3.4, 3.5 GRAPHS OF THE TRIGONOMETRIC FUNCTIONS

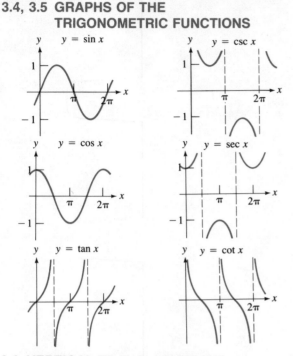

3.6 VERTICAL TRANSLATION AND PHASE SHIFT

The graph of $y = k + a \sin b(x + c)$ is identical to the graph of $y = a \sin bx$, except that it is translated

k units $\begin{Bmatrix} \text{up} \\ \text{down} \end{Bmatrix}$ if k is $\begin{Bmatrix} \text{positive} \\ \text{negative} \end{Bmatrix}$, and

$|c|$ units $\begin{Bmatrix} \text{left} \\ \text{right} \end{Bmatrix}$ if c is $\begin{Bmatrix} \text{positive} \\ \text{negative} \end{Bmatrix}$.

Similar statements are true for the other trigonometric functions.

4.2 FUNCTIONS OF TWO ANGLES

$\cos(A + B) = \cos A \cos B - \sin A \sin B$

$\cos(A - B) = \cos A \cos B + \sin A \sin B$

$\sin(A + B) = \sin A \cos B + \cos A \sin B$

$\sin(A - B) = \sin A \cos B - \cos A \sin B$

$\tan(A + B) = \dfrac{\tan A + \tan B}{1 - \tan A \tan B}$

$\tan(A - B) = \dfrac{\tan A - \tan B}{1 + \tan A \tan B}$

4.3 THE DOUBLE-ANGLE IDENTITIES

$\sin 2A = 2 \sin A \cos A$

$\cos 2A = \cos^2 A - \sin^2 A$

$\quad = 2 \cos^2 A - 1$

$\quad = 1 - 2 \sin^2 A$

$\tan 2A = \dfrac{2 \tan A}{1 - \tan^2 A}$

4.4 THE HALF-ANGLE IDENTITIES

$\cos \dfrac{A}{2} = \pm \sqrt{\dfrac{1 + \cos A}{2}}$

$\sin \dfrac{A}{2} = \pm \sqrt{\dfrac{1 - \cos A}{2}}$

$\tan \dfrac{A}{2} = \dfrac{\sin A}{1 + \cos A} = \dfrac{1 - \cos A}{\sin A}$

4.5 PRODUCT-TO-SUM AND SUM-TO-PRODUCT FORMULAS

$\sin A \cos B = \tfrac{1}{2}[\sin(A + B) + \sin(A - B)]$

$\cos A \sin B = \tfrac{1}{2}[\sin(A + B) - \sin(A - B)]$

$\sin A \sin B = \tfrac{1}{2}[\cos(A - B) - \cos(A + B)]$

$\cos A \cos B = \tfrac{1}{2}[\cos(A + B) + \cos(A - B)]$

$\sin A + \sin B = 2 \sin \tfrac{1}{2}(A + B) \cos \tfrac{1}{2}(A - B)$

$\sin A - \sin B = 2 \cos \tfrac{1}{2}(A + B) \sin \tfrac{1}{2}(A - B)$

$\cos A + \cos B = 2 \cos \tfrac{1}{2}(A + B) \cos \tfrac{1}{2}(A - B)$

$\cos A - \cos B = -2 \sin \tfrac{1}{2}(A + B) \sin \tfrac{1}{2}(A - B)$

4.6 SUMS OF FORM $A \sin x + B \cos x$

$A \sin x + B \cos x = k \sin(x + \phi)$

where $k = \sqrt{A^2 + B^2}$

and $\sin \phi = \dfrac{B}{\sqrt{A^2 + B^2}}$ and $\cos \phi = \dfrac{A}{\sqrt{A^2 + B^2}}$

5.3, 5.4 THE GRAPHS OF THE INVERSE TRIGONOMETRIC FUNCTIONS

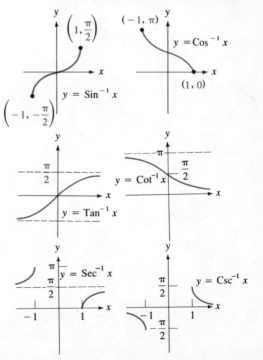

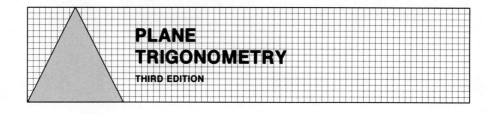

PLANE TRIGONOMETRY

THIRD EDITION

Books in the Gustafson and Frisk Series

BEGINNING ALGEBRA,   Second Edition
INTERMEDIATE ALGEBRA,   Second Edition
ALGEBRA FOR COLLEGE STUDENTS,   Second Edition
COLLEGE ALGEBRA,   Third Edition
PLANE TRIGONOMETRY,   Third Edition
COLLEGE ALGEBRA AND TRIGONOMETRY,   Second Edition
FUNCTIONS AND GRAPHS

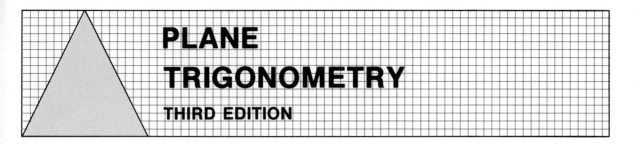

PLANE
TRIGONOMETRY
THIRD EDITION

R. David Gustafson
Rock Valley College

Peter D. Frisk
Rock Valley College

Brooks/Cole Publishing Company
Pacific Grove, California

Brooks/Cole Publishing Company
A Division of Wadsworth, Inc.

Printed in the United States of America

10   9   8   7   6   5   4

Library of Congress Cataloging-in-Publication Data

Gustafson, R. David (Roy David), [date]
    Plane trigonometry/R. David Gustafson, Peter D. Frisk.—3rd ed.
      p.   cm.
    Includes index.
    ISBN 0-534-09822-3
    1. Trigonometry, Plane.   I. Frisk, Peter D., [date].
  II. Title.
  QA533.G85   1988   88-19475
  516.2′4—dc 19   CIP

Sponsoring Editor: *Jeremy Hayhurst*
Editorial Assistant: *Virge Kelmser*
Production Editor: *Ellen Brownstein*
Manuscript Editor: *Brenda Griffing*
Permissions Editor: *Carline Haga*
Interior and Cover Design: *Roy R. Neuhaus*
Cover Photo: *Lee Hocker*
Art Coordinator: *Lisa Torri*
Interior Illustration: *Lori Heckleman*
Typesetting: *Polyglot, Singapore*
Cover Printing: *Lehigh Press Company, Pennsauken, NJ*
Printing and Binding: *R.R. Donnelley & Sons, Crawfordsville, IN*

To the teachers who most inspired our interest in mathematics:
  Theodosia Keeler
  Alice Iverson

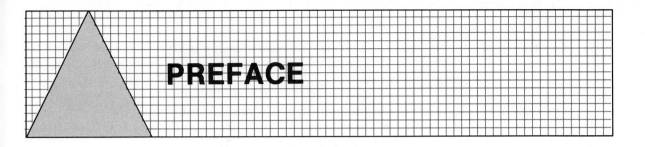

PREFACE

TO THE INSTRUCTOR

The basic philosophy of the third edition of *Plane Trigonometry* is unchanged from that of the highly successful second edition. The introduction of the trigonometric functions continues to be the popular angle-in-standard-position approach, with heavy emphasis on right-triangle applications. When using degrees to indicate the measurement of angles, we continue to emphasize their decimal forms. However, we now introduce minutes and seconds early in the course for those instructors who wish to use degrees-minutes-seconds.

Because a modern course in trigonometry must emphasize the trigonometric functions as functions with real number domains, we discuss the concept of radian measure early in the course, but not before the students need it. To convince students that the trigonometric functions can have real number domains, we include a section discussing the unit circle.

There are several major changes in the third edition:

1. The material on functions in Chapter 1 has been expanded and made slightly more rigorous. The horizontal line test to check for a one-to-one function is now included.
2. More application problems have been added throughout the text. Most exercise sets have been revised to allow for a greater range of problem solving skills.
3. In Chapter 5 more emphasis is given to writing general solutions to trigonometric equations.
4. A chapter on analytic geometry (Chapter 8) has been added for the many instructors who have requested it.
5. Appendix II has been expanded to include more work with base-10 and base-*e* exponential functions, more work with natural logarithms, and more applications of exponential and logarithmic functions.
6. Many sections have been rewritten to improve clarity.

We think that you will like this text because it contains the following features:

Review   Appendix I reviews the concepts from geometry that are necessary for success in trigonometry. Chapter 1 reviews the fundamental concept of functions. Each chapter concludes with a set of review exercises.

Many Exercises   The text includes more than 3000 exercises, with answers to the odd-numbered problems provided. All answers are provided for the review exercises.

Many Examples   The text includes more than 300 worked examples keyed to the exercise sets. By studying these examples, students can learn on their own, thereby easing the burden on the instructor.

Comprehensiveness   The topics of analytic geometry and logarithms are included for those instructors who want them. The material in *Plane Trigonometry* is very flexible and can be adapted to courses of various lengths.

Mathematical Honesty   The mathematical developments preserve the integrity of the mathematics, but they are not so rigorous as to confuse students.

Relevance   The text includes an abundant supply of application problems.

Instructor Support   For compiling individualized examinations, a computer-based question bank of test items with EXPTEST, a full-featured test-generating system for the IBM and compatible microcomputers, is available to those who adopt the text. Also available is an Instructor's Manual that gives answers to all even-numbered exercises.

We think your students will like the text because:

Informal   It is written for students to read. On the Fry Readability Test, the writing is at the tenth-grade level.

Worked Examples   There are more than 300 worked examples. Many of them include author's notes to explain important steps in the solution process.

Functional Use of Second Color   The text uses second color effectively—not just to highlight important definitions and theorems, but to point to terms and expressions that the instructor would point to in a classroom discussion.

Review Exercises   Each chapter concludes with a summary of key words, key ideas, and a set of review exercises.

Applications   Application problems abound throughout the text. The last section of most chapters emphasizes diverse applications that use the material discussed in the chapter.

Summary of Information   Key formulas and ideas from the text are listed inside the front and back covers for easy reference.

Student Support   Answers to the odd-numbered exercises and all answers to the review exercises are included in an appendix. Also available is a Student Solutions Manual, which provides complete solutions to the even-numbered exercises that are not multiples of four.

ORGANIZATION AND COVERAGE

Chapters 2 through 7 cover the topics usually associated with a course in trigonometry. Chapter 1 and Appendix I review the topics of function and basic geometry. Chapter 8 and Appendices II and III provide additional material for an expanded course.

The following diagram shows how the chapters and appendices are related.

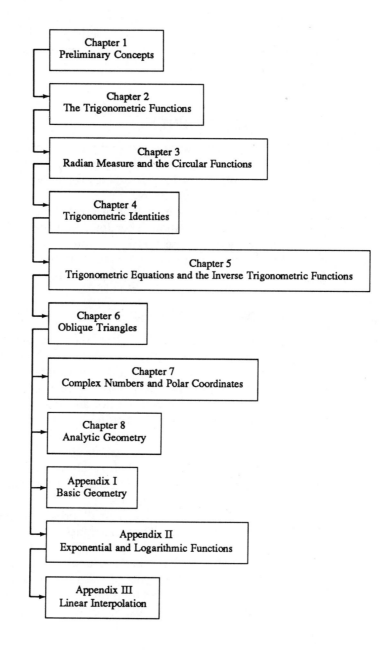

CALCULATORS

We encourage the use of calculators. In fact, they are essential to do the work in this text. All material is carefully marked where calculators should not be used.

ACCURACY

Dozens of mathematics teachers have reviewed either all or part of the text. We are grateful for their praise, constructive criticisms, and helpful suggestions. Both authors have independently worked every exercise, as have two professional problem checkers.

TO THE STUDENT

We have tried to write a text that you can read and understand. It provides an extensive number of worked examples and presents them in a way that will make sense to you. If you do not read the explanations carefully, however, much of the value of this text will be lost.

What you learn here will be of great value both in other course work and in your chosen occupation. Therefore, we suggest that you keep your text after completing this course. It is the one source of reference that will keep the material that you have learned here at your fingertips.

A WORD TO STUDENTS ABOUT CALCULATORS

Exercises and examples throughout the text assume that you will use a calculator. If you do not already own a calculator, now is the time to consider buying one. There are many types of calculators available and you should choose one that you will not outgrow. Inexpensive "checkbook" calculators that only add, subtract, multiply, and divide are not adequate for this course. For your work in mathematics, look for a scientific calculator capable of evaluating the trigonometric, exponential, and logarithmic functions, and that can display results in scientific notation. It should also be able to find powers, roots, reciprocals, and factorials, and save intermediate results in any of several addressable memories. Although the capability is not needed for this course, look for a calculator that also provides statistical functions.

Some calculators are really small, programmable computers and some provide graphical output. These machines are more powerful than simple calculators and are also more expensive. You will not need the extra power of a programmable calculator in this course but you might find it useful in later work.

There are two basic types of scientific calculators, depending on the way in which numbers and operations are entered. The most common is the algebraic calculator, but those that use Reverse Polish Notation (RPN) are also popular.

For example, to calculate the expression $\dfrac{5 \cdot 25}{17}$ on calculators of these two types, you would press these keys:

ALGEBRAIC calculator | 5 | × | 25 | ÷ | 17 | = |

RPN calculator | 5 | Enter | 25 | × | 17 | ÷ |

With either method, the result is 7.352941176...

Each model and each system has its special features and individual weaknesses. Whatever calculator you purchase, become familiar with its capabilities and quirks, and be sure to read its instruction manual carefully.

Calculators are intended to relieve the drudgery of doing mathematical calculations. Use your calculator freely when performing calculations. Know when an answer is unreasonable, thus indicating that an error has occurred. An answer is not correct just because a calculator produced it. Know also when *not* to use a calculator. Many results in mathematics depend not on calculation but on the logical relationships that link the concepts you are studying. In these situations, a calculator may provide insight but not solutions. No calculator can be a substitute for thinking.

We wish you well.

ACKNOWLEDGMENTS

We want to thank those who have reviewed the entire manuscript of *Plane Trigonometry* at various stages of its development.

Wilson Banks
Illinois State University

James Bolen
Tarrant County Junior College

John S. Cross
University of Northern Iowa

Grace De Velbiss
Sinclair Community College

Emily Dickinson
University of Arkansas

Russ Diprizio
Oakton Community College

Ray Edwards
Chabot College

Richard L. Francis
Southeast Missouri State University

Jerry Gustafson
Beloit College

Jerome Hahn
Bradley University

Douglas Hall
Michigan State University

David Hansen
Monterey Peninsula College

William Hinrichs
Rock Valley College

Arthur M. Hobbs
Texas A & M

Jack Hofer
California Polytechnic State University

Warren Jaech
Tacoma Community College

Diane Koenig
Rock Valley College

Marcus McWaters
University of Southern Florida

Eldon Miller
University of Mississippi

Stuart Mills
Louisiana State University

Juan P. Montemayor
Angelo State University

Gilbert W. Nelson
North Dakota State University

Anthony Peressini
University of Illinois

William D. Popejoy
University of Northern Colorado

Donald Sherbert
University of Illinois

L. Thomas Shiflett
Southwestern Missouri State University

Richard Slinkman
Bemidji State University

Jack Snyder
Sinclair Community College

Warren Strickland
Del Mar College

Lee Topham
North Harris County College

Robert J. Wisner
New Mexico State University

Robert Zink
Purdue University

We also thank the following people for their helpful suggestions: David Hinde, Darrell Ropp, and James Yarwood, all of Rock Valley College. We especially want to thank Diane Koenig for solving all of the problems and preparing the instructor's answer text and Jerry Frang, who prepared the *Student Study Guide*.

For their help in the production process, we express our thanks to Craig Barth, Jeremy Hayhurst, Ellen Brownstein, Brenda Griffing, Roy Neuhaus, Carline Haga, and Lisa Torri.

R. David Gustafson
Peter D. Frisk

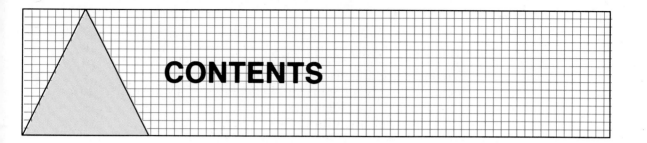

CONTENTS

1 PRELIMINARY CONCEPTS   1

1.1   The Rectangular Coordinate System and the Distance
        Formula   1
1.2   Functions and Relations   5
1.3   Inverses of Functions and Relations   11
        CHAPTER SUMMARY   18
        REVIEW EXERCISES   19

2 THE TRIGONOMETRIC FUNCTIONS   21

2.1   Functions of Trigonometric Angles   21
2.2   The Eight Fundamental Relationships of Trigonometric
        Functions   30
2.3   Trigonometric Functions of Certain Angles   38
2.4   Trigonometric Functions of Any Angle   46
2.5   Right Triangle Trigonometry   52
2.6   More Right Triangle Trigonometry   61
2.7   Introduction to Vectors   66
2.8   More Applications of the Trigonometric Functions   71
        CHAPTER SUMMARY   75
        REVIEW EXERCISES   77

3 RADIAN MEASURE AND THE CIRCULAR FUNCTIONS   79

3.1   Radian Measure   79
3.2   Linear and Angular Velocity   87
3.3   The Circular Functions   91
3.4   Graphs of Functions Involving Sin x and Cos x 96
3.5   Graphs of Functions Involving Tan x, Cot x, Csc x,
        and Sec x 103
3.6   Vertical and Horizontal Translations of the Trigonometric
        Functions   108
3.7   Graphs of Other Trigonometric Functions   114
3.8   Applications of Radian Measure   116
        CHAPTER SUMMARY   120
        REVIEW EXERCISES   121

4   TRIGONOMETRIC IDENTITIES   123

4.1   Trigonometric Identities   123
4.2   Identities Involving Sums and Differences of Two Angles   131
4.3   The Double-Angle Identities   137
4.4   The Half-Angle Identities   143
4.5   Sum-to-Product and Product-to-Sum Identities   150
4.6   Sums of the Form $A \sin x + B \cos x$ 156
4.7   Applications of the Trigonometric Identities   159
      CHAPTER SUMMARY   162
      REVIEW EXERCISES   164

**5   TRIGONOMETRIC EQUATIONS AND THE INVERSE
      TRIGONOMETRIC FUNCTIONS**   166

5.1   Trigonometric Equations   166
5.2   Introduction to the Inverse Trigonometric Relations   176
5.3   The Inverse Sine, Cosine, and Tangent Functions   179
5.4   The Inverse Cotangent, Secant, and Cosecant Functions
      (Optional)   189
5.5   Applications of Trigonometric Equations and Inverse
      Functions   194
      CHAPTER SUMMARY   197
      REVIEW EXERCISES   198

6   OBLIQUE TRIANGLES   201

6.1   The Law of Cosines   201
6.2   The Law of Sines   208
6.3   The Ambiguous Case: SSA   215
6.4   The Law of Tangents   220
6.5   Areas of Triangles   222
6.6   More on Vectors (Optional)   229
      CHAPTER SUMMARY   238
      REVIEW EXERCISES   240

7   COMPLEX NUMBERS AND POLAR COORDINATES   243

7.1   Complex Numbers   243
7.2   Graphing Complex Numbers   250
7.3   Trigonometric Form of a Complex Number   253
7.4   De Moivre's Theorem   257
7.5   Polar Coordinates   263
7.6   More on Polar Coordinates   270

7.7   Applications of Complex Numbers and Polar Coordinates   276
CHAPTER SUMMARY   278
REVIEW EXERCISES   279

8   **ANALYTIC GEOMETRY**   **281**

8.1   The Straight Line   282
8.2   The Circle   294
8.3   The Parabola   298
8.4   The Ellipse   305
8.5   The Hyperbola   312
8.6   Translation and Rotation of Axes   319
8.7   Polar Equations of the Conics   328
8.8   Parametric Equations   333
CHAPTER SUMMARY   341
REVIEW EXERCISES   344

APPENDIX I   **BASIC GEOMETRY**   **347**

I.1   Angles, Triangles, and the Pythagorean Theorem   347
I.2   Similar Triangles   353
I.3   Circles   356

APPENDIX II   **EXPONENTIAL AND LOGARITHMIC FUNCTIONS**   **360**

II.1   Exponential Functions   360
II.2   Logarithmic Functions   367
II.3   Applications of Base-10 and Base-*e* Logarithms   377
II.4   Exponential and Logarithmic Equations   382

APPENDIX III   **LINEAR INTERPOLATION**   **390**

APPENDIX IV   **TABLES**   **393**

Table A Values of the Trigonometric Functions   393
Table B Powers and Roots   397
Table C Base-10 Logarithms   398
Table D Base-*e* Logarithms   399

APPENDIX V   **ANSWERS TO SELECTED EXERCISES**   **400**

INDEX   **439**

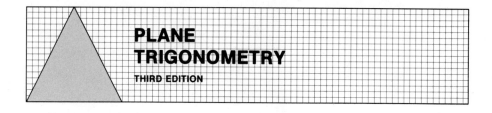

PLANE
TRIGONOMETRY

THIRD EDITION

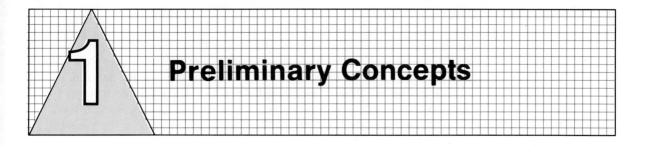

Preliminary Concepts

Many topics from algebra and geometry are applied in the study of trigonometry. The most basic of these ideas are discussed in Appendix I; this first chapter, however, treats other topics, the ones with which you might not be familiar.

1.1 THE RECTANGULAR COORDINATE SYSTEM AND THE DISTANCE FORMULA

It was the work of René Descartes (1596–1650) that merged algebra and geometry into a single unified subject. The genius of Descartes lay in his idea of a rectangular coordinate system. In such a system every point is assigned a pair of numbers as its unique address. Descartes's idea is based on two perpendicular number lines, one horizontal and one vertical. The horizontal number line (usually called the **x-axis**) and the vertical number line (usually called the **y-axis**) separate the geometric plane into four quadrants numbered as in Figure 1-1. The point at which the axes intersect, called the **origin**, is the point labeled 0 on each number line. The positive direction on the x-axis is to the right, the positive direction on the y-axis is upward, and the same unit of measurement is used on each axis.

To find (or plot) the point associated with the pair of real numbers $(-3, 2)$, for example, we start at the origin and count 3 units to the left and then 2 units up to locate point P as shown in Figure 1-2. Point P with **coordinates** $(-3, 2)$ lies in the

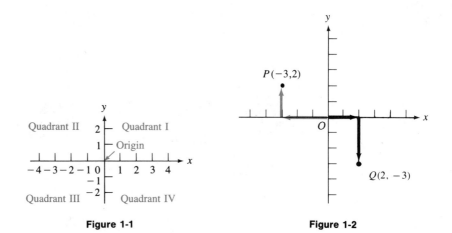

Quadrant II   Quadrant I
Origin
Quadrant III   Quadrant IV

Figure 1-1

$P(-3,2)$

$Q(2, -3)$

Figure 1-2

second quadrant. To find the point Q associated with the pair $(2, -3)$, we start at the origin and count 2 units to the right and then 3 units down. Point Q with coordinates $(2, -3)$ lies in the fourth quadrant. The order of the coordinates of a point is important. Point P with coordinates of $(-3, 2)$ is not the same as point Q with coordinates $(2, -3)$. For this reason, the pair of coordinates (x, y) is often called an **ordered pair**. The first number in the ordered pair is the x-coordinate and the second number is the y-coordinate.

It is always possible to find the distance between two points that are located on a rectangular coordinate system.

Definition.   If two points $P(x_1, y_1)$ and $Q(x_2, y_2)$ lie on a horizontal line, the distance $d(PQ)$ between them is the absolute value of the difference of the x-coordinate of Q and the x-coordinate of P. In symbols,

$$d(PQ) = |x_2 - x_1|$$

Example 1   Find the distance between $P(-4, -3)$ and $Q(2, -3)$.

Solution   Plot each point as in Figure 1-3 and note that PQ is a horizontal line segment. By definition, the distance between P and Q, represented by $d(PQ)$, is the absolute value of the difference of the x-coordinate of Q and the x-coordinate of P:

$$d(PQ) = |x_2 - x_1| = |2 - (-4)| = |2 + 4| = |6| = 6$$

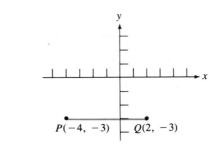

Figure 1-3

Definition.   If two points $P(x_1, y_1)$ and $Q(x_2, y_2)$ lie on a vertical line, the distance between them is the absolute value of the difference of the y-coordinate of Q and the y-coordinate of P. In symbols,

$$d(PQ) = |y_2 - y_1|$$

Example 2   Find the distance between $P(4, -3)$ and $Q(4, 7)$.

Solution   Plot the points as in Figure 1-4 and note that PQ is a vertical line segment. By definition, the distance between P and Q, represented by $d(PQ)$, is the absolute value of the difference of the y-coordinate of Q and the y-coordinate of P:

$$d(PQ) = |y_2 - y_1| = |7 - (-3)| = |10| = 10 \qquad \blacksquare$$

To find the distance between two points that do not lie on either a vertical or horizontal line, we use the distance formula.

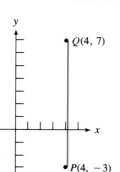

Figure 1-4

> **The Distance Formula.**   The distance between two points $P(x_1, y_1)$ and $Q(x_2, y_2)$ is given by the formula
>
> $$d(PQ) = \sqrt{(x_2 - x_1)^2 + (y_2 - y_1)^2}$$

Proof   We plot points $P(x_1, y_1)$ and $Q(x_2, y_2)$ and construct right triangle PRQ as in Figure 1-5. Because segment PR is horizontal and segment RQ is vertical, we have

$$d(PR) = |x_2 - x_1| \qquad \text{and} \qquad d(RQ) = |y_2 - y_1|$$

Because we know the lengths of the two legs of the right triangle, we can use the Pythagorean theorem to find the length of its hypotenuse, which is the distance $d(PQ)$.

$$d(PQ)^2 = |x_2 - x_1|^2 + |y_2 - y_1|^2$$

Because $|k|^2 = k^2$ for all numbers k, the previous result can be written as follows:

$$d(PQ)^2 = (x_2 - x_1)^2 + (y_2 - y_1)^2$$

or, by taking the positive square root of both sides, as follows:

1. $d(PQ) = \sqrt{(x_2 - x_1)^2 + (y_2 - y_1)^2}$

Equation 1 is called the **distance formula**, as noted above. It is an important formula and should be memorized.

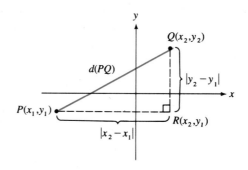

Figure 1-5   □

Example 3   Find the distance between points $P(-2, 3)$ and $Q(6, -5)$.

Solution   Use the distance formula and substitute -2 for x_1, 3 for y_1, 6 for x_2, and -5 for y_2:

$$\begin{aligned}
d(PQ) &= \sqrt{(x_2 - x_1)^2 + (y_2 - y_1)^2} \\
&= \sqrt{[6 - (-2)]^2 + (-5 - 3)^2} \\
&= \sqrt{8^2 + (-8)^2} \\
&= \sqrt{64 + 64} \\
&= \sqrt{64 \cdot 2} \\
&= \sqrt{64}\sqrt{2} \\
&= 8\sqrt{2}
\end{aligned}$$

The distance between points P and Q is $8\sqrt{2}$ units.   ■

■ EXERCISE 1.1

In Exercises 1–18, find the distance between points P and Q.

1. $P(2, 5)$ and $Q(7, 5)$

2. $P(2, -2)$ and $Q(10, -2)$

3. $P(2, 5)$ and $Q(2, 13)$

4. $P(-4, -2)$ and $Q(-4, 10)$

5. $Q(-5, -2)$ and $P(-5, 10)$

6. $Q(2, -4)$ and $P(-8, -4)$

7. $P(0, 0)$ and $Q(3, -4)$

8. $P(0, 0)$ and $Q(12, -35)$

9. $Q(5, 8)$ and $P(2, 4)$

10. $Q(8, 13)$ and $P(5, 9)$

11. $P(-2, -8)$ and $Q(3, 4)$

12. $P(-5, -2)$ and $Q(7, 3)$

13. $P(6, 8)$ and $Q(12, 16)$

14. $P(10, 4)$ and $Q(2, -2)$

15. $Q(-3, 5)$ and $P(-5, -5)$

16. $Q(2, -3)$ and $P(4, -8)$

17. $Q(-2, 4)$ and $P(2, -4)$

18. $Q(4, -3)$ and $P(4, 3)$

19. Show that a triangle with vertices at $A(-2, 4)$, $B(2, 8)$, and $C(6, 4)$ is an isosceles triangle (that is, a triangle with two sides of equal length).

20. Show that a triangle with vertices at $A(-2, 13)$, $B(-8, 9)$, and $C(-2, 5)$ is isosceles.

21. In Illustration 1, every point on the line CD is equidistant from points A and B. Use the distance formula to find the equation of line CD.

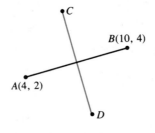

Illustration 1

22.   Use the distance formula to find the equation of a circle with center at $O(2, 4)$ and passing through $P(6, 9)$.

23.   What point on the x-axis is equidistant from points $A(-1, -3)$ and $B(6, 4)$?

24.   What two points on the y-axis are 13 units from the point $P(12, 2)$?

1.2   FUNCTIONS AND RELATIONS

Many equations that we discuss in algebra describe a correspondence between two variables. The equation $y = -3x + 1$, for example, describes a correspondence in which each number x determines exactly one value y. Such correspondences are called **functions**.

> **Definition.**   A **function** is a correspondence that assigns to each element of x of some nonempty set X a single value y of some set Y.
>
> The set X is called the **domain** of the function. The value y that corresponds to a particular x in the domain is called the **image** of x under the function. The set of all images of x is called the **range** of the function.

Because each value of y depends on some number x, we often call x the **independent variable** and y the **dependent variable**. Unless we indicate otherwise, the domain of a function is its implied domain—the set of all real numbers for which the function is defined.

By definition, a function f is a correspondence from the elements of one set X, to the elements of another set Y. We can visualize this correspondence with the aid of Figure 1-6. We represent the function f by drawing arrows from each element of set X to the corresponding element of set Y. If a function f assigns the element y_1 to the element x_1, we draw an arrow leaving x_1 and pointing to y_1. The set X, consisting of all elements from which arrows originate, is the domain of the function. The set Y, consisting of all elements to which arrows point, is the range.

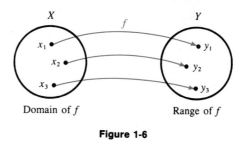

Figure 1-6

A correspondence is still a function if arrows point from several different elements of X to the same element y in the range as in Figure 1-7. In this correspondence, a single value y corresponds to each number x. However, if

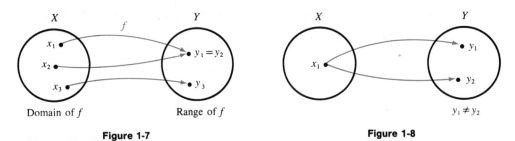

Figure 1-7   Figure 1-8

arrows leave some element x in the domain and point to several elements in the range, the correspondence is not a function because more than one value y corresponds to the number x. See Figure 1-8.

If to each value x there corresponds one or more values of y, we call the correspondence a **relation**.

Definition. A **relation** is a correspondence that assigns to each number x in a nonempty set X one or more values of y in a set Y.

The set X is called the **domain** of the relation. The **range** of the relation is the set containing all values of y that correspond to some number x in the domain.

Because of the preceding definitions, *a function is always a relation, but a relation is not necessarily a function.*

Example 1 Show that the equation $y = 2x - 3$ determines y to be a function of x. Find the domain and range of the function.

Solution To be a function, each real number x must determine *exactly one* value of y. To determine y in the equation $y = 2x - 3$, the number 3 must be subtracted from the product of 2 and x. Because for each number x this arithmetic gives a single result, only one value of y is produced. Thus, the equation does determine y to be a function of x.

Because x can be any real number, the domain of the function is its implied domain, which is the set of real numbers.

Because y can be any real number, the range of the function is also the set of real numbers.   ■

Example 2 Show that the equation $y^2 = 6x$ does not determine y to be a function of x. Give the domain and range of the relation.

Solution Let x be the number 6, for example. Then

$$y^2 = 6x = 6(6) = 36$$

and y could be either 6 or -6. Because more than one value of y corresponds to $x = 6$, the equation $y^2 = 6x$ does not determine y to be a function of x. However, it does determine a relation.

Because $y^2 = 6x$ and y^2 is always a nonnegative number, x must be a nonnegative number also. Thus, the domain of the relation is the set of nonnegative real numbers.

Because y can be any real number, the range is the set of all real numbers. ∎

Example 3   Show that the equation $y = \dfrac{8}{x + 2}$ determines y to be a function of x. Find its domain and range.

Solution   Since division by 0 is undefined, the denominator of the fraction $\dfrac{8}{x + 2}$ cannot be 0. Thus, x cannot be -2. Because each real number x other than -2 does determine *exactly one* value of y, the equation does determine y to be a function of x.

Since x can be any real number except -2, it follows that the domain of this function is the set of all real numbers except -2.

A fraction with a nonzero numerator cannot be 0. Thus, the range of this function is the set of all real numbers except 0. ∎

Example 4   Find the domain and the range of the relation determined by ordered pairs (x, y) such that $y^2 = x - 4$, and tell whether the relation is a function.

Solution   Because y^2 must be a nonnegative real number, the expression $x - 4$ must be nonnegative also, and x must be greater than or equal to 4. Thus, the domain of the relation is the set of all real numbers x such that $x \geq 4$.

Because y can be any real number, the range of the relation is the set of all real numbers.

Note that to some numbers x there correspond two values of y. For example, if $x = 13$, then y can be either 3 or -3. Thus, this relation is not a function. ∎

Function Notation

There is a notation that reinforces the idea that y is a function of x. If the variable y in a function is equal to some expression that involves the variable x, then y is often denoted as $f(x)$ and is read as "f of x." The symbol $f(a)$ indicates the value of y that results when a is substituted for x. For example, if $y = f(x) = 3x + 2$, then $f(4)$ is the value of y that is obtained when $x = 4$:

$$f(x) = 3x + 2$$
$$f(4) = 3(4) + 2$$
$$= 14$$

Likewise,

$$f(-2) = 3(-2) + 2 = -4$$

Example 5   The equation $y = f(x) = \dfrac{x+1}{x}$ defines y to be a function of x. Find the values **a.** $f(2)$, **b.** $f(h)$, and **c.** $f(0)$.

Solution   **a.** $f(2)$ is the value of y obtained when 2 is substituted for x:

$$f(2) = \frac{2+1}{2} = \frac{3}{2}$$

b. $f(h)$ is the value of y obtained when h is substituted for x:

$$f(h) = \frac{h+1}{h}$$

c. $f(0)$ is meaningless because 0 is not an acceptable replacement for x. The number 0 is not an element of the domain of f because division by 0 is not allowed.   ■

Example 6   Let $y = f(x) = x^2 - 2x - 3$. Find **a.** $f(a)$, **b.** $f(h)$, and **c.** $f(a + h)$.

Solution   **a.**   Because $f(a)$ is the value of y obtained when a is substituted for x, you have

$$f(a) = a^2 - 2a - 3$$

Likewise,

b. $f(h) = h^2 - 2h - 3$

c. $f(a + h) = (a + h)^2 - 2(a + h) - 3$
$$= a^2 + 2ah + h^2 - 2a - 2h - 3$$

Note that $f(a + h) \neq f(a) + f(h)$.   ■

Graphing Functions and Relations

If both the domain and the range of a relation are subsets of the set of real numbers, it is possible to graph the ordered pairs of the relation on a rectangular coordinate system. The graph consists of all points in the coordinate plane with coordinates $(x, y) = (x, f(x))$. The next two examples show how to graph a relation that is defined by an equation. In Example 7 the equation represents a function, and in Example 8 the equation does not.

Example 7   Graph the equation $y = x^2 - 10x + 24$.

Solution   Make a table of values of x and y and plot the points. For example, if $x = 2$, then

$y = 8$ because

$$y = f(x) = x^2 - 10x + 24$$
$$y = f(2) = 2^2 - 10(2) + 24$$
$$y = f(2) = 8$$

If $x = 7$, then $f(x) = f(7) = 3$. Because each number x determines a single value y, this equation determines y to be a function of x. Other pairs of coordinates are shown in Figure 1-9. Plot these points and join them with a smooth curve to obtain the parabola shown in the figure.

$y = x^2 - 10x + 24$

x	y
2	8
3	3
5	−1
7	3
8	8

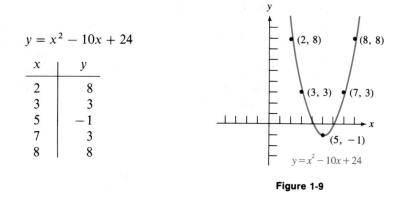

Figure 1-9

Example 8 Graph the equation $x^2 + y^2 = 13$.

Solution Make a table of values of x and y and plot the points. For example, if $x = 2$, then y is either 3 or -3. If $x = -2$, then y is again 3 or -3. Other pairs of co-ordinates are shown in Figure 1-10. Plot these points and join them with a smooth curve to obtain the circle shown in the figure. Because the equation $x^2 + y^2 = 13$ determines two values of y for each number x in the interval $-\sqrt{13} < x < \sqrt{13}$, it does not define a function.

$$x^2 + y^2 = 13$$

x	y
0	$\sqrt{13} \approx 3.6$
0	$-\sqrt{13} \approx -3.6$
2	3
2	−3
$\sqrt{13} \approx 3.6$	0
$-\sqrt{13} \approx -3.6$	0
−2	3
−2	−3

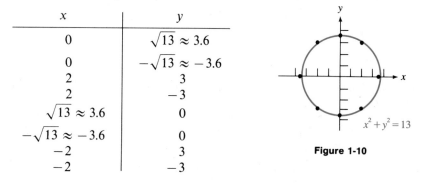

Figure 1-10

The symbol $\approx$ is read as "is approximately equal to."

Perhaps you know that any equation of the form $x^2 + y^2 = r^2$ has a circle for its graph, with center at the origin and radius of $|r|$. Thus, the graph of $x^2 + y^2 = 13$ is a circle with center at $(0, 0)$ and radius of $\sqrt{13}$. This is the circle shown in Figure 1-10.

The Vertical Line Test

A graph indicates pictorially whether a correspondence given by an equation represents a function. If any vertical line intersects a graph more than once, the correspondence set up by the graph cannot be a function. This is because to one number x, there would correspond more than one value of y. For example, the graph in Figure 1-11 cannot represent a function. This test to determine whether a graph represents a function is called the **vertical line test**. Because every vertical line in Figure 1-9 would intersect the graph exactly once, that graph and its equation determine a function. However, because some vertical lines drawn in Figure 1-10 would intersect the graph more than once, that graph and its equation do not determine a function.

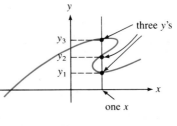

Figure 1-11

EXERCISE 1.2

In Exercises 1–10, tell whether the given relation determines y to be a function of x. Assume that x and y represent real numbers in each ordered pair (x, y).

1. $y = 3x - 4$

2. $y = -\frac{1}{2}x + 7$

3. $y = 3x^2 + 1$

4. $x = 2y^2 - 3$

5. $y = \pm\sqrt{x - 3}$

6. $y = \frac{x + 2}{2}$

7. $x = \frac{y - 3}{4}$

8. $x = \sqrt{y - 1}$

9. $x = |y|$

10. $y = |x|$

In Exercises 11–20, give the domain and the range of each function. Assume that x and y represent real numbers in each ordered pair (x, y).

11. $y = 2x + 3$

12. $y = \frac{1}{2}x - 4$

13. $y = \frac{4}{2 - x}$

14. $y = \frac{-6}{5x}$

15. $y = x^4$

16. $y = \frac{5}{x - 3}$

17. $y = |x|$

18. $y = \frac{1}{2}x^2 - 4$

19. $y = \sqrt{x - 2}$

20. $y = \sqrt{x^2 - 4}$

In Exercises 21–28, find the required value, if possible, when $f(x) = \dfrac{2x - 1}{x}$.

21. $f(2)$ **22.** $f(5)$ **23.** $f(-3)$ **24.** $f(-8)$

25. $f(0)$ **26.** $f(\frac{1}{2})$ **27.** $f(h)$ **28.** $f(a + h)$

In Exercises 29–36, find the required value, if possible, when $f(x) = \dfrac{\sqrt{x - 1}}{2}$. Assume that $f(x)$ must be a real number.

29. $f(1)$ **30.** $f(5)$ **31.** $f(0)$ **32.** $f(-2)$

33. $f(2a)$ **34.** $f(3s)$ **35.** $f(a + h)$ **36.** $f(a + h) - f(a)$

In Exercises 37–54, graph each equation. Use the vertical line test to decide whether each graph determines a function.

37. $y = 3x - 2$ **38.** $y = -2x + 3$ **39.** $y = |\frac{1}{2}x|$ **40.** $x = |2y|$

41. $y = x^2 + 2x + 1$ **42.** $y = -x^2 - 2x + 3$ **43.** $y^2 - 6y + 9 = x$ **44.** $y^2 + 4x = x - 4$

45. $xy = 12$ **46.** $xy = -6$ **47.** $x^2 + y^2 = 25$ **48.** $x^2 + y^2 = 16$

49. $x^2 - y^2 = 1$ **50.** $x^2 - y^2 = 9$ **51.** $y = x^3$ **52.** $x = y^3$

53. $y = \sqrt{x - 2}$ **54.** $y = \dfrac{1}{x}$

55. Use the distance formula to show that $x^2 + y^2 = r^2$ is the equation of a circle, which is centered at the origin and that has a radius of $|r|$ units.

56. Show that the graph of $y = \sqrt{25 - x^2}$ passes the vertical line test, but that the graph of $y^2 = 25 - x^2$ does not.

1.3   INVERSES OF FUNCTIONS AND RELATIONS

A nonempty set of ordered pairs (x, y) defines a relation because it determines a correspondence between values of x and y. For example, a relation R is defined by the set

$$R = \{(1, 10), (2, 20), (3, 30)\}$$ Read as "R is the set containing the ordered pairs (1, 10), (2, 20), and (3, 30)."

The domain of R is the set of first components of the ordered pairs and its range is the set of second components. Thus, the domain of R is $\{1, 2, 3\}$ and the range is $\{10, 20, 30\}$. Because only one second component y corresponds to each first component x of an ordered pair, the relation R is a function. However, the relation Q, where

$$Q = \{(5, 15), (5, 20), (6, 30), (7, 40)\}$$

with domain of $\{5, 6, 7\}$ and range $\{15, 20, 30, 40\}$, is not a function because two values of y (15 and 20) correspond to the first component 5.

Example 1   Let R be a relation defined by ordered pairs (x, y) such that $R = \{(2, 3), (5, 7), (6, 8), (9, 8)\}$. Give the domain and range of R and tell whether R is a function.

Solution   The domain of R is the set of first components of the ordered pairs. Thus, the domain is $\{2, 5, 6, 9\}$.

The range of R is the set of second components of the ordered pairs. Thus, the range is $\{3, 7, 8\}$.

Because only one value of y corresponds to each number x, the relation R is a function.   ■

Inverse Functions

If the components of each ordered pair in any given relation are interchanged, a new relation is formed that is called its **inverse relation**. For example, if R is the relation

$$R = \{(1, 10), (2, 20), (3, 30)\}$$

the inverse relation of R, denoted as R^{-1}, is

$$R^{-1} = \{(10, 1), (20, 2), (30, 3)\}$$

The -1 in the notation R^{-1} is not an exponent. It refers to the inverse of the relation. The symbol R^{-1} is read as "the inverse relation of R" or just "R inverse."

The domain of R^{-1} is $\{10, 20, 30\}$. This is also the range of R. The range of R^{-1} is $\{1, 2, 3\}$. This is also the domain of R. In general, we have the following definition.

> **Definition.**   If R is any relation and R^{-1} is the relation obtained from R by interchanging the components of each ordered pair of R, then R^{-1} is called the **inverse relation** of R.
> The domain of R^{-1} is the range of R, and the range of R^{-1} is the domain of R.

Figure 1-12 represents the relation

$$G = \{(2, 3), (2, 4), (3, 5), (4, 8)\}$$

where the numbers in the left-hand circle are the elements of the domain, and the numbers in the right-hand circle are elements of the range. The relation G is not a function because to the number 2 in the domain there corresponds both 3 and 4 in the range.

The inverse relation of G is G^{-1}, where

$$G^{-1} = \{(3, 2), (4, 2), (5, 3), (8, 4)\}$$

The relation G^{-1}, represented in Figure 1-13, is a function because to each element in the domain of G^{-1} there corresponds a single element in the range.

Figures 1-12 and 1-13 illustrate that the domain of a relation is the range of its inverse relation and that the range of a relation is the domain of its inverse relation.

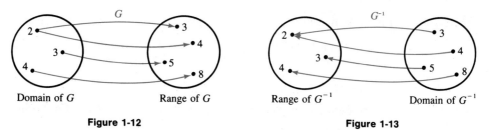

Domain of G Range of G Range of G^{-1} Domain of G^{-1}

Figure 1-12   **Figure 1-13**

An equation in x and y determines a set of ordered pairs. For example, if $x = 1$, 2, 3, 4, and 5, the equation $y = 6x$ determines the following set of ordered pairs:

$$T = \{(1, 6), (2, 12), (3, 18), (4, 24), (5, 30)\}$$

To form T^{-1}, we interchange the x- and y-coordinates to obtain

$$T^{-1} = \{(6, 1), (12, 2), (18, 3), (24, 4), (30, 5)\}$$

Example 2   The set of all pairs (x, y) determined by the equation $y = 4x + 2$ determines a function f. Find the inverse of function f and tell whether the inverse relation is a function.

Solution   To find the inverse relation of $y = 4x + 2$, interchange the variables x and y to obtain

1. $x = 4y + 2$

To decide whether this inverse relation is a function, solve Equation 1 for y as follows:

$$x = 4y + 2$$
$$x - 2 = 4y \qquad \text{Add } -2 \text{ to both sides.}$$
$$\frac{x - 2}{4} = y \qquad \text{Divide both sides by 4.}$$
$$y = \frac{x - 2}{4}$$

Because each number x that is substituted into the equation $y = \dfrac{x - 2}{4}$ gives a single value y, the inverse relation is a function.   ■

In Example 2 the inverse relation of the function $y = 4x + 2$ was found to be the function $y = \dfrac{x - 2}{4}$. In function notation, this inverse function can be denoted as follows:

$$f^{-1}(x) = \frac{x - 2}{4} \qquad \text{Read as ``f inverse of x is } \frac{x - 2}{4}.\text{''}$$

Example 3   Find the inverse relation of $y = f(x) = x^2 - 4$ and tell whether it is a function.

Solution   To find the inverse relation of $y = x^2 - 4$, interchange the variables x and y to obtain

1. $x = y^2 - 4$

To decide whether this inverse relation is a function, write Equation 1 as follows:

$$x = y^2 - 4$$
$$x + 4 = y^2 \qquad \text{Add 4 to both sides.}$$
$$y^2 = x + 4$$

If $x = 5$, for example, then $y^2 = 5 + 4 = 9$, and y could be either 3 or -3. Because the inverse relation contains pairs such as $(5, 3)$ and $(5, -3)$, in which two y values correspond to a single number x, the inverse relation $y^2 = x + 4$ is not a function.   ■

Example 4   Find the inverse of the function defined by $y = f(x) = x^3 + 3$ and graph both the function and its inverse on the same set of coordinate axes.

Solution   Find the inverse of the function defined by $y = x^3 + 3$ by interchanging x and y. The inverse relation is defined by the equation

$$x = y^3 + 3$$

Solve this equation for y by subtracting 3 from both sides to obtain

$$x - 3 = y^3$$

and finally, by taking the cube root of both sides to obtain

$$y = \sqrt[3]{x - 3}$$

Thus, $f^{-1}(x) = \sqrt[3]{x - 3}$. The graphs of the given function and its inverse appear in Figure 1-14. Note that the graph of $y = x$ is an axis of symmetry. When you

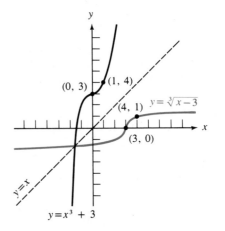

Figure 1-14

graph a function and its inverse on the same coordinate system, the line $y = x$ will always be an axis of symmetry because if the coordinates (a, b) satisfy an equation, the coordinates (b, a) will satisfy its inverse.   ■

Example 5   Find the inverse relation of $y = f(x) = \dfrac{2x}{x - 1}$ and tell whether the inverse is a function.

Solution   To find the inverse relation of $y = \dfrac{2x}{x - 1}$, interchange the variables x and y to obtain

1. $x = \dfrac{2y}{y - 1}$

To decide whether this inverse relation is a function, solve Equation 1 for y as follows:

$$x = \frac{2y}{y - 1}$$

$$x(y - 1) = 2y \qquad \text{Multiply both sides by } y - 1.$$

$$xy - x = 2y \qquad \text{Remove parentheses.}$$

$$xy - 2y = x \qquad \text{Add } -2y \text{ and } x \text{ to both sides.}$$

$$y(x - 2) = x \qquad \text{Factor out } y.$$

$$y = \frac{x}{x - 2} \qquad \text{Divide both sides by } x - 2.$$

Because each number x (except 2) that is substituted into this equation gives a single value y, the inverse relation is a function.   ■

If the inverse of a given function is also a function, the given function is said to be *one-to-one*. Stated another way, we have the following definition.

> **Definition.**   A function is called **one-to-one** if and only if each element in the range of the function corresponds to only one element in the domain of the function.

A test, called the **horizontal line test**, can be used to decide whether a function is one-to-one. If any horizontal line that intersects the graph of a function does so in *exactly one* point, the function is one-to-one. Otherwise, the function is not one-to-one.

Example 6   Use the horizontal line test to decide whether the functions defined by **a.** $y = x^3$ and **b.** $y = x^2 - 4$ are one-to-one.

Solution **a.** Graph the equation $y = x^3$ as in Figure 1-15a. Because each horizontal line that intersects the graph does so exactly once, each value of y corresponds to only one number x. Thus, the function defined by $y = x^3$ is one-to-one.

b. Graph the equation $y = x^2 - 4$ as in Figure 1-15b. Because many horizontal lines that intersect the graph do so twice, some values of y correspond to two numbers x. Thus, the function is not one-to-one.

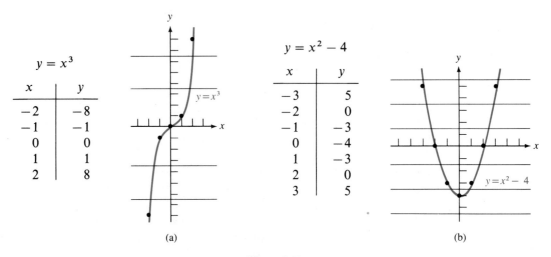

x	y
-2	-8
-1	-1
0	0
1	1
2	8

$y = x^3$

$y = x^2 - 4$

x	y
-3	5
-2	0
-1	-3
0	-4
1	-3
2	0
3	5

(a)   (b)

Figure 1-15

■

Example 7 Determine whether the function defined by $y = \dfrac{x + 2}{x}$ is a one-to-one function.

Solution Because to each number in the domain (x cannot be 0) there corresponds exactly one value y, the given equation determines a function. This function will be one-to-one if each possible value of y corresponds to exactly one number x. To see whether this is true, solve the equation for x as follows:

$$y = \frac{x + 2}{x}$$

$$xy = x + 2 \qquad \text{Multiply both sides by } x.$$

$$xy - x = 2 \qquad \text{Add } -x \text{ to both sides.}$$

$$x(y - 1) = 2 \qquad \text{Factor out } x.$$

$$x = \frac{2}{y - 1} \qquad \text{Divide both sides by } y - 1.$$

You can now see that as long as y does not equal 1, each value of y does determine exactly one number x. Thus, the function determined by the equation $y = \dfrac{x + 2}{x}$ is one-to-one.

■

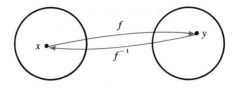

Figure 1-16

Figure 1-16 illustrates a fundamental property of a one-to-one function. If a function $y = f(x)$ changes some number x to some value y, the inverse function f^{-1} will change y back to x.

Example 8   The function $y = f(x) = x^2 + 3$ is not one-to-one. However, the function becomes one-to-one when the domain of f is restricted to some appropriate set, say the set of negative numbers. If the domain of f is so restricted, find the inverse of f and graph both the function and its inverse on the same set of coordinate axes.

Solution   To find the inverse of the function defined by $y = x^2 + 3$, where $x < 0$, interchange x and y and solve for y:

$$y = x^2 + 3 \qquad \text{where} \quad x < 0$$
$$x = y^2 + 3 \qquad \text{where} \quad y < 0 \qquad \text{Interchange } x \text{ and } y.$$
$$x - 3 = y^2 \qquad \text{where} \quad y < 0$$

Because y is less than zero, we can take the square root of both sides and obtain the function

$$y = -\sqrt{x - 3} \qquad \text{where} \quad y < 0$$

Thus, the inverse function is $f^{-1}(x) = -\sqrt{x - 3}$. The graphs of the two functions appear in Figure 1-17.

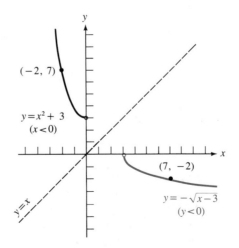

Figure 1-17

■ EXERCISE 1.3 ▬▬▬▬▬

In Exercises 1–6, give the domain and range of each relation and tell whether it is a function.

1. $\{(2, 1), (3, 2), (4, 3)\}$ **2.** $\{(1, 1), (4, 2), (9, 3)\}$ **3.** $\{(2, 3), (2, 4), (2, 5)\}$ **4.** $\{(3, 2), (4, 2), (8, 3)\}$

5. $\{(5, 1), (4, 1), (3, 1), (2, 1)\}$ **6.** $\{(3, 5)\}$

In Exercises 7–22, find the inverse of each relation and tell whether the inverse relation is a function.

7. $\{(1, 1), (2, 8), (3, 27), (4, 64)\}$ **8.** $\{(8, 8), (9, 9), (10, 10), (11, 11)\}$

9. $y = -2x - 2$ **10.** $y = 7x + 1$ **11.** $x + y = 3$ **12.** $2x - y = 4$

13. $y = x^2 + 2$ **14.** $y = 3x^2 - 1$ **15.** $y = \dfrac{2}{x - 4}$ **16.** $y = \dfrac{6}{2x + 1}$

17. $y = x^2 - 1$ **18.** $2y - 1 - x^2 = 0$ **19.** $y = |x - 2|$ **20.** $x = |y + 3|$

21. $x = \sqrt{y + 4}$ **22.** $y = \sqrt{2x - 3}$

In Exercises 23–26, tell whether each function is one-to-one.

23. $\{(1, 2), (2, 3), (3, 4)\}$ **24.** $\{(1, 1), (-1, 1), (2, 4), (-2, 4)\}$

25. $\{(1, 5), (2, 10), (3, 10), (4, 10)\}$ **26.** $\{(8, 2), (12, 3), (16, 4), (20, 5)\}$

In Exercises 27–40, graph each function and use the horizontal line test to decide whether each function is one-to-one.

27. $y = 2x - 3$ **28.** $y = -x + 4$ **29.** $y = x^2$ **30.** $y = (x + 3)^2$

31. $y = -x^3$ **32.** $y = (x - 5)^2$ **33.** $y = |2x|$ **34.** $y = |x + 3|$

35. $y = -\sqrt{x}$ **36.** $y = \sqrt[3]{x}$ **37.** $y = \dfrac{6}{x}$ **38.** $y = \sqrt{x - 2}$

39. $y = -x^3 + 2$ **40.** $y = x + x^2$

In Exercises 41–50, graph each relation and its inverse on one set of coordinate axes. What is the line of symmetry?

41. $y = -\frac{1}{2}x + 2$ **42.** $3x + 2y = 6$ **43.** $y = x^2 - 4$ **44.** $x = y^2 + 2$

45. $y = \sqrt{x - 1}$ **46.** $y = -\sqrt{x + 2}$

47. $y = x^2 - 1$, with $x \geq 0$ **48.** $y = x^2 + 2$, with $x < 0$

49. $y = |2x + 3|$ **50.** $y = \sqrt[3]{x}$

In Exercises 51–54, find the range of the function by finding the domain of the function's inverse.

51. $y = \dfrac{x + 3}{x}$ **52.** $y = \dfrac{x}{x + 3}$ **53.** $y = \dfrac{3x - 2}{x + 2}$ **54.** $y = \dfrac{2x + 1}{x - 2}$

▬▬▬▬▬▬ **CHAPTER SUMMARY** ▬▬▬▬▬▬

Key Words

coordinate system   (1.1)   *function*   (1.2)

distance between two points   (1.1)   *horizontal line test*   (1.3)

domain   (1.2)   *inverse relation*   (1.3)

one-to-one function   (1.3)   relation   (1.2)
origin   (1.1)   vertical line test   (1.2)
range   (1.2)   x- and y-axes   (1.1)

Key Ideas

(1.1)   If $P(x_1, y_1)$ and $Q(x_2, y_2)$ are points on a horizontal line, the distance between P and Q is $d(PQ) = |x_2 - x_1|$.

If $P(x_1, y_1)$ and $Q(x_2, y_2)$ are points on a vertical line, the distance between P and Q is $d(PQ) = |y_2 - y_1|$.

If $P(x_1, y_1)$ and $Q(x_2, y_2)$ are any two points on a line, then

$$d(PQ) = \sqrt{(x_2 - x_1)^2 + (y_2 - y_1)^2} \qquad \text{The distance formula}$$

(1.2)   If $y = f(x)$ is a function, then $f(a)$ represents the value of y when $x = a$.

If both the domain and the range of a relation are subsets of the real numbers, the relation can be graphed on a rectangular coordinate system.

(1.3)   If the components of each ordered pair in a relation R are interchanged, then R^{-1}, the inverse relation of R, is formed.

The graph of a relation is always symmetric to the graph of its inverse relation, with the line represented by $y = x$ as the axis of symmetry.

A function is one-to-one if and only if ordered pairs with different first components also have different second components.

The inverse relation of a one-to-one function is also a function.

REVIEW EXERCISES

In Review Exercises 1–12, find the distance between points P and Q.

1. $P(2, 5)$ and $Q(2, -2)$

2. $P(-2, -5)$ and $Q(-12, -5)$

3. $Q(5, 7)$ and $P(-12, 7)$

4. $Q(-3, -13)$ and $P(-3, 22)$

5. $P(0, 0)$ and $Q(3, 4)$

6. $P(0, 0)$ and $Q(12, 5)$

7. $P(2, 3)$ and $Q(5, 7)$

8. $P(-7, -6)$ and $Q(-4, -2)$

9. $Q(2, 8)$ and $P(7, 13)$

10. $Q(8, 2)$ and $P(13, 7)$

11. $Q(-2, -4)$ and $P(-9, 7)$

12. $Q(5, -4)$ and $P(-3, -2)$

13. Determine whether a triangle with vertices at $A(3, 1)$, $B(6, 2)$, and $C(4, 3)$ is isosceles.

14. Use the distance formula to show that points $A(0, 9)$, $B(2, 6)$, and $C(4, 3)$ all lie on the same line.

In Review Exercises 15–20, give the domain and range of the given relation. Assume that x and y represent real numbers in each ordered pair (x, y).

15. $\{(2, 1), (4, 2), (5, 6), (6, 8), (7, 7)\}$

16. $\{(1, 2), (-2, -2), (3, 3), (3, 4), (3, 5)\}$

17. $3y - 7x = 12$

18. $y = 2x^2 - 10$

19. $2y = \sqrt{x - 10}$

20. $y = x^2 + x^4$

In Review Exercises 21–26, tell whether the given relation is a function. Assume that x and y represent real numbers in each ordered pair (x, y).

21. $\{(1, 1), (1, 2), (2, 3), (2, 4)\}$

22. $\{(1, 1), (2, 1), (3, 1), (4, 1)\}$

23. $x^2 - y^2 = 17$

24. $y = x^4 + 2$

25. $x^2 + y^2 = 25$

26. $x^2 - y^2 = 1$

In Review Exercises 27–30, find the required value if $f(x) = x^2 + 2x - 4$.

27. $f(0)$

28. $f(-3)$

29. $f(h)$

30. $f(a + h) - f(a)$

In Review Exercises 31–34, graph each equation. Use the vertical line test to decide whether the graph represents a function.

31. $y = \frac{1}{2}x - 3$

32. $4x - y^2 = 3$

33. $x^2 - y = 4$

34. $xy = 6$

In Review Exercises 35–42, find the inverse relation of the given relation and tell whether it is a function.

35. $\{(2, 1), (3, 1), (4, 1)\}$

36. $\{(1, 10), (2, 20), (3, 30), (4, 40)\}$

37. $y = 7x + 12$

38. $y = \dfrac{x + 17}{4}$

39. $3y = x^2$

40. $xy = 36$

41. $y = \sqrt{2x - 1}$

42. $\sqrt{y - 4} = x$

In Review Exercises 43–46, graph each relation and its inverse relation on the same set of coordinate axes.

43. $y = 4x + 2$

44. $3x - y + 1 = 0$

45. $y = \dfrac{x^2}{2}$

46. $xy = 4$

In Review Exercises 47–52, tell whether the given function is one-to-one. Assume that x and y represent real numbers.

47. $3y = x - 2$

48. $y = x^2 - 3$

49. $y = x^4$

50. $y = \sqrt{x - 3}$

51. $y = x^3$

52. $y^3 = x$

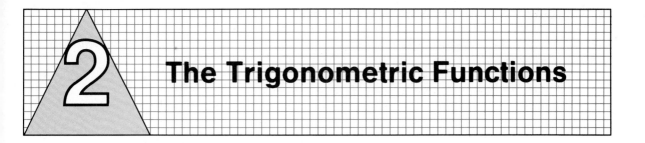

The Trigonometric Functions

The study of trigonometry is important for two reasons, one theoretical and one practical. The theoretical aspects of trigonometry appear in the development of many areas of mathematics, physics, biology, and chemistry. The practical uses of trigonometry appear in the fields of surveying, navigation, mechanical engineering, machine design, electrical engineering, and even carpentry.

2.1 FUNCTIONS OF TRIGONOMETRIC ANGLES

In geometry, an **angle** is defined as a figure formed by two rays originating from a common point, called the angle's **vertex**. A common measure of the size of an angle is the **degree**, denoted by the symbol °. An angle has a measure of 1° if it represents $\frac{1}{360}$ of one complete rotation.

We shall usually indicate fractional parts of a degree by using decimal fractions. For example,

$$\tfrac{1}{2}^\circ = 0.5^\circ \qquad \text{and} \qquad 33\tfrac{3}{4}^\circ = 33.75^\circ$$

Another method for expressing parts of a degree makes use of minutes, denoted by the symbol ′, and seconds, denoted by the symbol ″, where

$$1' = \frac{1}{60}(1^\circ) \qquad \text{and} \qquad 1'' = \frac{1}{60}(1') = \frac{1}{60}\left(\frac{1}{60}\right)(1^\circ) = \frac{1}{3600}(1^\circ)$$

Example 1 Change 37°24′18″ to decimal degrees.

Solution Because $24' = \frac{24}{60}(1^\circ)$, it follows that $24' = 0.4^\circ$. Because $18'' = \frac{18}{3600}(1^\circ)$, it follows that $18'' = 0.005^\circ$. Thus

$$37°24'18'' = 37^\circ + 0.4^\circ + 0.005^\circ = 37.405^\circ \qquad \blacksquare$$

Example 2 Change 83.41° to degrees-minutes-seconds.

Solution You must change 0.41° to minutes and seconds. Because 1° = 60′, you have

$$0.41(1)^\circ = (0.41)60'$$
$$0.41^\circ = 24.6'$$
$$= 24' + 0.6'$$

and because $1' = 60''$,

$$0.41° = 24' + (0.6)60''$$
$$= 24' + 36''$$

Thus, $83.41° = 83°24'36''$.   ■

In trigonometry, we shall define an angle in the following way.

Definition. Consider two rays with a common initial point O. A **trig-onometric angle** is the rotation required to move one ray, called the **initial side** of the angle, into coincidence with the other side, called the **terminal side** of the angle.

Angles that are rotations in a counterclockwise direction are considered to be positive. Angles that are rotations in a clockwise direction are considered to be negative.

This definition of a trigonometric angle is an extension of the definition of a geometric angle because one geometric angle can represent many different trigonometric angles. For example, the two rays forming the geometric angle in Figure 2-1 could represent any of the trigonometric angles in Figure 2-2.

To draw a trigonometric angle, we must indicate the rotation intended by including a curved arrow beginning at a point on the angle's initial side and ending at a point on the angle's terminal side.

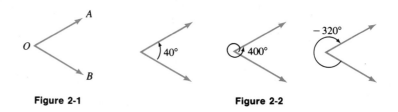

Figure 2-1   Figure 2-2

Definition. A trigonometric angle is said to be in **standard position** if and only if it is drawn on a rectangular coordinate system with its vertex at the origin and its initial side along the positive x-axis.

Such an angle is called a **first-**, **second-**, **third-**, or **fourth-quadrant angle** according to whether its terminal side lies in the first, second, third, or fourth quadrant. If the terminal side lies on the x- or y-axis, the angle is called a **quadrantal angle**.

Because each angle in Figure 2-3 has its initial side along the positive x-axis and its vertex at the origin, each one is in standard position.

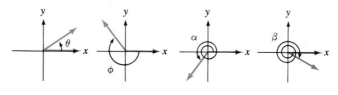

Figure 2-3

Angles in standard position are usually denoted by a letter, often a Greek letter, next to the curved arrow. Figure 2-3 illustrates first-quadrant angle θ (theta), second-quadrant angle ϕ (phi), third-quadrant angle α (alpha), and fourth-quadrant angle β (beta). Angles θ and α are positive angles, and angles ϕ and β are negative angles.

The angles shown in Figure 2-4 are not in standard position. In Figure 2-4a, the initial side of the angle is not on the positive x-axis. In Figure 2-4b, the vertex of the angle is not at the origin. In Figure 2-4c, the initial side of the angle is not on the positive x-axis.

> **Definition.**   If the terminal sides of two trigonometric angles in standard position coincide, the angles are called **coterminal**.

Because the three unequal trigonometric angles shown in Figure 2-5 are in standard position and share the same terminal side, they are coterminal.

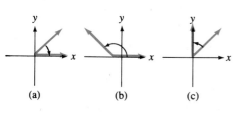

(a)   (b)   (c)

Figure 2-4

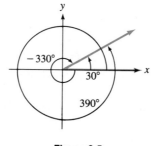

Figure 2-5

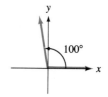

Figure 2-6

Example 3   Find three positive and three negative angles that are coterminal with an angle in standard position that measures 100°.

Solution   A trigonometric angle in standard position that measures 100° is shown in Figure 2-6. Three positive angles that are coterminal with it are shown in Figure 2-7. They are

$$100° + 360° \qquad \text{or} \qquad 460° \qquad (100° \text{ plus one revolution})$$
$$100° + 2(360°) \qquad \text{or} \qquad 820° \qquad (100° \text{ plus two revolutions})$$
$$100° + 3(360°) \qquad \text{or} \qquad 1180° \qquad (100° \text{ plus three revolutions})$$

Three negative angles that are coterminal with a 100° angle are also shown in Figure 2-7. They are

$$100° - 360° \qquad \text{or} \qquad -260° \qquad (100° \text{ minus one revolution})$$
$$100° - 2(360°) \qquad \text{or} \qquad -620° \qquad (100° \text{ minus two revolutions})$$
$$100° - 3(360°) \qquad \text{or} \qquad -980° \qquad (100° \text{ minus three revolutions})$$

As this example suggests, if n is a positive integer, then any angle of the form

$$100° \pm n360°$$

is coterminal with an angle of 100°.

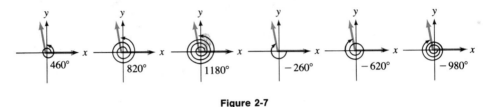

Figure 2-7   ∎

Trigonometric Functions

Most of the work in trigonometry is based on the following six trigonometric functions. The domain of each of these functions is a set of trigonometric angles, and the range is a set of real numbers.

Definition. Let θ be a trigonometric angle in standard position as shown in Figure 2-8. Let $P(x, y)$ be any point (except the origin) on the terminal side of angle θ. Then $r = \sqrt{x^2 + y^2}$ is the length of line segment OP. The **six trigonometric functions of the angle θ** are:

$\sin \theta = \dfrac{y}{r} \qquad$ read $\sin \theta$ as "sine theta"

$\cos \theta = \dfrac{x}{r} \qquad$ read $\cos \theta$ as "cosine theta"

$\tan \theta = \dfrac{y}{x} \qquad$ read $\tan \theta$ as "tangent theta"

$\csc \theta = \dfrac{r}{y} \qquad$ read $\csc \theta$ as "cosecant theta"

$\sec \theta = \dfrac{r}{x} \qquad$ read $\sec \theta$ as "secant theta"

$\cot \theta = \dfrac{x}{y} \qquad$ read $\cot \theta$ as "cotangent theta"

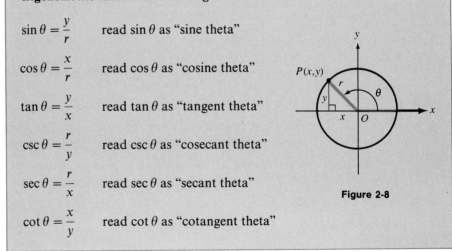

Figure 2-8

Remember that division by 0 is undefined. If $x = 0$, then $\tan \theta$ and $\sec \theta$ are undefined. If $y = 0$, then $\csc \theta$ and $\cot \theta$ are undefined.

The preceding definition implies that the values of trigonometric functions of coterminal angles are equal.

Example 4   The point $P(-3, 4)$ is on the terminal side of an angle ϕ. Draw angle ϕ in standard position and calculate the six trigonometric functions of ϕ.

Solution   Refer to Figure 2-9. The point $P(-3, 4)$ lies at a distance $r = \sqrt{(-3)^2 + 4^2} = \sqrt{25} = 5$ units from point O, and ϕ is a second-quadrant angle. From the definitions, you obtain

$$\sin \phi = \frac{y}{r} = \frac{4}{5}$$

$$\cos \phi = \frac{x}{r} = \frac{-3}{5} = -\frac{3}{5}$$

$$\tan \phi = \frac{y}{x} = \frac{4}{-3} = -\frac{4}{3}$$

$$\csc \phi = \frac{r}{y} = \frac{5}{4}$$

$$\sec \phi = \frac{r}{x} = \frac{5}{-3} = -\frac{5}{3}$$

$$\cot \phi = \frac{x}{y} = \frac{-3}{4} = -\frac{3}{4}$$

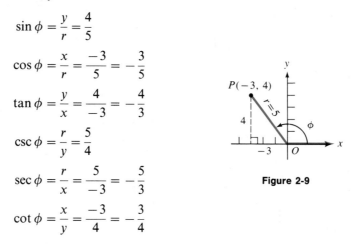

Figure 2-9

Any angle coterminal with angle ϕ will produce the same values for each of the six trigonometric functions. ∎

The following theorem enables us to use any point on the terminal side of angle θ (in standard position) to compute the six trigonometric functions of θ.

Theorem.   If angle θ is in standard position, and if P is some arbitrary point on the terminal side of angle θ (not the origin), then the values of the trigonometric functions of θ are independent of the choice of point P.

Proof   Let $P(x, y)$ and $P'(x', y')$ be two points on the terminal side of angle θ. Let $OP = r$ and $OP' = r'$. Points P, O, Q and P', O, Q' determine two triangles: triangle OPQ and triangle $OP'Q'$. See Figure 2-10. Since these triangles are similar, all ratios of corresponding sides are equal. In particular, we have

$$\sin \theta = \frac{y}{r} = \frac{y'}{r'}$$

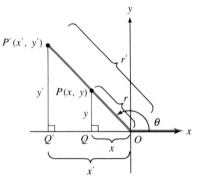

Figure 2-10

and

$$\cos \theta = \frac{x}{r} = \frac{x'}{r'}$$

This shows that the values of $\sin \theta$ and $\cos \theta$ are independent of the choice of point P. It can be shown in a similar manner that the values of the remaining four trigonometric functions are also independent of the choice of point P.   □

Example 5   Assume that $\sin \alpha = \frac{12}{13}$ and that α is in quadrant II. Find the values of the other five trigonometric functions.

Solution   Because any point on the terminal side of α can be used to compute $\cos \alpha$, choose a point P with a y-coordinate of 12 and $r = 13$, for then $\sin \alpha = \frac{y}{r} = \frac{12}{13}$ (see Figure 2-11). Then, substitute **12** for y and **13** for r in the equation $r = \sqrt{x^2 + y^2}$ and solve for x.

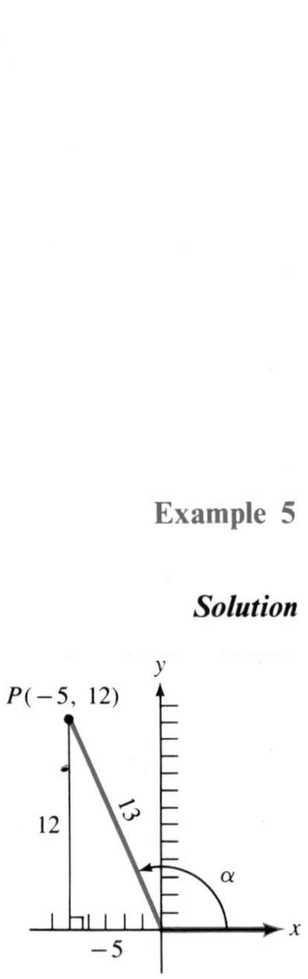

$$r = \sqrt{x^2 + y^2}$$
$$13 = \sqrt{x^2 + 12^2}$$
$$169 = x^2 + 144 \qquad \text{Square both sides.}$$
$$169 - 144 = x^2$$
$$25 = x^2$$
$$x = \pm 5$$

$P(-5, 12)$

12

13

α

-5

Figure 2-11

Because angle α is in quadrant II, the x-coordinate of P must be negative, so $x = -5$. Write the values of the other trigonometric functions by substituting the appropriate values of x, y, and r.

$$\cos \alpha = \frac{x}{r} = \frac{-5}{13} = -\frac{5}{13}$$

$$\tan \alpha = \frac{y}{x} = \frac{12}{-5} = -\frac{12}{5}$$

$$\csc \alpha = \frac{r}{y} = \frac{13}{12}$$

$$\sec \alpha = \frac{r}{x} = \frac{13}{-5} = -\frac{13}{5}$$

$$\cot \alpha = \frac{x}{y} = \frac{-5}{12} = -\frac{5}{12}$$

It is worthwhile to note in which quadrants each of the trigonometric functions is positive. Because r is always positive, the sine and cosecant functions are positive whenever y is positive. This occurs in quadrants I and II. Because x is positive in QI (quadrant I) and QIV (quadrant IV), the cosine and secant functions are positive in these quadrants also. The tangent and cotangent functions are positive when x and y agree in sign, and this happens in QI and QIII. Figure 2-12 will help you remember this information. The letters A, S, T, and C indicate the functions that are positive in the various quadrants. Some mathematics teachers remember these letters by fantasizing that All Students Take Calculus.

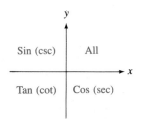

Figure 2-12

Example 6 Assume that angle β is not in QII and $\cos \beta = -\frac{3}{4}$. Find $\sin \beta$ and $\tan \beta$.

Solution Because $\cos \beta = -\frac{3}{4}$, angle β must be in QII or QIII. Because it is given that β is not in QII, it must be in QIII (see Figure 2-13). Choose point $P(x, y)$ so that $x = -3$ and $r = 4$ for $\cos \beta = -\frac{3}{4}$. Now apply the Pythagorean theorem to find y.

$$x^2 + y^2 = r^2$$
$$(-3)^2 + y^2 = 4^2$$
$$y^2 = 16 - 9$$
$$y^2 = 7$$
$$y = \pm\sqrt{7}$$

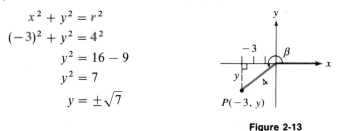

Figure 2-13

Choose y to be $-\sqrt{7}$ because y is negative in QIII. Once the values of x, y, and r have been determined, you can compute the required quantities:

$$\sin \beta = \frac{y}{r} = \frac{-\sqrt{7}}{4} = -\frac{\sqrt{7}}{4}$$

$$\tan \beta = \frac{y}{x} = \frac{-\sqrt{7}}{-3} = \frac{\sqrt{7}}{3}$$

The sine of β is negative and the tangent of β is positive in QIII, as expected.

Example 7   Assume that angle θ is not in QII and $\tan \theta = -\dfrac{40}{9}$. Find $\sin \theta$ and $\sec \theta$.

Solution   The tangent function is negative in QII and QIV. Because it is given that θ is not in QII, angle θ must be in QIV. Choose point $P(x, y)$ to be $P(9, -40)$ so that $\tan \theta = -\frac{40}{9}$ (see Figure 2-14). Thus, $y = -40$ and $x = 9$. Substitute these values into the equation $r = \sqrt{x^2 + y^2}$ to find r.

$$r = \sqrt{x^2 + y^2}$$
$$r = \sqrt{9^2 + (-40)^2}$$
$$r = \sqrt{1681}$$
$$r = 41$$

Finally, use the values of x, y, and r to compute the required quantities.

$$\sin \theta = \frac{y}{r} = \frac{-40}{41} = -\frac{40}{41}$$

$$\sec \theta = \frac{r}{x} = \frac{41}{9}$$

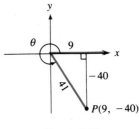

Figure 2-14

EXERCISE 2.1

In Exercises 1–4, change each angle to decimal degrees. Give each answer to the nearest thousandth.

1. $84°36'24''$ **2.** $25°48'6''$ **3.** $124°52'48''$ **4.** $356°12'20''$

In Exercises 5–8, change each angle to degrees-minutes-seconds.

5. $23.14°$ **6.** $73.87°$ **7.** $96.50°$ **8.** $224.48°$

In Exercises 9–20, tell whether the angle is in standard position. Indicate whether the angle is positive or negative.

9.

10.

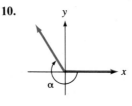

11.

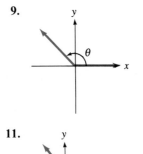

12.

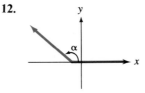

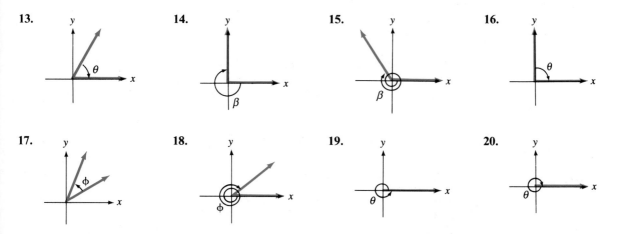

13.   14.   15.   16.

17.   18.   19.   20.

In Exercises 21–34, tell whether θ is a QI, QII, QIII, or QIV angle. Assume that θ is in standard position.

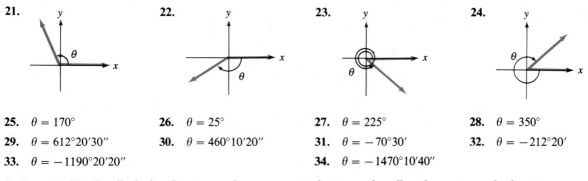

21.   22.   23.   24.

25. $\theta = 170°$
26. $\theta = 25°$
27. $\theta = 225°$
28. $\theta = 350°$

29. $\theta = 612°20'30''$
30. $\theta = 460°10'20''$
31. $\theta = -70°30'$
32. $\theta = -212°20'$

33. $\theta = -1190°20'20''$
34. $\theta = -1470°10'40''$

In Exercises 35–42, tell whether the given angles are coterminal. Assume that all angles are in standard position.

35. $40°, 400°$
36. $90°, -270°$

37. $135°, 270°$
38. $135°, -135°$

39. $740°, 380°$
40. $-340°, -700°$

41. $1035°30', -404°30'$
42. $10°40', -10°40'$

In Exercises 43–58, point P is on the terminal side of angle θ, which is in standard position. Draw a positive angle θ and calculate the six trigonometric functions of angle θ.

43. $P(3, 4)$
44. $P(5, 12)$
45. $P(-5, -12)$
46. $P(3, -4)$

47. $P(-9, 40)$
48. $P(9, -40)$
49. $P(1, 1)$
50. $P(-3, 3)$

51. $P(-3, 4)$
52. $P(-1, -1)$
53. $P(-4, 3)$
54. $P(24, 10)$

55. $P(3, 5)$
56. $P(-3, -4)$
57. $P(24, -10)$
58. $P(-5, 12)$

In Exercises 59–68, find the remaining trigonometric functions of angle θ. Assume that θ is positive and in standard position.

59. $\sin \theta = \dfrac{3}{5};$ θ in QI

60. $\tan \theta = 1;$ θ not in QI

61. $\cot \theta = \dfrac{5}{12}; \quad \cos \theta = -\dfrac{5}{13}$

62. $\cos \theta = \dfrac{\sqrt{5}}{5}; \quad \csc \theta = -\dfrac{\sqrt{5}}{2}$

63. $\sin \theta = -\dfrac{9}{41}; \quad \theta$ in QIV

64. $\tan \theta = -1: \quad \theta$ not in QII

65. $\sec \theta = -\dfrac{5}{3}; \quad \csc \theta = \dfrac{5}{4}$

66. $\tan \theta = -\dfrac{3}{5}; \quad \sin \theta = \dfrac{3\sqrt{34}}{34}$

67. $\tan \theta = -\dfrac{40}{9}; \quad \cos \theta = \dfrac{9}{41}$

68. $\sin \theta = -\dfrac{\sqrt{7}}{4}; \quad \cos \theta = -\dfrac{3}{4}$

In Exercises 69–74, explain why finding angle θ is impossible.

69. $\tan \theta = \dfrac{3}{4}$ and $\cot \theta = -\dfrac{4}{3}$

70. $\sin \theta = \dfrac{3}{4}$ and $\csc \theta = -\dfrac{4}{3}$

71. $\sin \theta = 2$

72. $\cos \theta = -\dfrac{1}{2}$ and θ in QIV

73. $\sec \theta = \dfrac{1}{2}$

74. $\csc \theta = 0$

2.2 THE EIGHT FUNDAMENTAL RELATIONSHIPS OF TRIGONOMETRIC FUNCTIONS

If the product of two numbers is 1, the numbers are called **reciprocals** of each other. For example, 2 and $\frac{1}{2}$ are reciprocals, as are $-\frac{3}{4}$ and $-\frac{4}{3}$. Such numbers as $\sqrt{2} + 1$ and $\sqrt{2} - 1$ are reciprocals also, because

$$(\sqrt{2} + 1)(\sqrt{2} - 1) = 1$$

Thus, the reciprocal of N does not always *look* like $\frac{1}{N}$.

The six trigonometric functions can be grouped in reciprocal pairs. For example, $\cos \theta$ and $\sec \theta$ are reciprocals because

$$\frac{x}{r} \quad \text{and} \quad \frac{r}{x}$$

are reciprocals. Likewise, $\sin \theta$ and $\csc \theta$ are reciprocals, as are $\tan \theta$ and $\cot \theta$.

Reciprocal Relationships. For any angle θ for which the functions are defined,

$$\sin \theta \csc \theta = 1 \qquad \sin \theta = \frac{1}{\csc \theta} \qquad \csc \theta = \frac{1}{\sin \theta}$$

$$\cos \theta \sec \theta = 1 \qquad \cos \theta = \frac{1}{\sec \theta} \qquad \sec \theta = \frac{1}{\cos \theta}$$

$$\tan \theta \cot \theta = 1 \qquad \tan \theta = \frac{1}{\cot \theta} \qquad \cot \theta = \frac{1}{\tan \theta}$$

There are many other algebraic relationships among the six trigonometric functions. These relationships are true only for those angles for which the functions are defined.

Example 1   Show that $\dfrac{\sin \theta}{\cos \theta} = \tan \theta$.

Solution   Using the definitions of the sine and cosine functions, you have

$$\frac{\sin \theta}{\cos \theta} = \frac{\dfrac{y}{r}}{\dfrac{x}{r}} = \frac{y}{r} \cdot \frac{r}{x} = \frac{y}{x} = \tan \theta \qquad ■$$

Example 2   Show that $\dfrac{\cos \theta}{\sin \theta} = \cot \theta$.

Solution   Using the definitions of the cosine and sine functions, you have

$$\frac{\cos \theta}{\sin \theta} = \frac{\dfrac{x}{r}}{\dfrac{y}{r}} = \frac{x}{r} \cdot \frac{r}{y} = \frac{x}{y} = \cot \theta \qquad ■$$

The six trigonometric functions involve the quantities x, y, and r, but, because of the Pythagorean theorem, these quantities are related by the formula $x^2 + y^2 = r^2$. This formula can be used to establish three more important relationships.

By agreement, $\sin^2\theta$ is standard notation for $(\sin \theta)^2$. Thus, $\sin^2\theta = \sin \theta \cdot \sin \theta$. Powers of the other trigonometric functions are denoted similarly.

Example 3   Use the result $y^2 + x^2 = r^2$ to show that $\sin^2\theta + \cos^2\theta = 1$.

Solution   Divide $y^2 + x^2 = r^2$ by r^2 to obtain

$$\frac{y^2}{r^2} + \frac{x^2}{r^2} = \frac{r^2}{r^2}$$

or

$$\left(\frac{y}{r}\right)^2 + \left(\frac{x}{r}\right)^2 = 1$$

Then use the definitions of the trigonometric functions to obtain

$$\sin^2\theta + \cos^2\theta = 1 \qquad ■$$

Example 4   Use the result $y^2 + x^2 = r^2$ to show that $\tan^2\theta + 1 = \sec^2\theta$.

Solution   Divide $y^2 + x^2 = r^2$ by x^2 and use the definitions of the trigonometric functions to obtain

$$\frac{y^2}{x^2} + \frac{x^2}{x^2} = \frac{r^2}{x^2}$$

$$\left(\frac{y}{x}\right)^2 + 1 = \left(\frac{r}{x}\right)^2$$

or

$$\tan^2\theta + 1 = \sec^2\theta \qquad \blacksquare$$

Example 5   Use the result $x^2 + y^2 = r^2$ to show that $\cot^2\theta + 1 = \csc^2\theta$.

Solution   Divide $x^2 + y^2 = r^2$ by y^2 to obtain

$$\frac{x^2}{y^2} + \frac{y^2}{y^2} = \frac{r^2}{y^2}$$

$$\left(\frac{x}{y}\right)^2 + 1 = \left(\frac{r}{y}\right)^2$$

or

$$\cot^2\theta + 1 = \csc^2\theta \qquad \blacksquare$$

The relationships developed so far are summarized in the following chart.

The Eight Fundamental Relationships.   For any trigonometric angle θ for which the functions are defined,

$$\sin\theta = \frac{1}{\csc\theta} \qquad\qquad \cos\theta = \frac{1}{\sec\theta}$$

$$\tan\theta = \frac{1}{\cot\theta} \qquad\qquad \tan\theta = \frac{\sin\theta}{\cos\theta}$$

$$\cot\theta = \frac{\cos\theta}{\sin\theta} \qquad\qquad \sin^2\theta + \cos^2\theta = 1$$

$$\tan^2\theta + 1 = \sec^2\theta \qquad \cot^2\theta + 1 = \csc^2\theta$$

Many other relationships can also be established.

Example 6   Show that $\tan\theta = \dfrac{\sec\theta}{\csc\theta}$.

Solution   Make use of the fundamental relationships of the trigonometric functions and proceed as follows.

$$\tan \theta = \frac{\sin \theta}{\cos \theta}$$

$$= \frac{\dfrac{1}{\csc \theta}}{\dfrac{1}{\sec \theta}}$$

$$= \frac{1}{\csc \theta} \cdot \frac{\sec \theta}{1}$$

$$= \frac{\sec \theta}{\csc \theta}$$ ■

Example 7 Show that $(\cot \theta \sec \theta + 1)(\cot \theta \sec \theta - 1) = \cot^2\theta$.

Solution Multiply $\cot \theta \sec \theta + 1$ and $\cot \theta \sec \theta - 1$ and proceed as follows:

$$(\cot \theta \sec \theta + 1)(\cot \theta \sec \theta - 1) = \cot^2\theta \sec^2\theta - 1$$

$$= \frac{\cos^2\theta}{\sin^2\theta} \cdot \frac{1}{\cos^2\theta} - 1 \qquad \cot^2\theta = \frac{\cos^2\theta}{\sin^2\theta} \quad \text{and} \quad \sec^2\theta = \frac{1}{\cos^2\theta}$$

$$= \frac{1}{\sin^2\theta} - 1 \qquad\qquad \text{Divide out } \cos^2\theta.$$

$$= \csc^2\theta - 1 \qquad\qquad \frac{1}{\sin^2\theta} = \csc^2\theta$$

$$= 1 + \cot^2\theta - 1 \qquad\qquad \csc^2\theta = 1 + \cot^2\theta$$

$$= \cot^2\theta$$ ■

Example 8 Assume that θ is in QII and that $\sin \theta = \frac{4}{5}$. Use the eight fundamental relationships of trigonometric functions to find the values of the other five trigonometric functions.

Solution Because the sine and cosecant of an angle are reciprocals, you have

$$\csc \theta = \frac{1}{\sin \theta} = \frac{1}{\frac{4}{5}} = \frac{5}{4}$$

Because $\sin^2\theta + \cos^2\theta = 1$, you have

$$\cos^2\theta = 1 - \sin^2\theta$$

$$\cos \theta = \pm\sqrt{1 - \sin^2\theta}$$

Because θ is in QII, $\cos \theta$ must be negative. Therefore, you have

$$\cos \theta = -\sqrt{1 - \left(\frac{4}{5}\right)^2} = -\sqrt{\frac{9}{25}} = -\frac{3}{5}$$

Because $\sec\theta$ is the reciprocal of $\cos\theta$, you have

$$\sec\theta = -\frac{5}{3}$$

Use the relationship $\tan\theta = \dfrac{\sin\theta}{\cos\theta}$ to find $\tan\theta$.

$$\tan\theta = \frac{\sin\theta}{\cos\theta} = \frac{\dfrac{4}{5}}{-\dfrac{3}{5}} = \frac{4}{5} \div \left(-\frac{3}{5}\right) = \frac{4}{5}\cdot\left(-\frac{5}{3}\right) = -\frac{4}{3}$$

Finally, $\cot\theta$ is the reciprocal of $\tan\theta$.

$$\cot\theta = -\frac{3}{4}$$ ∎

Example 9   Given that $\cos\theta = \frac{1}{3}$ and that $\tan\theta$ is negative, use the fundamental trigonometric relationships to find the values of the other five trigonometric functions.

Solution   First decide in which quadrant the terminal side of angle θ lies. Because $\cos\theta$ is positive, angle θ must be a QI or QIV angle. Because $\tan\theta$ is negative, angle θ must be a QII or QIV angle. Putting these facts together, you know that θ is a QIV angle. See Figure 2-15. From the eight basic trigonometric relationships, it follows that

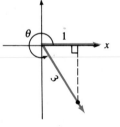

Figure 2-15

$$\sec\theta = \frac{1}{\cos\theta} = \frac{1}{\frac{1}{3}} = 3$$

Because $\tan^2\theta = \sec^2\theta - 1$, you have $\tan\theta = \pm\sqrt{\sec^2\theta - 1}$. Because $\tan\theta$ is given to be negative, substitute 3 for $\sec\theta$ in the equation $\tan\theta = -\sqrt{\sec^2\theta - 1}$ and simplify.

$$\tan\theta = -\sqrt{3^2 - 1}$$
$$= -\sqrt{8}$$
$$= -2\sqrt{2}$$

Because $\dfrac{\sin\theta}{\cos\theta} = \tan\theta$, it follows that $\sin\theta = \tan\theta\cos\theta$. To find $\sin\theta$, substitute $-2\sqrt{2}$ for $\tan\theta$ and $\frac{1}{3}$ for $\cos\theta$.

$$\sin\theta = -2\sqrt{2}\cdot\frac{1}{3}$$
$$= -\frac{2\sqrt{2}}{3}$$

To determine $\csc\theta$, find the reciprocal of $\sin\theta$.

$$\csc \theta = \frac{1}{\sin \theta}$$

$$= -\frac{3}{2\sqrt{2}}$$

$$= -\frac{3\sqrt{2}}{4} \qquad \text{Multiply numerator and denominator by } \sqrt{2} \text{ and simplify.}$$

Finally, because $\cot \theta$ is the reciprocal of $\tan \theta$, you have

$$\cot \theta = \frac{1}{-2\sqrt{2}}$$

$$= -\frac{\sqrt{2}}{4} \qquad \text{Multiply numerator and denominator by } \sqrt{2} \text{ and simplify.}$$

■

Trigonometric Functions of $(-\theta)$

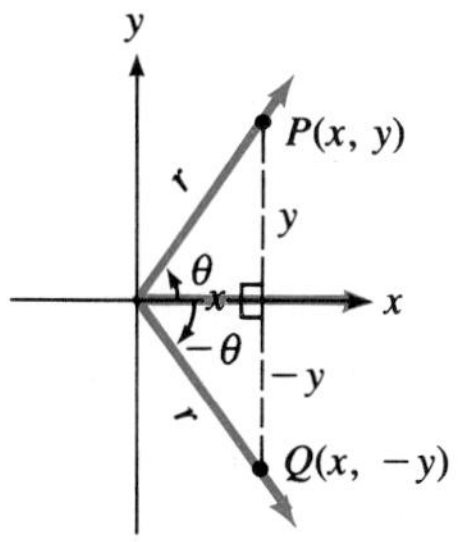

We consider three more basic relationships. Consider angles θ and $-\theta$, drawn in standard position as shown in Figure 2-16. Points P and Q are on the terminal sides of the respective angles. Let (x, y) be coordinates of point P, and let $(x, -y)$ be the coordinates of point Q. The trigonometric functions of these angles are

$$\begin{cases} \sin(-\theta) = \dfrac{-y}{r} \\[2mm] \sin \theta = \dfrac{y}{r} \end{cases} \quad \begin{cases} \cos(-\theta) = \dfrac{x}{r} \\[2mm] \cos \theta = \dfrac{x}{r} \end{cases} \quad \begin{cases} \tan(-\theta) = \dfrac{-y}{x} \\[2mm] \tan \theta = \dfrac{y}{x} \end{cases}$$

Figure 2-16

From these results, it follows that

$$\sin(-\theta) = -\sin \theta$$
$$\cos(-\theta) = \cos \theta$$
$$\tan(-\theta) = -\tan \theta$$

In the exercises you will be asked to verify these results for angles that are drawn in the second, third, and fourth quadrants.

Definition.   If $f(-x) = f(x)$ for all x, then f is called an **even function**. If $f(-x) = -f(x)$ for all x, then f is called an **odd function**.

Because $\cos(-\theta) = \cos \theta$, the cosine function is an even function. Because $\sin(-\theta) = -\sin \theta$ and $\tan(-\theta) = -\tan \theta$, the sine and tangent functions are odd functions.

Example 10 Show that $y = f(x) = x^2 + x^6$ is an even function.

Solution The function is an even function if and only if $f(-x) = f(x)$ for all x. Substitute $-x$ for x and simplify.

$$f(-x) = (-x)^2 + (-x)^6$$
$$= x^2 + x^6$$
$$= f(x) \quad \text{(for all } x)$$

Because $f(-x) = f(x)$ for all x, $f(x)$ is an even function. ∎

Example 11 Show that $y = f(\theta) = \cos \theta - \sin \theta$ is neither an even function nor an odd function.

Solution The function is an even function if and only if $f(-\theta) = f(\theta)$ for all θ. The function is an odd function if and only if $f(-\theta) = -f(\theta)$ for all θ. Calculate $f(-\theta)$ and $-f(\theta)$, and compare the results with $f(\theta)$.

$$f(\theta) = \cos \theta - \sin \theta$$
$$f(-\theta) = \cos(-\theta) - \sin(-\theta)$$
$$= \cos \theta + \sin \theta$$

and

$$-f(\theta) = -(\cos \theta - \sin \theta)$$
$$= -\cos \theta + \sin \theta$$

Because $f(-\theta) \neq f(\theta)$, the function $f(\theta) = \cos \theta - \sin \theta$ cannot be an even function. Because $f(-\theta) \neq -f(\theta)$, the function $f(\theta)$ cannot be an odd function. Hence, it is neither even nor odd. ∎

■ EXERCISE 2.2

In Exercises 1–10, use the letters x, y, and r and the definitions of the trigonometric functions to verify the following equations.

1. $\tan \theta \cos \theta = \sin \theta$

2. $\cot \theta \sin \theta = \cos \theta$

3. $\tan \theta = \dfrac{1}{\cos \theta \csc \theta}$

4. $\cot \theta = \cos \theta \csc \theta$

5. $\sin^2\theta + \sin^2\theta \cot^2\theta = 1$

6. $\cot \theta = \dfrac{\csc \theta}{\sec \theta}$

7. $\cot^2\theta + \sin^2\theta = \csc^2\theta - \cos^2\theta$

8. $\tan^2\theta + \cos^2\theta = \sec^2\theta - \sin^2\theta$

9. $\sin^2\theta + \dfrac{\cot^2\theta}{\csc^2\theta} = 1$

10. $\cos^2\theta + \dfrac{\tan^2\theta}{\sec^2\theta} = 1$

In Exercises 11–20, use one or more of the eight fundamental trigonometric relationships to verify the following equations.

11. $\tan \theta \cos \theta = \sin \theta$

12. $\cot \theta \sin \theta = \cos \theta$

13. $\tan \theta = \dfrac{1}{\cos \theta \csc \theta}$

14. $\cot \theta = \cos \theta \csc \theta$

15. $\sin^2\theta + \sin^2\theta \cot^2\theta = 1$

16. $\cot \theta = \dfrac{\csc \theta}{\sec \theta}$

17. $\cot^2\theta + \sin^2\theta = \csc^2\theta - \cos^2\theta$

18. $\tan^2\theta + \cos^2\theta = \sec^2\theta - \sin^2\theta$

19. $\sin^2\theta + \dfrac{\cot^2\theta}{\csc^2\theta} = 1$

20. $\cos^2\theta + \dfrac{\tan^2\theta}{\sec^2\theta} = 1$

In Exercises 21–32, use one or more of the eight fundamental trigonometric relationships to transform the left-hand side of each equation into the right-hand side.

21. $\dfrac{\sec \theta}{\csc \theta} = \tan \theta$

22. $\dfrac{\csc \theta}{\sec \theta} = \cot \theta$

23. $\cos \theta \tan \theta = \sin \theta$

24. $\sin \theta \cot \theta = \cos \theta$

25. $\dfrac{\sin^2\theta}{\sec^2\theta} + \sin^4\theta = \sin^2\theta$

26. $\dfrac{\cos^2\theta}{\csc^2\theta} + \cos^4\theta = \cos^2\theta$

27. $(\sec \theta + \tan \theta)(\sec \theta - \tan \theta) = 1$

28. $(\csc \theta + \cot \theta)(\csc \theta - \cot \theta) = 1$

29. $(1 + \cos \theta)(1 - \cos \theta) = \sin^2\theta$

30. $(1 + \sin \theta)(1 - \sin \theta) = \cos^2\theta$

31. $(\cos \theta \sec \theta + \cos \theta)(\cos \theta \sec \theta - \cos \theta) = \sin^2\theta$

32. $(\tan \theta \csc \theta + 1)(\tan \theta \csc \theta - 1) = \tan^2\theta$

In Exercises 33–42, use one or more of the eight fundamental trigonometric relationships to find the values of the remaining trigonometric functions.

33. θ in QI; $\sin \theta = \dfrac{4}{5}$

34. θ in QIII; $\sin \theta = -\dfrac{4}{5}$

35. θ in QII; $\cos \theta = -\dfrac{5}{13}$

36. θ in QIII; $\cos \theta = -\dfrac{5}{13}$

37. θ in QII; $\tan \theta = -\dfrac{4}{3}$

38. θ in QIV; $\tan \theta = -\dfrac{3}{4}$

39. θ in QI; $\cot \theta = \dfrac{9}{40}$

40. θ in QII; $\csc \theta = \dfrac{13}{5}$

41. $\sec \theta = -\dfrac{13}{5}$; θ in QIII

42. $\sec \theta = -\dfrac{13}{5}$; θ in QII

In Exercises 43–52, tell whether the given function is even, odd, or neither.

43. $y = x^3$

44. $y = x^2$

45. $y = x^4 + x^2$

46. $y = x^6 - 3x^4 + x$

47. $y = \cot \theta$

48. $y = \csc \theta + \sec \theta$

49. $y = \sin \theta + \cos \theta$

50. $y = \cos \theta + \sec \theta$

51. $y = 1 + x$

52. $y = -4 + \cos \theta$

53. Show that $y = \csc \theta$ is an odd function.

54. Show that $y = \cot \theta$ is an odd function.

55. Is the sum of two odd functions still odd?

56. Is the sum of two even functions still even?

57. Is the product of two odd functions still odd?

58. Is the product of two even functions still even?

59. Let θ be a second-quadrant angle. Draw a figure similar to Figure 2-16 and show that $\sin(-\theta) = -\sin \theta$, $\cos(-\theta) = \cos \theta$, and $\tan(-\theta) = -\tan \theta$.

60. Let θ be a third-quadrant angle. Draw a figure similar to Figure 2-16 and show that $\sin(-\theta) = -\sin\theta$, $\cos(-\theta) = \cos\theta$, and $\tan(-\theta) = -\tan\theta$.

61. Let θ be a fourth-quadrant angle. Draw a figure similar to Figure 2-16 and show that $\sin(-\theta) = -\sin\theta$, $\cos(-\theta) = \cos\theta$, and $\tan(-\theta) = -\tan\theta$.

62. If θ is a second-quadrant angle, is $\sin(-\theta)$ positive or negative?

63. If θ is a third-quadrant angle, is $\cos(-\theta)$ positive or negative?

64. Identify the product $\sin(-\theta)\cos(-\theta)\tan(-\theta)$ as positive or negative if θ is a second-quadrant angle.

65. Identify the product $\sin(-\theta)\cos(-\theta)\tan(-\theta)$ as positive or negative if θ is a third-quadrant angle.

66. Identify the product $\sin(-\theta)\cos(-\theta)\tan(-\theta)$ as positive or negative if θ is a fourth-quadrant angle.

2.3 TRIGONOMETRIC FUNCTIONS OF CERTAIN ANGLES

Certain angles are more common than others. For example, angles of 90° appear more often in geometric theorems than do angles of 89°. Angles of 30° and 45° receive more attention than angles of 31° and 43.5°. In this section, we will find values for the trigonometric functions of many of these common angles.

In Figure 2-17, an angle of 0° is placed in standard position with both its initial and terminal sides along the positive x-axis. The point $P(1, 0)$ lies on the terminal side and r, the distance of P from the origin, is 1 unit. Four of the six trigonometric functions of 0° are

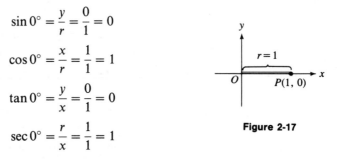

$$\sin 0° = \frac{y}{r} = \frac{0}{1} = 0$$

$$\cos 0° = \frac{x}{r} = \frac{1}{1} = 1$$

$$\tan 0° = \frac{y}{x} = \frac{0}{1} = 0$$

$$\sec 0° = \frac{r}{x} = \frac{1}{1} = 1$$

Figure 2-17

Values for $\cot 0°$ and $\csc 0°$ are undefined because division by 0 is undefined.

In a similar manner, the point $P(0, 1)$ lies on the terminal side of a 90° angle in standard position and at a distance $r = 1$ from the origin. See Figure 2-18. Again, four functions can be computed.

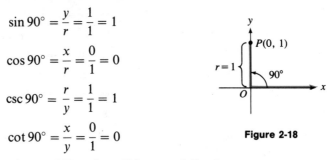

$$\sin 90° = \frac{y}{r} = \frac{1}{1} = 1$$

$$\cos 90° = \frac{x}{r} = \frac{0}{1} = 0$$

$$\csc 90° = \frac{r}{y} = \frac{1}{1} = 1$$

$$\cot 90° = \frac{x}{y} = \frac{0}{1} = 0$$

Figure 2-18

Values for $\tan 90°$ and $\sec 90°$ are undefined.

The point $P(-1, 0)$ lies on the terminal side of an angle of $180°$ that is placed in standard position. The distance of P from the origin is $r = 1$. Remember that r is always positive. See Figure 2-19. Four functions can be computed and the other two are undefined.

$$\sin 180° = \frac{y}{r} = \frac{0}{1} = 0$$

$$\cos 180° = \frac{x}{r} = \frac{-1}{1} = -1$$

$$\tan 180° = \frac{y}{x} = \frac{0}{-1} = 0$$

$$\sec 180° = \frac{r}{x} = \frac{1}{-1} = -1$$

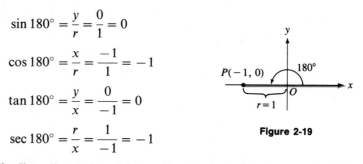

Figure 2-19

Finally, $P(0, -1)$ is a point on the terminal side of an angle of $270°$ placed in standard position. Again, four functions have values and two functions do not. See Figure 2-20.

$$\sin 270° = \frac{y}{r} = \frac{-1}{1} = -1$$

$$\cos 270° = \frac{x}{r} = \frac{0}{1} = 0$$

$$\cot 270° = \frac{x}{y} = \frac{0}{-1} = 0$$

$$\csc 270° = \frac{r}{y} = \frac{1}{-1} = -1$$

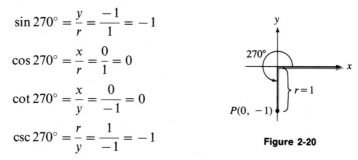

Figure 2-20

We summarize these results in a chart. This information should be memorized.

θ	$\sin \theta$	$\cos \theta$	$\tan \theta$	$\csc \theta$	$\sec \theta$	$\cot \theta$
$0°$	0	1	0	undefined	1	undefined
$90°$	1	0	undefined	1	undefined	0
$180°$	0	-1	0	undefined	-1	undefined
$270°$	-1	0	undefined	-1	undefined	0

Example 1   Find $\sin 990°$.

Solution   Because the sine of $990°$ is equal to the sine of any angle coterminal with $990°$, find an angle between $0°$ and $360°$ that is coterminal with $990°$. This can be done by repeatedly subtracting $360°$ from $990°$ until an angle less than $360°$ is found. Then, find the sine of that angle.

$$\sin 990° = \sin[990° - 2(360°)]$$
$$= \sin 270°$$
$$= -1$$

If $P(x, y)$ is a point on the terminal side of a 30° angle placed in standard position and r is the distance from P to the origin, then the triangle OPA shown in Figure 2-21 is a 30°–60° right triangle. Because the leg opposite the 30° angle measures half the length of the hypotenuse and the leg opposite the 60° angle is $\sqrt{3}$ times the leg opposite the 30° angle, it follows that $r = 2y$ and $x = \sqrt{3}y$. Thus,

$$\sin 30° = \frac{y}{r} = \frac{y}{2y} = \frac{1}{2}$$

$$\cos 30° = \frac{x}{r} = \frac{\sqrt{3}y}{2y} = \frac{\sqrt{3}}{2}$$

$$\tan 30° = \frac{y}{x} = \frac{y}{\sqrt{3}y} = \frac{1}{\sqrt{3}} = \frac{\sqrt{3}}{3}$$

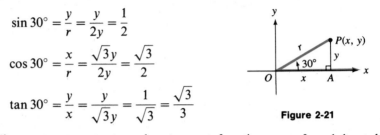

Figure 2-21

The cosecant, secant, and cotangent functions are found by taking the respective reciprocals of the sine, cosine, and tangent functions.

$$\csc 30° = \frac{1}{\sin 30°} = \frac{1}{\frac{1}{2}} = 2$$

$$\sec 30° = \frac{1}{\cos 30°} = \frac{1}{\frac{\sqrt{3}}{2}} = \frac{2}{\sqrt{3}} = \frac{2\sqrt{3}}{3}$$

$$\cot 30° = \frac{1}{\tan 30°} = \frac{1}{\frac{1}{\sqrt{3}}} = \sqrt{3}$$

A similar argument determines the trigonometric functions of a 60° angle. In Figure 2-22, triangle OPA is a 30°–60° right triangle with $r = 2x$ and $y = \sqrt{3}x$. Thus,

$$\sin 60° = \frac{y}{r} = \frac{\sqrt{3}x}{2x} = \frac{\sqrt{3}}{2}$$

$$\cos 60° = \frac{x}{r} = \frac{x}{2x} = \frac{1}{2}$$

$$\tan 60° = \frac{y}{x} = \frac{\sqrt{3}x}{x} = \sqrt{3}$$

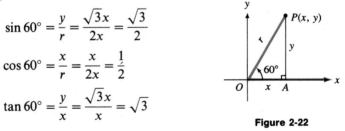

Figure 2-22

The following functions have values that are the reciprocals of the previous three.

$$\csc 60° = \frac{2}{\sqrt{3}} = \frac{2\sqrt{3}}{3}$$

$$\sec 60° = 2$$

$$\cot 60° = \frac{1}{\sqrt{3}} = \frac{\sqrt{3}}{3}$$

Because an angle of $45°$ is a base angle of the isosceles right triangle in Figure 2-23, it follows that $x = y$ and that $r = \sqrt{2}x = \sqrt{2}y$. Thus,

$$\sin 45° = \frac{y}{r} = \frac{y}{\sqrt{2}y}$$

$$= \frac{1}{\sqrt{2}} = \frac{\sqrt{2}}{2}$$

$$\cos 45° = \frac{x}{r} = \frac{x}{\sqrt{2}x}$$

$$= \frac{1}{\sqrt{2}} = \frac{\sqrt{2}}{2}$$

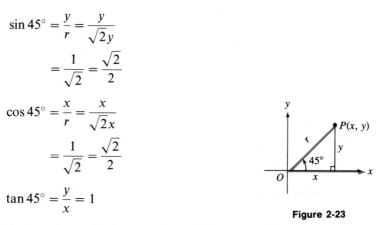

Figure 2-23

$$\tan 45° = \frac{y}{x} = 1$$

Taking the reciprocals of these values gives

$$\csc 45° = \sqrt{2}$$
$$\sec 45° = \sqrt{2}$$
$$\cot 45° = 1$$

We summarize all the results of this section in the following chart.

θ	$\sin\theta$	$\cos\theta$	$\tan\theta$	$\csc\theta$	$\sec\theta$	$\cot\theta$
$0°$	0	1	0	undefined	1	undefined
$30°$	$\dfrac{1}{2}$	$\dfrac{\sqrt{3}}{2}$	$\dfrac{\sqrt{3}}{3}$	2	$\dfrac{2\sqrt{3}}{3}$	$\sqrt{3}$
$45°$	$\dfrac{\sqrt{2}}{2}$	$\dfrac{\sqrt{2}}{2}$	1	$\sqrt{2}$	$\sqrt{2}$	1
$60°$	$\dfrac{\sqrt{3}}{2}$	$\dfrac{1}{2}$	$\sqrt{3}$	$\dfrac{2\sqrt{3}}{3}$	2	$\dfrac{\sqrt{3}}{3}$
$90°$	1	0	undefined	1	undefined	0
$180°$	0	-1	0	undefined	-1	undefined
$270°$	-1	0	undefined	-1	undefined	0

It is not necessary to memorize all of the values in the preceding chart. Many of them can be found by referring to the special right triangles in Figure 2-24 and using the definitions of the trigonometric functions.

By using the concept of a *reference triangle*, it is possible to determine the values of trigonometric functions of many more angles.

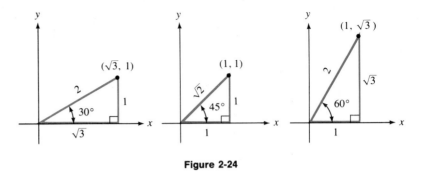

Figure 2-24

Example 2   Find the value of the six trigonometric functions of $\theta = 120°$.

Solution   Draw an angle of $120°$ in standard position and mark a point $P(x, y)$ on the terminal side at a distance r from the origin as in Figure 2-25. Draw segment PA so it is perpendicular to the x-axis. This forms right triangle OPA, which is called the **reference triangle**. Because P is in the second quadrant, x is negative and y is positive. In this triangle r is positive and triangle OPA is a $30°–60°$ right triangle, in which $r = -2x$ and $y = -\sqrt{3}x$. (Remember that if x is negative, $-2x$ and $-\sqrt{3}x$ are positive.) Thus,

$$\sin 120° = \frac{y}{r} = \frac{-\sqrt{3}x}{-2x} = \frac{\sqrt{3}}{2}$$

$$\cos 120° = \frac{x}{r} = \frac{x}{-2x} = -\frac{1}{2}$$

$$\tan 120° = \frac{y}{x} = \frac{-\sqrt{3}x}{x} = -\sqrt{3}$$

$$\csc 120° = \frac{1}{\sin 120°} = \frac{2}{\sqrt{3}} = \frac{2\sqrt{3}}{3}$$

$$\sec 120° = \frac{1}{\cos 120°} = -2$$

$$\cot 120° = \frac{1}{\tan 120°} = \frac{1}{-\sqrt{3}} = -\frac{\sqrt{3}}{3}$$

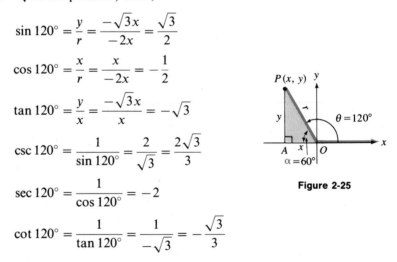

Figure 2-25

Note that angle α (not in standard position) formed by the terminal side of θ and the x-axis is $60°$. Also note that the value of the functions of $120°$ equal the values of the functions of a $60°$ angle, except for an occasional minus sign.   ■

Example 3   Find the values of the six trigonometric functions of $\theta = 225°$.

Solution   Draw angle θ in standard position. Place point $P(x, y)$ on the terminal side at a distance r from the origin and draw segment PA perpendicular to the x-axis to form the reference triangle OPA. See Figure 2-26. Because point P is in the third quadrant, x and y are both negative (and r is, of course, positive). Triangle OPA is an isosceles right triangle, so $x = y$ and $r = -\sqrt{2}x = -\sqrt{2}y$. Thus,

$$\sin 225° = \frac{y}{r} = \frac{y}{-\sqrt{2}y} = -\frac{1}{\sqrt{2}} = -\frac{\sqrt{2}}{2}$$

$$\cos 225° = \frac{x}{r} = \frac{x}{-\sqrt{2}x} = -\frac{1}{\sqrt{2}} = -\frac{\sqrt{2}}{2}$$

$$\tan 225° = \frac{y}{x} = 1$$

$$\csc 225° = \frac{1}{\sin 225°} = -\sqrt{2}$$

$$\sec 225° = \frac{1}{\cos 225°} = -\sqrt{2}$$

$$\cot 225° = \frac{1}{\tan 225°} = 1$$

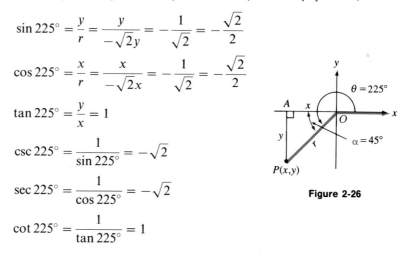

Figure 2-26

Note that the acute angle, angle α, between the terminal side of angle θ and the x-axis, is 45°. Also note that the six values of the functions of 225° equal the six values of the functions of 45°, except for sign.   ■

A pattern emerges from Examples 2 and 3. Associated with any angle θ is an acute angle in the reference triangle called the **reference angle**. The reference angle is that acute angle formed by the terminal side of angle θ and the x-axis. The reference angle is not always in standard position. However, if it were, it would be a first-quadrant angle with six positive values for the six trigonometric functions. The useful fact is this:

The six trigonometric functions of any angle θ are equal to those of the reference angle of θ, except possibly for sign.

The appropriate sign can be determined independently by considering the quadrant in which the terminal side of θ lies.

Example 4   Find the values of the six trigonometric functions of $\theta = 210°$.

Solution   Sketch an angle of 210° in standard position to determine in which quadrant the terminal side of θ lies. See Figure 2-27. Because $\theta = 210°$, the reference angle is 30°. Because 210° is a third-quadrant angle, only its tangent and its cotangent are positive. The other trigonometric functions of 210° are negative.

$$\sin 210° = -\sin 30° = -\frac{1}{2}$$

$$\cos 210° = -\cos 30° = -\frac{\sqrt{3}}{2}$$

$$\tan 210° = +\tan 30° = \frac{\sqrt{3}}{3}$$

Taking the reciprocals of the preceding results gives the values of the other functions.

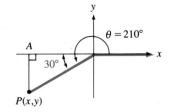

Figure 2-27

$$\csc 210° = \frac{1}{\sin 210°} = \frac{1}{-\dfrac{1}{2}} = -2$$

$$\sec 210° = \frac{1}{\cos 210°} = \frac{1}{-\dfrac{\sqrt{3}}{2}} = -\frac{2}{\sqrt{3}} = -\frac{2\sqrt{3}}{3}$$

$$\cot 210° = \frac{1}{\tan 210°} = \frac{1}{\dfrac{\sqrt{3}}{3}} = \frac{3}{\sqrt{3}} = \sqrt{3}$$ ■

Example 5   Find the values of the six trigonometric functions of $\theta = -405°$.

Solution   The sketch in Figure 2-28 determines the quadrant in which the angle θ terminates and also its reference angle. Recall that a negative angle has its rotation in a clockwise direction. Because an angle of $-405°$ is a fourth-quadrant angle, only its cosine and secant are positive.

$$\sin(-405°) = -\sin 45° = -\frac{\sqrt{2}}{2}$$

$$\cos(-405°) = +\cos 45° = +\frac{\sqrt{2}}{2}$$

$$\tan(-405°) = -\tan 45° = -1$$

Take the reciprocals of the preceding results to find the values of the other three functions.

$$\csc(-405°) = -\sqrt{2}$$
$$\sec(-405°) = +\sqrt{2}$$
$$\cot(-405°) = -1$$

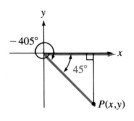

Figure 2-28

■

Example 6   If $\sin \theta = -\dfrac{1}{2}$ and $\cos \theta = \dfrac{\sqrt{3}}{2}$, find θ.

Solution   Because $\sin \theta$ is negative and $\cos \theta$ is positive, angle θ must be a QIV angle. See

Figure 2-29. Because

$$\sin 30° = \frac{1}{2} \quad \text{and} \quad \cos 30° = \frac{\sqrt{3}}{2}$$

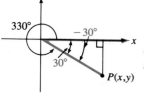

the reference angle must be 30°. There are infinitely many angles in QIV that have a reference angle of 30°. Two examples are $\theta = -30°$ and $\theta = 330°$. All such angles are of the form

$$330° \pm n \cdot 360°$$

Figure 2-29

where n is a nonnegative integer. ∎

■ EXERCISE 2.3

In Exercises 1–6, draw each angle in standard position and find the value of the sine, cosine, and tangent functions. **Do not use a calculator.**

1. 135°   **2.** 225°   **3.** 240°   **4.** 300°   **5.** 315°   **6.** 330°

In Exercises 7–12, draw each angle in standard position and find the value of the cosecant, secant, and cotangent functions. **Do not use a calculator.**

7. 225°   **8.** 135°   **9.** 300°   **10.** 240°   **11.** 330°   **12.** 315°

In Exercises 13–18, find the value of the sine, cosine, and tangent of the given angle. **Do not use a calculator.**

13. 390°   **14.** 480°   **15.** 510°   **16.** −690°   **17.** −45°   **18.** −150°

In Exercises 19–34, evaluate each expression. **Do not use a calculator.**

19. $\sin 0° + \cos 0° \tan 45°$

20. $\sin^2 90° + \cos 180° \tan 0°$

21. $\cos^2 90° + \cos 90° \sin^2 180°$

22. $\cos^2 0° + \sin^2 90° + \cot^2 90°$

23. $\sin^2 270° + \csc^2 270° + \cot^2 270°$

24. $\cos 180° \sin 180° - \tan^2 180°$

25. $\sin 30° \cos 60° - 2 \tan^2 60°$

26. $\sin^2 120° \cos 45° + \tan 45° \sin 90°$

27. $\sin 45° \cos 330° - \tan 150° \tan 60°$

28. $\cos 30° \tan 60° + \cos^2 45° \tan 45°$

29. $\csc^2 210° \sec 30° - \sec 315° \cot 60°$

30. $\csc 90° \csc 210° + \csc 45° \sin 135°$

31. $\sec 0° \tan 0° - \cot 30° \cot 45°$

32. $\cot^2 270° \csc 90° + \sec 60° \cot 30°$

33. $\csc^2 60° \sin^2 240°$

34. $\sec^2 45° \cos^2 315°$

In Exercises 35–48, use the given information to find values of θ, where $0° \leq \theta < 360°$. **Do not use a calculator.**

35. $\tan \theta = \frac{\sqrt{3}}{3}$; $\sin \theta = \frac{1}{2}$

36. $\tan \theta = -1$; $\cos \theta = \frac{-\sqrt{2}}{2}$

37. $\tan \theta = -\sqrt{3}$; $\cos \theta = \frac{1}{2}$

38. $\cot \theta = \sqrt{3}$; $\cos \theta = \frac{-\sqrt{3}}{2}$

39. $\sin \theta = -\frac{1}{2}$; $\sec \theta = \frac{-2\sqrt{3}}{3}$

40. $\cos \theta = \frac{\sqrt{3}}{2}$; $\csc \theta = 2$

41. $\tan \theta = -1$; $\sec \theta = \sqrt{2}$

42. $\tan \theta = -\sqrt{3}$; $\cos \theta = -\frac{1}{2}$

43. $\sec \theta = -\sqrt{2}; \quad \cot \theta = -1$

44. $\sin \theta = -1$

45. $\tan \theta$ is undefined.

46. $\csc \theta = -\sqrt{2}; \quad \cot \theta = -1$

47. $\cos \theta = \dfrac{\sqrt{3}}{2}; \quad \sin \theta = -\dfrac{1}{2}$

48. $\sin \theta = -\dfrac{\sqrt{3}}{2}; \quad \cos \theta = \dfrac{1}{2}$

In Exercises 49–60, use the given information to find all values of α, where possible, if $0° \le \alpha < 360°$.

49. $\sin \alpha = \dfrac{1}{2}$

50. $\cos \alpha = -\dfrac{\sqrt{3}}{2}$

51. $\tan \alpha = \dfrac{\sqrt{3}}{3}$

52. $\cot \alpha = -\sqrt{3}$

53. $\sec \alpha = -2$

54. $\csc \alpha = -2$

55. $\sin \alpha = -\dfrac{1}{2}; \alpha$ in QII

56. $\tan \alpha = -1; \alpha$ not in QII

57. $\cot \alpha = \sqrt{3}; \alpha$ not in QI

58. $\cos \alpha = \dfrac{1}{2}; \alpha$ in QIII

59. $\sin \alpha \cos \alpha = \dfrac{\sqrt{3}}{4}$

60. $\sin \alpha \cos \alpha = -\dfrac{\sqrt{3}}{4}$

2.4 TRIGONOMETRIC FUNCTIONS OF ANY ANGLE

Not every angle is one of the common angles considered in Section 2.3. We now discuss how to find values such as sin 23°, cos 313.27°, or tan 127°. One method requires a protractor, ruler, and compass to construct the **unit circle** (a circle with radius of 1 that is centered at the origin). Although this method is not accurate, it does reinforce the definitions of the trigonometric functions. A second method utilizes tables of values of trigonometric functions, such as Table A in Appendix IV. A third method is by far the easiest and most efficient. Use a calculator.

Example 1   Use a protractor, compass, and ruler to find approximate values for the sine, cosine, and tangent of an angle of 127°.

Solution   Extend the compass to a unit length as measured on the ruler and construct a circle with radius of 1. Use the protractor to draw an angle of 127° in standard position as in Figure 2-30a. Point P is the intersection of the circle and the terminal side of the 127° angle. Draw PA perpendicular to the x-axis. Because $OP = r = 1$,

$$\sin 127° = \frac{AP}{OP} = \frac{AP}{1} = AP$$

Carefully measure AP. That value, 0.8, is approximately equal to sin 127°. To find the cosine of 127°, also refer to Figure 2-30a, and note that

$$\cos 127° = \frac{OA}{OP} = \frac{OA}{1} = OA$$

Carefully measure OA and note that, because an angle of $127°$ is a second-quadrant angle, the directed distance from O to A is negative. That value, -0.6, is approximately equal to cos $127°$.

To find the tangent of $127°$, draw a tangent line to the unit circle at point B and extend it until it intersects with the extension of segment OP. See Figure 2-30b. The directed distance from O to B is -1, so

$$\tan 127° = \frac{BQ}{OB} = \frac{BQ}{-1} = -BQ$$

Carefully measure BQ. The negative of this measurement, -1.3, is approximately the value of tan $127°$.

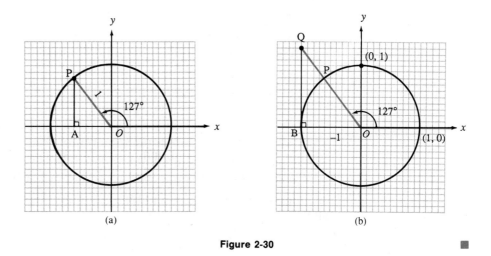

Figure 2-30

Example 2 Use Table A in Appendix IV to find the sine, cosine, tangent, and cotangent of 43.5°.

Solution In the **Degrees** column on the left-hand side of the table, locate the number 43.5. Move to the right in that row and read the entries in the columns headed by **Sin**, **Cos**, **Tan**, and **Cot**.

Radians	Degrees	Sin	Cos	Tan	Cot		
			...				
.7575	43.4	.6871	.7266	.9457	1.057	46.6	.8133
.7592	43.5	.6884	.7254	.9490	1.054	46.5	.8116
.7610	43.6	.6896	.7242	.9523	1.050	46.4	.8098
			...				
		Cos	**Sin**	**Cot**	**Tan**	**Degrees**	**Radians**

$$\sin 43.5° \approx 0.6884$$
$$\cos 43.5° \approx 0.7254$$
$$\tan 43.5° \approx 0.9490$$
$$\cot 43.5° \approx 1.054$$

■

Example 3   Use Table A in Appendix IV to find the sine, cosine, tangent, and cotangent of an angle of 46.5°.

Solution   When searching the **Degrees** column on the left-hand side of the table, you find that it ends at 45°. However, 46.5° can be found on the right-hand side in a column that is footed by **Degrees**. See the table in Example 2. Move to the left in the row containing 46.5° and read the entries in the columns footed by **Cos**, **Sin**, **Cot**, and **Tan**.

$$\sin 46.5° \approx 0.7254$$
$$\cos 46.5° \approx 0.6884$$
$$\tan 46.5° \approx 1.054$$
$$\cot 46.5° \approx 0.9490$$

■

The angles used in Examples 2 and 3 are **complementary angles** because $43.5° + 46.5° = 90°$. Note that the same row of the table was used in both examples. The table did double duty for these complementary angles because

$$\sin 43.5° = \cos 46.5° \qquad \text{and} \qquad \tan 43.5° = \cot 46.5°$$

are true statements. Because $\sin 43.5° = \cos 46.5°$, their reciprocals must be equal also. Therefore, we have

$$\csc 43.5° = \sec 46.5°$$

Definition.   The trigonometric functions of sine and cosine are called **cofunctions**. The tangent and cotangent functions are cofunctions, as are the secant and cosecant functions.

The sine of an acute angle θ is always equal to the cosine of the complement of θ. Likewise, the tangent of an acute angle θ is equal to the cotangent of the complement of θ, and the secant of an acute angle θ is equal to the cosecant of the complement of θ. These facts are summarized in the following theorem.

Theorem.   If θ is any acute angle, any trigonometric function of θ is equal to the cofunction of the complement of θ.

$$\sin \theta = \cos(90° - \theta) \qquad \text{and} \qquad \cos \theta = \sin(90° - \theta)$$
$$\tan \theta = \cot(90° - \theta) \qquad \text{and} \qquad \cot \theta = \tan(90° - \theta)$$
$$\csc \theta = \sec(90° - \theta) \qquad \text{and} \qquad \sec \theta = \csc(90° - \theta)$$

You will be asked to prove these cofunction relationships in the exercises. In a later chapter, you will see that these relationships can hold true for all values of θ.

Example 4 Use Table A in Appendix IV to find the values of the sine, cosine, tangent, and cotangent of an angle of 107°.

Solution Although an angle of 107° does not appear in the table, its reference angle, 73°, does. Remember that any trigonometric function of an angle in standard position can differ only in sign from that same trigonometric function of its reference angle. Because a 107° angle is a second-quadrant angle, only the sine and cosecant functions are positive. The rest are negative.

$$\sin 107° = +\sin 73° \approx \quad 0.9563$$
$$\cos 107° = -\cos 73° \approx -0.2924$$
$$\tan 107° = -\tan 73° \approx -3.271$$
$$\cot 107° = -\cot 73° \approx -0.3057 \qquad \blacksquare$$

Using Calculators

Only a few years ago tables provided the only practical way of finding values of trigonometric functions. However, the inexpensive pocket calculator has changed all that. Today the easiest and most efficient way to evaluate the trigonometric functions is to use a calculator.

Example 5 Use a calculator to find **a.** sin 23°, **b.** cos 313.27°, **c.** cos(−28.2°), and **d.** tan 90°.

Solution Set your calculator for degree measure of angles.

    a. To find sin 23°, enter the number 23 and press the $\boxed{\text{SIN}}$ key. The display should read .3907311285. To the nearest ten-thousandth,

$$\sin 23° = 0.3907$$

    b. To evaluate cos 313.27°, enter the number 313.27 and press the $\boxed{\text{COS}}$ key. The display should read 0.685437198. To the nearest ten-thousandth,

$$\cos 313.27° = 0.6854$$

    c. To evaluate cos(−28.2°), enter the number 28.2. Press the $\boxed{+/-}$ and $\boxed{\text{COS}}$ keys. The display should read .8813034521. To the nearest ten-thousandth,

$$\cos(-28.2°) = 0.8813$$

    d. Finally, to attempt to evaluate tan 90°, enter 90 and press the $\boxed{\text{TAN}}$ key. The display will either blink 9's, or read **ERROR**. This means that tan 90° is undefined. $\qquad \blacksquare$

Example 6 Find the value of sec 43°.

Solution   Neither tables nor calculators will allow you to evaluate the secant function directly. You must use the property that sec 43° is the reciprocal of cos 43°. To find sec 43° on a calculator, set your calculator for degrees, enter the number 43, and then press in order the $\boxed{\text{COS}}$ and $\boxed{1/x}$ keys. This gives the reciprocal of cos 43°, which is sec 43° ≈ 1.367327461. To the nearest ten-thousandth,

$$\sec 43° = 1.3673 \qquad \blacksquare$$

Example 7   If θ is an angle in QI and cos θ = 0.7660, find angle θ.

Solution   Angle θ can be found by using either a calculator or Table A.

If you use a calculator, be sure it is set for degrees. To find an angle whose cosine is known, most calculators provide an $\boxed{\text{INV}}$ (for inverse) key. To find an angle whose cosine is 0.7660, enter the number .7660 and press the $\boxed{\text{INV}}$ and $\boxed{\text{COS}}$ keys. If your calculator does not have an $\boxed{\text{INV}}$ key, consult your owner's manual. To the nearest tenth,

$$\theta = 40.0°$$

If you use Table A, find the column headed by **Cos** at the top of the page. Run your finger down the column, moving to successive pages if necessary, until you find the number .7660. Move to the left in that row to find the value of 40.0° in the degree column. $\qquad \blacksquare$

Example 8   If α is an acute angle and tan α = 5.671, find angle α.

Solution   Be sure your calculator is set for degrees. Enter the number 5.671 and press the $\boxed{\text{INV}}$ and $\boxed{\text{TAN}}$ keys in that order. To the nearest tenth of a degree,

$$\alpha = 80.0°$$

To use Table A, find the column footed by **Tan** at the bottom of the page. Run your finger up the column, moving to successive pages if necessary, to find the number 5.671. Move to the right to find the value of 80.0° in the degree column.

$\blacksquare$

Example 9   If θ is between 180° and 270° and sin θ = −0.9397, find angle θ.

Solution   See Figure 2-31. Any trigonometric function of θ has the same value as that same trigonometric function of θ's reference angle α, except possibly for sign. Enter the number .9397 in your calculator and press the $\boxed{\text{INV}}$ and $\boxed{\text{SIN}}$ keys. The display will give the value of the acute reference angle α. To the nearest tenth, $\alpha = 70.0°$. Because θ is a third-quadrant angle, add 180° to α to find angle θ. Thus,

$$\theta = 180° + 70.0°$$
$$= 250.0°$$

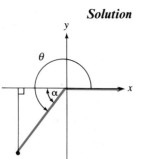

Figure 2-31   Table A can be used to find angle α if you do not have a calculator. $\blacksquare$

■ **EXERCISE 2.4** ■

In Exercises 1–4, use the unit circle in Illustration 1 to compute the sine, cosine, and tangent ratios of each angle.

1. 50°

2. 170°

3. 235°

4. 340°

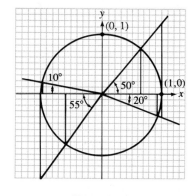

Illustration 1

In Exercises 5–10, draw the unit circle and use it to estimate the sine, cosine, and tangent ratios of each angle.

5. 12°   **6.** 72°   **7.** 115°   **8.** 175°   **9.** −260°   **10.** −310°

In Exercises 11–20, use Table A in Appendix IV to find the values of the sine, cosine, and tangent of each angle.

11. 17°   **12.** 29°   **13.** 62°   **14.** 82°   **15.** 119°   **16.** 121°

17. −233°   **18.** −182°   **19.** 1723°   **20.** 2811°

In Exercises 21–26, use a calculator to find the value of each function. Note that in each exercise, the second function is the cofunction of the first, and that the second angle is the complement of the first.

21. $\sin 20°$; $\cos 70°$ **22.** $\tan 9°$; $\cot 81°$ **23.** $\cos 5°$; $\sin 85°$ **24.** $\cot 64°$; $\tan 26°$

25. $\sec 84°$; $\csc 6°$ **26.** $\sec 12°$; $\csc 78°$

In Exercises 27–36, tell whether each statement is true for all acute angles A.

27. $\sin A = \sin(90° - A)$

28. $\cos A = \dfrac{1}{\sec A}$

29. $\sin^2 A + \sin^2(90° - A) = 1$

30. $\sin A = \cos(90° - A)$

31. $\tan A = \cot(90° - A)$

32. $\sin A = \dfrac{1}{\sec(90° - A)}$

33. $\cos A = \dfrac{1}{\csc(90° - A)}$

34. $\sin(90° - A) + \sin A = 1$

35. $\cot A = \dfrac{\cos(90° - A)}{\sin(90° - A)}$

36. $\cot A = \dfrac{\sin(90° - A)}{\cos(90° - A)}$

In Exercises 37–48, use a calculator to compute each value to four decimal places.

37. $\sin 23.1°$ **38.** $\sin 57.8°$ **39.** $\cos 133.7°$ **40.** $\cos 211.7°$

41. $\tan 223.5°$ **42.** $\tan(-223.5°)$ **43.** $\csc 312.4°$ **44.** $\csc 129.2°$

45. $\sec(-47.4°)$ **46.** $\sec 11.3°$ **47.** $\cot 640°36'$ **48.** $\cot 302°12'$

In Exercises 49–60, angle θ (0° ≤ θ < 360°) is in a given quadrant and the value of a trigonometric function is known. Use a calculator to find angle θ to the nearest tenth of a degree.

49. QI; $\tan \theta = 0.2493$ **50.** QI; $\sin \theta = 0.9986$

51. QII; $\cos \theta = -0.3420$ **52.** QII; $\cos \theta = -0.9063$

53. QIII; $\sin \theta = -0.4540$ **54.** QIII; $\cos \theta = -0.7193$

55. QIV; $\tan \theta = -5.6713$ **56.** QIV; $\sin \theta = -0.1908$

57. QI; $\csc \theta = 1.3250$ **58.** QII; $\sec \theta = -57.2987$

59. QIII; $\cot \theta = 1.1918$ **60.** QIV; $\csc \theta = -11.4737$

61. Devise a method to use the unit circle to find the secant, cosecant, and cotangent of an angle θ.

62. In Illustration 2, angle A is complementary to angle B. Prove that $\sin A = \cos B$, $\tan A = \cot B$, and $\csc A = \sec B$.

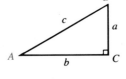

Illustration 2

2.5 RIGHT TRIANGLE TRIGONOMETRY

The word *trigonometry* means measuring triangles (trigons). Because it gives us the ability to determine all sides and all angles of a right triangle when only some are known, trigonometry is indispensable in astronomy, navigation, and surveying. In this section, we consider several applications of right triangle trigonometry.

Example 1   A right triangle ABC has an acute angle A that equals 27° and a hypotenuse c of length 14 feet. Solve the triangle by finding the other acute angle and the lengths of the two unknown sides.

Solution   Let the lengths of the two unknown sides of right triangle ABC be represented by a and b. Then, triangle ABC may be drawn in QI of a coordinate system as in Figure 2-32. In reference triangle ABC, the angle of 27° is the reference angle, $r = c = 14$, and point $B(x, y) = B(b, a)$. Because the triangle is a right triangle, angle A and angle B are complementary. Thus, angle $B = 63°$. The values of a and b can be found by using two trigonometric functions.

$$\sin 27° = \frac{a}{c} = \frac{a}{14}$$

$$\cos 27° = \frac{b}{c} = \frac{b}{14}$$

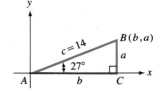

Use a calculator and solve these equations for a and b.

Figure 2-32

$$a = 14 \sin 27°$$
$$\approx 14(0.4539904997)$$
$$\approx 6.355866996$$
$$\approx 6.4$$

Similarly, we have

$$b = 14 \cos 27°$$
$$\approx 14(0.8910065242)$$
$$\approx 12.47409134$$
$$\approx 12$$

The remaining two sides of the triangle are 6.4 feet and 12 feet.   ■

Accuracy and Significant Digits

You may be curious why the answers in Example 1 were rounded off to two digits (called **significant digits**). If the hypotenuse were *exactly* 14 feet and the acute angle *exactly* 27°, rounding off would be unnecessary. More likely, however, 14 feet and 27° are only approximate. In that case, the impressive strings of decimal digits are unwarranted.

Example 2   In Example 1, the hypotenuse of the triangle was measured to two significant digits and the angle to the nearest degree. How many significant digits are appropriate in the answers?

Solution   Because the hypotenuse is measured to two significant digits as 14 feet, the actual value lies between 13.5 and 14.5 feet. The angle of 27° lies between 26.5° and 27.5°. Calculate the values of a for the worst possible cases. Using the smallest possible values for the hypotenuse and angle gives

$$a = 13.5 \sin 26.5° \approx 6.023670477$$

The complete answer from Example 1 is

$$a = 14 \sin 27° \approx 6.355866996$$

Using the largest possible values of the hypotenuse and angle gives

$$a = 14.5 \sin 27.5° \approx 6.695354892$$

The preceding results agree only in the units digit of 6. However impressive they might appear, all the digits to the right of the decimal points are meaningless. Yet because the hypotenuse is measured to two significant digits, convention allows rounding all sides to two significant digits also.   ■

Calculators routinely provide answers to 8, 10, or 12 figures. You must decide how many of these are significant. Resist the temptation of thinking "If I have them, I'll use them." A good rule of thumb for determining acceptable accuracy is provided in the following table.

Accuracy in measurements of sides	Accuracy in angles
Two significant digits	Nearest degree
Three significant digits	Nearest tenth of a degree or nearest 10′
Four significant digits	Nearest hundredth of a degree or nearest 1′

When solving triangles, remember that answers can be only as accurate as the least accurate of the given data. However, if a calculation requires several intermediate steps, *do not round off until you have the final answer.*

It is not always easy to decide how many significant digits a number has. For example, is the number 140 accurate to two or three significant digits? If the number is rounded to the nearest ten, the zero is merely a placeholder and 140 has two significant digits. On the other hand, if the number has been rounded to the nearest unit, the zero is significant and 140 has three significant digits. In this book, we will assume the greatest possible number of significant digits unless stated otherwise.

3234 has four significant digits.
104 has three significant digits.
140.00 has five significant digits.
0.00012 has two significant digits.
0.000120 has three significant digits.
0.0003 has one significant digit.
1.0003 has five significant digits.

Solving Right Triangles

To solve right triangles, it is helpful to view the definitions of the trigonometric functions in a different light. We place right triangle *ABC* on a coordinate system so it is the reference triangle for the acute angle *A*. See Figure 2-33. Let *a* be the length of *BC*, the side opposite angle *A*. Let *b* be the length of *AC*, the side

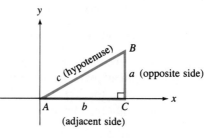

Figure 2-33

adjacent to angle A. Finally, let the hypotenuse have length c. The six trigonometric functions of the acute angle A can be defined as ratios involving sides of the right triangle ABC.

Definition. If angle A is an acute angle in right triangle ABC, then

$$\sin A = \frac{\text{opposite side}}{\text{hypotenuse}} \qquad \csc A = \frac{\text{hypotenuse}}{\text{opposite side}}$$

$$\cos A = \frac{\text{adjacent side}}{\text{hypotenuse}} \qquad \sec A = \frac{\text{hypotenuse}}{\text{adjacent side}}$$

$$\tan A = \frac{\text{opposite side}}{\text{adjacent side}} \qquad \cot A = \frac{\text{adjacent side}}{\text{opposite side}}$$

The terminology of "adjacent" and "opposite" sides is relative to acute angles of right triangles only. Note that a side adjacent to one acute angle of a right triangle is opposite the other acute angle and that the definitions above are a result of the earlier definitions of the six basic trigonometric functions.

If an observer looks up at an object (an airplane, the sun, or a cloud, perhaps), the angle that the observer's line of sight makes with the horizontal is called the **angle of elevation**. If the observer looks down to see the object, the angle made with the horizontal is called the **angle of depression**. See Figure 2-34.

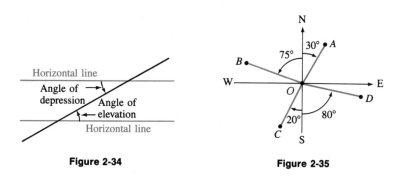

Figure 2-34

Figure 2-35

In nautical navigation and surveying, the concept of bearing is used. See Figure 2-35. The **bearing** of point A from point O (the observer) is the acute angle measured from the north–south line to the line segment OA. This bearing is denoted as N 30° E and is read as "north 30° east" or "30° east of north." The bearing of point B from O is N 75° W, the bearing of point C from O is S 20° W, and the bearing of point D from O is S 80° E.

Example 3 From a location 23.0 meters from the base of a flagpole, the angle of elevation to the top of the flagpole is 37°30′. How tall is the flagpole?

Solution   Consider the right triangle in Figure 2-36. The flagpole is the side opposite the 37°30′ angle and 23.0 meters is the length of the adjacent side. First, change 37°30′ to 37.5°, using the process shown in Section 2.1. Now, because the tangent of the angle involves the opposite and the adjacent sides, you have

$$\tan 37.5° = \frac{\text{opposite side}}{\text{adjacent side}}$$

$$= \frac{h}{23.0}$$

Solve for *h*.

$$h = 23.0 \tan 37.5°$$

$$\approx (23.0)(0.7673)$$

$$\approx 17.6479$$

$$\approx 17.6$$

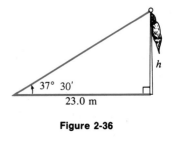

37° 30′

23.0 m

Figure 2-36

The flagpole is 17.6 meters tall. Note that the answer has been properly rounded to three significant digits, which is consistent with the accuracy of the data. ■

Example 4   A circus tightrope walker ascends from the ground to a platform 75 feet above the arena by walking a taut cable that is 92 feet long. At what angle is the cable from the horizontal?

Solution   Figure 2-37 helps determine the various parts of a triangle. You are given the lengths of the hypotenuse and a side, and must find angle θ. To do so, first use the sine ratio to find the value of sin θ.

$$\sin \theta = \frac{\text{opposite side}}{\text{hypotenuse}}$$

$$= \frac{75}{92}$$

$$\approx 0.8152$$

92 ft

75 ft

θ

Figure 2-37

If you look up .8152 in the body of Table A in Appendix IV, or enter .8152 and press $\boxed{\text{INV}}$ $\boxed{\text{SIN}}$ on a calculator, you will find that θ ≈ 55°. This result is accurate to the nearest degree and is consistent with the accuracy of the given sides. ■

Example 5   Perryville is 25.0 miles due south of Rock City, and Prairie Town is 90.0 miles due east of Rock City. What is the bearing of Prairie Town from Perryville?

Solution   See Figure 2-38. To find the bearing, find angle θ by using the tangent ratio.

$$\tan \theta = \frac{\text{opposite side}}{\text{adjacent side}}$$

$$= \frac{90}{25} = 3.6$$

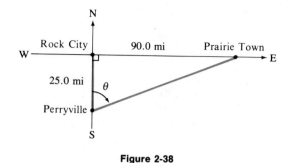

Figure 2-38

The angle with tangent of 3.6 is 74.5° (from either a calculator or the tables). Because the distances are accurate to three significant digits, the angle is given to the nearest tenth of a degree. The bearing of Prairie Town from Perryville is N 74.5° E. Incidentally, the bearing of Perryville from Prairie Town is S 74.5° W. ∎

Example 6 Two forest fire lookouts are on a north–south line, 17.2 miles apart. The bearing of a fire from lookout point A is S 27.0° W and the bearing of the fire from point B is N 63.0° W. How far from B is the fire?

Solution See Figure 2-39. Because the sum of the three angles of any triangle must be 180°, the angle at point F is 90°, so triangle AFB is a right triangle with hypotenuse AB. Find the distance that B is from the fire by using the cosine ratio.

$$\cos 63.0° = \frac{\text{adjacent side}}{\text{hypotenuse}}$$

$$= \frac{a}{17.2}$$

Solve for a.

$$a = 17.2 \cos 63.0°$$

$$\approx 7.81$$

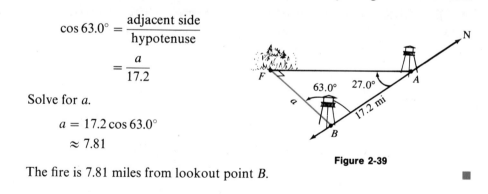

Figure 2-39

The fire is 7.81 miles from lookout point B. ∎

Example 7 A television tower stands on top of a building. From a point 75.3 feet from the base of the building, the angles of elevation to the top and the base of the tower are 60.1° and 47.4°, respectively. How tall is the tower?

Solution See Figure 2-40. You can find the total height H of the structure by using the tangent ratio.

$$\tan 60.1° = \frac{H}{75.3}$$

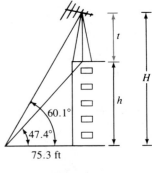

$$75.3(\tan 60.1°) = H$$
$$75.3(1.7391) \approx H$$
$$130.95 \approx H$$

The total height of the structure is approximately 130.95 feet.

You can find the height h of the building by using the tangent ratio.

$$\tan 47.4° = \frac{h}{75.3}$$

$$75.3(\tan 47.4°) = h$$
$$75.3(1.0875) \approx h$$
$$81.89 \approx h$$

Figure 2-40

The height of the building is approximately 81.89 feet.

The height t of the tower is the difference between the total height of the structure and the height of the building. Thus, you have

$$t = H - h$$
$$\approx 130.95 - 81.89$$
$$\approx 49.06$$

The height of the tower is approximately 49.1 feet. Note that the final answer is rounded off to the proper degree of accuracy. ■

■ EXERCISE 2.5

In Exercises 1–4, solve each triangle by finding the unknown sides and angles.

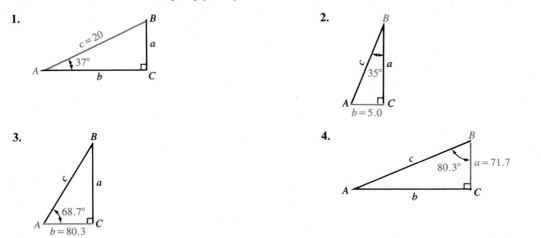

5. If the angle of elevation to the top of a flagpole from a point 40 feet from its base is 20°, find the height of the flagpole.

6. A person on the edge of a cliff looks down at a boat on a lake. The angle of depression of the person's line of sight is 10.0°, and the line-of-sight distance from the person to the boat is 555 feet. How high is the cliff?

7. A car drives up a long hill on a road that makes an angle of 7°30′ with the horizontal. How far will the car travel to reach the top if the horizontal distance traveled is 715 feet?

8. On an approach to an airport a plane is descending at an angle of 3°30′. How much altitude is lost as the plane travels a horizontal distance of 21.2 miles?

9. A plane loses 2750 feet in altitude as it travels a horizontal distance of 39,300 feet. What is its angle of descent to the nearest tenth of a degree?

10. A train rises 230 feet as it travels 1 mile up a steep grade. To the nearest tenth of a degree, what is its angle of ascent? (*Hint:* 5280 feet = 1 mile.)

11. The angle of depression from a point in the top of a tree to a point on the ground is 57°45′. Find the height of the tree if the line-of-sight distance from the top of the tree to the point on the ground is 34.23 feet.

12. An observer noted that the angle of elevation to a plane passing over a landmark was 32°40′. If the landmark was 1530 meters from the observer, what was the altitude of the plane?

13. A ship leaves from a port of call on a bearing of S 12.7° E. How far south has the ship traveled during a trip of 327 miles?

14. The bearing of Madison, Wisconsin, from Stevens Point, Wisconsin, is S 4.1° E. The distance between the cities is 108 miles. By how much is Madison east of Stevens Point?

15. A ship leaves port and sails 8800 kilometers due west. It then sails 4500 kilometers due south. To the nearest tenth of a degree, what is the ship's bearing from its port?

16. A ship is 3.3 miles from a lighthouse. It is also due north of a buoy that is 2.5 miles due east of the lighthouse. Find the bearing of the ship from the lighthouse.

17. Two lighthouses are on an east–west line. The bearing of a ship from one lighthouse is N 59° E, and the bearing from the other lighthouse is N 31° W. How far apart are the lighthouses if the ship is 5.0 miles from the first lighthouse?

18. Two lookout stations are on a north–south line. The bearing of a forest fire from one lookout is S 67° E, and the bearing of the fire from the second lookout is N 23° E. If the fire is 3.5 kilometers from the second lookout station, how far is the fire from the other lookout station?

19. A plane flying at 18,100 feet passes directly over an observer. Thirty seconds later, the observer notes that the plane's angle of elevation is 31.0°. How fast is the plane going in miles per hour? (*Hint:* 5280 feet = 1 mile.)

20. A plane flying horizontally at 650 miles per hour passes directly over a small city. One minute later the pilot notes that the angle of depression to that city is 13°. What is the plane's altitude in feet?

21. Use the information in Illustration 1 to compute the height of George Washington's face on Mount Rushmore.

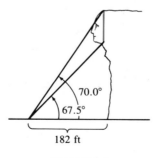

70.0°

67.5°

182 ft

Illustration 1

22. A boat is 537 meters from a lighthouse and has a bearing from the lighthouse of N 33.7° W. A second boat is 212 meters from the same lighthouse and has a bearing from the lighthouse of S 20.1° W. How many meters north of the second boat is the first?

23. A plane is flying at an altitude of 5120 feet. As it approaches an island, the navigator determines the angles of depression as in Illustration 2. What is the length of the island in feet and in miles?

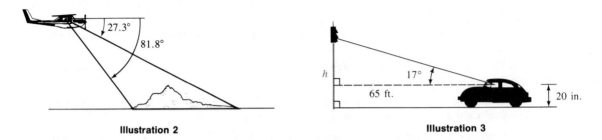

Illustration 2   **Illustration 3**

24. From an observation point 6500 meters from a launch site, an observer watches the vertical flight of a rocket. At one instant, the angle of elevation of the rocket is 15°. How far will the rocket ascend in the time it takes the angle of elevation to increase by 57°?

25. The sun visor in Sue's car prevents her from seeing objects more than 17° above the horizon. When her car is 65 feet from the intersection, Sue can barely see the bottom of the stoplight hanging above the roadway. How high is the stoplight? (See Illustration 3.)

26. The top of the stepladder in Illustration 4 is 8.0 feet above the floor, and the legs open at an angle θ. For stability, the legs of the ladder must be between 3.5 and 4.5 feet apart. What range of angles θ is possible?

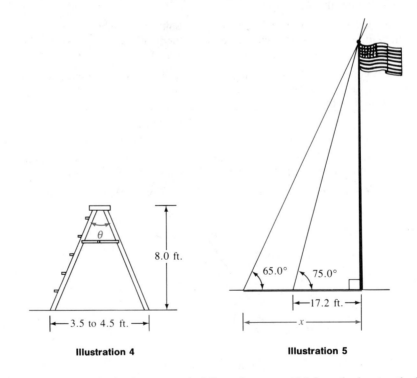

Illustration 4   **Illustration 5**

27. When the sun is 75.0° above the horizon, a vertical flagpole casts a 17.2 foot shadow on the horizontal ground. When the sun sinks to 65.0° above the horizon, how long is the shadow? (See Illustration 5.)

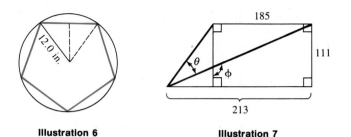

| Illustration 6 | Illustration 7 |

28. Find the perimeter of the regular pentagon inscribed in a circle with a radius of 12.0 inches. (See Illustration 6.)

29. Refer to Illustration 7 and find θ.

30. Refer to Illustration 7 and find ϕ.

2.6 MORE RIGHT TRIANGLE TRIGONOMETRY

We must work with two triangles simultaneously to solve many right triangle trigonometry problems.

Example 1 An observer A notices that the Space Needle of Seattle is due north, and the angle of elevation to its top is 44.4°. A second observer, B, 706 feet due east of A, notices that the bearing of the Space Needle is N 48.8° W. How tall is the Space Needle?

Solution In this problem you must consider two triangles: triangle PAB, which lies on the ground, and triangle APQ, which sits on its edge with the Space Needle as one of its sides. See Figure 2-41. Not enough information is given about triangle APQ to determine the height h from that triangle alone. If the distance x were known, however, h could be computed. From triangle APQ, you have

$$\tan 44.4° = \frac{h}{x} \quad \text{or} \quad h = x \tan 44.4°$$

The distance x may be found by working with the triangle on the ground. Triangle PAB is a right triangle with a right angle at A. Angle PBA is

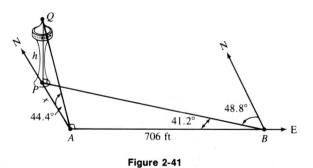

Figure 2-41

$90° − 48.8° = 41.2°$. Form the equation

$$\tan 41.2° = \frac{x}{706} \quad \text{or} \quad x = 706 \tan 41.2°$$

Putting these facts together gives

$$h = 706 \tan 41.2° \tan 44.4°$$
$$\approx 706(0.8754)(0.9793)$$
$$\approx 605.2391293$$
$$\approx 605$$

Seattle's Space Needle is approximately 605 feet tall.   ■

Example 2   Two Coast Guard lookouts, A and B, are on an east–west line, 21.3 kilometers apart. The bearing of a ship from lookout A is N 23.0° W, and the bearing of the ship from lookout B is N 65.0° E. How far is the ship from lookout B?

Solution   See Figure 2-42 in which triangle BAS is not a right triangle. However, if the perpendicular line segment SD of length h units is drawn from point S to side BA, then two right triangles are formed. If you let y represent the length of BD, then $21.3 − y$ will represent the length of DA. Then,

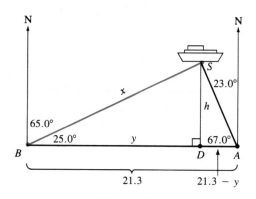

Figure 2-42

$$\tan 25.0° = \frac{h}{y} \quad \text{or} \quad h = y(\tan 25.0°)$$

and

$$\tan 67.0° = \frac{h}{21.3 − y} \quad \text{or} \quad h = (21.3 − y)\tan 67.0°$$

Setting the two values of h equal to each other and solving for y gives

$$y \tan 25.0° = (21.3 − y)\tan 67.0°$$
$$y \tan 25.0° = 21.3(\tan 67.0°) − y \tan 67.0°$$

$$y \tan 25.0° + y \tan 67.0° = 21.3 \tan 67.0°$$

$$y(\tan 25.0° + \tan 67.0°) = 21.3 \tan 67.0°$$

$$y = \frac{21.3 \tan 67.0°}{\tan 25.0° + \tan 67.0°}$$

$$\approx \frac{21.3(2.3559)}{0.4663 + 2.3559}$$

$$\approx 17.78$$

You can now use the cosine ratio to find x.

$$\cos 25.0° \approx \frac{17.78}{x}$$

$$x \approx \frac{17.78}{\cos 25.0°}$$

$$\approx \frac{17.78}{0.9063}$$

$$\approx 19.618$$

The ship is approximately 19.6 kilometers from lookout B. ■

Trigonometric functions can be used to relate the various sides and angles of geometric figures.

Example 3 An isosceles triangle has equal sides of k units and a vertex angle of α degrees. Express the length b of its base in terms of k and α.

Solution See Figure 2-43, in which CD is the perpendicular drawn from the vertex of the vertex angle to the base. Because this perpendicular bisects the vertex angle, you have

$$\sin \frac{\alpha}{2} = \frac{x}{k}$$

$$x = k \sin \frac{\alpha}{2}$$

Because the perpendicular CD also bisects the base, you have

1. $b = 2x$

Finally, substitute $k \sin \frac{\alpha}{2}$ for x in Equation 1 to obtain

$$b = 2k \sin \frac{\alpha}{2}$$

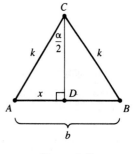

Figure 2-43 ■

Example 4   A regular polygon has n equal sides, each of length a. The radius of the circumscribed circle is R. Express a as a function of n and R.

Solution   See Figure 2-44, in which point O is the center of an n-sided regular polygon with each side a units long. Because the number of degrees in one complete revolution is $360°$, and because an n-sided regular polygon has n equal central angles, angle $BOC = \frac{360°}{n}$. The perpendicular from point O to side BC bisects angle BOC and side BC. Thus, you have

$$BE = \frac{a}{2}$$

$$\text{angle } BEO = 90°$$

and

$$\text{angle } BOE = \frac{1}{2}\left(\frac{360°}{n}\right) = \frac{180°}{n}$$

Because triangle BOE is a right triangle, you can write

$$\sin(\text{angle } BOE) = \frac{BE}{OB}$$

After substituting values for the above quantities, you have

$$\sin\left(\frac{180°}{n}\right) = \frac{\frac{a}{2}}{R} \quad \text{or} \quad a = 2R\sin\left(\frac{180°}{n}\right) \qquad \blacksquare$$

Figure 2-44

EXERCISE 2.6

1. Compute the height of the Sears Tower using the information given in Illustration 1.

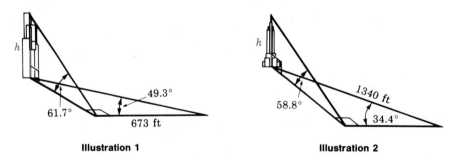

Illustration 1   **Illustration 2**

2. Compute the height of the Empire State Building using the information given in Illustration 2.

3. Bill and Paula, standing on the same side of and in line with the Washington Monument, are looking at its top. The angle of elevation from Bill's position is $34.1°$, and the angle of elevation from Paula's position is $60.0°$. If Bill and Paula stand on level ground and are 500 feet apart, how tall is the monument?

4. Use the information given in Illustration 3 to compute the height of the figure part of the Statue of Liberty.

5. Use the information given in Illustration 4 to compute the height of the Gateway Arch in St. Louis.

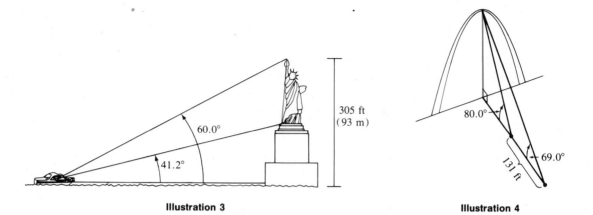

Illustration 3   Illustration 4

6. The bearing of point *C* from point *A* is N 25.0° E, and the bearing of point *C* from point *B* is N 40.0° W. If *A* and *B* are on an east–west line and 200 kilometers apart, how far is *B* from *C*?

7. Find the height of the triangle shown in Illustration 5.

8. Find the length of side *AC* shown in Illustration 5.

9. Gail is standing 152 feet due east of a television tower. She then travels along a bearing of S 60.0° W. How tall is the tower if the angle of elevation to its top from her closest position to the tower is 58.2°?

10. Jeffrey stands due south of the Eiffel Tower and notes that the angle of elevation to its top is 60.0°. His brother Grant, standing 298 feet due east of Jeff, notes an angle of elevation of 56.9°. How tall is the tower? (*Hint:* Consider a triangle on the ground and use the Pythagorean theorem.)

11. Two tangents are drawn from a point *P* to a circle of radius *r* as in Illustration 6. The angle between the tangents is *θ*. How long is each tangent?

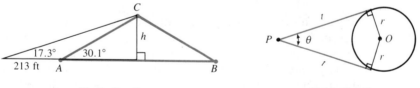

Illustration 5   Illustration 6

12. A regular polygon has *n* equal sides, each of length *a*. The radius of the inscribed circle is *r*. Express *a* as a function of *n* and *r*.

13. An isosceles triangle has a base angle of *α* and a side of *k* centimeters. Express the length *b* of its base in terms of *α* and *k*.

14. At noon a ship left port on a bearing of N *α* E and steamed *k* kilometers. At the same time another ship left port on a bearing of N *α* W and steamed *k* kilometers. Express the distance *d* between the two ships in terms of *α* and *k*.

15. Point *A* is *k* meters from a building. From point *A* the angle of elevation to the top of the building is *α*, and the angle of depression to the base of the building is *β*. Express the height *H* of the building in terms of *k*, *α*, and *β*.

16. A ladder *k* feet long reaches a height of *H* feet on the side of a building, and the ladder makes an angle of *α* with the horizontal. The ladder slips so that it reaches a height of only *h* feet on the side of the building. It then makes an angle of *β* with the horizontal. Express the distance *d* that the ladder has come down the building in terms of *k*, *α*, and *β*.

17. From an observation point D meters from a launch site, an observer watches the vertical flight of a rocket. At one instant, the angle of elevation of the rocket is θ, and at a later instant, ϕ. How far has the rocket traveled during that time?

18. Two helicopters are hovering at an altitude of d feet. The pilot of one helicopter observes a crate being dropped from the other aircraft. At one instant the pilot measures the angle of depression of the crate to be α. When the crate hits the ground, the pilot measures the angle of depression to be β. How far has the crate fallen in that time?

19. Two tall buildings are separated by a distance of d meters. From the top of the shorter building, the angle of elevation to the top of the taller building is α, and the angle of depression to the base of the taller building is β. Express the height a of the shorter building in terms of d and β.

20. Express the height b of the taller building described in Exercise 19 in terms of d, α, and a.

2.7 INTRODUCTION TO VECTORS

Quantities that have only magnitude are called **scalar quantities**. However, some quantities have both magnitude and direction. For example, the flight of an airplane is described by both a speed and a direction. Quantities with both magnitude and direction are called **vector quantities** and are represented by mathematical entities called **vectors**.

> **Definition.** A **vector** is a directed line segment. The direction of the vector is indicated by the angle it makes with some convenient reference line.
>   The **norm**, or **magnitude**, of the vector is the length of the line segment. If a vector is denoted by **v**, the norm of the vector is denoted by |**v**|.

Any two directed line segments with the same length and the same direction are regarded as **equal vectors**. When drawing diagrams containing vectors, it is common to position them in the most convenient location. A force of 30 pounds, for example, exerted in a northwesterly direction, may be represented by the directed line segment in Figure 2-45. The length of the arrow is 30 units and because it is the terminal side of a 135° angle, it points northwest.

A wind blowing from the east at 20 miles per hour can be represented by the vector in Figure 2-46.

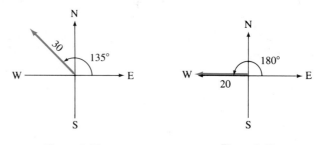

Figure 2-45   Figure 2-46

An airplane flying 350 miles per hour on a **heading** (an intended direction of travel measured clockwise from the north line) of 240° can be represented by the vector in Figure 2-47.

Vector quantities can be added, but the process must take into account both their norms and directions. Two forces of 40 pounds, for example, exerted on the same object might not combine to be an 80-pound force. If they acted in opposite directions, the net force would be zero.

Vector quantities can be added by using the **parallelogram law**. If two vectors originating at a common point are adjacent sides of a parallelogram, their vector sum, called the **resultant vector**, is the vector represented by the diagonal of the parallelogram that is drawn from the common point. In Figure 2-48, the sum of vectors **AB** and **AD** is the vector **AC**, given by the diagonal of the parallelogram *ABCD*. Note that vector **DC** has the same magnitude and direction as vector **AB**.

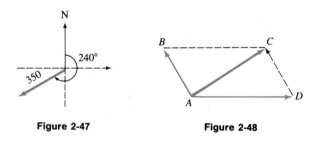

Figure 2-47   **Figure 2-48**

Example 1   A boat capable of a speed of 8.0 miles per hour in still water attempts to go directly across a river flowing at 3.0 miles per hour. By what angle is the boat pushed off its intended path? What is the effective speed of the boat?

Solution   The two given velocities can be represented by vectors as in Figure 2-49. The direction the boat travels, called its **course**, is represented by a vector sum, the diagonal of the rectangle. Angle θ specifies the direction in which the boat is forced to travel, and the length of OP specifies the effective speed of the boat. Because AP is also 3.0 units, $\tan \theta = \frac{3}{8}$. From this relationship, you can calculate that $\theta = 21°$. Thus, the river pushes the boat approximately 21° off its intended path. Use the Pythagorean theorem to find the length of OP, hence the speed of the boat.

$$OP = \sqrt{8^2 + 3^2}$$
$$= \sqrt{64 + 9}$$
$$= \sqrt{73}$$
$$\approx 8.5$$

Figure 2-49

The effective speed of the boat is approximately 8.5 miles per hour.

Example 2   An airplane capable of a speed of 270 miles per hour in still air sets a heading of 75°. A strong wind is blowing in the direction of 165° and forces the plane onto a course that is due east. What is the velocity of the wind, and what is the ground speed (the speed relative to the ground) of the plane?

Solution   The velocities involved are represented in Figure 2-50. Vector **w** represents the wind velocity and vector **v** represents the resultant velocity, or **ground speed**, of the plane. Because $165° - 75° = 90°$, the vector parallelogram is a rectangle with angle α equal to 90°. Because each triangle formed by the diagonal is a right triangle, you can write

$$\tan 15° = \frac{|\mathbf{w}|}{270}$$

Multiplying both sides by 270 and simplifying gives

$$|\mathbf{w}| = 270 \tan 15°$$

$$|\mathbf{w}| \approx 72$$

Also, you have

Figure 2-50

$$\cos 15° = \frac{270}{|\mathbf{v}|}$$

$$|\mathbf{v}| \cos 15° = 270 \qquad \text{Multiply both sides by } |\mathbf{v}|.$$

$$|\mathbf{v}| = \frac{270}{\cos 15°} \qquad \text{Divide both sides by } \cos 15°.$$

$$|\mathbf{v}| \approx 280 \qquad \text{Simplify.}$$

The wind speed is approximately 72 miles per hour, and the ground speed of the plane (the resultant of the plane's air speed and the wind speed) is approximately 280 miles per hour.   ■

It is possible to separate a single vector into several components. Suppose that a car weighing 3000 pounds is parked on a hill. It is pulled directly downward by gravity with a force of 3000 pounds. Part of this force appears as a tendency to roll the car down the hill, and another part presses the car against the road. Just how the 3000 pounds is apportioned depends on the angle of the hill. If there were no hill, there would be no tendency to roll. If the hill were very steep, only a small force would hold the car to the road. The weight of the car is said to be **resolved** into two components—one directed down the hill, and the other directed into the hill. These vectors obey the parallelogram law. See Figure 2-51. Note that the

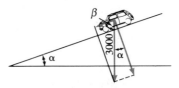

Figure 2-51

angle of the hill, α, is also the angle between two of the vectors because both of these angles are complementary to angle β.

Example 3   A 3000-pound car sits on a 23.0° incline. What force is required to prevent the car from rolling down the hill? With what force is it held to the roadway?

Solution   You must find the magnitudes of vectors **t** and **n** in Figure 2-52. Because the figure *OACB* is a rectangle, the opposite sides are equal and angle $B = 90°$. Hence, you have

$$\sin 23.0° = \frac{|t|}{3000}$$

$$|t| = 3000 \sin 23.0°$$

$$|t| \approx 1170$$

Also, you have

$$\cos 23.0° = \frac{|n|}{3000}$$

$$|n| = 3000 \cos 23.0°$$

$$|n| \approx 2760$$

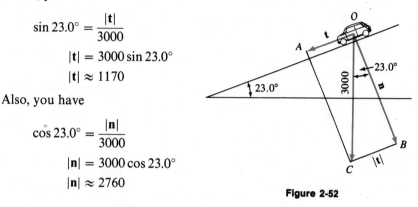

Figure 2-52

A force of approximately 1170 pounds is required to prevent rolling, and a force of approximately 2760 pounds keeps the car on the hill. Note that because the angles are to the nearest tenth of a degree, only three-place accuracy is permitted for the sides. In this example, the lengths of the vectors have only three significant digits.   ■

EXERCISE 2.7

1. A boat capable of a speed of 6 miles per hour in still water attempts to go directly across a river. As the boat crosses the river, it drifts 30° from its intended path. How strong is the current? What is the effective speed of the boat?

2. A boat capable of a speed of 11 miles per hour in still water attempts to go directly across a river with a current of 5.6 miles per hour. By what angle is the boat pushed off its intended path? What is the effective speed of the boat?

3. Laura can row a boat 0.5 mile per hour in still water. She attempts to row straight across a river that has a current of 1 mile per hour. If she must row for 2 hours to cross the river, by what angle is the current pushing her off her intended path? Give your answer to the nearest degree.

4. A boat attempts to go directly across a river with a current of 3.7 miles per hour. The current causes the boat to drift 23° from its intended path. How far will the boat travel if the trip takes 10 minutes?

5. A plane has a heading of 260.0° and is flying at 357 miles per hour. If a southerly wind causes the plane's course (the direction it is actually going) to be due west, find the ground speed of the plane (its speed relative to the ground).

6. A plane has an air speed of 411 miles per hour and a heading of 90.0°. A wind from the north is blowing at 31.0 miles per hour. By how many degrees is the plane blown off its heading? Find the ground speed of the plane (its speed relative to the ground).

7. A plane leaves an airport with a heading of 45.0° and an air speed of 201 miles per hour. At the same time, another

plane leaves the same airport with a heading of 135.0° and an air speed of 305 miles per hour. At the end of 2 hours, what is the bearing of the first plane from the second? Assume no wind.

8. A rifle that fires bullets having a muzzle velocity of 4100 feet per second is fixed at an angle of elevation of 32°. What is the horizontal component of the bullet's velocity?

9. A 317-pound weight is hanging from the ceiling on a long rope. A man pushes horizontally against the weight to rotate the rope through an angle of 11.2°. What resultant force is being counteracted by the rope?

10. A plane leaves an airport with a heading of 170°. At the same time, a second plane leaves the same airport with a heading of 260°. One hour later, the first plane is 1300 miles directly southeast of the second plane. How fast is the first plane going? Assume no wind.

11. What force is required to keep a 2210-pound car from rolling down a ramp that makes a 10.0° angle with the horizontal?

12. A force of 25 pounds is necessary to hold a barrel in place on a ramp that makes an angle of 7° with the horizontal. How much does the barrel weigh?

13. A vehicle presses against a roadway with a force of 1100 pounds. How much does the vehicle weigh if the roadway has an 18.0° grade?

14. A board will break if it is subjected to a force greater than 350 pounds. Will the board hold a 450-pound piano supported by a single dolly as it slides up the board and into a truck? Assume that the board makes an angle of 35.0° with the horizontal.

15. A garden tractor weighing 351 pounds is being driven up a ramp onto a trailer. If the tractor presses against the ramp with a 341-pound force, what angle does the ramp make with the horizontal?

16. If a force of 21.3 pounds is necessary to keep a 50.1-pound barrel from rolling down an inclined plane, what angle does the inclined plane make with the horizontal?

17. A 201-pound force is directed due east. What force, directed due north, is needed to produce a resultant force of 301 pounds? What angle is formed by the vectors representing the 201- and the 301-pound forces?

18. A 312-pound force is directed due west. What force, directed exactly southeast, would cause the resultant force to be directed due south?

19. It requires 35 pounds of force to keep two children from sliding down the chute shown in Illustration 1. If one child weighs 110 pounds and the other 51 pounds, what angle does the chute make with the horizontal?

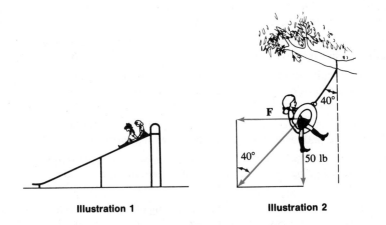

Illustration 1   Illustration 2

20. A 50-pound girl is playing in a tire swing hanging from the limb of a large tree. See Illustration 2. To get her started,

a friend pulls the swing backward until it makes an angle of 40° with the vertical. What horizontal force F is required to hold the swing in this position?

21. A weight of 160 pounds is supported by a wire as in Illustration 3. Find the vertical and horizontal components of force F_1. (*Hint:* The vertical component of F_1 supports half the weight.)

22. A weight of 220 pounds is supported by a cable as shown in Illustration 4. Find the vertical and horizontal components of force F_1.

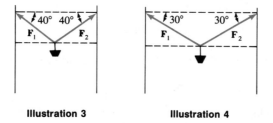

Illustration 3   Illustration 4

In Exercises 23 and 24, you must draw some lines to create right triangles.

23. A plane leaves an airport at 12:00 noon with a heading of 60.0° and an air speed of 451 miles per hour. One hour later, another plane leaves the same airport with a heading of 290.0° and an air speed of 611 miles per hour. What is the bearing of the first plane from the second at 2:00 P.M.? Assume no wind.

24. A plane leaves an airport at 3:00 P.M. with a heading of 70.0° and an air speed of 512 miles per hour. Two hours later, another plane leaves the same airport with a heading of 100.0° and an air speed of 621 miles per hour. How far apart are the two planes at 6:00 P.M.? Assume no wind.

In Exercises 25 and 26, recall that the diagonals of a rhombus are perpendicular and bisect each other.

25. Two forces of 30 pounds each make an angle of 30° with each other. What is the magnitude of their resultant?

26. Two equal forces of f pounds each make an angle of θ with each other. What is the magnitude of their resultant?

2.8   MORE APPLICATIONS OF THE TRIGONOMETRIC FUNCTIONS

In this chapter, we have discussed how to use trigonometry to calculate forces, directions, velocities, and distances. These are important applications of trigonometry to navigation, mechanics, and surveying, but they represent only a small fraction of the possible uses of trigonometry. Here are four more.

1. Slope of a Straight Line

If a nonvertical line passes through points $P(x_1, y_1)$ and $Q(x_2, y_2)$, the **slope** of the line is defined as

$$\text{slope of } PQ = \frac{\text{rise}}{\text{run}} = \frac{y_2 - y_1}{x_2 - x_1} \quad (x_2 \neq x_1)$$

In the right triangle PQR in Figure 2-53, $y_2 - y_1$ is the length of the side opposite angle α, and $x_2 - x_1$ is the length of the side adjacent to angle α. The slope,

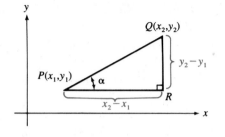

Figure 2-53

therefore, is the ratio of the opposite side to the adjacent side, and

slope of $PQ = \tan \alpha$

Angle α, measured counterclockwise from the horizontal line segment, is called the **angle of inclination** of line PQ. When $\alpha = 90°$, $\tan \alpha$ is undefined. This indicates that a vertical line has no slope.

2. Optics

Light travels more rapidly in air than in water or glass. When light passes from one substance into another in which it slows down, its path is bent. This is called **refraction** and is the principle behind camera lenses, eye glasses, and the sparkle of a diamond.

Suppose, in Figure 2-54, that a beam of light is passing from air into glass. Angle i is called the **angle of incidence**, and angle r the **angle of refraction**. The two angles are related by a simple formula named after the Dutch astronomer Willebrord Snell (1591–1629):

$$\frac{\sin i}{\sin r} = n$$

where n is a constant, called the **refractive index**, dependent upon the optical properties of the glass. If, for example, light striking a piece of glass at a 10° angle is refracted to an angle of 6°, the refractive index is

$$
\begin{aligned}
n &= \frac{\sin i}{\sin r} \\
&= \frac{\sin 10°}{\sin 6°} \\
&\approx \frac{0.1736}{0.1045} \\
&\approx 1.7
\end{aligned}
$$

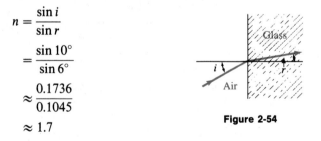

Figure 2-54

The refractive index of glass ranges from 1.5 to 1.9, whereas the refractive index of diamond is approximately 2.4. This is why diamonds sparkle more than rhinestones.

3. Electrical Engineering

Certain components of an electronic circuit behave in ways that are conveniently analyzed by using trigonometric functions. For example, a resistor and an inductor connected as a series network will oppose the flow of an alternating current. A measure of this opposition is called the network's **impedance**. The impedance, z, is determined by the *reactance*, X, of the inductor, and the *resistance*, R, of the resistor. The three quantities z, R, and X represent the sides of the right triangle in Figure 2-55. These quantities are related to each other by the Pythagorean theorem as follows:

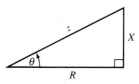

Figure 2-55

$$z^2 = X^2 + R^2$$

The angle θ shown in the figure is called the **phase angle**, and from the figure we can see that

$$\tan \theta = \frac{X}{R}$$

Example 1   An alternating current generator G and an impedance z, consisting of a resistance R and an inductive reactance X, are connected as in Figure 2-56. If $R = 300$ ohms and $X = 400$ ohms, find the impedance z and the phase angle θ.

Solution   The circuit and the impedance triangle shown in the figure will help you determine both z and θ. By the Pythagorean theorem, you have

$$z^2 = X^2 + R^2$$

By substituting **300** for R and **400** for X and simplifying, you obtain

$$z = \sqrt{300^2 + 400^2}$$
$$= \sqrt{250,000}$$
$$= 500$$

Thus, the impedance is 500 ohms.

To find the phase angle θ, use the fact that

$$\tan \theta = \frac{X}{R}$$
$$= \frac{400}{300}$$
$$\approx 1.3333$$

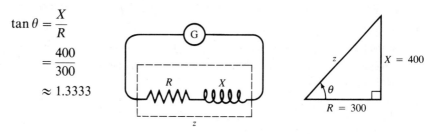

Figure 2-56

Thus, angle θ is approximately 53.1°.

4. The Seasons

In answer to the question "Why is summer hotter than winter?" many people respond "Because the earth is closer to the sun in the summer." However, this is not true. In fact, the earth is closest to the sun in December. Summers and winters are caused by the angle at which the sun's energy hits the earth. This angle varies as the earth moves along its orbit because the axis of the earth is tilted by approximately 23°30'. In the summer, sunlight approaches the earth at angles close to 90°; but in the winter, the sunlight is not nearly as direct.

In the northern hemisphere at noon on June 21 (the summer solstice), the sun is as far north as it will ever get; this is the first day of summer. Six months later, at noon on December 22, the sun appears 47° farther south in the sky; this is the beginning of winter. The heat energy from the sun that reaches the earth is proportional to the cosine of the angle between the sun's rays and the vertical. Because this angle is greater during the winter months, less energy reaches us and the days are cold.

Example 2   At noon on the first day of summer in Havana, Cuba, the sun is directly overhead. How much less energy reaches Havana at noon on the first day of winter? See Figure 2-57.

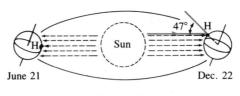

June 21   Dec. 22

Figure 2-57

Solution   The energy E that reaches the earth is proportional to the cosine of the angle of incidence. In June, $E = k \cos 0° = k$, while in December, $E = k \cos 47° \approx 0.68k$. Havana receives about $1 - 0.68 = 0.32$, or 32% less energy on the first day of winter than on the first day of summer. ∎

EXERCISE 2.8

1. A line passes through the points $(1, 3)$ and $(3, 5)$. What is its angle of inclination?

2. A line passes through the points $(0, -1)$ and $(-2\sqrt{3}, 1)$. What is its angle of inclination?

3. A beam of light enters a block of ice at the angle of incidence of 8.0° and is refracted to 6.1°. What is the index of refraction of the ice?

4. A clear liquid, tetrachloroethylene (used in dry cleaning), has an index of refraction equal to that of a certain brand of glass. Would you be able to see a piece of that glass submerged in the liquid? How could you distinguish real diamonds from a handful of fakes?

5. An impedance consists of a resistance of 120 ohms and an inductive reactance of 50.0 ohms. Find the impedance.

6. Use the information in Exercise 5 to find the phase angle.

7. An impedance consists of a resistance of 200 ohms in series with an unknown inductance. If the impedance measures 290 ohms, what is the phase angle?

8. Use the information in Exercise 7 to find the inductive reactance.

9. The impedance of a series resistor–inductor network is measured as 240 ohms, and the phase angle as 60.0°. Find the resistance R.

10. Use the information in Exercise 9 to find the inductive reactance X.

In Exercises 11 and 12, assume that the amount of the sun's energy reaching the earth is directly proportional to the cosine of the angle of incidence.

11. Find the amount of the sun's energy that reaches earth if the angle of incidence is 30°.

12. If the sun's energy reaching earth is 90% of what it would be if the sun were directly overhead, what is the angle of incidence?

13. A photographer aims his lights directly at a painting he wishes to photograph. How much less light will hit the painting if he lowers his lights by 20°? See Illustration 1. Assume that the light remains the same distance from the painting.

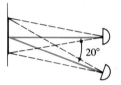

Illustration 1

CHAPTER SUMMARY

Key Words

angle (2.1)	initial side of a
angle of depression (2.5)	trigonometric angle (2.1)
angle of elevation (2.5)	minutes (2.1)
angle of incidence (2.8)	odd functions (2.2)
angle of refraction (2.8)	phase angle (2.8)
bearing (2.5)	quadrant (2.1)
cofunctions (2.4)	quadrantal angle (2.1)
complementary angles (2.4)	reciprocals (2.2)
cosecant θ (2.1)	reference angle (2.3)
cosine θ (2.1)	reference triangle (2.3)
cotangent θ (2.1)	refraction (2.8)
coterminal angles (2.1)	refractive index (2.8)
equal vectors (2.7)	resultant (2.7)
even functions (2.2)	secant θ (2.1)
ground speed (2.7)	seconds (2.1)
heading (2.7)	significant digits (2.5)
impedance (2.8)	sine θ (2.1)

slope of a line (2.8)

standard position of a
   trigonometric angle (2.1)

tangent θ (2.1)

terminal side of a
   trigonometric angle (2.1)

trigonometric angle (2.1)

unit circle (2.4)

vectors (2.7)

vertex of an angle (2.1)

Key Ideas

(2.1) $1' = \dfrac{1}{60}(1°)$ and $1'' = \dfrac{1}{3600}(1°)$

If $P(x, y)$ is any point on the terminal side of an angle θ in standard position and at a distance r from the origin, then

$$\sin \theta = \frac{y}{r} \qquad \cos \theta = \frac{x}{r} \qquad \tan \theta = \frac{y}{x}$$

$$\csc \theta = \frac{r}{y} \qquad \sec \theta = \frac{r}{x} \qquad \cot \theta = \frac{x}{y}$$

(2.2)   The eight fundamental relationships are:

$$\sin \theta = \frac{1}{\csc \theta} \qquad\qquad \cos \theta = \frac{1}{\sec \theta}$$

$$\tan \theta = \frac{1}{\cot \theta} \qquad\qquad \tan \theta = \frac{\sin \theta}{\cos \theta}$$

$$\cot \theta = \frac{\cos \theta}{\sin \theta} \qquad\qquad \sin^2\theta + \cos^2 \theta = 1$$

$$\tan^2\theta + 1 = \sec^2 \theta \qquad \cot^2\theta + 1 = \csc^2\theta$$

Three other important relationships are:

$$\sin(-\theta) = -\sin \theta \qquad \cos(-\theta) = \cos \theta \qquad \tan(-\theta) = -\tan \theta$$

(2.3)   The six trigonometric functions of any angle θ are equal to those of the reference angle of θ, except possibly for sign.

	0°	30°	45°	60°	90°	180°	270°
$\sin \theta$	0	$\dfrac{1}{2}$	$\dfrac{\sqrt{2}}{2}$	$\dfrac{\sqrt{3}}{2}$	1	0	-1
$\cos \theta$	1	$\dfrac{\sqrt{3}}{2}$	$\dfrac{\sqrt{2}}{2}$	$\dfrac{1}{2}$	0	-1	0
$\tan \theta$	0	$\dfrac{\sqrt{3}}{3}$	1	$\sqrt{3}$	—	0	—

(2.4)   If θ is any acute angle, then any trigonometric function of θ is equal to the cofunction of the complement of θ.

(2.5)   When solving right triangles, your answers can be only as accurate as the least accurate of the given data.

If θ is an acute angle in a right triangle, then

$$\sin \theta = \frac{\text{opposite side}}{\text{hypotenuse}} \qquad \csc \theta = \frac{\text{hypotenuse}}{\text{opposite side}}$$

$$\cos \theta = \frac{\text{adjacent side}}{\text{hypotenuse}} \qquad \sec \theta = \frac{\text{hypotenuse}}{\text{adjacent side}}$$

$$\tan \theta = \frac{\text{opposite side}}{\text{adjacent side}} \qquad \cot \theta = \frac{\text{adjacent side}}{\text{opposite side}}$$

(2.6)   Many times, auxiliary lines can be drawn to form right triangles. These right triangles can then be solved by using the methods of Section 2.5.

(2.7)   Vectors can be used to describe quantities that have both magnitude and direction. Vector quantities are added by using a parallelogram law.

REVIEW EXERCISES

In Review Exercises 1–4, tell whether the given angle is in standard position.

1.

2.

3.

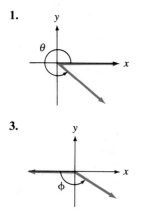

4.

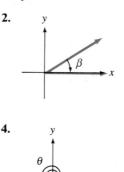

5.   Are angles of 310° and −410° coterminal?

6.   Are angles of 190°24′ and 820°24′ coterminal?

In Review Exercises 7–10, find the values of the remaining trigonometric functions of angle θ, which is in standard position.

7. $\sin \theta = \dfrac{-7}{10}$; θ in QIII

8. $\tan \theta = \dfrac{7}{9}$; θ not in QI

9. $\cos \theta = \dfrac{-7}{10}$; θ in QII

10. $\cot \theta = \dfrac{-9}{8}$; θ in QIV

11. Use the eight fundamental relationships to show that $\dfrac{1}{\sec\theta} = \sin\theta\cot\theta$.

12. Use the eight fundamental relationships to show that $\cos\theta\csc\theta = \cot\theta$.

In Review Exercises 13–16, evaluate each trigonometric expression. **Do not use a calculator.**

13. $\sin 45° \cos 30°$

14. $\cos 120° \tan 135°$

15. $\tan^2 225° \cos^2 30° \sin^2 300°$

16. $\sec 30° \csc 30° + \sec 330° \csc 330°$

In Review Exercises 17–20, find the value of the sine, cosine, and tangent of the given angle. **Do not use a calculator.**

17. $930°$

18. $1380°$

19. $-300°$

20. $-585°$

In Review Exercises 21–24, draw the unit circle and use it to estimate the sine, cosine, and tangent ratios of the given angles.

21. $15°$

22. $160°$

23. $265°$

24. $340°$

In Review Exercises 25–30, use Table A in Appendix IV to evaluate the sine, cosine, and tangent of the given angle. Check your work with a calculator.

25. $15°$ 26. $160°$ 27. $265°$ 28. $340°$ 29. $-160°$ 30. $-340°$

In Review Exercises 31–36, use Table A in Appendix IV to find angle α $(0° \le \alpha < 360°)$, which is in the given quadrant. Check your work with a calculator.

31. QII; $\sin\alpha = 0.8746$ 32. QIII; $\tan\alpha = 0.6009$ 33. QIV; $\cos\alpha = 0.7314$ 34. QI; $\sec\alpha = 1.871$

35. QII; $\cot\alpha = -0.1763$ 36. QIV; $\csc\alpha = -1.046$

37. From a location 32.1 meters from the base of a flagpole, the angle of elevation to its top is α. Find α if the flagpole is 10.0 meters tall.

38. The angle of depression from a window in a building to a point on the ground is $17.7°$. If the point on the ground is 187 feet from the base of the building, how high is the observer?

39. Owatonna, Minnesota, is approximately 55 miles due south of Minneapolis, and the bearing of Winona, Minnesota, from Minneapolis is about S 44° E. How far is Winona from Owatonna if Owatonna is due west of Winona?

40. Assume that the bearing of South Bend, Indiana, from Fort Wayne, Indiana, is N 48° W and that the distance between the cities is about 71 miles. Further assume that South Bend is due north of Indianapolis and that the bearing of Fort Wayne from Indianapolis is N 21° E. How far is Fort Wayne from Indianapolis?

41. A barrel weighing 60 pounds rests on a ramp that makes an angle of $10°$ with the horizontal. How much force is necessary to keep the barrel from rolling down the ramp?

42. A 2500-pound car rests on a hill. A force of 500 pounds is required to keep the car from rolling down the hill. How steep is the grade?

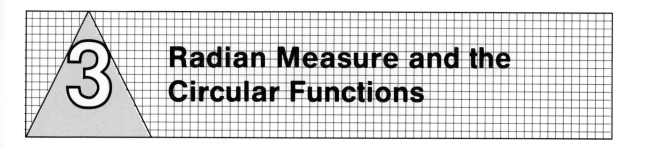

Radian Measure and the Circular Functions

We have seen that angles can be measured in degrees. In this chapter, we will introduce and use a different unit of angle measurement: the radian.

3.1 RADIAN MEASURE

People of ancient times noticed that the annual progression of the seasons appeared to repeat in 360-day cycles. Accordingly, they divided their calendar of the year's circle into 360 days. It was reasonable, then, to divide the circle itself into 360 equal parts, and degree measure of angles was developed. Although calendars were revised when closer observation showed that a year contains approximately 365 days, the 360-degree circle persists to this day. While still common, degree measure is not always convenient to use. In higher mathematics a different unit, called a **radian**, is used to express angular measure.

> **Definition.** Consider the circle with radius r in Figure 3-1. The measure of central angle θ is **1 radian** if and only if θ intercepts an arc whose length is r.

Figure 3-1

Because the circumference of a circle is $2\pi r$ and each arc with length r that is marked off on the circumference determines an angle of 1 radian,

$$\frac{2\pi r}{r} = 2\pi$$

is the number of radians in one complete revolution. Thus,

2π radians = 360°

Dividing both sides of this equation by 2 leads to the fundamental relationship between radians and degrees.

π radians = 180°

This relationship is very important and should be memorized.

79

Example 1 Change **a.** 30°, **b.** 45°, **c.** 60°, **d.** 225°, **e.** −315°, and **f.** $D°$ to radians.

Solution

a.
$$180° = \pi \text{ radians}$$
$$\frac{1}{6}(180°) = \frac{1}{6}(\pi \text{ radians})$$
$$30° = \frac{\pi}{6} \text{ radians}$$

b.
$$180° = \pi \text{ radians}$$
$$\frac{1}{4}(180°) = \frac{1}{4}(\pi \text{ radians})$$
$$45° = \frac{\pi}{4} \text{ radians}$$

c.
$$180° = \pi \text{ radians}$$
$$\frac{1}{3}(180°) = \frac{1}{3}(\pi \text{ radians})$$
$$60° = \frac{\pi}{3} \text{ radians}$$

d.
$$180° = \pi \text{ radians}$$
$$1° = \frac{\pi}{180} \text{ radians}$$
$$225(1°) = 225\left(\frac{\pi}{180} \text{ radians}\right)$$
$$225° = \frac{5}{4}\pi \text{ radians}$$

e.
$$180° = \pi \text{ radians}$$
$$1° = \frac{\pi}{180} \text{ radians}$$
$$-315(1°) = -315\left(\frac{\pi}{180} \text{ radians}\right)$$
$$-315° = -\frac{7\pi}{4} \text{ radians}$$

f.
$$180° = \pi \text{ radians}$$
$$1° = \frac{\pi}{180} \text{ radians}$$
$$D(1°) = D\left(\frac{\pi}{180} \text{ radians}\right)$$
$$D° = \frac{\pi D}{180} \text{ radians}$$ ■

To find the number of degrees in 1 radian, we proceed as follows.

$$\pi \text{ radians} = 180°$$
$$1 \text{ radian} = \frac{180°}{\pi} \qquad \text{Divide both sides by } \pi.$$
$$\approx 57.3°$$

Hence, 1 radian is approximately 57.3°.

Example 2 Change to degrees:

a. $\frac{2}{3}\pi$ radians, **b.** $\frac{5}{7}\pi$ radians, **c.** −7 radians, and **d.** R radians.

Solution

a.
$$\pi \text{ radians} = 180°$$
$$\frac{2}{3}(\pi \text{ radians}) = \frac{2}{3}(180°)$$
$$\frac{2\pi}{3} \text{ radians} = 120°$$

b.
$$\pi \text{ radians} = 180°$$
$$\frac{5}{7}(\pi \text{ radians}) = \frac{5}{7}(180°)$$
$$\frac{5\pi}{7} \text{ radians} = \frac{900°}{7}$$

c. π radians $= 180°$

$$\frac{1}{\pi}(\pi \text{ radians}) = \frac{1}{\pi}(180°)$$

$$1 \text{ radian} = \frac{180°}{\pi}$$

$$-7(1 \text{ radian}) = -7\left(\frac{180°}{\pi}\right)$$

$$-7 \text{ radians} = -\frac{1260°}{\pi}$$

d. π radians $= 180°$

$$1 \text{ radian} = \frac{180°}{\pi}$$

$$R(1 \text{ radian}) = R\left(\frac{180°}{\pi}\right)$$

$$R \text{ radians} = \frac{R180°}{\pi}$$

■

The following table gives the degree measure and the corresponding radian measure of five common angles.

Degree measure	0°	30°	45°	60°	90°
Radian measure	0	$\dfrac{\pi}{6}$	$\dfrac{\pi}{4}$	$\dfrac{\pi}{3}$	$\dfrac{\pi}{2}$

The information given in this table helps us to convert angles such as 120° and 225° from degree measure to radian measure in the following way.

$$120° = 2(60°)$$
$$= 2\left(\frac{\pi}{3}\right) = \frac{2\pi}{3}$$
$$225° = 5(45°)$$
$$= 5\left(\frac{\pi}{4}\right) = \frac{5\pi}{4}$$

We can also use this information to change angles such as $\frac{7\pi}{4}$ and $\frac{11\pi}{6}$ from radian measure to degree measure.

$$\frac{7\pi}{4} = 7\left(\frac{\pi}{4}\right)$$
$$= 7(45°) = 315°$$
$$\frac{11\pi}{6} = 11\left(\frac{\pi}{6}\right)$$
$$= 11(30°) = 330°$$

Figure 3-2 shows the degree measure and the corresponding radian measure of all angles from 0° to 360° that are multiples of 30° or multiples of 45°.

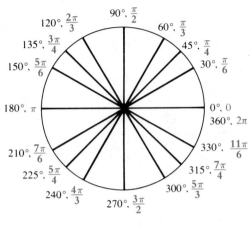

Figure 3-2

Arc Length

Radian measure permits easy calculation of arc length in a circle. Suppose that the central angle in Figure 3-3 is in radians. Because the arc length s in a circle is the same fractional part of the circle's circumference as the angle θ is of one complete revolution, we can set up the proportion

$$\frac{s}{2\pi r} = \frac{\theta}{2\pi}$$

and solve it for s:

$$2\pi s = 2\pi r\theta$$

$$s = r\theta$$

Figure 3-3

We summarize this important result.

Formula for Arc Length.   If a central angle θ in a circle with radius r is in radians, the length s of its intercepted arc is given by the formula

$$s = r\theta$$

The radian measure of a central angle θ of a circle is the ratio of its intercepted arc to the circle's radius. In a circle of radius 5 inches, for example, a central angle that intercepts an arc of 10 inches is $\frac{10 \text{ in.}}{5 \text{ in.}}$, or 2 radians. In a circle of radius 8 centimeters, a central angle that intercepts an arc of 12 centimeters is $\frac{12 \text{ cm}}{8 \text{ cm}}$, or $\frac{3}{2}$ radians. In a circle of radius r, a central angle that intercepts an arc of length s is $\frac{s}{r}$ radians. In general, if θ is a central angle in radians,

$$\theta = \frac{s}{r}$$

where s is the length of the intercepted arc and r is the radius of the circle.

Example 3   If a central angle θ of a circle with a radius of 5 centimeters is 80°, find the length of the intercepted arc.

Solution   First change θ to radians.

$$180° = \pi \text{ radians}$$

$$1° = \frac{\pi}{180} \text{ radians}$$

$$80° = \frac{4\pi}{9} \text{ radians} \qquad \text{Multiply both sides by 80.}$$

Because the radius of the circle is 5 centimeters and $\theta = \dfrac{4\pi}{9}$, substitute 5 for r and $\dfrac{4\pi}{9}$ for θ in the formula $s = r\theta$, and simplify.

$$s = 5\left(\frac{4\pi}{9}\right) = \frac{20\pi}{9} \approx 6.981$$

Thus, the arc length is approximately 6.981 centimeters.   ■

Area of a Sector of a Circle

We can find the area of a **sector** of a circle if the central angle of the sector is measured in radians. See Figure 3-4. We set up the proportion that arc s is to the circumference of the circle as the area A of the sector is to the area of the circle

$$\frac{s}{2\pi r} = \frac{A}{\pi r^2}$$

and solve it for A.

$$2\pi r A = s\pi r^2$$

$$A = \frac{s\pi r^2}{2\pi r}$$

$$A = \frac{1}{2} rs$$

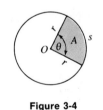

Figure 3-4

Because $s = r\theta$, we can substitute $r\theta$ for s.

$$A = \frac{1}{2} rr\theta \qquad A = \frac{1}{2} r^2 \theta$$

Formula for the Area of a Sector.   If a sector of a circle with radius r has a central angle θ in radians, then the area of the sector is given by the formula

$$A = \frac{1}{2} r^2 \theta$$

Example 4   A sector of a circle has a central angle of 50.0° and an area of 605 square centimeters. Find the radius of the circle.

Solution   Change the angle of 50.0° to radians.

$$180° = \pi \text{ radians}$$

$$1° = \frac{\pi}{180} \text{ radians}$$

$$50.0° = \frac{5\pi}{18} \text{ radians}$$

Because the values of A and θ are known, substitute those values into the formula $A = \frac{1}{2} r^2 \theta$ and solve for r.

$$A = \frac{1}{2} r^2 \theta$$

$$605 = \frac{1}{2} r^2 \frac{5\pi}{18}$$

$$\frac{605(2)(18)}{5\pi} = r^2$$

$$1386.56 \approx r^2$$

$$\sqrt{1386.56} \approx r$$

$$37.2 \approx r$$

The radius of the circle is approximately 37.2 centimeters.   ∎

Example 5   In a circle a sector is formed by an arc of length 15π meters intercepted by a central angle measuring $\frac{5\pi}{6}$ radians. Find the area of the sector.

Solution   First find the radius of the circle by substituting 15π for s and $\frac{5\pi}{6}$ for θ in the formula $s = r\theta$ and solving for r.

$$s = r\theta$$

$$15\pi = r\left(\frac{5\pi}{6}\right)$$

$$15\pi\left(\frac{6}{5\pi}\right) = r$$

$$r = 18$$

Then find the area of the sector by substituting 18 for r and $\frac{5\pi}{6}$ for θ in the formula $A = \frac{1}{2} r^2 \theta$ and simplifying.

$$A = \frac{1}{2}r^2\theta$$

$$= \frac{1}{2}(18)^2\left(\frac{5\pi}{6}\right)$$

$$= 135\pi$$

The area of the sector is 135π square meters. ■

Example 6 An automobile is traveling north as it enters a curve. After it has traveled 1530 feet around the curve, the road straightens and the automobile heads northwest. What is the radius of the curve?

Solution From Figure 3-5, the change in direction of the automobile is 45° (from north to northwest). Because angle α is supplementary to both the 45° angle and to θ, angle θ must be 45°. Before using the formula $s = r\theta$, angle θ must be expressed in radians: $\theta = \frac{\pi}{4}$.

$$s = r\theta$$

$$1530 = r\left(\frac{\pi}{4}\right)$$

$$r = \frac{4(1530)}{\pi}$$

$$r \approx 1948$$

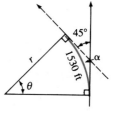

Figure 3-5

The radius of the curve is approximately 1948 feet. ■

Example 7 A pulley with a diameter of exactly 8 inches drives another pulley with diameter of exactly 6 inches. If the larger pulley turns through one revolution, through what angle does the smaller pulley turn?

Solution The radii of the two pulleys are exactly 4 and 3 inches, respectively. When the larger pulley turns through one revolution (2π radians), a point on the edge of the pulley moves

$$s = r\theta = 4(2\pi) = 8\pi \text{ inches}$$

If the belt does not slip, that motion is transferred to the smaller pulley, which turns through an angle

$$\theta = \frac{s}{r} = \frac{8\pi}{3} \approx 8.38 \text{ radians}$$ ■

Example 8 Scott knows that the moon is about 237,000 miles from the earth, but he has forgotten its diameter. If the angle between his lines of sight to either side of the moon is 0.52°, how can Scott estimate its diameter? Give the answer to three significant digits.

Solution   Because the moon is so far from the earth, the length of the arc AB (with the earth as its center, and intercepted by a diameter of the moon) is a good estimate of the length of the diameter. See Figure 3-6. Because the moon is approximately 237,000 miles from the earth, side EB of triangle EAB is approximately 237,000 miles long as well. Change $0.52°$ to radians and use the formula $s = r\theta$.

$$\pi \text{ radians} = 180°$$

$$\frac{\pi}{180} \text{ radians} = 1°$$

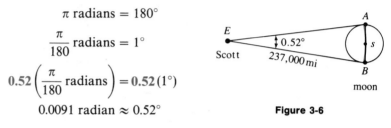

$$0.52\left(\frac{\pi}{180} \text{ radians}\right) = 0.52\,(1°)$$

$$0.0091 \text{ radian} \approx 0.52°$$

Figure 3-6

Substitute $237,000$ for r and 0.0091 for θ into the formula $s = r\theta$ to get

$$s = r\theta \approx 237,000\,(0.0091) \approx 2160$$

Scott's estimate of the diameter of the moon is 2160 miles.   ■

■ EXERCISE 3.1

In Exercises 1–12, change each angle to radians.

1. 15°	**2.** 75°	**3.** 120°	**4.** 150°	**5.** 210°	**6.** 240°
7. 300°	**8.** 330°	**9.** 780°	**10.** 660°	**11.** −520°	**12.** −880°

In Exercises 13–24, each angle is expressed in radians. Change each angle to degrees.

13. $\dfrac{3\pi}{4}$	**14.** 3π	**15.** $\dfrac{5\pi}{2}$	**16.** $\dfrac{7\pi}{3}$	**17.** $\dfrac{4\pi}{3}$	**18.** $\dfrac{11\pi}{6}$
19. 6	**20.** 8	**21.** −10	**22.** −5	**23.** 12.5	**24.** −15.3

In Exercises 25–36, find the values of the trigonometric functions. All angles are measured in radians. **Do not use a calculator or tables.**

25. $\sin\dfrac{\pi}{3}$	**26.** $\cos\dfrac{\pi}{6}$	**27.** $\tan\dfrac{\pi}{4}$	**28.** $\sin\pi$
29. $\cos\dfrac{-3\pi}{4}$	**30.** $\tan\dfrac{-5\pi}{4}$	**31.** $\sin\dfrac{7\pi}{6}$	**32.** $\sin\dfrac{10\pi}{3}$
33. $\csc\dfrac{11\pi}{6}$	**34.** $\sec\dfrac{7\pi}{4}$	**35.** $\cot\left(-\dfrac{5}{3}\pi\right)$	**36.** $\csc\left(-\dfrac{5}{3}\pi\right)$

37. Find the radius of a circle if a central angle of 25° intercepts an arc of 17 centimeters.

38. Find the central angle in radians that intercepts an arc of 10 centimeters in a circle with diameter of 10 centimeters.

39. Find the central angle in degrees that intercepts a 10-inch arc on a 15-inch diameter wheel.

40. If the radius of a circle is 3 meters, find the central angle in degrees that intercepts an arc of 6 meters.

In Exercises 41–44, use 3960 miles for the radius of the earth. Consider this estimate to be accurate to three figures.

41.   The latitude of Manchester, New Hampshire, is 43.0° N. How far is Manchester from the equator? See Illustration 1.

42.   The latitude of Seattle, Washington, is 47.6° N. How far is Seattle from the equator? See Illustration 1.

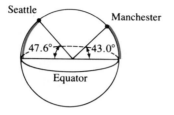

Illustration 1

43.   St. Louis, Missouri, is 2670 miles north of the equator. Find its latitude.

44.   Pittsburgh, Pennsylvania, is 2800 miles north of the equator. Find its latitude.

45.   Find the area, to the nearest hundredth, of a sector of a circle if the sector has a central angle of exactly 30° and the circle has a radius of exactly 20 units.

46.   If a circle contains a sector with central angle of $\frac{2\pi}{3}$ and area of 30 square meters, find the diameter of the circle.

47.   The diameter of the moon is approximately 2160 miles. How far is the center of the moon from earth if the central angle from a point on earth intercepting a diameter on the moon is 0.56°? Give the answer to the nearest thousand.

48.   A regular octagon is inscribed in a circle 30 centimeters in diameter. How long is an arc intercepted by one of the sides of the octagon?

49.   A railroad track curves along a 17° arc of a circle. If the radius of the circle is 250 meters, how long is the track?

50.   The earth is approximately 93 million miles from the sun. In one day, the earth moves through an arc of 0.986°. How many miles does the earth travel through space in one week?

51.   A **nautical mile** is defined as $\frac{1}{60}°$ of arc on the equator. If the radius of the earth is 3960 statute miles, find the number of statute miles contained in one nautical mile. Give your answer to the nearest hundredth.

52.   A ship steams 500 nautical miles. How far has the ship gone in statute miles? Give your answer to the nearest hundredth. (*Hint:* See Exercise 51.)

3.2   LINEAR AND ANGULAR VELOCITY

The question "How fast is it moving?" may be answered in two ways, depending on what "it" is. "How fast is that train moving?" might be answered "70 miles per hour" or "90 feet per second." These answers indicate the train's **linear velocity,*** a measure of how far (miles, feet, meters, and so on) the train will travel in one unit of time (hour, minute, second, and so on). It is not necessary that linear velocity be in a straight line; a car could travel at 55 miles per hour around a curve.

---

* The words *linear velocity* and *speed* are used interchangeably; the word *velocity* alone denotes a vector quantity, as used in Section 2.7.

The question "How fast is that phonograph record turning?" might be answered "$33\frac{1}{3}$ revolutions per minute" or "45 revolutions per minute." These answers indicate the record's **angular velocity**, a measure of the angle (radian, degree, or revolutions) through which it rotates in one unit of time (hour, minute, second, and so on).

Example 1   What is the angular velocity of the earth about its axis?

Solution   Because the earth rotates once in 24 hours, its angular velocity is one revolution per 24 hours, or $\frac{1}{24}$ revolution per hour. Since one revolution is 2π radians, the earth's angular velocity is also 2π radians per 24 hours, or $\frac{\pi}{12}$ radian per hour. Other possible answers are one revolution per day, 15 degrees per hour, or 2π radians per day.   ■

There are situations in which linear velocity and angular velocity are related. For example, the question "How fast is the tire of that car moving?" might be answered in two ways. Because the tire moves with the car, it has linear velocity. Because the tire is spinning, it has angular velocity as well. Since for a given linear velocity, small motorcycle tires spin faster than the bigger tires of a semitrailer, one suspects that linear velocity, angular velocity, and the radius of the wheel are all related.

If a wheel of radius r rolls a distance s without slipping, any point on the circumference of the wheel also moves a distance s, measured along the arc of the wheel. See Figure 3-7. The wheel rotates through an angle θ, and

$s = r\theta$ (θ in radians)

If this movement is accomplished in a length of time, t, then s/t is the linear velocity of the wheel, and θ/t is the angular velocity. Dividing both sides of the equation $s = r\theta$ by t gives the formula that relates linear and angular velocity.

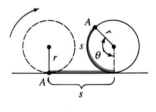

Figure 3-7

$$\frac{s}{t} = r\left(\frac{\theta}{t}\right)$$

If we denote the linear velocity s/t by v, and the angular velocity θ/t by ω (the Greek letter omega), the relation between linear and angular velocity is

$v = r\omega$

where ω is in units of radians per unit time.

Example 2   Find the linear velocity of a point on the earth's equator in miles per hour.

Solution   Assume that the radius of the earth is 4000 miles. From Example 1, the angular velocity of the earth can be expressed as $\omega = \frac{\pi}{12}$ radian/hour. To find the linear velocity, use the formula $v = r\omega$, substitute **4000** for r and $\frac{\pi}{12}$ for ω, and simplify.

$$v = r\omega$$

$$= 4000\left(\frac{\pi}{12}\right)$$

$$\approx 1047$$

Because 4000 miles is the radius of the earth to two-place accuracy, round the answer to two-place accuracy also. The linear velocity of a point on the earth's equator is approximately 1000 miles per hour. ∎

Example 3   A bicycle with 24-inch wheels is traveling down a road at 10 miles per hour. Find the angular velocity of the wheels in revolutions per minute.

Solution   Because the radius of each wheel is 12 inches and the angular velocity is to be given in revolutions per minute, change 10 miles per hour to units of inches per minute.

$$10\frac{\text{mi}}{\text{hr}} \cdot 63360\frac{\text{in.}}{\text{mi}} \cdot \frac{1}{60}\frac{\text{hr}}{\text{min}} = 10560\frac{\text{in.}}{\text{min}}$$

Substitute 10560 for v and 12 for r in the formula $v = r\omega$, and solve for ω.

$$v = r\omega$$

$$10560 = 12\omega$$

$$880\frac{\text{rad}}{\text{min}} = \omega$$

To find ω in revolutions per minute, multiply 880 radians/minute by $\frac{1}{2\pi}$ revolution per radian.

$$\omega = 880\frac{\text{rad}}{\text{min}} \cdot \frac{1}{2\pi}\frac{\text{rev}}{\text{rad}}$$

$$\omega = \frac{440}{\pi}\frac{\text{rev}}{\text{min}}$$

$$\omega \approx 140 \text{ rpm}$$ ∎

Example 4   An 8-inch diameter pulley drives a 6-inch diameter pulley. If the larger pulley makes 15 revolutions per second, find the angular velocity of the smaller pulley in revolutions per second.

Solution   The angular velocity of the drive pulley is

$$15\frac{\text{rev}}{\text{sec}} \cdot 2\pi\frac{\text{rad}}{\text{rev}} = 30\pi\frac{\text{rad}}{\text{sec}}$$

If you assume that the belt that connects the two pulleys does not slip, the linear velocities of points on either circumference are the same—the product $r_1\omega_1$ for one pulley is equal to the product $r_2\omega_2$ for the second pulley. Thus,

$$r_1\omega_1 = v = r_2\omega_2$$

$$4(30\pi) = 3(\omega_2) \qquad \text{Substitute 4 for } r_1, 30\pi \text{ for } \omega_1, \text{ and 3 for } r_2.$$

$$\frac{4(30\pi)}{3} = \omega_2$$

$$40\pi = \omega_2$$

The angular velocity of the smaller pulley is 40π radians per second. To convert to revolutions per second, multiply by $\frac{1}{2\pi}$ revolution/radian and simplify.

$$40\pi \frac{\text{rad}}{\text{sec}} \cdot \frac{1}{2\pi} \frac{\text{rev}}{\text{rad}} = 20 \frac{\text{rev}}{\text{sec}}$$

The angular velocity of the smaller pulley is 20 revolutions per second.   ■

■ EXERCISE 3.2

In Exercises 1–6, find the angular velocity of the object in the unit specified.

1. The minute hand of a clock in radians per hour.

2. The second hand of a clock in radians per second.

3. The minute hand of a clock in radians per second.

4. The earth in its orbit in radians per month.

5. The moon in its orbit in radians per day. Assume that the moon circles the earth in 29.5 days.

6. A phonograph record turning at $33\frac{1}{3}$ revolutions per minute in radians per second.

7. A car is traveling 60 miles per hour (approximately 88 feet per second). How fast are its 30-inch diameter tires spinning in revolutions per second?

8. A Volkswagen is traveling 30 miles per hour. Find the angular velocity of its 22-inch diameter tires in revolutions per minute.

9. A large truck is traveling 65 miles per hour. How fast are its 20-inch radius tires spinning in revolutions per minute?

10. The 30-inch diameter tires of a bus are turning at 400 revolutions per minute. What is the linear velocity of the bus in miles per hour?

11. The 27-inch diameter tires of a bicycle are turning at the rate of 125 revolutions per minute. What is the linear velocity of the bicycle in feet per second?

12. A wheel that is driven by a belt is making 1 revolution per second. If the wheel is 6 inches in diameter, what is the linear velocity of the belt in feet per second?

13. A wheel that is 8.0 centimeters in diameter is driven by a belt with a linear velocity of 40 centimeters per second. How many revolutions per minute is the wheel making?

14. An idler pulley 3 inches in diameter is making 95 revolutions per minute. What is the linear velocity of the belt driving the pulley in inches per second?

15. A belt drives two pulleys. One pulley is 10 inches in diameter and the other is 12 inches in diameter. How many revolutions per minute is the large pulley making if the small pulley is turning at 20 revolutions per minute?

16. A belt drives two wheels that are 15 and 20 inches in diameter. If the 20-inch wheel is turning at 10 revolutions per minute, how fast is the other wheel turning in revolutions per minute?

17. The 10-inch diameter sprocket of a ten-speed bike drives a 3-inch diameter sprocket that is attached to the 27-inch diameter rear wheel. If the pedals make 30 revolutions per minute, how fast is the bike traveling in miles per hour?

18. What is the linear velocity due to the rotation of the earth of Green Bay, Wisconsin (latitude of 44.5° N)? Assume the radius of the earth to be 3960 miles. See Illustration 1.

19. What is the linear velocity due to the rotation of the earth of Miami, Florida (latitude of 25.8° N)? Assume the radius of the earth to be 3960 miles. See Illustration 2.

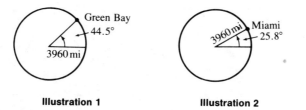

Illustration 1   **Illustration 2**

20. What is the latitude of Pocatello, Idaho, if its linear velocity due to the rotation of the earth is 759 miles per hour? Assume the radius of the earth to be 3960 miles.

21. What is the latitude of Salem, Oregon, if its linear velocity due to the rotation of the earth is 734 miles per hour? Assume the radius of the earth to be 3960 miles.

22. If a gear of radius R_1 drives a gear of radius R_2, show that the angular velocity of the driven gear is R_1/R_2 times that of the driving gear. See Illustration 3.

23. Show that, if the gears in Illustration 3 are separated by an idler gear of any radius, the angular velocity of the driven gear is still R_1/R_2 times that of the driving gear. See Illustration 4.

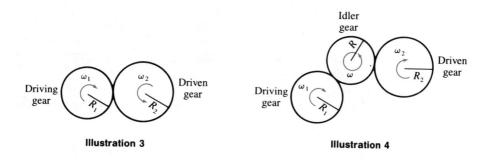

Illustration 3   **Illustration 4**

3.3   THE CIRCULAR FUNCTIONS

In Chapter 2 we defined the trigonometric functions as functions of an angle in standard position. We now present an alternative definition that makes use of the **unit circle**, the circle that is centered at the origin and has a radius of 1 unit.

Definition. Let θ be a central angle of the unit circle shown in Figure 3-8. Let $P(x, y)$ be the point where the terminal side of angle θ intersects the circle. Then, $r = OP = 1$. The **six trigonometric functions of angle θ** are

$$\sin \theta = y$$

$$\cos \theta = x$$

$$\tan \theta = \frac{y}{x} \qquad (x \neq 0)$$

$$\csc \theta = \frac{1}{y} \qquad (y \neq 0)$$

$$\sec \theta = \frac{1}{x} \qquad (x \neq 0)$$

$$\cot \theta = \frac{x}{y} \qquad (y \neq 0)$$

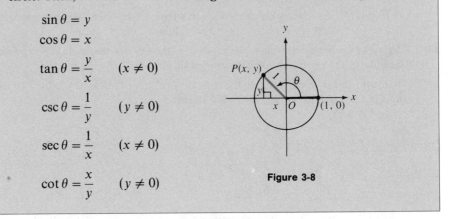

Figure 3-8

The unit-circle definition of the trigonometric functions is a special case of the angle-in-standard-position definition. In the unit-circle definition, the radius r is required to be 1 unit in length. In the angle-in-standard-position definition, r can be any positive number.

The domain of the function defined by the equation $y = f(x)$ is the set of all admissible values of x and the range is the set of all values of y. For virtually all functions previously studied, both the domain and the range of the function have been subsets of the real numbers. A real number x was fed into the function and the function produced a corresponding real number y. Thus far, however, the trigonometric functions have been different. The domain of $y = \sin \theta$, for example, has been a set of angles, rather than a set of real numbers. One advantage of the unit-circle definition is subtle but important. The unit-circle approach allows us to bring the trigonometric functions into conformity with other functions—that is, to show that the trigonometric functions can be thought of as functions with real-number domains.

The argument begins by noting that any real number can represent the length of exactly one arc on the unit circle. If t is a positive number, we can find the arc of

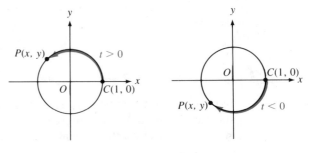

Figure 3-9a   Figure 3-9b

length t by measuring a distance t in a counterclockwise direction along an arc of the unit circle beginning at the point $C(1,0)$. This determines arc CP of length t. See Figure 3-9a. If t is a negative number, we can find the arc of length t by measuring a distance $|t|$ in a clockwise direction along an arc of the unit circle beginning at the point $C(1,0)$. This determines arc CP of length $|t|$. See Figure 3-9b.

In each case, marking off the arc determines the unique point P with coordinates (x, y) that corresponds to the real number t. Now suppose that we let t be any real number and place an arc of length $|t|$ on the unit circle as in Figure 3-10. From Section 3.1 we know that, if s is an arc intercepted by a central angle θ and θ is measured in radians, we have

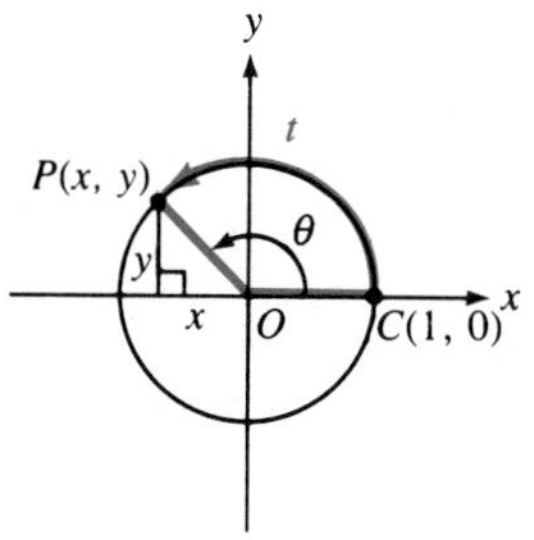

$$s = r\theta$$

We can substitute t for s and 1 for r in the formula $s = r\theta$ to obtain

$$t = \theta$$

or

Figure 3-10

$$\theta = t$$

Thus, when the measure of an arc on the unit circle is the real number t, then t is also the radian measure of the central angle determined by that arc. This gives

$$\sin\theta = \sin t \qquad \csc\theta = \csc t$$
$$\cos\theta = \cos t \qquad \sec\theta = \sec t$$
$$\tan\theta = \tan t \qquad \cot\theta = \cot t$$

where θ is an angle measured in radians and t is a real number. Thus, we can think of each trigonometric expression as being either a trigonometric function of an angle measured in radians or as a trigonometric function of a real number. The important point is this: *The trigonometric functions can now be thought of as functions that have domains and ranges that are subsets of the real numbers.*

Example 1 Find **a.** $\sin\dfrac{\pi}{3}$, **b.** $\cos\dfrac{5\pi}{2}$, **c.** $\tan\dfrac{3\pi}{4}$, and **d.** $\csc\dfrac{7\pi}{6}$.

Solution **a.** $\sin\dfrac{\pi}{3} = \sin\left(\dfrac{\pi}{3}\,\text{rad}\right) = \dfrac{\sqrt{3}}{2}$

  b. $\cos\dfrac{5\pi}{2} = \cos\left(\dfrac{5\pi}{2}\,\text{rad}\right) = \cos\left(\dfrac{\pi}{2}\,\text{rad}\right) = 0$

  c. $\tan\dfrac{3\pi}{4} = \tan\left(\dfrac{3\pi}{4}\,\text{rad}\right) = -\tan\left(\dfrac{\pi}{4}\,\text{rad}\right) = -1$

  d. $\csc\dfrac{7\pi}{6} = \csc\left(\dfrac{7\pi}{6}\,\text{rad}\right) = \dfrac{1}{\sin\left(\dfrac{7\pi}{6}\,\text{rad}\right)} = -2$ ∎

Example 2   Find the coordinates (x, y) of the point P on the unit circle that correspond to the real numbers **a.** π, **b.** $\dfrac{\pi}{6}$, and **c.** $\dfrac{-5\pi}{4}$.

Solution   **a.**   From the definitions of the circular functions, you have

$$x = \cos \theta$$
$$y = \sin \theta$$

But, if θ is measured in radians, you also have

$$x = \cos t$$
$$y = \sin t$$

where t is the real number that is the radian measure of central angle θ. Thus, the coordinates (x, y) of the point P that correspond to the real number π are

$$x = \cos \pi = \cos(\pi \text{ rad}) = -1$$
$$y = \sin \pi = \sin(\pi \text{ rad}) = 0$$

Thus, point P has coordinates $(-1, 0)$.

Because the circumference of the unit circle is 2π, the number π is the measure of half that circumference. From Figure 3-11, you can see that the coordinates of the point P must be $(-1, 0)$.

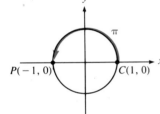

Figure 3-11

b.   The point P on the unit circle that corresponds to $\frac{\pi}{6}$ has coordinates of

$$x = \cos \frac{\pi}{6} = \cos\left(\frac{\pi}{6} \text{ rad}\right) = \frac{\sqrt{3}}{2}$$

$$y = \sin \frac{\pi}{6} = \sin\left(\frac{\pi}{6} \text{ rad}\right) = \frac{1}{2}$$

Thus, the point P has coordinates $\left(\dfrac{\sqrt{3}}{2}, \dfrac{1}{2}\right)$.

c.   The point P on the unit circle that corresponds to $\frac{-5\pi}{4}$ has coordinates of

$$x = \cos\left(\frac{-5\pi}{4}\right) = \cos\left(\frac{-5\pi}{4} \text{ rad}\right) = -\frac{\sqrt{2}}{2}$$

$$y = \sin\left(\frac{-5\pi}{4}\right) = \sin\left(\frac{-5\pi}{4} \text{ rad}\right) = \frac{\sqrt{2}}{2}$$

Thus, the point P has coordinates $\left(-\dfrac{\sqrt{2}}{2}, \dfrac{\sqrt{2}}{2}\right)$.   ∎

Example 3   Find the coordinates of the point P on the unit circle that correspond to the real numbers **a.** 7 and **b.** -1.37.

Solution   **a.**   The coordinates of the point P are

$$x = \cos 7 = \cos(7 \text{ rad})$$
$$y = \sin 7 = \sin(7 \text{ rad})$$

Use a calculator, set in radian mode, to determine that

$$x \approx 0.7539 \qquad \text{and} \qquad y \approx 0.6570$$

Thus, point P has approximate coordinates of $(0.7539, 0.6570)$.

b.   The coordinates of the point P are

$$x = \cos(-1.37) = \cos(-1.37 \text{ rad}) \approx 0.1994$$
$$y = \sin(-1.37) = \sin(-1.37 \text{ rad}) \approx -0.9799$$

Thus, point P has approximate coordinates of $(0.1994, -0.9799)$.

■ EXERCISE 3.3

In Exercises 1–12, evaluate each given expression. **Do not use a calculator.**

1. $\sin \dfrac{\pi}{6}$

2. $\cos \dfrac{\pi}{6}$

3. $\cos\left(-\dfrac{5\pi}{6}\right)$

4. $\sin\left(-\dfrac{5\pi}{6}\right)$

5. $\tan \dfrac{5\pi}{4}$

6. $\cot \dfrac{5\pi}{4}$

7. $\csc \dfrac{5\pi}{6}$

8. $\sec \dfrac{5\pi}{6}$

9. $\sin\left(-\dfrac{8\pi}{3}\right)$

10. $\cos \dfrac{8\pi}{3}$

11. $\tan \dfrac{15\pi}{4}$

12. $\csc \dfrac{7\pi}{3}$

In Exercises 13–24, use a calculator to evaluate each given expression to four decimal places. Remember to set your calculator in radian mode.

13. $\sin 2$

14. $\sin 3$

15. $\cos 8$

16. $\tan 5$

17. $\sin \dfrac{3}{\pi}$

18. $\cos\left(-\dfrac{3}{\pi}\right)$

19. $\sec 5$

20. $\csc 4$

21. $\cot 1$

22. $\cot(-1)$

23. $\sin(3 + \pi)$

24. $\cos(3 - \pi)$

In Exercises 25–48, find the coordinates of the point P on the unit circle that correspond to each given real number. **Do not use a calculator.**

25. $\dfrac{3\pi}{2}$

26. $-\dfrac{\pi}{2}$

27. $-\pi$

28. 2π

29. 3π

30. 4π

31. $-\dfrac{7\pi}{2}$

32. $\dfrac{9\pi}{2}$

33. $\dfrac{\pi}{4}$

34. $-\dfrac{\pi}{4}$

35. $-\dfrac{3\pi}{4}$

36. $\dfrac{5\pi}{4}$

37. $\dfrac{\pi}{3}$

38. $-\dfrac{\pi}{6}$

39. $-\dfrac{2\pi}{3}$

40. $\dfrac{2\pi}{3}$

41. $\dfrac{4\pi}{3}$

42. $-\dfrac{4\pi}{3}$

43. $-\dfrac{5\pi}{3}$

44. $\dfrac{5\pi}{6}$

45. $\dfrac{11\pi}{6}$

46. $\dfrac{17\pi}{6}$

47. $\dfrac{23\pi}{6}$

48. $-\dfrac{17\pi}{6}$

49.   Show that the area of a sector of the unit circle is given by the formula $A = \frac{1}{2}\theta$, where θ is the central angle of the sector measured in radians.

50. Show that the area of triangle OAR in Illustration 1 is given by the formula

$$A = \tfrac{1}{2}\cos\theta\sin\theta$$

51. Show that the area of triangle OTP in Illustration 1 is given by the formula

$$A = \tfrac{1}{2}\tan\theta$$

52. Use the inequality

area of $\triangle OAR \le$ area of sector $OTR \le$ area of $\triangle OTP$

Illustration 1

to show that the ratio $\frac{\sin\theta}{\theta}$ approaches 1 as θ approaches 0. Note that θ must be a real number for this ratio to have meaning. This fact is extremely important in calculus. (*Hint:* Refer to Exercises 49–51 and to Illustration 1.)

3.4   GRAPHS OF FUNCTIONS INVOLVING SIN X AND COS X

In this section we will discuss the graphs of the sine and cosine functions. A study of these graphs will emphasize some properties of the sine and cosine functions that are not obvious from their equations.

We begin by noting that the sine of an angle θ is equal to the sine of any angle that is coterminal with angle θ. Likewise, the cosine of an angle θ is equal to the cosine of any angle that is coterminal with angle θ. Therefore, for any real number x,

$$\sin x = \sin(x \pm 2\pi) = \sin(x \pm 4\pi) = \cdots$$
$$\cos x = \cos(x \pm 2\pi) = \cos(x \pm 4\pi) = \cdots$$

The sine and cosine functions are called **periodic functions** because, as x increases, the values of $\sin x$ and $\cos x$ repeat in a predictable way.

> **Definition.**   A function f is said to be **periodic** with period p if p is the smallest positive number for which
>
> $$f(x) = f(x + p)$$
>
> for all x in the domain of f.

The sine function has a period of 2π because $\sin x = \sin(x + 2\pi)$ for all x, and $\sin x = \sin(x + p)$ is true for no positive number less than 2π. To graph the sine function, we find points (x, y) that satisfy the equation $y = \sin x$ and plot them on a rectangular coordinate system. Remember that x, often called the **argument of the function**, can be thought of either as an angle measured in radians or as a real number.

To graph the equation $y = \sin x$, we refer to the table of values in Figure 3-12. These values show how the values of $\sin x$ vary as x increases from 0 to 2π.

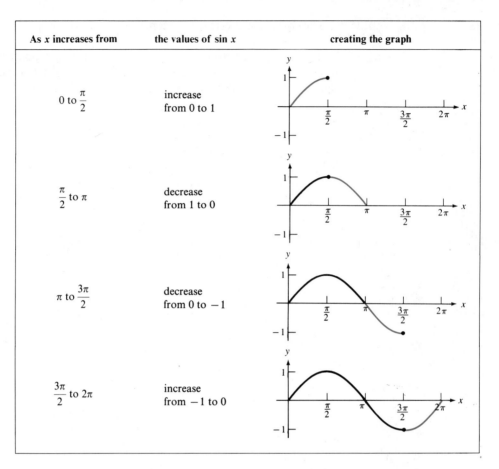

As *x* increases from	the values of sin *x*	creating the graph
0 to $\frac{\pi}{2}$	increase from 0 to 1	
$\frac{\pi}{2}$ to π	decrease from 1 to 0	
π to $\frac{3\pi}{2}$	decrease from 0 to -1	
$\frac{3\pi}{2}$ to 2π	increase from -1 to 0	

$$y = \sin x$$

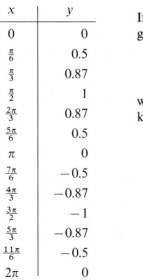

x	*y*
0	0
$\frac{\pi}{6}$	0.5
$\frac{\pi}{3}$	0.87
$\frac{\pi}{2}$	1
$\frac{2\pi}{3}$	0.87
$\frac{5\pi}{6}$	0.5
π	0
$\frac{7\pi}{6}$	-0.5
$\frac{4\pi}{3}$	-0.87
$\frac{3\pi}{2}$	-1
$\frac{5\pi}{3}$	-0.87
$\frac{11\pi}{6}$	-0.5
2π	0

If we plot the pairs of values $(x, \sin x)$ shown in Figure 3-12, we can draw the graph of $y = \sin x$, often called a **sine wave**. We note that, for all *x*,

$$-1 \le \sin x \le 1$$

The graph of $y = \sin x$ does not start at 0, nor does it end at 2π. The sine wave goes on forever in both directions. Since the function is periodic, if we know its behavior through one period, we know its behavior everywhere.

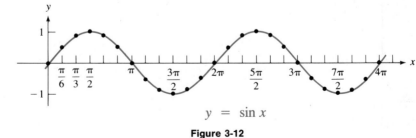

$$y = \sin x$$

Figure 3-12

The cosine function also has a period of 2π because $\cos x = \cos(x + 2\pi)$ for all x, and $\cos x = \cos(x + p)$ is true for no positive number less than 2π. Because $\cos 0 = 1$, we shall draw its graph beginning at the point $(0, 1)$. To graph the equation $y = \cos x$, we refer to the table of values accompanying Figure 3-13, which shows what happens to the values of $\cos x$ as x increases from 0 to 2π.

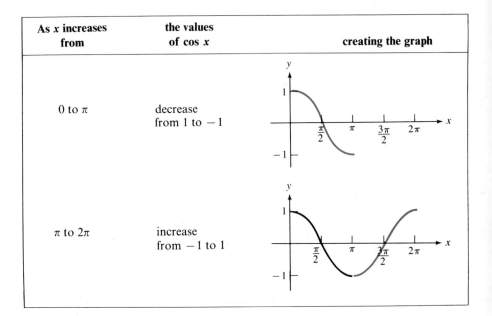

As x increases from	the values of cos x	creating the graph
0 to π	decrease from 1 to -1	
π to 2π	increase from -1 to 1	

$y = \cos x$

x	y
0	1
$\frac{\pi}{6}$	0.87
$\frac{\pi}{3}$	0.5
$\frac{\pi}{2}$	0
$\frac{2\pi}{3}$	-0.5
$\frac{5\pi}{6}$	-0.87
π	-1
$\frac{7\pi}{6}$	-0.87
$\frac{4\pi}{3}$	-0.5
$\frac{3\pi}{2}$	0
$\frac{5\pi}{3}$	0.5
$\frac{11\pi}{6}$	0.87
2π	1

The graph of $y = \cos x$ appears in Figure 3-13. We note that, for all x,

$$-1 \le \cos x \le 1$$

$y = \cos x$

Figure 3-13

Example 1   Graph the function defined by $y = 3 \sin x$.

Solution   The values of $3 \sin x$ can be found by multiplying each value of $\sin x$ by 3. Because the values of $\sin x$ are between -1 and 1, the values of $3 \sin x$ must be between -3 and 3 as x increases from 0 to 2π. The way the values of $3 \sin x$ vary is shown in the following chart.

As *x* increases from	the values of 3 sin *x*
0 to $\dfrac{\pi}{2}$	increase from 0 to 3
$\dfrac{\pi}{2}$ to $\dfrac{3\pi}{2}$	decrease from 3 to -3
$\dfrac{3\pi}{2}$ to 2π	increase from -3 to 0

The graph of $y = 3 \sin x$, shown in Figure 3-14, can be found by stretching the graph of $y = \sin x$ vertically by a factor of 3.

$y = 3 \sin x$

x	y
0	0
$\frac{\pi}{2}$	3
π	0
$\frac{3\pi}{2}$	-3
2π	0

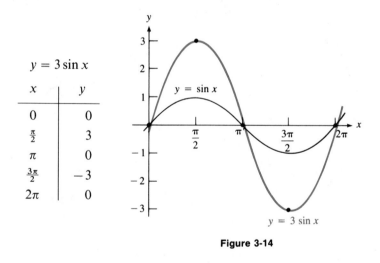

Figure 3-14   ∎

Example 2   Graph $y = 2 \cos x$ and $y = -2 \cos x$ on the same set of coordinate axes.

Solution   The values of $2 \cos x$ can be found by multiplying each value of $\cos x$ by 2. As x increases from 0 to π, the values of $2 \cos x$ decrease from 2 to -2. As x increases from π to 2π, they increase from -2 back to 2. Thus, the graph of $y = 2 \cos x$ can be found by stretching the graph of $y = \cos x$ vertically by a factor of 2.

The graph of $y = -2 \cos x$ can be found by reflecting the graph of $y = 2 \cos x$ in the x-axis. Both graphs appear in Figure 3-15.

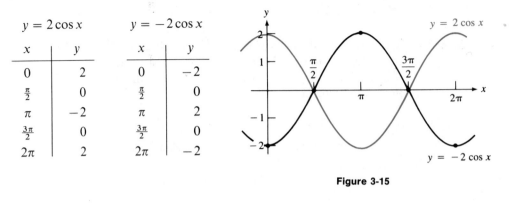

$y = 2\cos x$

x	y
0	2
$\frac{\pi}{2}$	0
π	-2
$\frac{3\pi}{2}$	0
2π	2

$y = -2\cos x$

x	y
0	-2
$\frac{\pi}{2}$	0
π	2
$\frac{3\pi}{2}$	0
2π	-2

Figure 3-15

Example 3 Graph $y = \sin 2x$.

Solution Because the graph of $y = \sin x$ completes one cycle as x increases from 0 to 2π, the graph of $y = \sin 2x$ must complete one cycle as $2x$ increases from 0 to 2π, or as x increases from 0 to π. Thus, the period of the graph of $y = \sin 2x$ is π, and the sine wave will oscillate twice as fast as the graph of $y = \sin x$. The graph of $y = \sin 2x$ appears in Figure 3-16. Note that the coefficient of x is a number greater than 1 and that the curve is compressed horizontally.

$y = \sin 2x$

x	y
0	0
$\frac{\pi}{4}$	1
$\frac{\pi}{2}$	0
$\frac{3\pi}{4}$	-1
π	0

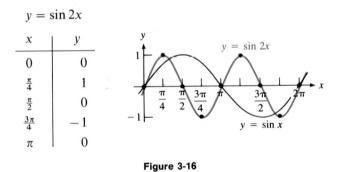

Figure 3-16

Example 4 Graph $y = \cos\frac{1}{3}x$.

Solution Because the graph of $y = \cos x$ completes one cycle as x increases from 0 to 2π, the graph of $y = \cos\frac{1}{3}x$ must complete one cycle as $\frac{1}{3}x$ increases from 0 to 2π, or as x increases from 0 to 6π. Thus, the period of the graph of $y = \cos\frac{1}{3}x$ is 6π, and the graph will oscillate one-third as fast as the graph of $y = \cos x$. The graph of $y = \cos\frac{1}{3}x$ appears in Figure 3-17. Note that the coefficient of x is a positive number less than 1 and that the curve is stretched horizontally.

$y = \cos\frac{1}{3}x$

x	y
0	1
$\frac{3\pi}{2}$	0
3π	-1
$\frac{9\pi}{2}$	0
6π	1

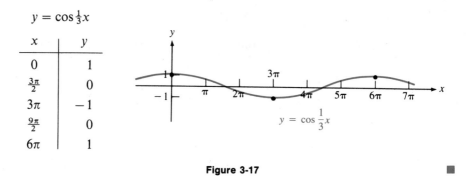

Figure 3-17

Example 5   Graph $y = \cos \pi x$.

Solution   Because the graph of $y = \cos x$ completes one cycle as x increases from 0 to 2π, the graph of $y = \cos \pi x$ must complete one cycle as πx increases from 0 to 2π, or as x increases from 0 to 2. Thus, the period of the graph of $y = \cos \pi x$ is 2. The graph of $y = \cos \pi x$ appears in Figure 3-18. Note that the coefficient of x is a number greater than 1 and that the curve is compressed horizontally.

$y = \cos \pi x$

x	y
0	1
$\frac{1}{2}$	0
1	-1
$\frac{3}{2}$	0
2	1
$\frac{5}{2}$	0
3	-1
$\frac{7}{2}$	0
4	1
$\frac{9}{2}$	0
5	-1
$\frac{11}{2}$	0
6	1

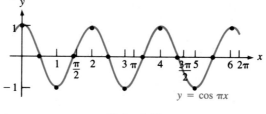

Figure 3-18

Period and Amplitude

One cycle of the graph of $y = a \sin bx$ is completed as $|bx|$ increases from 0 to 2π, or as x increases from 0 to $|\frac{2\pi}{b}|$. Thus, the period of the graph of $y = a \sin bx$ is $|\frac{2\pi}{b}|$.

Because $-1 \leq \sin bx \leq 1$, the largest value that can be attained by $y = a \sin bx$ is $|a| \cdot 1$, or just $|a|$. This value is called the **amplitude** of the graph of $y = a \sin bx$.

A similar argument applies to determine the period and the amplitude of the graph of $y = a \cos bx$.

Period and Amplitude.   The **period** of the graph of $\begin{cases} y = a \sin bx \\ y = a \cos bx \end{cases}$ is $\left|\frac{2\pi}{b}\right|$ and the **amplitude** is $|a|$.

Example 6   Graph $y = 5 \sin 7x$.

Solution   The amplitude is 5 and the period is $\frac{2\pi}{7}$. One cycle of the graph of $y = 5 \sin 7x$ appears in Figure 3-19.

$$y = 5 \sin 7x$$

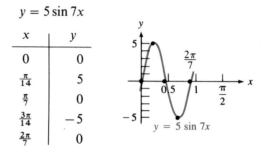

x	y
0	0
$\frac{\pi}{14}$	5
$\frac{\pi}{7}$	0
$\frac{3\pi}{14}$	-5
$\frac{2\pi}{7}$	0

Figure 3-19

Example 7   Graph $y = 2 \cos \frac{1}{2}x$.

Solution   The amplitude is 2 and the period is $\frac{2\pi}{\frac{1}{2}}$, or 4π. Because $2 \cos \frac{1}{2}x$ is zero when

$$x = \ldots, -\pi, \pi, 3\pi, \ldots$$

the graph intersects the x-axis at these points. One cycle of the graph appears in Figure 3-20.

$$y = 2 \cos \tfrac{1}{2}x$$

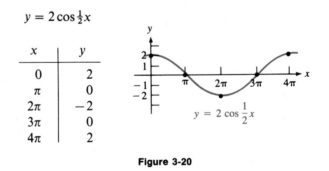

x	y
0	2
π	0
2π	-2
3π	0
4π	2

Figure 3-20

▬ EXERCISE 3.4

*In Exercises 1–16, find the amplitude and period of each function. **Do not draw the graph.***

1. $y = 2 \sin x$ **2.** $y = 3 \cos x$ **3.** $y = \cos 9x$ **4.** $y = -\sin 11x$

5. $y = \sin \frac{1}{3}x$ **6.** $y = \cos \frac{1}{4}x$ **7.** $y = -\cos 0.2x$ **8.** $y = \sin 0.25x$

9. $y = 3\sin\dfrac{1}{2}x$

10. $y = \dfrac{1}{2}\cos\dfrac{1}{3}x$

11. $y = -\dfrac{1}{2}\cos\pi x$

12. $y = 17\sin 2\pi x$

13. $y = 3\sin 2\pi x$

14. $y = 8\cos\pi x$

15. $y = -\dfrac{1}{3}\sin\dfrac{3x}{\pi}$

16. $y = -\dfrac{5}{3}\cos\dfrac{x}{\pi}$

In Exercises 17–26, graph each pair of functions over the indicated interval.

17. $y = \sin x$ and $y = 2\sin x, 0 \le x \le 2\pi$

18. $y = \cos x$ and $y = -3\cos x, 0 \le x \le 2\pi$

19. $y = \cos x$ and $y = -\dfrac{1}{3}\cos x, 0 \le x \le 2\pi$

20. $y = \sin x$ and $y = \dfrac{1}{2}\sin x, 0 \le x \le 2\pi$

21. $y = \sin x$ and $y = \sin 2x, 0 \le x \le 2\pi$

22. $y = \cos x$ and $y = \cos 3x, 0 \le x \le 2\pi$

23. $y = \cos x$ and $y = \cos\dfrac{1}{3}x, 0 \le x \le 6\pi$

24. $y = \sin x$ and $y = \sin\dfrac{1}{2}x, 0 \le x \le 4\pi$

25. $y = \sin x$ and $y = \sin\pi x, 0 \le x \le 2\pi$

26. $y = \cos x$ and $y = \cos\pi x, 0 \le x \le 2\pi$

In Exercises 27–40, graph each given function over an interval that is at least one period long.

27. $y = 3\cos x$

28. $y = 4\sin x$

29. $y = -\sin x$

30. $y = -\cos x$

31. $y = \cos 2x$

32. $y = \sin 3x$

33. $y = -\sin\dfrac{x}{4}$

34. $y = \cos\dfrac{x}{4}$

35. $y = 3\sin\pi x$

36. $y = -2\cos\pi x$

37. $y = \dfrac{1}{2}\cos 4x$

38. $y = -4\sin 3x$

39. $y = -4\sin\dfrac{x}{2}$

40. $y = \dfrac{1}{3}\cos\dfrac{x}{2}$

3.5   GRAPHS OF FUNCTIONS INVOLVING TAN X, COT X, CSC X, AND SEC X

Because the reference angles of x and $x + \pi$ shown in Figure 3-21 are equal and because the tangents of angles in nonadjacent quadrants agree in sign, we have

$$\tan x = \tan(x + \pi)$$

for all x. Furthermore, the number π is the smallest positive number p for which $\tan x = \tan(x + p)$. Therefore, the period of the tangent function is π.

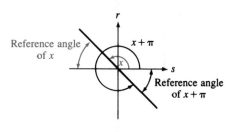

Figure 3-21

$y = \tan x$

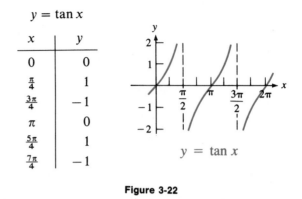

x	y
0	0
$\frac{\pi}{4}$	1
$\frac{3\pi}{4}$	-1
π	0
$\frac{5\pi}{4}$	1
$\frac{7\pi}{4}$	-1

Figure 3-22

Because $\tan 0 = 0$, the graph of $y = \tan x$ passes through the origin as in Figure 3-22. As x gets very close to $\frac{\pi}{2}$, the value of $|\tan x|$ becomes very large. Since a value for $\tan \frac{\pi}{2}$ does not exist, the graph of $y = \tan x$ cannot intersect the line $x = \frac{\pi}{2}$. However, the graph of $y = \tan x$ does approach the vertical line $x = \frac{\pi}{2}$ as x approaches $\frac{\pi}{2}$. The vertical line $x = \frac{\pi}{2}$ is called an **asymptote**. Other vertical asymptotes of the tangent function are the vertical lines $x = \frac{3\pi}{2}$, $x = \frac{5\pi}{2}$, and so on. The graph of $y = \tan x$ approaches these lines, but it never touches them.

For all x between $\frac{\pi}{2}$ and π, the values of $\tan x$ are negative, returning to 0 when $x = \pi$. The graph of $y = \tan x$ is shown in Figure 3-22. In the exercises, you will be asked to explain why the tangent function has no defined amplitude.

The cotangent function, whose graph is shown in Figure 3-23, has a period of π, with vertical asymptotes at multiples of π. In the exercises, you will be asked to explain why the cotangent function has no defined amplitude.

$y = \cot x$

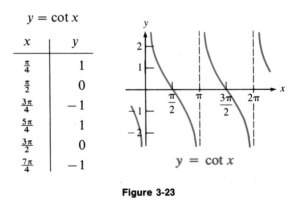

x	y
$\frac{\pi}{4}$	1
$\frac{\pi}{2}$	0
$\frac{3\pi}{4}$	-1
$\frac{5\pi}{4}$	1
$\frac{3\pi}{2}$	0
$\frac{7\pi}{4}$	-1

Figure 3-23

Because $\csc x$ is the reciprocal of $\sin x$, it has the same period as $\sin x$ and is undefined whenever $\sin x$ is 0—at $x = 0$, at $x = \pi$, at $x = 2\pi$, and so on. These values determine the vertical asymptotes for the graph of $y = \csc x$. The graph of

$y = \csc x$

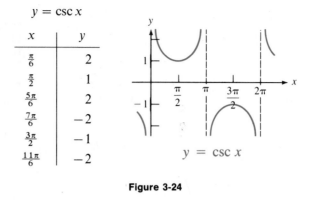

x	y
$\frac{\pi}{6}$	2
$\frac{\pi}{2}$	1
$\frac{5\pi}{6}$	2
$\frac{7\pi}{6}$	-2
$\frac{3\pi}{2}$	-1
$\frac{11\pi}{6}$	-2

Figure 3-24

$y = \csc x$ appears in Figure 3-24. In the exercises, you will be asked to explain why the cosecant function has no defined amplitude.

Because $\sec x$ is the reciprocal of $\cos x$ and $\cos x$ has a period of 2π, the graph of $y = \sec x$, shown in Figure 3-25, has a period of 2π. There is no value for the secant function at $x = \frac{\pi}{2}$, at $x = \frac{3\pi}{2}$, and so on because the cosine of these values is 0. Thus, the vertical lines $x = \frac{\pi}{2}$, $x = \frac{3\pi}{2}$,... are the vertical asymptotes for the graph of $y = \sec x$. In the exercises, you will be asked to explain why the secant function has no defined amplitude.

$y = \sec x$

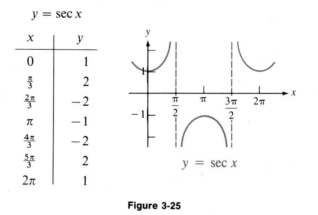

x	y
0	1
$\frac{\pi}{3}$	2
$\frac{2\pi}{3}$	-2
π	-1
$\frac{4\pi}{3}$	-2
$\frac{5\pi}{3}$	2
2π	1

Figure 3-25

Example 1   Graph $y = \tan 3x$.

Solution   The graph of $y = \tan 3x$, shown in Figure 3-26, intersects the x-axis when $\tan 3x = 0$. This is true when $3x = 0$, when $3x = \pi$, when $3x = 2\pi$, and so on. Hence, the graph intersects the x-axis at $x = 0$, at $x = \frac{\pi}{3}$, at $x = \frac{2\pi}{3}$, and so on. The graph of $y = \tan 3x$ completes one cycle as $3x$ increases from 0 to π—that is, as x itself increases from 0 to $\frac{\pi}{3}$. A value for $\tan 3x$ is undefined at $3x = \frac{\pi}{2}$ or at $x = \frac{\pi}{6}$. The period is $\frac{\pi}{3}$, and the vertical asymptotes are $x = \frac{\pi}{6}$, $x = \frac{\pi}{2}$, $x = \frac{5\pi}{6}$, and so on.

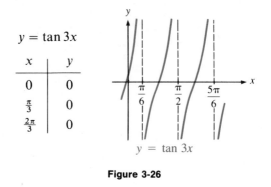

$y = \tan 3x$

x	y
0	0
$\frac{\pi}{3}$	0
$\frac{2\pi}{3}$	0

$y = \tan 3x$

Figure 3-26

Example 2   Graph $y = 2 \csc x$ on the interval from 0 to 2π.

Solution   Because $\csc x$ is the reciprocal of $\sin x$, it has the same period as $\sin x$ and is undefined whenever $\sin x$ is 0. Thus, in the interval from 0 to 2π, a value for $2 \csc x$ is undefined at $x = 0$, at $x = \pi$, and at $x = 2\pi$. These values determine vertical asymptotes for the graph of $y = 2 \csc x$. A table of values and the graph of $y = 2 \csc x$ appear in Figure 3-27.

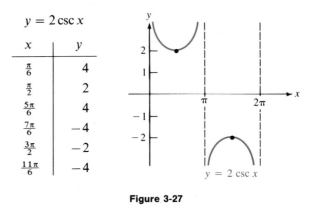

$y = 2 \csc x$

x	y
$\frac{\pi}{6}$	4
$\frac{\pi}{2}$	2
$\frac{5\pi}{6}$	4
$\frac{7\pi}{6}$	-4
$\frac{3\pi}{2}$	-2
$\frac{11\pi}{6}$	-4

$y = 2 \csc x$

Figure 3-27

Example 3   Graph $y = -2 \sec 2x$ on the interval from 0 to 2π.

Solution   Because $\sec 2x$ is the reciprocal of $\cos 2x$, it has the same period as $\cos 2x$. Thus, the period of $y = -2 \sec 2x$ is $\frac{2\pi}{2}$ or π. Furthermore, a value for $\sec 2x$ is undefined whenever $\cos 2x$ is 0. In the interval from 0 to 2π, this occurs when $x = \frac{\pi}{4}$, when $x = \frac{3\pi}{4}$, when $x = \frac{5\pi}{4}$, and when $x = \frac{7\pi}{4}$. These values determine the vertical asymptotes for the graph of $y = -2 \sec 2x$. The graph of $y = -2 \sec 2x$ appears in Figure 3-28.

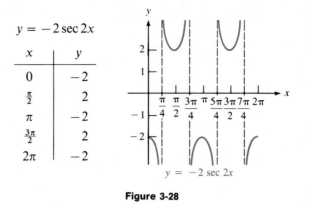

$y = -2\sec 2x$

x	y
0	-2
$\frac{\pi}{2}$	2
π	-2
$\frac{3\pi}{2}$	2
2π	-2

Figure 3-28

We summarize the results of this section.

Period and Amplitude.   The **period** of

$$\left.\begin{cases} y = a\tan bx \\ y = a\cot bx \end{cases}\right\} \quad \text{is} \quad \left|\frac{\pi}{b}\right|.$$

The **period** of

$$\left.\begin{cases} y = a\csc bx \\ y = a\sec bx \end{cases}\right\} \quad \text{is} \quad \left|\frac{2\pi}{b}\right|.$$

The tangent, cotangent, cosecant, and secant functions have no defined amplitude.

EXERCISE 3.5

In Exercises 1–18, give the period of each of the given functions. ***Do not draw the graph.***

1. $y = 3\tan x$

2. $y = 2\csc x$

3. $y = \dfrac{1}{2}\sec x$

4. $y = \dfrac{1}{3}\cot x$

5. $y = \dfrac{1}{3}\tan 3x$

6. $y = \dfrac{1}{2}\sec 2x$

7. $y = -2\csc \pi x$

8. $y = -3\tan \pi x$

9. $y = 3\sec \dfrac{x}{3}$

10. $y = -2\tan \dfrac{\pi x}{3}$

11. $y = \dfrac{7}{2}\cot \dfrac{2\pi x}{3}$

12. $y = -\dfrac{2}{3}\csc \dfrac{\pi x}{3}$

13. $y = 3\csc \dfrac{\pi x}{2}$

14. $y = -4\sec \dfrac{2\pi x}{5}$

15. $y = -\cot \dfrac{x}{2\pi}$

16. $y = 7\csc \dfrac{x}{4\pi}$

17. $y = -\dfrac{2}{5}\sec \dfrac{3x}{\pi}$

18. $y = \dfrac{7}{9}\cot \dfrac{2x}{\pi}$

In Exercises 19–30, graph each given function over the indicated interval.

19. $y = 2 \tan x, \quad \dfrac{-\pi}{2} < x < \dfrac{3\pi}{2}$

20. $y = 2 \csc x, \quad 0 < x < 2\pi$

21. $y = -3 \sec x, \quad 0 \le x \le 2\pi$

22. $y = -\csc 2x, \quad 0 < x < \dfrac{3\pi}{2}$

23. $y = \cot 2x, \quad 0 < x < \pi$

24. $y = \sec 3x, \quad 0 \le x \le \dfrac{2\pi}{3}$

25. $y = -2 \tan \dfrac{x}{2}, \quad 0 < x < 2\pi$

26. $y = -3 \csc \dfrac{x}{2}, \quad 0 < x < 4\pi$

27. $y = 2 \sec 2x, \quad 0 \le x \le 2\pi$

28. $y = 2 \csc 2x, \quad 0 < x < 2\pi$

29. $y = -2 \cot \dfrac{\pi}{4} x, \quad 0 < x < 4$

30. $y = -2 \sec \dfrac{\pi}{4} x, \quad 0 \le x \le 8$

31. Explain why amplitude has no meaning when discussing the tangent, cotangent, secant, and cosecant functions.

32. For what values of x, if any, are the values of $\sin x$ and $\csc x$ equal?

33. For what values of x, if any, are the values of $\cos x$ and $\sec x$ equal?

34. For what values of x, if any, are the values of $\tan x$ and $\cot x$ equal?

3.6 VERTICAL AND HORIZONTAL TRANSLATIONS OF THE TRIGONOMETRIC FUNCTIONS

In applied work we often encounter graphs of trigonometric functions that have been shifted in either a vertical or a horizontal direction. Such shifts are called **translations**. The graph of a function can be shifted vertically (a vertical translation) by adding a constant value to the function. The graph of a function can be shifted horizontally (a horizontal translation) by adding a constant value to the argument of the function. The first two examples illustrate vertical translations and the rest involve horizontal translations.

Example 1   Graph $y = 2 + \cos x$.

Solution   You already know that $y = \cos x$ has a period of 2π, has values that lie between -1 and 1, and has x-intercepts at

$$x = \frac{\pi}{2}, \quad x = \frac{3\pi}{2}, \quad x = \frac{5\pi}{2}, \dots$$

The values of $2 + \cos x$ will be similar to the values of $\cos x$, except that each will be increased by 2. Thus, the graph of $y = 2 + \cos x$ has a period of 2π and has values that lie between $2 + (-1)$ and $2 + 1$, or between 1 and 3. The curve will intersect the line $y = 2$ at $x = \frac{\pi}{2}$, at $x = \frac{3\pi}{2}$, at $x = \frac{5\pi}{2}$, and so on. A table of values and the graph of $y = 2 + \cos x$ are shown in Figure 3-29.

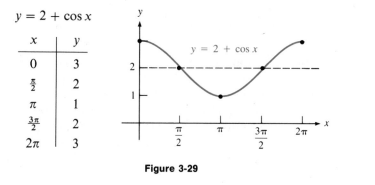

$y = 2 + \cos x$

x	y
0	3
$\frac{\pi}{2}$	2
π	1
$\frac{3\pi}{2}$	2
2π	3

Figure 3-29

Example 2   Graph $y = -3 + \tan\dfrac{x}{2}$.

Solution   The function $y = \tan\frac{x}{2}$ has a period of $\pi/\frac{1}{2}$, or 2π, and has vertical asymptotes at

$$x = \pi, \quad x = 3\pi, \quad x = 5\pi,\ldots$$

The values of $-3 + \tan\frac{x}{2}$ will be similar to the values of $\tan\frac{x}{2}$, except that each value of $\tan\frac{x}{2}$ will be decreased by 3. The graph of $y = -3 + \tan\frac{x}{2}$ appears in Figure 3-30.

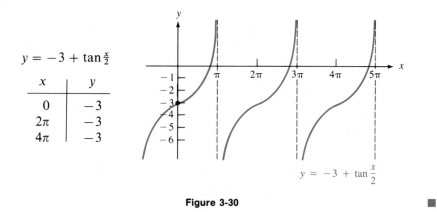

$y = -3 + \tan\frac{x}{2}$

x	y
0	-3
2π	-3
4π	-3

$$y = -3 + \tan\frac{x}{2}$$

Figure 3-30

The two preceding examples suggest the following facts.

Vertical Translations.   The graph of $y = k + a \sin bx$ is identical to the graph of $y = a \sin bx$, except that it is translated $|k|$ units

$$\begin{Bmatrix} \text{up} \\ \text{down} \end{Bmatrix} \text{ if } k \text{ is } \begin{Bmatrix} \text{positive} \\ \text{negative} \end{Bmatrix}$$

An analogous statement is true for each of the other trigonometric functions.

Example 3   Graph $y = \sin\left(x + \dfrac{\pi}{6}\right)$ and $y = \sin x$ on the same set of coordinate axes.

Solution   One complete cycle of $y = \sin x$ is described as x increases from 0 to 2π. Similarly, one cycle of $y = \sin(x + \frac{\pi}{6})$ is completed when $x + \frac{\pi}{6}$ increases from 0 to 2π—that is, when x itself increases from $-\frac{\pi}{6}$ to $(2\pi - \frac{\pi}{6})$. At $x = -\frac{\pi}{6}$, $\sin(x + \frac{\pi}{6})$ is zero. The graph of $y = \sin(x + \frac{\pi}{6})$ looks just like the graph of $y = \sin x$, except that it is translated to the left by a distance of $\frac{\pi}{6}$. The graph appears in Figure 3-31.

$$y = \sin(x + \tfrac{\pi}{6})$$

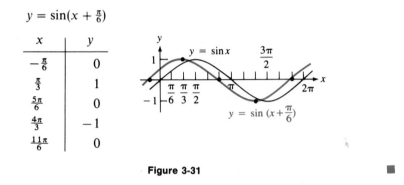

x	y
$-\frac{\pi}{6}$	0
$\frac{\pi}{3}$	1
$\frac{5\pi}{6}$	0
$\frac{4\pi}{3}$	-1
$\frac{11\pi}{6}$	0

Figure 3-31   ■

The distance that a graph is translated to the left or to the right is called the **phase shift** of the graph.

Phase Shift.   The graph of $y = a \sin b(x + c)$ is identical to the graph of $y = a \sin bx$, except that it is translated $|c|$ units to the

$$\begin{Bmatrix} \text{left} \\ \text{right} \end{Bmatrix} \text{ if } c \text{ is } \begin{Bmatrix} \text{positive} \\ \text{negative} \end{Bmatrix}.$$

The number $|c|$ is called the **phase shift** of the graph. An analogous statement is true for $y = a \cos b(x + c)$.

Example 4   Graph $y = 3\cos\left(2x - \dfrac{\pi}{3}\right)$.

Solution   Rewrite

$$y = 3\cos\left(2x - \frac{\pi}{3}\right)$$

in the form $y = a \cos b(x + c)$ by factoring out a 2 from the binomial $2x - \frac{\pi}{3}$. Then

$$y = 3\cos\left(2x - \frac{\pi}{3}\right) = 3\cos 2\left(x + \frac{-\pi}{6}\right)$$

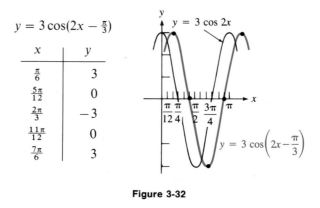

$$y = 3\cos(2x - \tfrac{\pi}{3})$$

x	y
$\frac{\pi}{6}$	3
$\frac{5\pi}{12}$	0
$\frac{2\pi}{3}$	-3
$\frac{11\pi}{12}$	0
$\frac{7\pi}{6}$	3

Figure 3-32

From the equation $y = 3\cos 2(x + \tfrac{-\pi}{6})$, you can read that the amplitude is 3, the period is $\frac{2\pi}{2} = \pi$, and the graph looks like that of $y = 3\cos 2x$. However, it is translated $\frac{\pi}{6}$ units to the right. With $y = 3\cos 2x$ included for reference, the graph appears in Figure 3-32.

Example 5   Graph $y = \tan\left(x - \dfrac{\pi}{4}\right)$.

Solution   One complete cycle of $y = \tan x$ is described as x increases from 0 to π. Thus, one cycle of $y = \tan(x - \tfrac{\pi}{4})$ is completed when $x - \tfrac{\pi}{4}$ increases from 0 to π: that is, when x itself increases from $\tfrac{\pi}{4}$ to $\tfrac{5\pi}{4}$. At $x = \tfrac{\pi}{4}$, the value of $\tan(x - \tfrac{\pi}{4})$ is zero. The graph of $y = \tan(x - \tfrac{\pi}{4})$ looks just like the graph of $y = \tan x$, except that it is translated to the right by a distance of $\tfrac{\pi}{4}$. The graph appears in Figure 3-33.

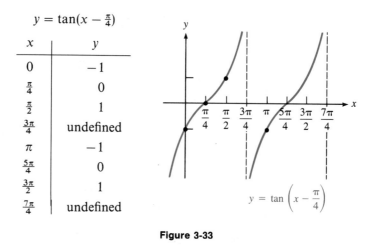

$$y = \tan(x - \tfrac{\pi}{4})$$

x	y
0	-1
$\frac{\pi}{4}$	0
$\frac{\pi}{2}$	1
$\frac{3\pi}{4}$	undefined
π	-1
$\frac{5\pi}{4}$	0
$\frac{3\pi}{2}$	1
$\frac{7\pi}{4}$	undefined

Figure 3-33

The following statement summarizes the facts involving phase shifts of the tangent and the cotangent functions.

Phase Shift.   The graph of

$$\begin{cases} y = a \tan b(x + c) \\ y = a \cot b(x + c) \end{cases}$$

is identical to the graph of $\begin{cases} y = a \tan bx \\ y = a \cot bx \end{cases}$

except that it is translated $|c|$ units to the $\begin{cases} \text{left} \\ \text{right} \end{cases}$ if c is $\begin{cases} \text{positive} \\ \text{negative} \end{cases}$.

The number $|c|$ is called the **phase shift** of the graph.

Example 6   Graph $y = \csc\left(3x - \dfrac{\pi}{2}\right)$.

Solution   Factor 3 out of the binomial $3x - \frac{\pi}{2}$ to write the equation in the form $y = \csc b(x + c)$.

$$y = \csc\left(3x - \frac{\pi}{2}\right)$$

$$= \csc 3\left[x + \left(-\frac{\pi}{6}\right)\right]$$

As with the sine function, the period of the cosecant function is $\frac{2\pi}{b}$ or in this case $\frac{2\pi}{3}$. The graph looks like that of $y = \csc 3x$, but it is translated $\frac{\pi}{6}$ units to the right. The graph appears in Figure 3-34.

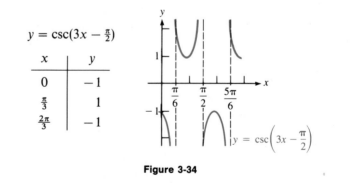

$y = \csc(3x - \frac{\pi}{2})$

x	y
0	-1
$\frac{\pi}{3}$	1
$\frac{2\pi}{3}$	-1

Figure 3-34

The following statement summarizes the facts involving phase shifts of the cosecant and secant functions.

Phase Shift.   The graph of

$$\begin{cases} y = a\csc b(x + c) \\ y = a\sec b(x + c) \end{cases}$$

is identical to the graph of $\begin{cases} y = a\csc bx \\ y = a\sec bx \end{cases}$

except that it is translated $|c|$ units to the $\begin{cases} \text{left} \\ \text{right} \end{cases}$ if c is $\begin{cases} \text{positive} \\ \text{negative} \end{cases}$.

The number $|c|$ is called the **phase shift** of the graph.

■ **EXERCISE 3.6**

In Exercises 1–12, give the number of units and the direction (either up or down) that each trigonometric function has been shifted. Also give the period. **Do not draw the graph.**

1. $y = 2 + \sin x$

2. $y = -4 + \cos x$

3. $y = \tan x - 1$

4. $y = \csc x + 3$

5. $y = 7 + 9\sec 5x$

6. $y = 7 + 9\cot 5x$

7. $y = 3 - \sin x$

8. $y = -2 - \cos x$

9. $y + 5 = \csc 2x$

10. $y + 5 = \cot 2x$

11. $y = 2(3 + \tan \pi x)$

12. $y = -4(1 - \sec \pi x)$

In Exercises 13–30, give the period and the phase shift (including direction), if any, of each given function. **Do not draw the graph.**

13. $y = \sin\left(x - \dfrac{\pi}{3}\right)$

14. $y = \cos\left(x + \dfrac{\pi}{4}\right)$

15. $y = \cos\left(x + \dfrac{\pi}{6}\right)$

16. $y = -\sin\left(x - \dfrac{\pi}{2}\right)$

17. $y + 2 = 3\cos 2\pi x$

18. $y = 3\sin\dfrac{2x}{\pi}$

19. $y = \tan(x - \pi)$

20. $y = \csc\left(x + \dfrac{\pi}{6}\right)$

21. $y = -\sec\left(x + \dfrac{\pi}{4}\right)$

22. $y = -2\sec\left(x - \dfrac{\pi}{3}\right)$

23. $y = \sin(2x + \pi)$

24. $y = \cos(2x - \pi)$

25. $y = \tan\left(\dfrac{\pi x}{2} + \dfrac{\pi}{4}\right)$

26. $y = \csc\left(\dfrac{2\pi x}{3} + \dfrac{\pi}{9}\right)$

27. $y = 2\sec\left(\dfrac{1}{3}x - 6\pi\right)$

28. $y = 2\cot\left(\dfrac{\pi}{10} + \dfrac{x}{5}\right)$

29. $2y = 3\cot\left(7x - \dfrac{21}{2}\pi\right)$

30. $17y = \sec\left(\dfrac{x}{5} + \dfrac{\pi}{4}\right)$

In Exercises 31–46, graph each function through at least one period.

31. $y = -4 + \sin x$

32. $y + 2 = \tan\dfrac{x}{2}$

33. $y = 3 - \sec x$

34. $y = 1 + \csc x$

35. $y = \cot\dfrac{x}{2} - 2$

36. $y = 1 - 2\cos x$

37. $y = \sin\left(x + \dfrac{\pi}{2}\right)$

38. $y = -\cos\left(x - \dfrac{\pi}{2}\right)$

39. $y = \tan\left(x - \dfrac{\pi}{2}\right)$

40. $y = \csc\left(x + \dfrac{\pi}{4}\right)$

41. $y = \cos(2x + \pi)$

42. $y = \sin(3x - \pi)$

43. $y = \sec\left(3x + \dfrac{\pi}{2}\right)$ **44.** $y = \tan\left(\dfrac{x}{2} - \dfrac{\pi}{2}\right)$ **45.** $y - 1 = \csc\left(2x - \dfrac{\pi}{6}\right)$ **46.** $y + 2 = \cot\left(\dfrac{x}{3} - \dfrac{\pi}{2}\right)$

3.7 GRAPHS OF OTHER TRIGONOMETRIC FUNCTIONS

There are other graphs of trigonometric functions that are of interest.

Example 1   Graph $y = \sin^4 x$ for all x from 0 to 2π.

Solution   The function can be graphed by taking the fourth power of each y value of $y = \sin x$. Because the exponent is the even number 4, the graph of $y = \sin^4 x$ will never go below the x-axis. The graph is shown in Figure 3-35. Note that the shape of this curve differs from the shape of the sine curve.

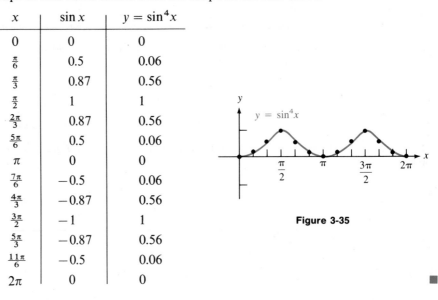

x	$\sin x$	$y = \sin^4 x$
0	0	0
$\frac{\pi}{6}$	0.5	0.06
$\frac{\pi}{3}$	0.87	0.56
$\frac{\pi}{2}$	1	1
$\frac{2\pi}{3}$	0.87	0.56
$\frac{5\pi}{6}$	0.5	0.06
π	0	0
$\frac{7\pi}{6}$	-0.5	0.06
$\frac{4\pi}{3}$	-0.87	0.56
$\frac{3\pi}{2}$	-1	1
$\frac{5\pi}{3}$	-0.87	0.56
$\frac{11\pi}{6}$	-0.5	0.06
2π	0	0

Figure 3-35

Example 2   Graph $y = x + \cos x$.

Solution   It would be possible to make an extensive table of values, plot the points, and draw the curve. However, it is easier to use your knowledge of the functions $y = x$ and $y = \cos x$ and sketch each one separately as in Figure 3-36. Then pick several numbers x such as x_1, and add the values of x_1 and $\cos x_1$ together to obtain the y value of a point on the graph of $y = x + \cos x$. For example, if $x_1 = 0$, the value of $\cos x_1 = 1$. Thus, the point $(0, 0 + 1)$ or $(0, 1)$ is on the graph of $y = x + \cos x$.

As another example, if $x_1 = 1$, the value of $\cos x_1 \approx 0.54$. Thus, the point $(1, 1 + 0.54)$ or $(1, 1.54)$ is on the graph of $y = x + \cos x$.

It is easy to use a compass to add these values of y. Set your compass so that it spans segment BA and transfer that length to form segment CD, locating point D on the desired graph. However, if $x_1 = \pi$, the value of $\cos x_1$ is negative. In this case, you must subtract the magnitudes of the y values. Do this by using a compass to transfer segment MN to position PQ to locate point Q on the desired graph. Repeat this process of adding y values to determine several points. Then join them with a smooth curve to obtain the graph of $y = x + \cos x$.

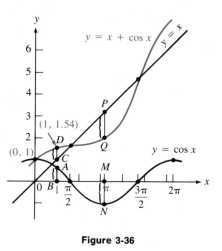

Figure 3-36   ■

Example 3   Graph $y = \cos x + 2 \sin 2x$.

Solution   Proceed as in Example 2. For values of x between 0 and $\frac{\pi}{2}$ such as $\frac{\pi}{4}$, the values of $\cos x$ and $2 \sin 2x$ are both positive. In this case, you must add the y values. Do this by transferring segments such as MN to new positions such as PQ. This locates point Q, which is on the graph of $y = \cos x + 2 \sin 2x$.

For values of x that are between π and $\frac{3\pi}{2}$, such as $\frac{5\pi}{4}$, the y value of $2 \sin 2x$ is positive while the y value of $\cos x$ is negative. In this case, subtract the magnitudes of the y values of the two functions. Do this by transferring segments such as AB to new positions such as CD to locate point D on the graph of $y = \cos x + 2 \sin 2x$. See Figure 3-37.

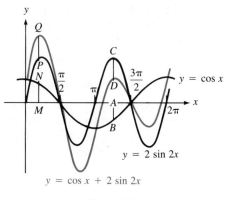

Figure 3-37   ■

The process illustrated in Examples 2 and 3 is called **addition of ordinates**.

■ EXERCISE 3.7

Graph each of the following functions. A calculator will be helpful.

1. $y = \cos^4 x$
2. $y = -\tan^4 x$
3. $y = -\sin^3 x$
4. $y = \cos^3 x$

5. $y = \sin^{10} x$
6. $y = \cot^{10} x$
7. $y = \sin^{100} x$
8. $y = \cos^{101} x$

9. $y^3 = \sin x$
10. $y^3 = \tan x$
11. $y = \cos^2 3x$
12. $y = -\sin^2 \dfrac{x}{2}$

13. $y = \sin x + \cos x$
14. $y = 2\sin x + \cos x$
15. $y = \sin \dfrac{x}{2} + \sin x$
16. $y = x + \sin x$

17. $y = -x + \cos x$
18. $y = x - \cos x$
19. $y = 2\sin x + 2\cos \dfrac{x}{2}$
20. $y = \sin^2 x + \cos^2 x$

21. $y = 2\sin x \cos x$
22. $y = \cos^2 x - \sin^2 x$
23. $y = |\sin x|$
24. $y = |\cos x|$

25. $y = \sin x^2$
26. $y = \sin \dfrac{1}{x}, x \neq 0$
27. $y = \sin \sqrt{x}, x \geq 0$
28. $y = \cos \sqrt{x}, x \geq 0$

3.8 APPLICATIONS OF RADIAN MEASURE

The trigonometric functions have many applications when their domains are considered to be the set of real numbers. We will discuss three of these that come from the fields of electronics and mechanics.

1. Alternating Current

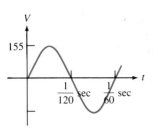

Figure 3-38

The humming sound of a fluorescent light fixture or a toy train transformer is caused by electric current reversing itself 120 times each second. The voltage available at a wall outlet is described by the formula

$$V = V_0 \sin(2\pi f t)$$

where V_0 is approximately 155 volts, f is the frequency, and t is time. In the United States, the frequency is 60 cycles per second (or 60 hertz), and time is measured in seconds. The graph of $V = V_0 \sin(2\pi f t) = 155 \sin(120\,\pi t)$ is given in Figure 3-38. When $t = \frac{1}{60}$, $V = 155 \sin(120\pi \cdot \frac{1}{60}) = 155 \sin 2\pi = 0$. One cycle is completed in $\frac{1}{60}$ of a second, so there are 60 cycles a second.

2. Simple Harmonic Motion

If an object is suspended from a ceiling by a spring, and then pushed up and released, it will start to bounce. See Figure 3-39. Its position above (positive) or below (negative) the equilibrium position is given by the formula involving another circular function:

1. $$y = A \cos\left(\sqrt{\dfrac{k}{m}}\, t\right)$$

The coefficient A, called the **amplitude**, represents the distance the object is pushed above the equilibrium position. The constant k, called the **spring constant**, depends on the stiffness of the spring, and m is the mass of the object. The variable t represents the time in seconds since the object was released. The first cycle is completed when

$$\sqrt{\frac{k}{m}}\,t = 2\pi \quad \text{or when} \quad t = \sqrt{\frac{m}{k}}\,2\pi$$

The value

$$\sqrt{\frac{m}{k}}\,2\pi$$

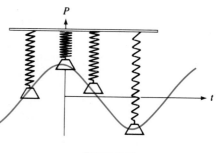

is called the **period of oscillation**, p, and indicates the number of seconds per complete cycle.

The **frequency of the oscillation** indicates the number of cycles per second. The frequency, f, is the reciprocal of the period, p.

Figure 3-39

$$f = \frac{1}{p}$$

Thus, the frequency of the oscillation is

$$f = \frac{1}{p} = \frac{1}{\sqrt{\frac{m}{k}}\,2\pi} = \frac{1}{2\pi}\sqrt{\frac{k}{m}}\,\text{cycles/sec}$$

Example 1   A mass of 4 grams, attached to a spring with a spring constant of 9 grams per centimeter, is raised above the equilibrium position and released. Find the period and the frequency of the oscillation.

Solution   The period p of the oscillation is given by

$$p = \sqrt{\frac{m}{k}}\,2\pi$$

Substituting 4 for m and 9 for k gives

$$p = \sqrt{\frac{m}{k}}\,2\pi = \sqrt{\frac{4}{9}}\,2\pi = \frac{2}{3}\cdot 2\pi = \frac{4\pi}{3}$$

It takes the mass $\frac{4\pi}{3}$ seconds, or about 4.2 seconds, to return to its starting point. Because the frequency of the oscillation is the reciprocal of the period, you have

$$f = \frac{1}{p} = \frac{3}{4\pi}$$

The frequency is $\frac{3}{4\pi}$, or approximately 0.24 hertz (cycle per second).   ■

Any motion described by an equation such as Equation 1, which is of the form

$$y = A \cos kt$$

is called **simple harmonic motion**. We give two more examples of such motion.

A pendulum swinging through a small arc makes an angle with the vertical, where the angle, θ, is given by the formula

$$\theta = A \cos \sqrt{\frac{g}{l}} \, t$$

The coefficient A is the maximum amplitude of the swing, l is the length of the pendulum, and g is a constant related to the force of gravity. The motion described by the pendulum is simple harmonic motion.

If a floating object is pushed under water and then released, it will oscillate as it bobs up and down. For certain shapes, the position y relative to the equilibrium position is given by the equation

$$y = A \cos \sqrt{\frac{k}{w}} \, t$$

where A is the maximum displacement from equilibrium, w is the object's weight, and k depends on the density of the water and the force of gravity. The bobbing action of such a floating object also describes simple harmonic motion.

3. Sound Waves

Figure 3-40

When a tuning fork is struck, its tongs vibrate at a rate that depends on its dimensions and the material of which it is made. See Figure 3-40. The vibrations of the fork set the surrounding air into vibration, and it is these tiny periodic variations in pressure that our ears perceive as a musical tone.

The vibrations of a tuning fork and the surrounding air are described by the equation

$$y = A \sin (2\pi f t)$$

where f is the frequency, t is the time, and A is the amplitude of the oscillations.

The first cycle is completed when $2\pi f t = 2\pi$, or when $t = \frac{1}{f}$. The reciprocal of the frequency is the period—the time it takes to complete one cycle.

Sound travels through air at approximately 1100 feet per second. The distance that sound can travel in the time of one cycle is $1100 \cdot \frac{1}{f}$ feet. This distance is called the **wavelength** of that sound in air.

The rate at which the pressure varies determines the frequency, or pitch, of the tone. In the United States, 440 pressure variations per second is the pitch recognized as musical A. Middle C on the piano is 261.6 variations per second. Physicists use a different scale of pitch. For them, middle C is 256 variations per second. If the frequency of a tone is doubled, the pitch of the tone is one octave higher. Human ears are sensitive to frequencies between about 30 and 20,000 vibrations per second, although the ability to hear tones in the top end of that range usually decreases with age.

A vibrating tuning fork is very quiet if you hold it in your hand. However, if you touch it to a surface such as a tabletop, the surface becomes a sounding board and makes the weak vibrations of the tuning fork much louder. This is because the larger surface is capable of transferring the vibrations to a larger volume of air.

A thorough knowledge of the nature of sound and of vibrations in both solid materials and air is needed for the proper design of the sounding board of such musical instruments as the violin, piano, and guitar. This is because the vibrations of the sounding board also add distinctive colorations to the tone. This is one reason why a grand piano's tone is fuller than that of a spinet.

The analysis of these more complex tones still involves the sine function, but in the form of

$$y = A \sin(2\pi f t) + A_1 \sin(2\pi f_1 t) + A_2 \sin(2\pi f_2 t) + \cdots$$

where $f_1, f_2, \ldots$ and $A_1, A_2, \ldots$ are the frequencies and amplitudes of the overtones present in the original sound.

EXERCISE 3.8

1. In some European countries, the power distribution network operates at a frequency of 50 hertz. If V_0 is 310 volts in Europe, express V as a function of t.

2. A radio station operates on a federally assigned frequency of 1400 kilohertz (1 kilohertz is 1000 hertz). If radio waves travel at 300 million meters per second, what is the wavelength of the radio station's signal?

3. A spring with a spring constant of 6 newtons per meter hangs from a ceiling and supports a mass of 24 kilograms. The mass is pulled to a starting point a few centimeters below the equilibrium position and released. How long will it take to return to the starting point?

4. What is the frequency of the oscillation in Exercise 3?

5. A spring supports a mass of 12 grams. The frequency of its oscillation is $\frac{1}{\pi}$ hertz (cycles per second). What is the spring constant? The units of your answer will be dynes per centimeter.

6. For a given spring, what is the effect on the period of its oscillation if a mass suspended by the spring is doubled?

7. A pendulum 1 meter long is set to swinging through a small amplitude. Let $g = 9.8$ meters per second per second. Find the period and the frequency of the oscillation.

8. A cubical box floats on its side and oscillates with a period of $\frac{1}{3}$ second. What is its weight? Let $k = 200,000$ pounds per second per second.

9. One tine of a tuning fork takes 0.0035 second to complete one vibration. What is the frequency of the tone?

10. A tuning fork vibrates at a frequency of 256 cycles per second. How long does it take to complete one cycle?

11. Construct the graph of $y = A \sin(2\pi f t)$.

12. Construct the graph of $y = A \cos kt$.

CHAPTER SUMMARY

Key Words

addition of ordinates   (3.7)

amplitude   (3.4)

angular velocity   (3.2)

arc length   (3.1)

argument of a function   (3.4)

asymptote   (3.5)

circular functions   (3.3)

frequency of an oscillation   (3.8)

linear velocity   (3.2)

period   (3.4)

periodic function   (3.4)

phase shift   (3.4)

radian   (3.1)

sector of a circle   (3.1)

simple harmonic motion   (3.8)

Key Ideas

(3.1)   If a central angle θ of a circle with radius of r intercepts an arc of length s, then the radian measure of angle θ is s/r.

π radians $= 180°$

If a central angle θ in a circle with radius r is in radians, the length s of the intercepted arc is given by the formula $s = r\theta$.

The area of a sector of a circle with radius r and a central angle θ (measured in radians) is given by the formula $A = \frac{1}{2}r^2\theta$.

(3.2)   Linear and angular velocity of a point on a rotating wheel are related by the formula $v = r\omega$, where $v = s/t$, $\omega = \theta/t$, and r is the radius of the wheel.

(3.3)   It is possible to think of trigonometric functions as functions with domains and ranges that are real numbers.

(3.4)   The period of $y = a \sin bx$ and $y = a \cos bx$ is $\left|\dfrac{2\pi}{b}\right|$. The amplitude of each of these functions is $|a|$.

(3.5)   The period of $y = a \tan bx$ and $y = a \cot bx$ is $\left|\dfrac{\pi}{b}\right|$.

The period of $y = a \csc bx$ and $y = a \sec bx$ is $\left|\dfrac{2\pi}{b}\right|$.

(3.6)   The graph of $y = k + a \sin bx$ is identical to the graph of $y = a \sin bx$, except that it is translated $|k|$ units

$$\left.\begin{matrix} \text{up} \\ \text{down} \end{matrix}\right\} \text{ if } k \text{ is } \left\{\begin{matrix} \text{positive} \\ \text{negative} \end{matrix}\right\}$$

An analogous statement is true for each of the other trigonometric functions.

The graph of $y = a \sin b(x + c)$ is identical to the graph of $y = a \sin bx$, except that it is translated $|c|$ units to the

$$\left\{ \begin{matrix} \text{left} \\ \text{right} \end{matrix} \right\} \text{ if } c \text{ is } \left\{ \begin{matrix} \text{positive} \\ \text{negative} \end{matrix} \right\}$$

An analogous statement is true for each of the other trigonometric functions.

(3.7)   Functions that are sums of simpler functions can be graphed by using the method of addition of ordinates.

REVIEW EXERCISES

In Review Exercises 1–4, change each angle to radians.

1. $105°$

2. $325°$

3. $318°$

4. $-105°$

In Review Exercises 5–8, each angle is expressed in radians. Change each angle to degrees.

5. $\dfrac{19}{6}\pi$

6. $-\dfrac{5}{6}\pi$

7. 7π

8. 8

In Review Exercises 9–12, find the value of the given function without using a calculator. Then, check your work with a calculator.

9. $\sin\dfrac{5\pi}{6}$

10. $\cos\left(-\dfrac{13}{6}\pi\right)$

11. $\tan\left(-\dfrac{\pi}{3}\right)$

12. $\csc\dfrac{\pi}{6}$

13. The latitude of Springfield, Illinois, is $39.8°$ N. How far is Springfield from the equator? Use 3960 miles as an estimate of the radius of the earth.

14. Des Moines, Iowa, is approximately 2870 miles north of the equator. If the radius of the earth is about 3960 miles, find the latitude of Des Moines.

15. Find the area of a sector of a circle if the sector has a central angle of $15°$ and the circle has a radius of 12 centimeters.

16. Find the angular velocity of the earth in radians per second.

17. A truck is traveling 55 miles per hour. How fast are the 32-inch diameter tires spinning in revolutions per minute?

18. A vehicle has 40-inch diameter tires that are making 100 revolutions per minute. How fast is the vehicle going in miles per hour?

In Review Exercises 19 and 20, find the coordinates of the point P on the unit circle that corresponds to each given real number. **Do not use a calculator.**

19. $\dfrac{7\pi}{6}$

20. $\dfrac{13\pi}{4}$

In Review Exercises 21–28, use a calculator to find the values of the given functions. Give answers to the nearest ten-thousandth.

21. $\sin(5\pi)$

22. $\cos 7$

23. $\tan(2 + \pi)$

24. $\csc 3\pi$

25. $\sec(-1)$

26. $\cot\left(\dfrac{5\pi}{3}\right)$

27. $\sin(\pi^3)$

28. $\cos\sqrt{\pi}$

In Review Exercises 29–32, find the amplitude and the period of the given function.

29. $y = 4\sin 3x$

30. $y = \dfrac{\cos 4x}{8}$

31. $y = -\dfrac{1}{3}\cos\dfrac{x}{3}$

32. $y = 0.875\sin\dfrac{1}{4}x$

In Review Exercises 33–36, find the vertical shift and phase shift, if any, of each of the given functions.

33. $y = 2 + \tan x$

34. $y = \csc\left(2x + \dfrac{\pi}{3}\right)$

35. $y - 4 = 3\sin\left(\dfrac{x}{7} + \dfrac{3}{2}\right)$

36. $y = -\cos\left(\dfrac{1}{5}x - \dfrac{1}{2}\pi\right) - 1$

In Review Exercises 37–50, graph the given function.

37. $y = 4\sin x$

38. $y = 0.5\cos x$

39. $y = \cos\dfrac{x}{4}$

40. $y = \tan\dfrac{\pi x}{3}$

41. $y = 3 + \sin x$

42. $y = -2 + \tan x$

43. $y = 2\sin\left(x - \dfrac{5\pi}{6}\right)$

44. $y = \tan\left(x - \dfrac{2\pi}{3}\right)$

45. $y = \dfrac{1}{2}\tan\left(x + \dfrac{\pi}{4}\right)$

46. $y = 2\sin 3x$

47. $y = 2\sin^2 x$

48. $y = \dfrac{1}{2}\cos^2 x$

49. $y = 2\cos x + \sin\dfrac{x}{2}$

50. $y = \dfrac{x}{2} + \cos\dfrac{x}{2}$

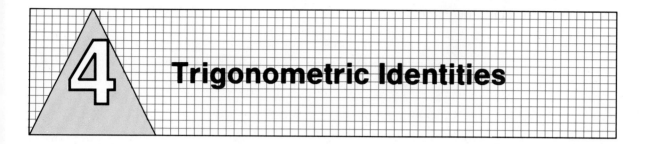

4 Trigonometric Identities

We have used trigonometry to solve problems such as finding heights of flagpoles and distances across rivers. However, trigonometry is useful in another, more theoretical way. Because the trigonometric functions are related to each other and to certain algebraic expressions, they are often used to transform difficult and awkward mathematical expressions into more manageable ones. To explore some of these relationships and to gain facility in manipulating trigonometric expressions, we turn our attention to trigonometric identities.

4.1 TRIGONOMETRIC IDENTITIES

Because no numbers satisfy the equation $x = x + 1$, it is called an **impossible equation**. An equation such as $x + 3 = 5$ for which some numbers are solutions and others are not is called a **conditional equation**. An equation such as

$$x + 3 = x + 3$$

for which all numbers are solutions is called an **identity**. Because this chapter deals with identities, we emphasize the following definition.

> **Definition.** An **identity** is an equation that is true for every value of its variable(s) for which each side of the equation is defined.

The equations

$$3(x + 2) = 3x + 6 \quad \text{and} \quad x^2 - y^2 = (x + y)(x - y)$$

for example, are true for all values of x and y. Thus, they are identities. In Section 2.2, we discussed several fundamental trigonometric identities. Because they are so fundamental, we repeat them here using the variable x.

The Fundamental Identities.   For any number x for which the functions are defined,

$$\sin x = \frac{1}{\csc x} \qquad \cos x = \frac{1}{\sec x}$$

$$\tan x = \frac{1}{\cot x} \qquad \tan x = \frac{\sin x}{\cos x}$$

$$\cot x = \frac{\cos x}{\sin x} \qquad \sin^2 x + \cos^2 x = 1$$

$$\tan^2 x + 1 = \sec^2 x \qquad \cot^2 x + 1 = \csc^2 x$$

These relationships and the rules of algebra will be used to show that many more equations are identities. In doing so, we usually work with one side of the equation only, manipulating it until it is transformed into the other side. Success at verifying identities comes only with practice. The work may be frustrating at times, but the skill is indispensable later.

Here are some suggestions that will make the work easier. They do not, however, guarantee success, because each identity presents its own unique challenge.

1. Memorize the fundamental identities given above. Whenever you see one side of one of these identities, the other side should come to mind immediately.
2. Start with the more complicated side of the equation, and try to transform it into the other side.
3. Use the rules of algebra. Trigonometric fractions may be added, multiplied, and simplified just as the fractions of algebra are handled. Trigonometric expressions may also be factored and similar terms combined.
4. As you work on one side of the equation, always keep an eye on the other side. It is easier to hit a target that you can see.
5. It is sometimes helpful to begin by changing all functions into sine and cosine functions.
6. If the numerator or denominator of a fraction contains a factor of $1 + \sin x$ (or $1 - \sin x$), consider multiplying both the numerator and the denominator by the conjugate $1 - \sin x$ (or $1 + \sin x$). This operation creates a factor of $1 - \sin^2 x$, which may be replaced with $\cos^2 x$. The same idea applies to factors of $1 \pm \cos x$, $\sec x \pm 1$, and $\sec x \pm \tan x$.

As we work with one side of the equation and transform it into the other side, we can use the two-column format illustrated in Examples 1–3.

Example 1   Verify that $\dfrac{\tan x}{\sin x} = \sec x$ is an identity.

Solution Work on the left-hand side because it is the more complicated.

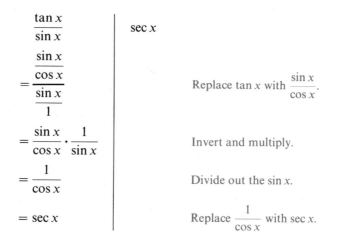

$$\frac{\tan x}{\sin x} \qquad\qquad \sec x$$

$$= \frac{\dfrac{\sin x}{\cos x}}{\dfrac{\sin x}{1}} \qquad\qquad \text{Replace } \tan x \text{ with } \frac{\sin x}{\cos x}.$$

$$= \frac{\sin x}{\cos x} \cdot \frac{1}{\sin x} \qquad\qquad \text{Invert and multiply.}$$

$$= \frac{1}{\cos x} \qquad\qquad \text{Divide out the } \sin x.$$

$$= \sec x \qquad\qquad \text{Replace } \frac{1}{\cos x} \text{ with } \sec x.$$

Because the left-hand side has been transformed into the right-hand side, the identity is verified. ∎

Example 2 Verify the identity $\tan x + \cot x = \csc x \sec x$.

Solution Work with the left-hand side of this equality and change $\tan x$ and $\cot x$ into expressions containing $\sin x$ and $\cos x$. Then proceed as follows.

$$\tan x + \cot x \qquad\qquad \csc x \sec x$$

$$= \frac{\sin x}{\cos x} + \frac{\cos x}{\sin x}$$

$$= \frac{\sin x \sin x}{\sin x \cos x} + \frac{\cos x \cos x}{\sin x \cos x} \qquad\qquad \text{Change each fraction to a} \\ \text{fraction with a common} \\ \text{denominator.}$$

$$= \frac{\sin^2 x + \cos^2 x}{\sin x \cos x} \qquad\qquad \text{Add the fractions.}$$

$$= \frac{1}{\sin x \cos x} \qquad\qquad \text{Replace } \sin^2 x + \cos^2 x \text{ with } 1.$$

$$= \frac{1}{\sin x} \cdot \frac{1}{\cos x} \qquad\qquad \text{Rewrite as two fractions.}$$

$$= \csc x \sec x$$

Because the left-hand side has been transformed into the right-hand side, the identity is verified. ∎

Example 3   Verify that $1 + \tan x = \sec x(\cos x + \sin x)$.

Solution   Applying the distributive property to the right-hand side and remembering that $\sec x \cos x = 1$ is the strategy:

$$1 + \tan x \quad \bigg| \quad \sec x(\cos x + \sin x)$$

$$= \sec x \cos x + \sec x \sin x \qquad \text{Remove parentheses.}$$

$$= 1 + \sec x \sin x \qquad \text{Recall that } \sec x \cos x = 1.$$

$$= 1 + \frac{\sin x}{\cos x} \qquad \text{Recall that } \sec x = \frac{1}{\cos x}.$$

$$= 1 + \tan x \qquad \text{Recall that } \frac{\sin x}{\cos x} = \tan x.$$

The identity is verified.   ■

In Examples 4–6, we shall transform one side of the equation into the other without using a two-column format.

Example 4   Verify that $(1 + \tan^2 x)\cos^2 x = 1$.

Solution   Work on the left-hand side. Change $\tan^2 x$ into an expression involving sines and cosines, use the distributive property to remove parentheses, and proceed as follows:

$$(1 + \tan^2 x)\cos^2 x = \left(1 + \frac{\sin^2 x}{\cos^2 x}\right)\cos^2 x \qquad \text{Recall that } \tan^2 x = \frac{\sin^2 x}{\cos^2 x}.$$

$$= \cos^2 x + \sin^2 x \qquad \text{Remove parentheses.}$$

$$= 1 \qquad \text{Recall that } \cos^2 x + \sin^2 x = 1.$$

The identity is verified.   ■

Example 5   Verify that $\cos^4 x - \sin^4 x = 1 - 2\sin^2 x$.

Solution   The left-hand side is the difference of two squares and can be factored. Because your target—the expression on the right—involves only the sine function, eliminate the cosine function from the left-hand side and proceed as follows:

$$\cos^4 x - \sin^4 x = (\cos^2 x + \sin^2 x)(\cos^2 x - \sin^2 x)$$

$$= 1 \cdot (\cos^2 x - \sin^2 x)$$

$$= \cos^2 x - \sin^2 x$$

$$= 1 - \sin^2 x - \sin^2 x \qquad \text{Remember that } \cos^2 x = 1 - \sin^2 x.$$

$$= 1 - 2\sin^2 x$$

The identity is verified.   ■

Example 6   Verify that $\dfrac{1}{\sec x - \tan x} - \dfrac{1}{\sec x + \tan x} = 2\tan x$.

Solution Work on the left-hand side, find the common denominator for the fractions, and add them.

$$\frac{1}{\sec x - \tan x} - \frac{1}{\sec x + \tan x} = \frac{(\sec x + \tan x)}{(\sec x - \tan x)(\sec x + \tan x)} - \frac{(\sec x - \tan x)}{(\sec x + \tan x)(\sec x - \tan x)}$$

$$= \frac{\sec x + \tan x - \sec x + \tan x}{\sec^2 x - \tan^2 x}$$

$$= \frac{2 \tan x}{1}$$

$$= 2 \tan x$$

The identity is verified. ∎

In Examples 1–6, all the work done in verifying each identity was performed on one side of the equation only. Occasionally, if the equation seems to warrant it, mathematicians work on both sides of the identity independently until each side has been transformed into a common third expression. When following this strategy, it is important that each step in the process be reversible so that either side of the identity can be derived from the other side. It is also important to note that each side must be worked on independently. It is incorrect to multiply or divide both sides of an equation whose truth you are trying to establish by an expression containing a variable, because the resulting equation might not be equivalent to the given equation.

Example 7 Verify that $\dfrac{1 - \cos x}{1 + \cos x} = (\csc x - \cot x)^2$.

Solution In this example, change the left-hand side and the right-hand side of the equation to a common third expression. The left-hand side can be changed as follows:

$$\frac{1 - \cos x}{1 + \cos x} = \frac{(1 - \cos x)(1 - \cos x)}{(1 + \cos x)(1 - \cos x)}$$

$$= \frac{(1 - \cos x)^2}{1 - \cos^2 x}$$

$$= \frac{(1 - \cos x)^2}{\sin^2 x}$$

$$= \left(\frac{1 - \cos x}{\sin x}\right)^2$$

The right-hand side can be changed as follows:

$$(\csc x - \cot x)^2 = \left(\frac{1}{\sin x} - \frac{\cos x}{\sin x}\right)^2$$

$$= \left(\frac{1 - \cos x}{\sin x}\right)^2$$

Each side has been transformed independently into the expression

$$\left(\frac{1 - \cos x}{\sin x}\right)^2$$

Because each step is reversible, it follows that the given equation is an identity. ∎

Example 8   For what values of x is $\sqrt{1 - \cos^2 x} = \sin x$?

Solution   Take the square root of both sides of the identity $1 - \cos^2 x = \sin^2 x$.

$$\sqrt{1 - \cos^2 x} = \sqrt{\sin^2 x}$$
$$\sqrt{1 - \cos^2 x} = |\sin x|$$

The equation $\sqrt{1 - \cos^2 x} = |\sin x|$ is an identity, but because $\sin x$ is sometimes negative, $\sqrt{1 - \cos^2 x} = \sin x$ is *not* an identity. If you think of x as a real number from 0 to 2π, $\sin x$ is nonnegative for $0 \le x \le \pi$. Hence, $\sqrt{1 - \cos^2 x} = \sin x$ only if x is a number from 0 to π, from 2π to 3π, from 4π to 5π, and so on. ∎

When verifying identities, always remember to write the variable associated with a trigonometric function. The notation "$\cos x$" represents a numerical value, but the notation "cos" is meaningless.

▬ EXERCISE 4.1

SET A

In Exercises 1–30, indicate whether the statement is an identity. If so, verify it. If not, explain why. Remember that if an equation is false for one value of its variable, it cannot be an identity.

1. $\sin x + \sin x = 2 \sin x$

2. $\sin x + \sin(\pi - x) = 0$

3. $\sin \dfrac{1}{x} = \csc x$

4. $\sec = \dfrac{1}{\cos}$

5. $\sin x + \cos x = 1$

6. $\sec^2 x - \tan^2 x = 1$

7. $(\sin x + \cos x)^2 = 1 + 2 \sin x \cos x$

8. $(\sec x - \tan x)^2 = 1$

9. $\cos^2 x + \sin^4 x = 1$

10. $\sqrt{\sin^2 x + \cos^2 x} = \sin x + \cos x$

11. $\dfrac{1}{\sin} = \csc$

12. $\dfrac{1}{\tan x} = \dfrac{\cos x}{\sin x}$

13. $\dfrac{1}{\sin x} = \cos x$

14. $\tan \alpha \cos \alpha \csc \alpha = 1$

15. $\cot \beta \sec \beta \sin \beta = 1$

16. $\sin^2 \alpha + \cos^2 \beta = 1$

17. $1 + \tan^2 x = \sec^2 y$

18. $\sqrt{1 - \sin^2 x} = |\cos x|$

19. $\sqrt{1 - \sin^2 x} = \cos x$

20. $\sqrt{1 - \cos^2 x} = \sin x$

21. $\sin^4 x - \cos^4 x = \sin^2 x - \cos^2 x$

22. $\sin^2 x + 2 \sin x + 1 = (\sin x + 1)^2$

23. $(1 + \sin x)(1 - \sin x) = \cos^2 x$

24. $(1 - \cos x)(1 + \cos x) = \sin^2 x$

25. $\sec^2 z + 1 = \tan^2 z$

26. $\sin(60° - x) = \sin 60° - \sin x$

27. $\cos(\pi + x) + \cos x = \pi$

28. $\tan(45° + x) - \tan x = 0$

29. $\sin 2x = 2 \sin x$

30. $\cos 2x = 2 \cos x$

In Exercises 31–80, verify each identity.

31. $\dfrac{\tan x}{\sec x} = \sin x$

32. $\dfrac{\cot x}{\csc x} = \cos x$

33. $\dfrac{1 - \cos^2 x}{\sin x} = \sin x$

34. $\dfrac{1 - \sin^2 x}{\cos x} = \cos x$

35. $\tan x \csc x = \sec x$

36. $\cot x \sec x = \csc x$

37. $\dfrac{1 + \tan^2 x}{\sec^2 x} = 1$

38. $\dfrac{1 + \cot^2 x}{\csc^2 x} = 1$

39. $\dfrac{\sin x}{\csc x} + \dfrac{\cos x}{\sec x} = 1$

40. $1 - \dfrac{\cos x}{\sec x} = \sin^2 x$

41. $\dfrac{\sin x}{\cos^2 x - 1} = -\csc x$

42. $\dfrac{\cos x}{\sin^2 x - 1} = -\sec x$

43. $\dfrac{1}{1 - \sin^2 x} = 1 + \tan^2 x$

44. $\dfrac{1}{1 - \cos^2 x} = 1 + \cot^2 x$

45. $\sin^2 x + \cos^2 x = \sin^2 x \csc^2 x$

46. $\dfrac{1 - \sin^2 x}{1 - \cos^2 x} = \cot^2 x$

47. $\dfrac{1 - \cos^2 x}{1 - \sin^2 x} = \tan^2 x$

48. $\sin^2 x + \cos^2 x = \cos^2 x \sec^2 x$

49. $\dfrac{1 - \sin^2 x}{1 + \tan^2 x} = \cos^4 x$

50. $\tan x \sin x = \dfrac{\csc x}{\cot x + \cot^3 x}$

51. $\dfrac{\cos x(\cos x + 1)}{\sin x} = \cos x \cot x + \cot x$

52. $\cos^2 x \csc x - \csc x = -\sin x$

53. $\sin^2 x \sec x - \sec x = -\cos x$

54. $(\sin x - \cos x)(1 + \sin x \cos x) = \sin^3 x - \cos^3 x$

55. $\sin^2 x - \tan^2 x = -\sin^2 x \tan^2 x$

56. $\dfrac{1 + \cot x}{\csc x} = \sin x + \cos x$

57. $\dfrac{1 - \csc x}{\cot x} = \tan x - \sec x$

58. $\dfrac{\cos^2 x - \tan^2 x}{\sin^2 x} = \cot^2 x - \sec^2 x$

59. $\dfrac{1}{\sec x - \tan x} = \tan x + \sec x$

60. $\csc^2 x + \sec^2 x = \csc^2 x \sec^2 x$

61. $\dfrac{\cos x}{\cot x} + \dfrac{\sin x}{\tan x} = \sin x + \cos x$

62. $\dfrac{\cos x}{1 + \sin x} = \sec x - \tan x$

63. $\dfrac{\cos x}{1 - \sin x} = \dfrac{1 + \sin x}{\cos x}$

64. $\cos^2 x + \sin x \cos x = \dfrac{\cos x(\cot x + 1)}{\csc x}$

65. $(\sin x + \cos x)^2 + (\sin x - \cos x)^2 = 2$

66. $\dfrac{\sec x + 1}{\tan x} = \dfrac{\tan x}{\sec x - 1}$

67. $\dfrac{\cos x - \cot x}{\cos x \cot x} = \dfrac{\sin x - 1}{\cos x}$

68. $\cos^4 x - \sin^4 x = 2\cos^2 x - 1$

69. $\dfrac{1}{1 + \sin x} + \dfrac{1}{1 - \sin x} = 2\sec^2 x$

70. $\sqrt{\dfrac{1 - \sin x}{1 + \sin x}} = \sec x - \tan x$ with $0 < x < \dfrac{\pi}{2}$

71. $\sqrt{\dfrac{1 - \sin x}{1 + \sin x}} = \dfrac{1 - \sin x}{\cos x}$ with $0 < x < \dfrac{\pi}{2}$

72. $-\sqrt{\dfrac{\sec x - 1}{\sec x + 1}} = \dfrac{1 - \sec x}{\tan x}$ with $0 < x < \dfrac{\pi}{2}$

73. $\sqrt{\dfrac{\csc x - \cot x}{\csc x + \cot x}} = \dfrac{\sin x}{1 + \cos x}$ with $0 < x < \dfrac{\pi}{2}$

74. $\dfrac{\cos x}{1 - \sin x} - \dfrac{1}{\cos x} = \tan x$

75. $\dfrac{1}{\sec x(1 + \sin x)} = \sec x(1 - \sin x)$

76. $\dfrac{\cot x - \cos x}{\cot x \cos x} = \dfrac{1 - \sin x}{\cos x}$

77. $\dfrac{1}{\tan x(\csc x + 1)} = \tan x(\csc x - 1)$

78. $\dfrac{(\cos x + 1)^2}{\sin^2 x} = 2\csc^2 x + 2\csc x \cot x - 1$

79. $\dfrac{\csc x}{\sec x - \csc x} = \dfrac{\cot^2 x + \csc x \sec x + 1}{\tan^2 x - \cot^2 x}$

80. $(\sin x + \cos x)^2 - (\sin x - \cos x)^2 = 4\sin x \cos x$

SET B

In Exercises 81–100, verify each identity.

81. $\dfrac{\sin^2 x \tan x + \sin^2 x}{\sec x - 2\sin x \tan x} = \dfrac{\sin^2 x}{\cos x - \sin x}$

82. $\dfrac{\sin x + \cos x + 1}{\sin x + \cos x - 1} = \csc x \sec x + \csc x + \sec x + 1$

83. $\dfrac{\sec x + \tan x}{\csc x + \cot x} = \dfrac{\cot x - \csc x}{\tan x - \sec x}$

84. $\dfrac{\cos x + \sin x}{\cos x - \sin x} = \dfrac{\csc x + 2\cos x}{\csc x - 2\sin x}$

85. $\dfrac{3\cos^2 x + 11\sin x - 11}{\cos^2 x} = \dfrac{3\sin x - 8}{1 + \sin x}$

86. $\dfrac{3\sin^2 x + 5\cos x - 5}{\sin^2 x} = \dfrac{3\cos x - 2}{1 + \cos x}$

87. $\dfrac{\cos x + \sin x + 1}{\cos x - \sin x - 1} = \dfrac{1 + \cos x}{-\sin x}$

88. $\dfrac{1 + \sin x + \cos x}{1 + \sin x - \cos x} = \dfrac{\sin x}{1 - \cos x}$

89. $\dfrac{1 + \sin x}{\cos x} = \dfrac{\sin x - \cos x + 1}{\sin x + \cos x - 1}$

90. $\dfrac{\tan x + \sec x + 1}{\tan x + \sec x - 1} = \dfrac{1 + \cos x}{\sin x}$

91. $\dfrac{\csc x + 1 + \cot x}{\csc x + 1 - \cot x} = \dfrac{1 + \cos x}{\sin x}$

92. $\dfrac{3\cot x \csc x - 2\csc^2 x}{\csc^2 x + \cot x \csc x} - 3 = -5(\csc^2 x - \csc x \cot x)$

93. $\dfrac{\sin^3 x \cos x + \cos x - \sin^2 x \cos x - \cos x \sin x}{\cos^4 x} = \sec x - \tan x$

94. $\dfrac{\sin x + \sin x \cos x - \sin x \cos^2 x - \sin x \cos^3 x}{\cos^4 x - 2\cos^2 x + 1} = \csc x + \cot x$

95. $\dfrac{1}{\sec x + \csc x - \sec x \csc x} = \dfrac{\sin x + \cos x + 1}{2}$

96. $\dfrac{2(\sin x + 1)}{1 + \cot x + \csc x} = \sin x + 1 - \cos x$

97. $\dfrac{2\sin x \cos x + \cos x - 2\sin x - 1}{-\sin^2 x} = \dfrac{2\sin x + 1}{\cos x + 1}$

98. $-4\cos^2 x - 3\sin^2 x = \tan^2 x \cos^2 x - 4$

99. $\sec^2 x(\tan x + 3\sec x)(\tan x - \sec x) = 2\tan x \sec^3 x - 2\sec^4 x - \sec^2 x$

100. $\dfrac{\csc x + 1}{\csc x - 1} + \dfrac{\tan x - \sec x}{\tan x + \sec x} = \dfrac{4\tan x}{\cos x}$

4.2   IDENTITIES INVOLVING SUMS AND DIFFERENCES OF TWO ANGLES

We often encounter expressions that contain a trigonometric function of either a sum or a difference of two angles. It is tempting to believe that the expression $\cos(A + B)$, for example, is equal to $\cos A + \cos B$. However, this is *not* true, as the following work shows.

$$\cos(30° + 60°) = \cos 90° = 0$$

$$\cos 30° + \cos 60° = \frac{\sqrt{3}}{2} + \frac{1}{2} = \frac{\sqrt{3} + 1}{2}$$

Hence, we have

$$\cos(30° + 60°) \neq \cos 30° + \cos 60°$$

In general, a trigonometric function of the sum (or difference) of two angles is not equal to the sum (or difference) of the trigonometric functions of each angle. It is possible, however, to develop formulas for finding a trigonometric function of the sum (or difference) of two angles. We will use the distance formula to derive a formula to evaluate the expression $\cos(A + B)$.

We draw angles A and $-B$ in standard position on the unit circle as in Figure 4-1, locate point R on the circle so that angle POR is equal to angle B, and form triangles POQ and ROS. Because these two isosceles triangles have equal vertex angles, they are congruent. Hence, $d(RS) = d(PQ)$. The coordinates of points P, Q, R, and S are:

P:	$(\cos A, \sin A)$
Q:	$(\cos[-B], \sin[-B]) = (\cos B, -\sin B)$
R:	$(\cos[A + B], \sin[A + B])$
S:	$(1, 0)$

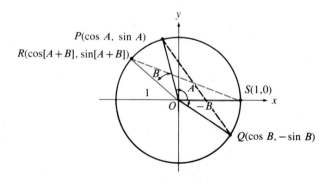

Figure 4-1

Because $d(RS) = d(PQ)$, their squares are also equal. Hence, $[d(RS)]^2 = [d(PQ)]^2$. By the distance formula, we have

$$[\cos(A + B) - 1]^2 + [\sin(A + B) - 0]^2$$
$$= (\cos A - \cos B)^2 + (\sin A + \sin B)^2$$

or

$$\cos^2(A + B) - 2\cos(A + B) + 1 + \sin^2(A + B)$$
$$= \cos^2 A - 2\cos A \cos B + \cos^2 B + \sin^2 A + 2\sin A \sin B + \sin^2 B$$

Because $\cos^2\theta + \sin^2\theta = 1$ for any angle θ, the preceding equation can be written as follows.

$$1 - 2\cos(A + B) + 1 = 1 - 2\cos A \cos B + 1 + 2\sin A \sin B$$
$$2 - 2\cos(A + B) = 2 - 2\cos A \cos B + 2\sin A \sin B$$
$$-2\cos(A + B) = -2\cos A \cos B + 2\sin A \sin B$$

Dividing both sides of the equation above by -2 gives the identity*

$$\cos(A + B) = \cos A \cos B - \sin A \sin B$$

Example 1   Find $\cos 75°$ without using tables or a calculator.

Solution   Note that $\cos 75° = \cos(45° + 30°)$ and use the formula for the cosine of the sum of two angles.

$$\cos(A + B) = \cos A \cos B - \sin A \sin B$$
$$\cos(45° + 30°) = \cos 45° \cos 30° - \sin 45° \sin 30°$$
$$= \frac{\sqrt{2}}{2} \cdot \frac{\sqrt{3}}{2} - \frac{\sqrt{2}}{2} \cdot \frac{1}{2}$$
$$\cos 75° = \frac{\sqrt{6} - \sqrt{2}}{4}$$

∎

The identity for the cosine of the difference of two angles follows from the fact that $A - B = A + (-B)$.

$$\cos(A - B) = \cos[A + (-B)]$$
$$= \cos A \cos(-B) - \sin A \sin(-B)$$

Because $\cos(-B) = \cos B$ and $\sin(-B) = -\sin B$, this result can be rewritten as

---

* Keep in mind that this formula and all others developed in this chapter are also true if A and B represent real numbers.

follows:

$$\cos(A - B) = \cos A \cos B + \sin A \sin B$$

Example 2   Find the value of $\cos\dfrac{\pi}{12}$ without using tables or a calculator.

Solution   Because $\dfrac{\pi}{12} = \dfrac{\pi}{4} - \dfrac{\pi}{6}$, it follows that

$$\cos\frac{\pi}{12} = \cos\left(\frac{\pi}{4} - \frac{\pi}{6}\right)$$

$$= \cos\frac{\pi}{4}\cos\frac{\pi}{6} + \sin\frac{\pi}{4}\sin\frac{\pi}{6}$$

$$= \frac{\sqrt{2}}{2} \cdot \frac{\sqrt{3}}{2} + \frac{\sqrt{2}}{2} \cdot \frac{1}{2}$$

$$= \frac{\sqrt{6} + \sqrt{2}}{4}$$

∎

In Section 2.4, you learned that if θ is an acute angle, any trigonometric function of θ is equal to the cofunction of the complement of θ. This property may be extended to all angles. To this end, we find a value for $\cos(90° - \theta)$.

$$\cos(90° - \theta) = \cos 90° \cos \theta + \sin 90° \sin \theta$$

$$= 0 \cdot \cos \theta + 1 \cdot \sin \theta$$

$$= \sin \theta$$

Thus, for any angle θ, we have

1. $$\sin \theta = \cos(90° - \theta)$$

Now note that $\theta = 90° - (90° - \theta)$ and find a value for $\cos \theta$.

$$\cos \theta = \cos[90° - (90° - \theta)]$$

$$= \cos 90° \cos(90° - \theta) + \sin 90° \sin(90° - \theta)$$

$$= \sin(90° - \theta)$$

Thus, for any angle θ, we have

2. $$\cos \theta = \sin(90° - \theta)$$

Equations 1 and 2 show that the sine of any angle θ is equal to the cosine of $(90° - \theta)$ and that the cosine of any angle θ is equal to the sine of $(90° - \theta)$. These facts can be used to develop a formula for $\sin(A + B)$.

If $A + B$ is substituted for θ in Equation 1, we have

$$\sin(A + B) = \cos[90° - (A + B)]$$
$$= \cos[(90° - A) - B]$$
$$= \cos(90° - A)\cos B + \sin(90° - A)\sin B$$

Because $\cos(90° - A) = \sin A$ and $\sin(90° - A) = \cos A$, it follows that

$$\sin(A + B) = \sin A \cos B + \cos A \sin B$$

To find a formula for $\sin(A - B)$, proceed as follows.

$$\sin(A - B) = \sin[A + (-B)]$$
$$= \sin A \cos(-B) + \cos A \sin(-B)$$
$$= \sin A \cos B - \cos A \sin B$$

This gives the identity

$$\sin(A - B) = \sin A \cos B - \cos A \sin B$$

Example 3   Derive a formula for $\sin(A + B + C)$.

Solution

$$\sin(A + B + C) = \sin[(A + B) + C]$$
$$= \sin(A + B)\cos C + \cos(A + B)\sin C$$
$$= (\sin A \cos B + \cos A \sin B)\cos C$$
$$+ (\cos A \cos B - \sin A \sin B)\sin C$$
$$= \sin A \cos B \cos C + \cos A \sin B \cos C$$
$$+ \cos A \cos B \sin C - \sin A \sin B \sin C \qquad ∎$$

To find identities for the tangent of the sum and difference of two angles, we make use of the identity

$$\tan x = \frac{\sin x}{\cos x}$$

and substitute $A + B$ for x.

$$\tan(A + B) = \frac{\sin(A + B)}{\cos(A + B)}$$
$$= \frac{\sin A \cos B + \cos A \sin B}{\cos A \cos B - \sin A \sin B}$$

To simplify this result, we divide both the numerator and denominator of the fraction by cos A cos B.

$$\tan(A + B) = \frac{\dfrac{\sin A \cos B}{\cos A \cos B} + \dfrac{\cos A \sin B}{\cos A \cos B}}{\dfrac{\cos A \cos B}{\cos A \cos B} - \dfrac{\sin A \sin B}{\cos A \cos B}}$$

$$= \frac{\tan A + \tan B}{1 - \tan A \tan B}$$

This gives the identity

3.
$$\tan(A + B) = \frac{\tan A + \tan B}{1 - \tan A \tan B}$$

If we substitute $-B$ for B in Equation 3 and simplify, we obtain an identity for the tangent of the difference of two angles.

$$\tan(A - B) = \frac{\tan A - \tan B}{1 + \tan A \tan B}$$

Example 4   Given that $\sin \alpha = \dfrac{12}{13}$, α in QI, $\cos \beta = -\dfrac{4}{5}$, and β in QII, find $\sin(\beta - \alpha)$.

Solution   Because $\sin \alpha = \frac{12}{13}$ and α is in QI, you can draw Figure 4-2a and determine that $\cos \alpha = \frac{5}{13}$. Because $\cos \beta = -\frac{4}{5}$ and β is in QII, you can draw Figure 4-2b and determine that $\sin \beta = \frac{3}{5}$. You can substitute $\frac{3}{5}$ for $\sin \beta$, $\frac{5}{13}$ for $\cos \alpha$, $-\frac{4}{5}$ for $\cos \beta$, and $\frac{12}{13}$ for $\sin \alpha$ in the following identity and simplify.

$$\sin(\beta - \alpha) = \sin \beta \cos \alpha - \cos \beta \sin \alpha$$

$$= \frac{3}{5}\left(\frac{5}{13}\right) - \left(-\frac{4}{5}\right)\left(\frac{12}{13}\right)$$

$$= \frac{15}{65} + \frac{48}{65}$$

$$= \frac{63}{65}$$

Thus, you have

$$\sin(\beta - \alpha) = \frac{63}{65}$$

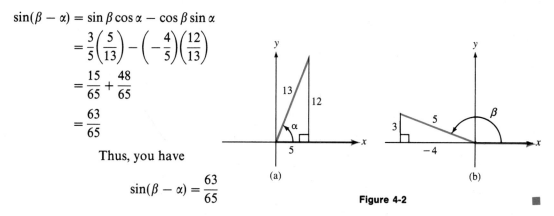

(a)   (b)

Figure 4-2

Example 5   Verify the identity $\tan(\theta + 45°) = \dfrac{1 + \tan\theta}{1 - \tan\theta}$.

Solution   Use the formula for the tangent of the sum of two angles.

$$\tan(\theta + 45°) = \frac{\tan\theta + \tan 45°}{1 - \tan\theta \tan 45°}$$

$$= \frac{\tan\theta + 1}{1 - \tan\theta}$$

$$= \frac{1 + \tan\theta}{1 - \tan\theta}$$

The identity is verified.   ∎

EXERCISE 4.2

1. Find $\sin 195°$ from the trigonometric functions of $45°$ and $150°$.

2. Find $\cos 195°$ from the trigonometric functions of $45°$ and $150°$.

3. Find $\tan 195°$ from the trigonometric functions of $225°$ and $30°$.

4. Find $\tan 165°$ from the trigonometric functions of $210°$ and $45°$.

5. Find $\cos\dfrac{11\pi}{12}$ from the trigonometric functions of $\dfrac{\pi}{6}$ and $\dfrac{3\pi}{4}$.

6. Find $\sin\dfrac{11\pi}{12}$ from the trigonometric functions of $\dfrac{\pi}{6}$ and $\dfrac{3\pi}{4}$.

7. Find $\cos\dfrac{19\pi}{12}$ from the trigonometric functions of $\dfrac{11\pi}{6}$ and $\dfrac{\pi}{4}$.

8. Find $\tan\dfrac{19\pi}{12}$ from the trigonometric functions of $\dfrac{11\pi}{6}$ and $\dfrac{\pi}{4}$.

In Exercises 9–20, find each value by using an identity involving a trigonometric function of two angles.

9. $\sin 75°$ 10. $\cos 285°$ 11. $\tan 105°$ 12. $\sin 255°$ 13. $\sin 105°$ 14. $\cos 255°$

15. $\sin\dfrac{\pi}{12}$ 16. $\sin\dfrac{7\pi}{12}$ 17. $\tan\dfrac{5\pi}{12}$ 18. $\cos\dfrac{13\pi}{12}$ 19. $\cos\dfrac{21\pi}{12}$ 20. $\tan\dfrac{23\pi}{12}$

In Exercises 21–28, find each value.

21. $\sec 75°$ 22. $\csc 105°$ 23. $\cot(-165°)$ 24. $\sec 435°$

25. $\csc\dfrac{7\pi}{12}$ 26. $\sec\dfrac{5\pi}{12}$ 27. $\sec\left(-\dfrac{\pi}{12}\right)$ 28. $\cot\dfrac{29\pi}{12}$

29. Show that $\sin(60° + \theta) = \dfrac{\sqrt{3}}{2}\cos\theta + \dfrac{1}{2}\sin\theta$.

30. Show that $\cos\left(\dfrac{\pi}{2} + x\right) = -\sin x$.

31. Show that $\tan(\pi + x) = \tan x$.

32. Show that $\sin\left(\dfrac{3\pi}{2} - x\right) = -\cos x$.

33. Show that $\cos(\pi - x) = -\cos x$.

34. Show that $\tan\left(\dfrac{\pi}{4} - x\right) = \dfrac{1 - \tan x}{1 + \tan x}$.

In Exercises 35–42, express each quantity as a single function of one angle.

35. $\sin 10° \cos 30° + \cos 10° \sin 30°$

36. $\cos 20° \cos 30° - \sin 20° \sin 30°$

37. $\dfrac{\tan 75° + \tan 40°}{1 - \tan 75° \tan 40°}$

38. $\sin 100° \cos 80° - \cos 100° \sin 80°$

39. $\cos 120° \cos 40° + \sin 120° \sin 40°$

40. $\dfrac{\tan 37° - \tan 125°}{1 + \tan 37° \tan 125°}$

41. $\sin x \cos 2x + \sin 2x \cos x$

42. $\cos 2y \cos 3y - \sin 2y \sin 3y$

43. Given that $\sin \alpha = \frac{3}{5}$, α in QII, and $\cos \beta = -\frac{12}{13}$, and β in QII, find $\sin(\alpha + \beta)$ and $\cos(\alpha - \beta)$.

44. Given that $\sin \alpha = \frac{7}{25}$ with α in QI, and $\sin \beta = \frac{15}{17}$ with β in QI, find $\sin(\alpha - \beta)$ and $\cos(\alpha + \beta)$.

45. Given that $\tan \alpha = -\frac{5}{12}$ with α in QII, and $\tan \beta = \frac{15}{8}$ with β in QI, find $\tan(\alpha + \beta)$ and $\tan(\alpha - \beta)$.

46. Given that $\sin \alpha = \frac{12}{13}$ with α in QI, and $\sin(\alpha + \beta) = \frac{24}{25}$ with $\alpha + \beta$ in QII, find $\sin \beta$ and $\cos \beta$.

47. Given that $\cos \beta = -\frac{15}{17}$ with β in QII, and $\sin(\alpha - \beta) = -\frac{24}{25}$ with $\alpha - \beta$ in QIV, find $\sin \alpha$ and $\cos \alpha$.

48. Given that $\sin(\alpha + \beta) = \frac{3}{5}$ with $\alpha + \beta$ in QI, and $\cos(\alpha - \beta) = \frac{12}{13}$ with $\alpha - \beta$ in QIV, find $\sin 2\alpha$ and $\cos 2\beta$.

In Exercises 49–64, verify each identity.

49. $\sin(30° + \theta) - \cos(60° + \theta) = \sqrt{3} \sin \theta$

50. $\sin(60° + \theta) - \cos(30° + \theta) = \sin \theta$

51. $\sin(30° + \theta) + \cos(60° + \theta) = \cos \theta$

52. $\sin(A + B) - \sin(A - B) = 2 \cos A \sin B$

53. $\cos(A + B) + \cos(A - B) = 2 \cos A \cos B$

54. $\sin(A + B) \sin(A - B) = \sin^2 A - \sin^2 B$

55. $\cos(A + B) \cos(A - B) = \cos^2 A + \cos^2 B - 1$

56. $\cot(A + B) = \dfrac{\cot A \cot B - 1}{\cot A + \cot B}$

57. $\cot(A - B) = \dfrac{\cot A \cot B + 1}{\cot B - \cot A}$

58. $\cos(A + B) - \cos(A - B) = -2 \sin A \sin B$

59. $\sin(A + B) + \sin(A - B) = 2 \sin A \cos B$

60. $-\tan A = \cot\left(A + \dfrac{\pi}{2}\right)$

61. $\dfrac{\tan A + \tan B}{1 + \tan A \tan B} = \dfrac{\sin(A + B)}{\cos(A - B)}$

62. $\dfrac{\cos(A + B)}{\sin(A - B)} = \dfrac{\cot A - \tan B}{1 - \cot A \tan B}$

63. $\dfrac{\cos(A - B)}{\sin(A + B)} = \dfrac{1 + \tan A \tan B}{\tan A + \tan B}$

64. $\dfrac{\sin A \cos B + \cos A \sin B}{\sin A \cos B - \cos A \sin B} = \dfrac{1 + \cot A \tan B}{1 - \cot A \tan B}$

65. Derive a formula for $\cos(A + B + C)$.

66. Derive a formula for $\sin(A - B - C)$.

67. Derive a formula for $\cos(A - B - C)$.

68. Derive a formula for $\tan(A + B + C)$.

4.3   THE DOUBLE-ANGLE IDENTITIES

If angles A and B are equal, the formulas for the sine of the sum of two angles and the cosine of the sum of two angles can be transformed into formulas called the **double-angle identities**. For example, if we let $A = B$ in the identity

$\sin(A + B) = \sin A \cos B + \cos A \sin B$ and simplify, we obtain

$$\sin(A + A) = \sin A \cos A + \cos A \sin A$$

$\sin 2A = 2 \sin A \cos A$

Similarly, if we let $A = B$ in the identity $\cos(A + B) = \cos A \cos B - \sin A \sin B$ and simplify, we obtain

$$\cos(A + A) = \cos A \cos A - \sin A \sin A$$

1.

$\cos 2A = \cos^2 A - \sin^2 A$

There are two variations of this identity for $\cos 2A$. To obtain one of them, we substitute $1 - \cos^2 A$ for $\sin^2 A$ in Equation 1, and simplify.

$$\cos 2A = \cos^2 A - (1 - \cos^2 A)$$

$\cos 2A = 2 \cos^2 A - 1$

To obtain the other, we substitute $1 - \sin^2 A$ for $\cos^2 A$ in Equation 1, and simplify.

$$\cos 2A = 1 - \sin^2 A - \sin^2 A$$

$\cos 2A = 1 - 2 \sin^2 A$

To find the double-angle identity for the tangent function, we let $A = B$ in the identity

$$\tan(A + B) = \frac{\tan A + \tan B}{1 - \tan A \tan B}$$

and simplify.

$$\tan(A + A) = \frac{\tan A + \tan A}{1 - \tan A \tan A}$$

$\tan 2A = \dfrac{2 \tan A}{1 - \tan^2 A}$

Example 1   Simplify the expression $\sin 5A \cos 5A$.

Solution   Because the given expression is one-half of $2 \sin 5A \cos 5A$, it is one-half of $\sin 2(5A)$.

$$\sin 5A \cos 5A = \frac{1}{2}(2 \sin 5A \cos 5A)$$

$$= \frac{1}{2} \sin 2(5A)$$

$$= \frac{1}{2} \sin 10A$$

∎

Example 2   Use a double-angle identity to evaluate $\tan 2(60°)$.

Solution   Use the identity $\tan 2A = \dfrac{2 \tan A}{1 - \tan^2 A}$ and substitute $60°$ for A.

$$\tan 2A = \frac{2 \tan A}{1 - \tan^2 A}$$

$$\tan 2(60°) = \frac{2 \tan(60°)}{1 - \tan^2(60°)}$$

$$= \frac{2\sqrt{3}}{1 - 3}$$

$$= \frac{2\sqrt{3}}{-2}$$

$$= -\sqrt{3}$$

∎

Example 3   Find the **a.** sine, **b.** cosine, and **c.** tangent of 2θ if $\cos \theta = \dfrac{12}{13}$ and θ is in QIV.

Solution   Use the information that $\cos \theta = \frac{12}{13}$ and that θ is in QIV to sketch the angle in standard position as in Figure 4-3. From the figure, it follows that

$$\sin \theta = \frac{-5}{13}$$

$$= -\frac{5}{13}$$

$$\tan \theta = \frac{-5}{12}$$

$$= -\frac{5}{12}$$

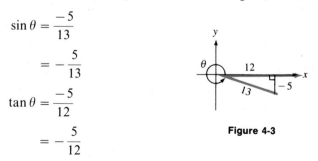

Figure 4-3

Use the double-angle identities to find the functions of 2θ.

a. $\sin 2\theta = 2 \sin \theta \cos \theta = 2\left(-\dfrac{5}{13}\right)\left(\dfrac{12}{13}\right) = -\dfrac{120}{169}$

b. $\cos 2\theta = \cos^2\theta - \sin^2\theta = \left(\dfrac{12}{13}\right)^2 - \left(-\dfrac{5}{13}\right)^2 = \dfrac{119}{169}$

c. $\tan 2\theta = \dfrac{2\left(-\frac{5}{12}\right)}{1 - \left(-\frac{5}{12}\right)^2} = \dfrac{-\frac{10}{12}}{1 - \frac{25}{144}} = \dfrac{-\frac{5}{6}}{\frac{119}{144}} = -\dfrac{5}{\cancel{6}} \cdot \dfrac{\overset{24}{\cancel{144}}}{119} = -\dfrac{120}{119}$ ∎

Example 4 Verify the identity $\cos 2A = \cos^4 A - \sin^4 A$.

Solution Because the right-hand side factors, work on that side.

$$\cos^4 A - \sin^4 A = (\cos^2 A + \sin^2 A)(\cos^2 A - \sin^2 A)$$
$$= 1 \cdot (\cos^2 A - \sin^2 A)$$
$$= \cos 2A$$ ∎

Example 5 Verify the identity $\cos A \sin 2A = 2 \sin A - 2 \sin^3 A$.

Solution Although the right-hand side looks more complicated, work with the other side because it involves a double angle.

$$\cos A \sin 2A = \cos A (2 \sin A \cos A)$$
$$= 2 \cos^2 A \sin A$$
$$= 2(1 - \sin^2 A) \sin A$$
$$= 2 \sin A - 2 \sin^3 A$$ ∎

Example 6 Verify the identity $\cos 2A \sec^2 A = 2 - \sec^2 A$.

Solution Work on the left-hand side to eliminate the double angle.

$$\cos 2A \sec^2 A = (2 \cos^2 A - 1) \sec^2 A$$
$$= 2 \cos^2 A \sec^2 A - \sec^2 A$$
$$= 2 - \sec^2 A$$ ∎

Example 7 Verify that $\sin 2A \sec^2 A = 2 \tan A$.

Solution Work on the left-hand side of the equation.

$$\sin 2A \sec^2 A = 2 \sin A \cos A \sec^2 A$$
$$= 2 \sin A \sec A$$
$$= 2 \tan A$$ ∎

Example 8 Verify the identity $2 \cot 2A = \cot A - \tan A$.

Solution   Work with the left-hand side of the equation.

$$2 \cot 2A = 2\left(\frac{1}{\tan 2A}\right)$$

$$= 2\left(\frac{1 - \tan^2 A}{2 \tan A}\right)$$

$$= \frac{1 - \tan^2 A}{\tan A}$$

$$= \frac{1}{\tan A} - \frac{\tan^2 A}{\tan A}$$

$$= \cot A - \tan A$$

■ EXERCISE 4.3

In Exercises 1–26, write each expression in terms of a single trigonometric function of twice the given angle.

1. $2 \sin \alpha \cos \alpha$

2. $2 \cos^2 \alpha - 1$

3. $2 \sin 3\theta \cos 3\theta$

4. $2 \cos^2 2A - 1$

5. $\cos^2 \beta - \sin^2 \beta$

6. $1 - 2 \sin^2 \beta$

7. $2 \cos^2 \dfrac{\beta}{2} - 1$

8. $\sin 5\theta \cos 5\theta$

9. $4 \sin \theta \cos \theta$

10. $\cos^2 \dfrac{\theta}{2} - \sin^2 \dfrac{\theta}{2}$

11. $4 \sin^2 2\theta \cos^2 2\theta$

12. $2 - 4 \sin^2 6B$

13. $\cos^2 \alpha - \dfrac{1}{2}$

14. $2 - 4 \sin^2 \dfrac{\alpha}{4}$

15. $\cos^2 9\theta - \sin^2 9\theta$

16. $4 \sin 4B \cos 4B$

17. $4 \sin^2 5\theta \cos^2 5\theta$

18. $\dfrac{2 \tan A}{1 - \tan^2 A}$

19. $\dfrac{2 \tan 4C}{1 - \tan^2 4C}$

20. $3 - 6 \cos^2 6x$

21. $\dfrac{\tan \dfrac{A}{2}}{\dfrac{1}{2} - \dfrac{1}{2} \tan^2 \dfrac{A}{2}}$

22. $\dfrac{2 \tan \alpha}{2 - \sec^2 \alpha}$

23. $\cos^4 4x - \sin^4 4x$

24. $\dfrac{1}{2} \sec \theta \csc \theta$

25. $1 - 2 \cos^2 5x$

26. $\dfrac{1 - \tan^2 3x}{2 \tan 3x}$

In Exercises 27–38, use a double-angle identity to find the value of the given expression. Then, check your work by evaluating the expression directly.

27. $\sin 2(30°)$

28. $\cos 2(30°)$

29. $\tan 2(45°)$

30. $\cot 2(135°)$

31. $\cos 2(240°)$

32. $\sin 2(150°)$

33. $\cot 2(225°)$

34. $\tan 2(315°)$

35. $\sin 2\left(\dfrac{\pi}{3}\right)$

36. $\cos 2\left(\dfrac{7\pi}{6}\right)$

37. $\cos 2\left(\dfrac{11\pi}{6}\right)$

38. $\sin 2\left(\dfrac{5\pi}{3}\right)$

In Exercises 39–50, find the exact value of the sine, cosine, and tangent of 2θ using the given information.

39. $\sin \theta = \dfrac{12}{13}$ and θ is in QI

40. $\cos \theta = \dfrac{5}{13}$ and θ is in QIV

41. $\tan \theta = \dfrac{12}{5}$ and θ is in QIII

42. $\sin \theta = \dfrac{3}{5}$ and θ is in QII

43. $\cos \theta = -\dfrac{4}{5}$ and θ is in QIII

44. $\cot \theta = -\dfrac{15}{8}$ and θ is in QIV

45. $\sin \theta = -\dfrac{24}{25}$ and θ is in QIV

46. $\cos \theta = -\dfrac{7}{25}$ and θ is in QIII

47. $\sin \theta = \dfrac{40}{41}$ and $\cos \theta$ is positive

48. $\sin \theta = -\dfrac{40}{41}$ and $\tan \theta$ is negative

49. $\cos \theta = \dfrac{40}{41}$ and $\tan \theta$ is negative

50. $\cos \theta = 0$

In Exercises 51–78, verify each identity.

51. $\dfrac{\tan 2x}{\sin 2x} = \sec 2x$

52. $\dfrac{\cot 2x}{\cos 2x} = \csc 2x$

53. $2 \csc 2A = \sec A \csc A$

54. $\sin 2A + 2 \sin A = \dfrac{-2 \sin^3 A}{\cos A - 1}$

55. $\cos 2x - \dfrac{\sin 2x}{\cos x} + 2 \sin^2 x = 1 - 2 \sin x$

56. $2 \cos^4 \theta + 2 \sin^2 \theta \cos^2 \theta - 1 = \cos 2\theta$

57. $\sin 2A - 2 \sin A = -\dfrac{2 \sin^3 A}{\cos A + 1}$

58. $2 \cos A - 1 = \dfrac{\cos 2A + \cos A}{\cos A + 1}$

59. $2 \sin^4 \theta + 2 \sin^2 \theta \cos^2 \theta - 1 = -\cos 2\theta$

60. $\sin 2\theta - \tan \theta = \cos 2\theta \tan \theta$

61. $2 \cos A + 1 = \dfrac{\cos 2A - \cos A}{\cos A - 1}$

62. $(\sin A + \cos A)^2 = 1 + \sin 2A$

63. $\sec 2\theta = \dfrac{\tan \theta + \cot \theta}{\cot \theta - \tan \theta}$

64. $\sin 4x = 4 \cos x \sin x - 8 \cos x \sin^3 x$

65. $4 \csc^2 2A = \sec^2 A + \csc^2 A$

66. $1 - \dfrac{1}{2} \sin 2x = \dfrac{\sin^3 x + \cos^3 x}{\sin x + \cos x}$

67. $\cos 4x = 8 \cos^4 x - 8 \cos^2 x + 1$

68. $-\tan 2x = \dfrac{2}{\tan x - \cot x}$

69. $2 \tan 2x = \dfrac{\sin x - \cos x}{\sin x + \cos x} - \dfrac{\sin x + \cos x}{\sin x - \cos x}$

70. $\sin^4 \theta = \dfrac{\cos 4\theta}{8} - \dfrac{\cos 2\theta}{2} + \dfrac{3}{8}$

71. $\dfrac{\sin x + \sin 2x}{\cos x - \cos 2x} = \dfrac{\sec x + 2}{\csc x - \cot x + \tan x}$

72. $\dfrac{\cos x - \sin x}{\cos 2x} = \dfrac{\cos x + \sin x}{1 + \sin 2x}$

73. $\dfrac{1 - \tan^2 x}{1 + \tan^2 x} = \cos 2x$

74. $\dfrac{\sec^2 x}{\sec 2x} = 2 - \sec^2 x$

75. $\dfrac{1 + \tan x}{1 - \tan x} = \dfrac{\cos 2x}{1 - \sin 2x}$

76. $\cos 2x = \dfrac{1}{\tan 2x \tan x + 1}$

77. $\tan A \sin 2A + \cos 2A = 1$

78. $\dfrac{2(\sin A - 1)}{\sec^2 A} + 1 = \sin 2A \cos A - \cos 2A$

4.4   THE HALF-ANGLE IDENTITIES

If the trigonometric functions of angle A are known, the formulas of Section 4.3 allow us to find the functions of $2A$. The identities in this section will enable us to find the trigonometric functions of $\frac{1}{2}A$.

We begin by solving the identity $\cos 2\theta = 2\cos^2\theta - 1$ for $\cos\theta$.

$$\cos 2\theta = 2\cos^2\theta - 1$$

$$2\cos^2\theta = 1 + \cos 2\theta$$

$$\cos^2\theta = \frac{1 + \cos 2\theta}{2}$$

$$\cos\theta = \pm\sqrt{\frac{1 + \cos 2\theta}{2}}$$

With an appropriate choice of the sign preceding the radical, this identity is true for all values of θ. To derive the first **half-angle identity**, we let $\theta = \frac{A}{2}$.

1.
$$\cos\frac{A}{2} = \pm\sqrt{\frac{1 + \cos A}{2}}$$

The sign preceding the radical in Equation 1 is determined by the quadrant in which $\frac{A}{2}$ lies.

In a similar fashion, we solve $\cos 2\theta = 1 - 2\sin^2\theta$ for $\sin\theta$ and let $\theta = \frac{A}{2}$ to obtain the half-angle identity for the sine function.

$$\cos 2\theta = 1 - 2\sin^2\theta$$

$$2\sin^2\theta = 1 - \cos 2\theta$$

$$\sin^2\theta = \frac{1 - \cos 2\theta}{2}$$

$$\sin\theta = \pm\sqrt{\frac{1 - \cos 2\theta}{2}}$$

$$\sin\frac{A}{2} = \pm\sqrt{\frac{1 - \cos A}{2}}$$

Again, the + or − sign is chosen by the quadrant in which $\frac{A}{2}$ lies.

Example 1   Use a half-angle identity to find $\sin 15°$.

Solution   Because $15°$ is $\frac{1}{2}(30°)$, it follows that

$$\sin 15° = \sin\frac{30°}{2} = +\sqrt{\frac{1 - \cos 30°}{2}}$$

$$= +\sqrt{\frac{1 - \dfrac{\sqrt{3}}{2}}{2}}$$

$$= +\sqrt{\frac{2 - \sqrt{3}}{4}}$$

$$= +\frac{\sqrt{2 - \sqrt{3}}}{2}$$

The $+$ sign is chosen because $15°$ is a first-quadrant angle and the sine of a first-quadrant angle is positive.   ∎

Example 2   Use a half-angle identity to find $\cos\dfrac{7\pi}{12}$.

Solution   Because $\dfrac{7\pi}{12}$ is $\dfrac{1}{2}\cdot\dfrac{7\pi}{6}$, it follows that

$$\cos\frac{7\pi}{12} = \cos\frac{1}{2}\cdot\frac{7\pi}{6} = -\sqrt{\frac{1 + \cos\dfrac{7\pi}{6}}{2}}$$

$$= -\sqrt{\frac{1 - \dfrac{\sqrt{3}}{2}}{2}}$$

$$= -\frac{\sqrt{2 - \sqrt{3}}}{2}$$

Here, the $-$ sign is chosen because $\frac{7\pi}{12}$ is a second-quadrant angle and the cosine of a second-quadrant angle is negative.   ∎

To derive the half-angle tangent identities, we will show that

$$\tan x = \frac{\sin 2x}{1 + \cos 2x}$$

and then substitute $\frac{A}{2}$ for x.

$$\frac{\sin 2x}{1 + \cos 2x} = \frac{2 \sin x \cos x}{1 + (2 \cos^2 x - 1)}$$

$\sin 2x = 2 \sin x \cos x$ and
$\cos 2x = 2 \cos^2 x - 1.$

$$= \frac{2 \sin x \cos x}{2 \cos^2 x}$$

$$= \frac{\sin x}{\cos x}$$

$$= \tan x$$

Thus,

$$\tan x = \frac{\sin 2x}{1 + \cos 2x}$$

After substituting $\frac{A}{2}$ for x and simplifying, we have

2.
$$\tan \frac{A}{2} = \frac{\sin A}{1 + \cos A}$$

By multiplying both the numerator and denominator of Equation 2 by $1 - \cos A$, we obtain another version of the half-angle tangent identity.

$$\tan \frac{A}{2} = \frac{\sin A (1 - \cos A)}{(1 + \cos A)(1 - \cos A)}$$

$$= \frac{\sin A(1 - \cos A)}{1 - \cos^2 A}$$

$$= \frac{\sin A(1 - \cos A)}{\sin^2 A}$$

$$= \frac{1 - \cos A}{\sin A}$$

Thus,

$$\tan \frac{A}{2} = \frac{1 - \cos A}{\sin A}$$

Example 3   Use the identity $\tan \dfrac{A}{2} = \dfrac{1 - \cos A}{\sin A}$ to find $\tan \dfrac{\pi}{8}$.

Solution Because $\dfrac{\pi}{8}$ is $\dfrac{1}{2}\left(\dfrac{\pi}{4}\right)$, it follows that

$$\tan\frac{\pi}{8} = \tan\frac{\dfrac{\pi}{4}}{2} = \frac{1 - \cos\dfrac{\pi}{4}}{\sin\dfrac{\pi}{4}}$$

$$= \frac{1 - \dfrac{\sqrt{2}}{2}}{\dfrac{\sqrt{2}}{2}}$$

$$= \frac{2 - \sqrt{2}}{\sqrt{2}} \qquad \text{Multiply numerator and denominator by } \sqrt{2}$$
$$\text{and simplify.}$$

$$= \sqrt{2} - 1 \qquad\qquad\qquad\qquad\qquad\qquad\qquad \blacksquare$$

Example 4 Use the identity $\tan\dfrac{A}{2} = \dfrac{\sin A}{1 + \cos A}$ to find $\tan 157.5°$.

Solution Because $157.5°$ is $\dfrac{315°}{2}$, then

$$\tan 157.5° = \tan\frac{315°}{2} = \frac{\sin 315°}{1 + \cos 315°}$$

$$= \frac{-\dfrac{\sqrt{2}}{2}}{1 + \dfrac{\sqrt{2}}{2}}$$

$$= \frac{-\dfrac{\sqrt{2}}{2}}{\dfrac{2 + \sqrt{2}}{2}}$$

$$= \frac{-\sqrt{2}}{2 + \sqrt{2}}$$

$$= \frac{-\sqrt{2}}{(2 + \sqrt{2})} \cdot \frac{(2 - \sqrt{2})}{(2 - \sqrt{2})}$$

$$= \frac{-\sqrt{2}(2 - \sqrt{2})}{2}$$

$$= 1 - \sqrt{2}$$

This number is negative, as it must be for the tangent of a QII angle. $\blacksquare$

Example 5   Find the **a.** sine, **b.** cosine, and **c.** tangent of $\dfrac{\theta}{2}$, if $\sin\theta = \dfrac{3}{5}$ and $\dfrac{\pi}{2} < \theta < \pi$.

Solution   Use the information that $\sin\theta = \frac{3}{5}$ and that θ is in QII to sketch the angle in standard position as in Figure 4-4. The value of $\cos\theta$ can be read from the figure.

$$\cos\theta = -\frac{4}{5}$$

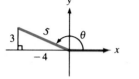

Figure 4-4

Use the half-angle identities to find the functions of $\dfrac{\theta}{2}$.

a. $\sin\dfrac{\theta}{2} = \sqrt{\dfrac{1 - \left(\dfrac{-4}{5}\right)}{2}} = \sqrt{\dfrac{9}{10}} = \dfrac{3\sqrt{10}}{10}$

b. $\cos\dfrac{\theta}{2} = \sqrt{\dfrac{1 + \left(\dfrac{-4}{5}\right)}{2}} = \sqrt{\dfrac{1}{10}} = \dfrac{\sqrt{10}}{10}$

c. $\tan\dfrac{\theta}{2} = \dfrac{\sin\dfrac{\theta}{2}}{\cos\dfrac{\theta}{2}} = \dfrac{\dfrac{3\sqrt{10}}{10}}{\dfrac{\sqrt{10}}{10}} = 3$

Choose the radicals to be positive, because if $\dfrac{\pi}{2} < \theta < \pi$, then $\dfrac{\theta}{2}$ is in QI.   ∎

Example 6   Write $\dfrac{\sin(-20A)}{-\cos 20A - 1}$ as a trigonometric function of $10A$.

Solution   Because $\sin(-20A) = -\sin 20A$, it follows that

$$\frac{\sin(-20A)}{-\cos 20A - 1} = \frac{-\sin 20A}{-(1 + \cos 20A)} = \frac{\sin 20A}{1 + \cos 20A}$$

One of the identities for $\tan\dfrac{\theta}{2}$ is

$$\frac{\sin\theta}{1 + \cos\theta} = \tan\frac{\theta}{2}$$

Hence, you have

$$\frac{\sin 20A}{1 + \cos 20A} = \tan\frac{20A}{2}$$

$$= \tan 10A$$

∎

Example 7   Verify the identity $\tan \dfrac{\theta}{2} = \csc \theta - \cot \theta$.

Solution   Work on the left-hand side to eliminate the half-angle.

$$\tan \frac{\theta}{2} = \frac{1 - \cos \theta}{\sin \theta}$$

$$= \frac{1}{\sin \theta} - \frac{\cos \theta}{\sin \theta}$$

$$= \csc \theta - \cot \theta \qquad \blacksquare$$

Example 8   Verify the identity $2 \sin^2 \dfrac{x}{2} \tan x = \tan x - \sin x$.

Solution   Work on the left-hand side to eliminate the half-angle.

$$2 \sin^2 \frac{x}{2} \tan x = 2 \cdot \frac{1 - \cos x}{2} \cdot \frac{\sin x}{\cos x}$$

$$= \frac{\sin x - \sin x \cos x}{\cos x}$$

$$= \frac{\sin x}{\cos x} - \sin x$$

$$= \tan x - \sin x \qquad \blacksquare$$

Example 9   Verify the identity $\sec^2 \dfrac{x}{2} = \dfrac{2 \tan \dfrac{x}{2}}{\sin x}$.

Solution   This time work on the right-hand side.

$$\frac{2 \tan \dfrac{x}{2}}{\sin x} = \frac{2 \cdot \dfrac{\sin x}{1 + \cos x}}{\sin x} \qquad \text{Replace } \tan \frac{x}{2} \text{ with } \frac{\sin x}{1 + \cos x}.$$

$$= \frac{2}{1 + \cos x} \qquad \text{Divide numerator and denominator by } \sin x.$$

$$= \frac{1}{\dfrac{1 + \cos x}{2}} \qquad \text{Divide numerator and denominator by 2.}$$

$$= \frac{1}{\cos^2 \dfrac{x}{2}}$$

$$= \sec^2 \frac{x}{2} \qquad \blacksquare$$

▪ EXERCISE 4.4

In Exercises 1–12, use half-angle identities to find the required values. **Do not use a calculator.**

1. $\cos 15°$ **2.** $\tan 15°$ **3.** $\tan 105°$ **4.** $\sin 105°$ **5.** $\sin \dfrac{\pi}{8}$ **6.** $\cos \dfrac{\pi}{8}$

7. $\cos \dfrac{\pi}{12}$ **8.** $\sin \dfrac{\pi}{12}$ **9.** $\tan 165°$ **10.** $\cos 165°$ **11.** $\cot \dfrac{5\pi}{4}$ **12.** $\tan \dfrac{7\pi}{4}$

In Exercises 13–24, use the given information to find the exact value of the sine, cosine, and tangent of $\frac{\theta}{2}$. Assume that $0° \le \theta < 360°$.

13. $\sin \theta = \dfrac{3}{5}$; θ in QI

14. $\cos \theta = \dfrac{12}{13}$; θ in QI

15. $\tan \theta = \dfrac{4}{3}$; θ in QIII

16. $\tan \theta = \dfrac{3}{4}$; θ in QIII

17. $\cos \theta = \dfrac{8}{17}$; θ in QIV

18. $\sin \theta = -\dfrac{7}{25}$; θ in QIII

19. $\cot \theta = \dfrac{40}{9}$; θ in QI

20. $\sec \theta = -\dfrac{41}{40}$; θ in QII

21. $\csc \theta = \dfrac{17}{8}$; θ in QII

22. $\csc \theta = -\dfrac{5}{3}$; θ in QIV

23. $\sec \theta = \dfrac{3}{2}$; θ in QIV

24. $\cos \theta = -0.1$; θ in QIII

In Exercises 25–42, write each expression as a single trigonometric function of half the given angle.

25. $\sqrt{\dfrac{1 + \cos 30°}{2}}$ **26.** $\sqrt{\dfrac{1 - \cos 30°}{2}}$ **27.** $\sqrt{\dfrac{1 - \cos 80°}{2}}$ **28.** $\sqrt{\dfrac{1 + \cos 80°}{2}}$

29. $\sqrt{\dfrac{1 - \cos 400°}{2}}$ **30.** $\sqrt{\dfrac{1 + \cos 350°}{2}}$ **31.** $\sqrt{\dfrac{1 + \cos 550°}{2}}$ **32.** $\sqrt{\dfrac{1 - \cos 200°}{2}}$

33. $\dfrac{1 - \cos 200°}{\sin 200°}$

34. $\dfrac{\sin 50°}{1 + \cos 50°}$

35. $\csc 80° - \cot 80°$

36. $\dfrac{2 \tan 140°}{\sin 280°}$ (280° is the given angle)

37. $\dfrac{\sin 2\pi}{1 + \cos 2\pi}$

38. $\dfrac{1 - \cos 4\theta}{\sin 4\theta}$

39. $\dfrac{1 - \cos \dfrac{x}{2}}{\sin \dfrac{x}{2}}$

40. $\left(\dfrac{1 - \cos 2x}{\sin 2x}\right)^2$

41. $\dfrac{\sin 10A}{1 + \cos 10A}$

42. $\left(\dfrac{\sin 4x}{1 + \cos 4x}\right)^2$

In Exercises 43–56, verify each identity.

43. $\sin^2\dfrac{\theta}{2} = \dfrac{1}{2}(1 - \cos\theta)$

44. $\cos^2\dfrac{\theta}{2} = \dfrac{1}{2}(1 + \cos\theta)$

45. $\sec^2\dfrac{\theta}{2} = \dfrac{2}{1 + \cos\theta}$

46. $\csc^2\dfrac{\theta}{2} = \dfrac{2}{1 - \cos\theta}$

47. $\cot\dfrac{\theta}{2} = \dfrac{1 + \cos\theta}{\sin\theta}$

48. $\cot\dfrac{\theta}{2} = \dfrac{\sin\theta}{1 - \cos\theta}$

49. $\csc^2\dfrac{\theta}{2} = 2\csc^2\theta + 2\cot\theta\csc\theta$

50. $\sin^2\dfrac{\theta}{2} = \dfrac{\sec\theta - 1}{2\sec\theta}$

51. $-\sec B = \dfrac{\sec^2\dfrac{B}{2}}{\sec^2\dfrac{B}{2} - 2}$

52. $\tan\left(\dfrac{\pi}{4} + \dfrac{\theta}{2}\right) = \dfrac{1 + \cos\theta + \sin\theta}{1 + \cos\theta - \sin\theta}$

53. $\csc\theta = \dfrac{1}{2}\csc\dfrac{\theta}{2}\sec\dfrac{\theta}{2}$

54. $\dfrac{1}{2}\sin x\tan\dfrac{x}{2}\csc^2\dfrac{x}{2} = 1$

55. $\tan\dfrac{B}{2}\cos B + \tan\dfrac{B}{2} = \sin B$

56. $\left(\cos\dfrac{\alpha}{2} - \sin\dfrac{\alpha}{2}\right)^2 = 1 - \sin\alpha$

4.5 SUM-TO-PRODUCT AND PRODUCT-TO-SUM IDENTITIES

The identities discussed in this section are needed whenever it is necessary to convert a product of two trigonometric functions into a sum, or a sum into a product.

If the identities for $\sin(x + y)$ and $\sin(x - y)$ are added, some of these new identities result.

1. $\sin(x + y) = \sin x\cos y + \cos x\sin y$
2. $\sin(x - y) = \sin x\cos y - \cos x\sin y$

Adding Equations 1 and 2 causes the $\cos x\sin y$ term to drop out. The result is shown in Equation 3.

3. $\sin(x + y) + \sin(x - y) = 2\sin x\cos y$

We divide both sides of Equation 3 by 2 to get

$$\sin x\cos y = \frac{1}{2}[\sin(x + y) + \sin(x - y)]$$

This identity is used to convert a product of the sine and cosine functions into a sum.

Example 1 Calculate the value of $\sin 67.5° \cos 22.5°$.

Solution Let $x = 67.5°$, $y = 22.5°$, and use the boxed identity on page 150.

$$\sin x \cos y = \frac{1}{2}[\sin(x + y) + \sin(x - y)]$$

$$\sin 67.5° \cos 22.5° = \frac{1}{2}[\sin(67.5° + 22.5°) + \sin(67.5° - 22.5°)]$$

$$= \frac{1}{2}(\sin 90° + \sin 45°)$$

$$= \frac{1}{2}\left(1 + \frac{\sqrt{2}}{2}\right)$$

$$= \frac{1}{2} + \frac{\sqrt{2}}{4}$$ ∎

We now develop a formula to convert the sum of two sines into a product. To do so, we let $A = x + y$ and $B = x - y$, and solve for x and y. We first solve

$$\begin{cases} A = x + y \\ B = x - y \end{cases}$$

for x by adding the equations.

$$A + B = 2x$$

$$x = \frac{1}{2}(A + B)$$

To find y, we subtract one equation from the other.

$$A - B = 2y$$

$$y = \frac{1}{2}(A - B)$$

We substitute

$$A \text{ for } x + y, \qquad B \text{ for } x - y, \qquad \frac{A + B}{2} \text{ for } x, \qquad \text{and} \qquad \frac{A - B}{2} \text{ for } y$$

in Equation 3 to obtain

$$\sin A + \sin B = 2\sin\frac{A + B}{2}\cos\frac{A - B}{2}$$

Example 2 Find the value of $\sin 75° + \sin 15°$.

Solution Use the identity $\sin A + \sin B = 2\sin\dfrac{A+B}{2}\cos\dfrac{A-B}{2}$ and let $A = 75°$ and $B = 15°$.

$$\sin A + \sin B = 2\sin\frac{A+B}{2}\cos\frac{A-B}{2}$$

$$\sin 75° + \sin 15° = 2\sin\frac{75° + 15°}{2}\cos\frac{75° - 15°}{2}$$

$$= 2\sin\frac{90°}{2}\cos\frac{60°}{2}$$

$$= 2\sin 45° \cos 30°$$

$$= 2\cdot\frac{\sqrt{2}}{2}\cdot\frac{\sqrt{3}}{2}$$

$$= \frac{\sqrt{6}}{2}$$

∎

Example 3 Verify the identity $\sin 3\theta + \sin\theta = 2\sin 2\theta\cos\theta$.

Solution Use the identity $\sin A + \sin B = 2\sin\dfrac{A+B}{2}\cos\dfrac{A-B}{2}$ and let $A = 3\theta$ and $B = \theta$. Work on the left-hand side.

$$\sin 3\theta + \sin\theta = 2\sin\frac{3\theta + \theta}{2}\cos\frac{3\theta - \theta}{2}$$

$$= 2\sin\frac{4\theta}{2}\cos\frac{2\theta}{2}$$

$$= 2\sin 2\theta\cos\theta$$

∎

If Equation 2 above is subtracted from Equation 1, more identities result.

1. $\qquad\sin(x + y) = \sin x\cos y + \cos x\sin y$
2. $\qquad\sin(x - y) = \sin x\cos y - \cos x\sin y$

_____

3. $\sin(x + y) - \sin(x - y) = 2\cos x\sin y$

Dividing both sides of Equation 3 by 2 gives a new identity.

$$\cos x\sin y = \frac{1}{2}[\sin(x + y) - \sin(x - y)]$$

This identity is used to convert a product into a difference. If we let

$$x + y = A \qquad \text{and} \qquad x - y = B$$

then we can solve the system of equations to obtain

$$x = \frac{A + B}{2} \quad \text{and} \quad y = \frac{A - B}{2}$$

We can substitute the values for $x + y$, $x - y$, x, and y into Equation 3 to obtain the next identity.

$$\sin A - \sin B = 2 \cos \frac{A + B}{2} \sin \frac{A - B}{2}$$

This formula is used to convert the difference of the sines of two angles into a product.

If the formulas for the cosines of the sum and the difference of two angles are added, still more identities result.

4. $\begin{cases} \cos(x + y) = \cos x \cos y - \sin x \sin y \\ \cos(x - y) = \cos x \cos y + \sin x \sin y \end{cases}$
5.

After adding, the $\sin x \sin y$ terms drop out, as shown in Equation 6.

6. $\cos(x + y) + \cos(x - y) = 2 \cos x \cos y$

We divide both sides of Equation 6 by 2 to obtain a new identity.

$$\cos x \cos y = \frac{1}{2}[\cos(x + y) + \cos(x - y)]$$

This identity is used when a product of cosines needs to be changed to a sum. To derive an identity to convert from sums to products, we let $x + y = A$ and $x - y = B$. Then

$$x = \frac{A + B}{2} \quad \text{and} \quad y = \frac{A - B}{2}$$

We use Equation 6 and substitute to get

$$\cos A + \cos B = 2 \cos \frac{A + B}{2} \cos \frac{A - B}{2}$$

We obtain the final identities when we subtract Equation 4 from Equation 5 and make the usual substitutions.

$$\sin x \sin y = \frac{1}{2}[\cos(x - y) - \cos(x + y)]$$

$$\cos A - \cos B = -2\sin\frac{A + B}{2}\sin\frac{A - B}{2}$$

For easy reference, all the preceding formulas are listed together as follows.

The product-to-sum formulas:

$$\sin A \cos B = \frac{1}{2}[\sin(A + B) + \sin(A - B)]$$

$$\cos A \sin B = \frac{1}{2}[\sin(A + B) - \sin(A - B)]$$

$$\sin A \sin B = \frac{1}{2}[\cos(A - B) - \cos(A + B)]$$

$$\cos A \cos B = \frac{1}{2}[\cos(A + B) + \cos(A - B)]$$

The sum-to-product formulas:

$$\sin A + \sin B = 2\sin\frac{A + B}{2}\cos\frac{A - B}{2}$$

$$\sin A - \sin B = 2\cos\frac{A + B}{2}\sin\frac{A - B}{2}$$

$$\cos A + \cos B = 2\cos\frac{A + B}{2}\cos\frac{A - B}{2}$$

$$\cos A - \cos B = -2\sin\frac{A + B}{2}\sin\frac{A - B}{2}$$

Example 4   Write $\cos 2\theta + \cos 6\theta$ as a product of two functions.

Solution   Substitute 2θ for A and 6θ for B in the identity for $\cos A + \cos B$, and simplify.

$$\cos 2\theta + \cos 6\theta = 2\cos\frac{2\theta + 6\theta}{2}\cos\frac{2\theta - 6\theta}{2}$$

$$= 2\cos 4\theta\cos(-2\theta)$$

$$= 2\cos 4\theta\cos 2\theta$$ Remember that $\cos(-2\theta) = \cos 2\theta$.   ∎

Example 5   Verify the identity $\tan 3A = \dfrac{\sin 4A + \sin 2A}{\cos 4A + \cos 2A}$.

Solution   Work on the right-hand side of the equation.

$$\frac{\sin 4A + \sin 2A}{\cos 4A + \cos 2A} = \frac{2 \sin \dfrac{4A + 2A}{2} \cos \dfrac{4A - 2A}{2}}{2 \cos \dfrac{4A + 2A}{2} \cos \dfrac{4A - 2A}{2}}$$

$$= \frac{2 \sin 3A \cos A}{2 \cos 3A \cos A}$$

$$= \frac{\sin 3A}{\cos 3A}$$

$$= \tan 3A$$

EXERCISE 4.5

In Exercises 1–10, express each quantity as a sum or a difference and find its value. **Do not use a calculator or tables.**

1. $\cos 75° \cos 15°$

2. $\sin 15° \cos 75°$

3. $\sin 165° \sin 105°$

4. $\sin 165° \sin 75°$

5. $\cos 22.5° \cos 67.5°$

6. $\cos 105° \sin 15°$

7. $\cos \dfrac{19\pi}{12} \sin \dfrac{5\pi}{12}$

8. $\sin \dfrac{5\pi}{12} \cos \dfrac{13\pi}{12}$

9. $\cos \dfrac{7\pi}{12} \cos \dfrac{5\pi}{12}$

10. $\cos \dfrac{7\pi}{12} \sin \dfrac{13\pi}{12}$

In Exercises 11–20, express each quantity as a product and find its value. **Do not use a calculator or tables.**

11. $\cos 75° + \cos 15°$

12. $\sin 75° + \sin 15°$

13. $\sin 165° - \sin 105°$

14. $\sin 15° - \sin 75°$

15. $\cos 165° - \cos 105°$

16. $\cos 105° - \cos 15°$

17. $\sin \dfrac{\pi}{12} + \sin \dfrac{5\pi}{12}$

18. $\sin \dfrac{5\pi}{12} - \sin \dfrac{13\pi}{12}$

19. $\cos \dfrac{7\pi}{12} + \cos \dfrac{5\pi}{12}$

20. $\sin \dfrac{7\pi}{12} + \sin \dfrac{13\pi}{12}$

In Exercises 21–30, express each quantity as a sum or a difference.

21. $\sin 40° \sin 30°$

22. $\sin 75° \cos 70°$

23. $\cos \dfrac{5\pi}{7} \cos \dfrac{\pi}{7}$

24. $\sin \dfrac{4\pi}{5} \cos \dfrac{3\pi}{5}$

25. $\cos 5\theta \cos 3\theta$

26. $\sin 3\theta \sin 2\theta$

27. $\sin A \cos \dfrac{A}{2}$

28. $\cos 3B \sin B$

29. $\sin \dfrac{x}{2} \sin \dfrac{x}{3}$

30. $\cos \dfrac{x}{3} \cos x$

In Exercises 31–40, express each quantity as a product.

31. $\sin 50° + \sin 30°$

32. $\sin 75° - \sin 70°$

33. $\cos \dfrac{7\pi}{9} - \cos \dfrac{5\pi}{9}$

34. $\cos \dfrac{3\pi}{5} + \cos \dfrac{\pi}{5}$

35. $\cos 5\theta + \cos 3\theta$ **36.** $\cos 7\theta - \cos \theta$ **37.** $\sin A - \sin \dfrac{A}{2}$ **38.** $\sin 5B + \sin B$

39. $\cos \dfrac{x}{2} + \cos \dfrac{x}{3}$ **40.** $\cos x - \cos \dfrac{x}{2}$

In Exercises 41–50, verify each identity.

41. $\dfrac{\sin A + \sin B}{\sin A - \sin B} = \tan\dfrac{1}{2}(A + B)\cot\dfrac{1}{2}(A - B)$ **42.** $\dfrac{\sin A + \sin B}{\cos A + \cos B} = \tan\dfrac{1}{2}(A + B)$

43. $\dfrac{\sin A + \sin B}{\cos A - \cos B} = -\cot\dfrac{1}{2}(A - B)$ **44.** $\dfrac{\cos A + \cos B}{\sin A - \sin B} = \cot\dfrac{1}{2}(A - B)$

45. $\dfrac{\cos A + \cos 5A}{\cos A - \cos 5A} = \cot 3A \cot 2A$ **46.** $\sin^2 A - \sin^2 B = \sin(A + B)\sin(A - B)$

47. $\cos^2 A - \cos^2 B = \sin(B + A)\sin(B - A)$ **48.** $\cos 2A(1 + 2\cos A) = \cos A + \cos 2A + \cos 3A$

49. $2\cos 5A \sin 2A = \sin 7A - \sin 3A$ **50.** $\cot 7A \cot 5A = \dfrac{\cos 12A + \cos 2A}{\cos 2A - \cos 12A}$

In Exercises 51–54, assume that $A + B + C = 180°$. Verify each identity.

51. $\sin A + \sin B + \sin C = 4\cos\dfrac{A}{2}\cos\dfrac{B}{2}\cos\dfrac{C}{2}$ **52.** $\cos A + \cos B - \cos C = 4\cos\dfrac{A}{2}\cos\dfrac{B}{2}\sin\dfrac{C}{2} - 1$

53. $\tan A + \tan B + \tan C = \tan A \tan B \tan C$ **54.** $\cos A + \cos B + \cos C = 1 + 4\sin\dfrac{A}{2}\sin\dfrac{B}{2}\sin\dfrac{C}{2}$

4.6 SUMS OF THE FORM $A \sin X + B \cos X$

The graph of $y = A \sin x + B \cos x$ is a sine curve of the form $y = k \sin(x + \phi)$. The amplitude k and the phase shift angle ϕ are determined by the values of A and B. The following discussion will establish this result and provide the proper values for k and ϕ.

From the terms of the expression $A \sin x + B \cos x$, we factor out a common factor of $\sqrt{A^2 + B^2}$. This is not difficult if we realize that the product

$$\sqrt{A^2 + B^2} \cdot \frac{A}{\sqrt{A^2 + B^2}}$$

is just A, and that the product

$$\sqrt{A^2 + B^2} \cdot \frac{B}{\sqrt{A^2 + B^2}}$$

is just B. Thus, we have

1. $A \sin x + B \cos x = \sqrt{A^2 + B^2}\left(\dfrac{A}{\sqrt{A^2 + B^2}}\sin x + \dfrac{B}{\sqrt{A^2 + B^2}}\cos x\right)$

Because the sum of the squares of the coefficients

$$\frac{A}{\sqrt{A^2 + B^2}} \quad \text{and} \quad \frac{B}{\sqrt{A^2 + B^2}}$$

is 1, one of these coefficients is sin ϕ and one is cos ϕ, for some angle ϕ. If ϕ is an angle such that

$$\sin \phi = \frac{B}{\sqrt{A^2 + B^2}} \quad \text{and} \quad \cos \phi = \frac{A}{\sqrt{A^2 + B^2}}$$

then Equation 1 becomes

$$A \sin x + B \cos x = \sqrt{A^2 + B^2} \,(\cos \phi \sin x + \sin \phi \cos x)$$
$$= \sqrt{A^2 + B^2} \sin(x + \phi)$$

In summary, we have the following theorem.

Theorem. $A \sin x + B \cos x = k \sin(x + \phi)$, where $k = \sqrt{A^2 + B^2}$ and ϕ is any angle for which

$$\sin \phi = \frac{B}{\sqrt{A^2 + B^2}} \quad \text{and} \quad \cos \phi = \frac{A}{\sqrt{A^2 + B^2}}$$

Example 1   Express $3 \sin x + 4 \cos x$ in the form $k \sin(x + \phi)$.

Solution   Begin by evaluating k.

$$k = \sqrt{A^2 + B^2} = \sqrt{3^2 + 4^2} = \sqrt{25} = 5$$

Here ϕ is an angle such that $\sin \phi = \frac{4}{5} = 0.8$ and $\cos \phi = \frac{3}{5} = 0.6$. Use a calculator to determine that $\phi \approx 53.1°$. Thus, you have

$$3 \sin x + 4 \cos x \approx 5 \sin(x + 53.1°)$$

■

Example 2   Express $5 \sin 3x - 12 \cos 3x$ as a single expression involving the sine function only.

Solution   The expression $5 \sin 3x - 12 \cos 3x$ can be written as $k \sin(3x + \phi)$, where

$$k = \sqrt{5^2 + (-12)^2} = \sqrt{25 + 144} = \sqrt{169} = 13$$

The angle ϕ is such that $\sin \phi = -\frac{12}{13} \approx -0.9231$ and $\cos \phi = \frac{5}{13} \approx 0.3846$. Because $\sin \phi$ is negative and $\cos \phi$ is positive, ϕ must lie in the fourth quadrant; $\phi \approx -67.4°$. Hence, you have

$$5 \sin 3x - 12 \cos 3x \approx 13 \sin(3x - 67.4°)$$

■

Example 3   Express $3 \sin \theta - \sqrt{3} \cos \theta$ in the form $k \cos(\theta + \phi)$.

Solution   First, express the quantity in the form $k \sin(\theta + \phi)$.

$$k = \sqrt{3^2 + (-\sqrt{3})^2} = \sqrt{9 + 3} = \sqrt{12} = 2\sqrt{3}$$

$$\sin \phi = \frac{-\sqrt{3}}{2\sqrt{3}} = -\frac{1}{2}$$

$$\cos \phi = \frac{3}{2\sqrt{3}} = \frac{\sqrt{3}}{2}$$

Because $\sin \phi$ is negative and $\cos \phi$ is positive, ϕ is a fourth-quadrant angle. A fourth-quadrant angle with a sine of $-\frac{1}{2}$ is $-30°$. Hence, you have

$$3 \sin \theta - \sqrt{3} \cos \theta = 2\sqrt{3} \sin(\theta - 30°)$$

Because $\sin x = \cos(90° - x)$, it follows that

$$3 \sin \theta - \sqrt{3} \cos \theta = 2\sqrt{3} \cos[90° - (\theta - 30°)]$$
$$= 2\sqrt{3} \cos(120° - \theta)$$

Because $\cos(120° - \theta) = \cos[-(120° - \theta)] = \cos(\theta - 120°)$, this can be written as follows:

$$3 \sin \theta - \sqrt{3} \cos \theta = 2\sqrt{3} \cos(\theta - 120°)$$ ■

Example 4   Graph the function $y = \cos x - \sin x$.

Solution   Note that $\cos x - \sin x$ or $-\sin x + \cos x$ can be changed to $k \sin(x + \phi)$ with

$$k = \sqrt{(-1)^2 + 1^2} = \sqrt{2}, \quad \sin \phi = \frac{1}{\sqrt{2}} = \frac{\sqrt{2}}{2}, \quad \text{and} \quad \cos \phi = \frac{-1}{\sqrt{2}} = -\frac{\sqrt{2}}{2}$$

Because $\sin \phi$ is positive and $\cos \phi$ is negative, angle ϕ is a second-quadrant angle. The second-quadrant angle with a sine of $\frac{\sqrt{2}}{2}$ is $\frac{3\pi}{4}$ radians. Hence, $y = \cos x - \sin x$ is equivalent to

$$y = \sqrt{2} \sin\left(x + \frac{3\pi}{4}\right)$$

The graph of this function (see Figure 4-5) is a simple sine curve with amplitude of $\sqrt{2}$ and a phase shift of $\frac{3\pi}{4}$ to the left. Note that the function $y = \cos x - \sin x$ can be graphed using the method of addition of ordinates, which was discussed in Section 3.6.

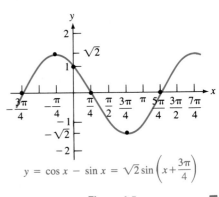

$$y = \cos x - \sin x = \sqrt{2} \sin\left(x + \frac{3\pi}{4}\right)$$

Figure 4-5   ■

■ EXERCISE 4.6

In Exercises 1–12, write each expression in the form $k \sin(x + \phi)$.

1. $6 \sin x + 8 \cos x$ **2.** $12 \sin x + 5 \cos x$ **3.** $6 \sin x - 8 \cos x$ **4.** $12 \sin x - 5 \cos x$

5. $2 \sin x + \cos x$ **6.** $\sin x - \cos x$ **7.** $\sin x + \cos x$ **8.** $2 \sin x - \cos x$

9. $-\sin x + 5 \cos x$ **10.** $\sqrt{3} \sin x + 3 \cos x$ **11.** $\sqrt{3} \sin x - 3 \cos x$ **12.** $-\sin x - 5 \cos x$

13. Verify that the sum of the squares of the coefficients of $\sin x$ and $\cos x$ in Equation 1 is 1.

14. If, in the development of this section, the assignments of $\sin \phi$ and $\cos \phi$ had been interchanged, what would the results have been?

In Exercises 15–20, use the method of Example 4 to graph each function.

15. $y = \sin x + \cos x$ **16.** $y = \sin x - \cos x$

17. $y = \sin x - \sqrt{3} \cos x$ **18.** $y = \sqrt{3} \sin x + \cos x$

19. $y = \sin 2x + \sqrt{3} \cos 2x$ **20.** $y = \sin 2x - \sqrt{3} \cos 2x$

4.7 APPLICATIONS OF THE TRIGONOMETRIC IDENTITIES

The results of this chapter have made it clear that the circular functions are interrelated in many ways. So far, the work has been theoretical, proving given identities and developing new ones. We now turn our attention to three applications that use trigonometric identities.

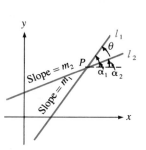

Figure 4-6

1. Finding the Angle Between Two Lines

If lines l_1 and l_2, with angles of inclination α_1 and α_2 and slopes m_1 and m_2, intersect at point P (see Figure 4-6), then the angle θ from l_2 to l_1 is the difference of the angles of inclination $\alpha_1 - \alpha_2$.

$$\theta = \alpha_1 - \alpha_2$$

This angle, θ, can be found by first finding $\tan \theta$. By an identity of Section 4.2, it follows that

$$\tan \theta = \tan(\alpha_1 - \alpha_2)$$

$$= \frac{\tan \alpha_1 - \tan \alpha_2}{1 + \tan \alpha_1 \tan \alpha_2}$$

Recall that the tangent of the angle of inclination of a line is its rise divided by its run—that is, the slope of that line. Hence, we have

$$\tan \alpha_1 = m_1$$

and

$$\tan \alpha_2 = m_2$$

and

$$\tan \theta = \frac{m_1 - m_2}{1 + m_1 m_2} \qquad \text{Remember that } \theta \text{ is the angle from } l_2 \text{ to } l_1.$$

Example 1   Find the angle from the line $y = 3x + 1$ to the line $y = -2x + 3$.

Solution   Let l_1 be the line $y = -2x + 3$ and l_2 be the line $y = 3x + 1$. Because the slope of $y = mx + b$ is m, the slope of $l_1 = -2$ and the slope of $l_2 = 3$. See Figure 4-7. Because $m_1 = -2$ and $m_2 = 3$, you have

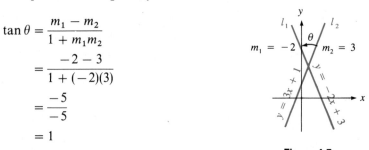

$$\tan \theta = \frac{m_1 - m_2}{1 + m_1 m_2}$$

$$= \frac{-2 - 3}{1 + (-2)(3)}$$

$$= \frac{-5}{-5}$$

$$= 1$$

Figure 4-7

The only angle less than $180°$ whose tangent is 1 is $45°$. Hence, the angle from the line $y = 3x + 1$ to the line $y = -2x + 3$ is $45°$. ∎

2. Simple Harmonic Motion

If a weight, hanging by a spring from the ceiling, is pushed upward a small distance and released, it starts to bounce. Its position y at any time t can be found by the formula

$$y = A \cos (2\pi f t)$$

This formula was discussed in Section 3.8.

If, as the weight is released, it is given a push upward, it will still continue to bounce. However, its position depends on time in a more complicated way.

$$y = A \sin (2\pi f t) + B \cos (2\pi f t)$$

where A, B, and f are dependent on the characteristics of the spring, the upward displacement, and the initial push.

Example 2   A weight is attached to a certain spring and is pushed to start it bouncing. Suppose the position y depends on t according to the formula

$$y = 3 \sin 6t + 4 \cos 6t$$

What is the amplitude of the oscillation?

Solution   By the identity of Section 4.6, it follows that

$$A \sin x + B \cos x = \sqrt{A^2 + B^2} \sin(x + \phi)$$
$$y = 3 \sin 6t + 4 \cos 6t = \sqrt{3^2 + 4^2} \sin(6t + \phi)$$
$$y = 5 \sin(6t + \phi)$$

The oscillation may still be described by the sine function. Because the amplitude is 5, the weight is at most 5 units above (or below) the equilibrium position. ∎

3. Beat Frequencies

It is difficult to recognize that two tones are distinct if they are very close in pitch and sounded in succession. If they are sounded together, however, the two tones beat against each other. The combined sound has a noticeable warble, slow or rapid, depending on how far apart the initial frequencies are. A piano tuner listens for these lower beat frequencies and tries to eliminate them by tightening or loosening the individual piano strings.

If two equally loud tones of frequencies f_1 and f_2 are sounded together, the combined amplitude A is

$$A = k\sin(2\pi f_1 t) + k\sin(2\pi f_2 t)$$
$$= k[\sin(2\pi f_1 t) + \sin(2\pi f_2 t)]$$

where k is the amplitude of each of the individual tones. By the identity

$$\sin x + \sin y = 2\cos\frac{x-y}{2}\sin\frac{x+y}{2}$$

the combined amplitude can be written in a different form:

$$A = k(\sin 2\pi f_1 t + \sin 2\pi f_2 t)$$
$$= k\left(2\cos\frac{2\pi f_1 t - 2\pi f_2 t}{2}\sin\frac{2\pi f_1 t + 2\pi f_2 t}{2}\right)$$
$$= 2k\cos\left[2\pi\left(\frac{f_1 - f_2}{2}\right)t\right]\sin\left[2\pi\left(\frac{f_1 + f_2}{2}\right)t\right]$$

The factor

$$\sin 2\pi\left(\frac{f_1 + f_2}{2}\right)t$$

represents a sound of frequency $(f_1 + f_2)/2$, which is the average of the frequencies of the two original tones. The other factor

$$\cos 2\pi\left(\frac{f_1 - f_2}{2}\right)t$$

represents a time-varying amplitude, changing at the frequency $(f_1 - f_2)/2$, or half of the difference of the original frequencies. It is this factor that describes the slow warble of volume of the combined sound, because its frequency $(f_1 - f_2)/2$ is quite low when f_1 and f_2 are close to each other. The graph of the sound formed by two tones that are close in pitch appears in Figure 4-8.

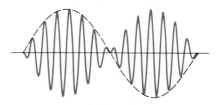

Figure 4-8

Example 3   If the C below middle C on a piano is depressed, a felt-covered hammer strikes two strings. If they are not carefully tuned, the tones beat with each other and the note sounds sour. A piano tuner counts 20 pulsations of sound in a 10-second time interval. How far apart are the frequencies of the two piano strings?

Solution   Figure 4-8 represents one cycle of the time-varying amplitude but two pulsations of sound. Twenty pulsations in 10 seconds is two pulsations per second, or a beat frequency of 1 cycle per second.

$$\frac{f_1 - f_2}{2} = 1$$

$$f_1 - f_2 = 2$$

The two strings are tuned at 2 cycles per second apart.   ■

EXERCISE 4.7

1. Find the angle from the line $y = 2x - 5$ to the line $y = 3x + 1$.

2. Find the angle from $y = 3x - 8$ to $y = 5x$.

3. Show that if lines l_1 and l_2 are perpendicular and each has a nonzero slope, then

$$m_1 = -\frac{1}{m_2}$$

4. If in Example 1 in the text, l_1 had been $y = 3x + 1$ and l_2 had been $y = -2x + 3$, what would have been the angle between the lines?

5. What is the amplitude of the oscillation described by $y = 5 \sin 3t + 12 \cos 3t$?

6. What is the amplitude of the oscillation described by $y = 6 \sin \pi t - 8 \cos \pi t$?

7. A piano tuner counts 50 pulsations of sound in 20 seconds. How far apart are the frequencies of two piano strings that the tuner has struck?

8. Two tuning forks, cut for frequencies of 2000 and 2003 hertz (cycles per second), respectively, are struck and simultaneously touched to a sounding board. How many pulsations of the resulting sound will be heard each second?

CHAPTER SUMMARY

Key Words

beat frequency   (4.7)   identity   (4.1)
conditional equation   (4.1)   impossible equation   (4.1)

Key Ideas

(4.1)   Identities are verified by manipulating one side of an identity algebraically until it is transformed so that it is identical to the other side. An alternative method is to work on both sides of an identity independently until each side is transformed into some common third expression.

(4.2)   **Identities involving sums and differences of two angles:**

$$\cos(A + B) = \cos A \cos B - \sin A \sin B$$
$$\cos(A - B) = \cos A \cos B + \sin A \sin B$$
$$\sin \theta = \cos(90° - \theta)$$
$$\cos \theta = \sin(90° - \theta)$$
$$\sin(A + B) = \sin A \cos B + \cos A \sin B$$
$$\sin(A - B) = \sin A \cos B - \cos A \sin B$$
$$\tan(A + B) = \frac{\tan A + \tan B}{1 - \tan A \tan B}$$
$$\tan(A - B) = \frac{\tan A - \tan B}{1 + \tan A \tan B}$$

(4.3)   **The double-angle identities:**

$$\sin 2A = 2 \sin A \cos A$$
$$\cos 2A = \cos^2 A - \sin^2 A$$
$$\cos 2A = 2 \cos^2 A - 1$$
$$\cos 2A = 1 - 2 \sin^2 A$$
$$\tan 2A = \frac{2 \tan A}{1 - \tan^2 A}$$

(4.4)   **The half-angle identities:**

$$\cos \frac{A}{2} = \pm \sqrt{\frac{1 + \cos A}{2}}$$
$$\sin \frac{A}{2} = \pm \sqrt{\frac{1 - \cos A}{2}}$$
$$\tan \frac{A}{2} = \frac{\sin A}{1 + \cos A}$$
$$\tan \frac{A}{2} = \frac{1 - \cos A}{\sin A}$$

(4.5)   **Product-to-sum and sum-to-product identities:**

$$\sin A \cos B = \frac{1}{2}[\sin(A + B) + \sin(A - B)]$$

$$\cos A \sin B = \frac{1}{2}[\sin(A + B) - \sin(A - B)]$$

$$\sin A \sin B = \frac{1}{2}[\cos(A - B) - \cos(A + B)]$$

$$\cos A \cos B = \frac{1}{2}[\cos(A + B) + \cos(A - B)]$$

$$\sin A + \sin B = 2 \sin \frac{A + B}{2} \cos \frac{A - B}{2}$$

$$\sin A - \sin B = 2 \cos \frac{A + B}{2} \sin \frac{A - B}{2}$$

$$\cos A + \cos B = 2 \cos \frac{A + B}{2} \cos \frac{A - B}{2}$$

$$\cos A - \cos B = -2 \sin \frac{A + B}{2} \sin \frac{A - B}{2}$$

(4.6)   **Sums of the form $A \sin x + B \cos x$:**

$A \sin x + B \cos x = k \sin(x + \phi)$, where $k = \sqrt{A^2 + B^2}$ and ϕ is any angle for which

$$\sin \phi = \frac{B}{\sqrt{A^2 + B^2}} \quad \text{and} \quad \cos \phi = \frac{A}{\sqrt{A^2 + B^2}}$$

REVIEW EXERCISES

In Review Exercises 1–10, find each value by using trigonometric functions of $300°$ *and* $45°$.

1. $\sin 345°$ 2. $\cos 345°$ 3. $\tan 345°$ 4. $\sin 255°$

5. $\cos 255°$ 6. $\tan 255°$ 7. $\sin 600°$ 8. $\cos 600°$

9. $\tan 600°$ 10. $\sin 22.5°$

11. Evaluate $\cos 22.5°$.   12. Evaluate $\tan 22.5°$.

In Review Exercises 13–24, express each quantity as a single function of one angle, and simplify if possible.

13. $\sin 20° \cos 51° + \cos 20° \sin 51°$

14. $\dfrac{\tan 20° - \tan 51°}{1 + \tan 20° \tan 51°}$

15. $\cos 35° \cos 15° + \sin 35° \sin 15°$

16. $\cos \dfrac{3\pi}{11} \cos \dfrac{\pi}{11} - \sin \dfrac{3\pi}{11} \sin \dfrac{\pi}{11}$

17. $\sin \dfrac{2\pi}{7} \cos \dfrac{\pi}{7} - \cos \dfrac{2\pi}{7} \sin \dfrac{\pi}{7}$

18. $2 \cos 17° \sin 17°$

19. $2 \cos^2 17° - 1$

20. $\dfrac{2 \tan 217°}{1 - \tan^2 217°}$

21. $4 \sin \dfrac{3\pi}{8} \cos \dfrac{3\pi}{8}$

22. $1 - 2 \sin^2 \dfrac{\pi}{3}$

23. $\cos^4 \dfrac{\pi}{13} - \sin^4 \dfrac{\pi}{13}$

24. $8 \sin^2 \dfrac{\theta}{2} \cos^2 \dfrac{\theta}{2}$

In Review Exercises 25–36, write the right-hand side of the indicated formula.

25. $\sin(\theta + \alpha) =$ 26. $\cos(\theta + \alpha) =$ 27. $\tan(\theta + \alpha) =$ 28. $\sin(\theta - \alpha) =$

29. $\cos(\theta - \alpha) =$ 30. $\tan(\theta - \alpha) =$ 31. $\sin 2\theta =$ 32. $\cos 2\theta =$

33. $\tan 2\theta =$ 34. $\sin \dfrac{\theta}{2} =$ 35. $\cos \dfrac{\theta}{2} =$ 36. $\tan \dfrac{\theta}{2} =$

37. Show that $\cos(60° + \theta) = \dfrac{1}{2}(\cos \theta - \sqrt{3} \sin \theta)$.

38. Show that $\sin\left(\dfrac{3\pi}{2} - \theta\right) = -\cos \theta$.

39. Show that $\tan(180° - \theta) = -\tan \theta$.

40. Show that $\sin(120° + \theta) = \dfrac{1}{2}(\sqrt{3} \cos \theta - \sin \theta)$.

41. Show that $\cos(300° - \theta) = \dfrac{1}{2}(\cos \theta - \sqrt{3} \sin \theta)$.

42. Show that $\tan\left(\dfrac{\pi}{4} + \theta\right) = \dfrac{1 + \tan \theta}{1 - \tan \theta}$.

43. Show that $\sin x = \dfrac{\sin 2x}{2 \cos x}$.

44. Show that $\cos \theta = \pm\sqrt{\sin^2\theta + \cos 2\theta}$.

In Review Exercises 45–50, use the given information to find the sine, cosine, and tangent of 2θ.

45. $\cos \theta = -\dfrac{12}{13}$; θ in QII

46. $\sin \theta = -\dfrac{5}{13}$; θ in QIII

47. $\tan \theta = -\dfrac{20}{21}$; θ in QIV

48. $\cos \theta = \dfrac{21}{29}$; θ in QI

49. $\sin \theta = -\dfrac{4}{5}$; θ in QIII

50. $\cot \theta = -\dfrac{3}{4}$; θ in QII

In Review Exercises 51–56, use the given information to find the sine, cosine, and tangent of $\frac{\theta}{2}$. Assume that $0° \le \theta < 360°$.

51. $\cos \theta = \dfrac{12}{13}$; θ in QI

52. $\sin \theta = -\dfrac{5}{13}$; θ in QIV

53. $\tan \theta = -\dfrac{5}{12}$; θ in QIV

54. $\cos \theta = -\dfrac{3}{5}$; θ in QII

55. $\sin \theta = \dfrac{4}{5}$; θ in QII

56. $\cot \theta = \dfrac{3}{4}$; θ in QIII

*In Review Exercises 57–60, evaluate each of the products. **Do not use a calculator or tables.***

57. $\sin 285° \cos 15°$

58. $\cos 285° \sin 15°$

59. $\sin \dfrac{5\pi}{4} \sin \dfrac{\pi}{12}$

60. $\cos \dfrac{\pi}{12} \cos \dfrac{5\pi}{4}$

In Review Exercises 61–64, express each quantity as a product.

61. $\sin 5° + \sin 7°$

62. $\sin 312° - \sin 140°$

63. $\cos \dfrac{3\pi}{5} - \cos \dfrac{\pi}{5}$

64. $\cos \dfrac{2\pi}{7} + \cos \dfrac{3\pi}{7}$

In Review Exercises 65–68, express each quantity as a product and find its value without using a calculator or tables.

65. $\sin 285° + \sin 15°$

66. $\sin 15° - \sin 285°$

67. $\cos \dfrac{5\pi}{12} - \cos \dfrac{\pi}{12}$

68. $\cos \dfrac{\pi}{12} + \cos \dfrac{5\pi}{12}$

In Review Exercises 69–72, verify each identity.

69. $\cot^2 x \sec x = \dfrac{\csc x}{\tan x}$

70. $\cos x(\sec x + \cos x) = 2 - \sin^2 x$

71. $\sec^2 x(\csc x - \sin x) = \csc x$

72. $\dfrac{1}{\sec \theta - \tan \theta} + \dfrac{1}{\sec \theta + \tan \theta} = 2 \sec \theta$

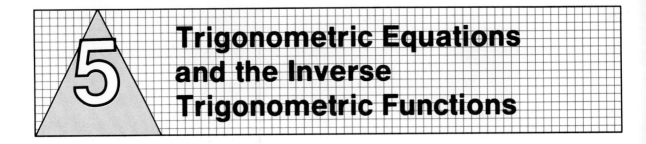

Trigonometric Equations and the Inverse Trigonometric Functions

We have discussed equations that are identities. We now consider trigonometric equations that are true for some, but not all, values of the variable.

5.1 TRIGONOMETRIC EQUATIONS

The equation $\sin 2x = 2 \sin x$ is not an identity because it is false for some values of x. It is true, however, for other values of x. For example, the equation is true if $x = 0$, as the following check shows:

$$\sin 2x = 2 \sin x$$
$$\sin 2(0) \overset{?}{=} 2 \sin 0$$
$$\sin 0 \overset{?}{=} 2 \cdot 0$$
$$0 = 0$$

Other values of x, such as π, 2π, and 3π, also satisfy the equation. The process of finding all such values that satisfy the equation is called **solving the equation**. Solving a trigonometric equation is similar to solving an equation in algebra. We combine terms, add terms to both sides of the equation, factor, and so on. However, we must also use the trigonometric formulas and identities discussed in the preceding chapters. We begin by providing the solution of the equation $\sin 2x = 2 \sin x$.

Example 1 Solve $\sin 2x = 2 \sin x$ for all values of x, where x is a real number.

Solution Use the identity for $\sin 2x$ to write the left-hand side of the equation as $2 \sin x \cos x$.

$$\sin 2x = 2 \sin x$$
1. $$2 \sin x \cos x = 2 \sin x$$

Do not divide both sides of Equation 1 by $\sin x$ because a solution might be lost if you divided both sides of the equation by an expression that could be zero. Instead, divide both sides of the equation by 2, subtract $\sin x$ from both sides, and proceed as follows:

$$\sin x \cos x = \sin x \qquad \text{Divide both sides of Equation 1 by 2.}$$

$$\sin x \cos x - \sin x = 0 \qquad \text{Subtract } \sin x \text{ from both sides.}$$

$$\sin x(\cos x - 1) = 0 \qquad \text{Factor out } \sin x.$$

$$\sin x = 0 \quad \text{or} \quad \cos x - 1 = 0 \qquad \text{Set each factor equal to 0.}$$

$$\cos x = 1$$

The solutions of the equation $\sin x = 0$ are

$$0, \ \pm\pi, \ \pm 2\pi, \ \pm 3\pi, \ldots$$

Each solution of $\sin x = 0$ has the form $x = n\pi$, for some integer n.
The solutions of the equation $\cos x = 1$ are

$$0, \ \pm 2\pi, \ \pm 4\pi, \ \pm 6\pi, \ldots$$

Each solution of $\cos x = 1$ has the form $x = 2n\pi$, for some integer n. Thus, the solutions of the equation $\sin 2x = 2\sin x$ are

$$n\pi \qquad \text{or} \qquad 2n\pi$$

or just

$$n\pi$$

for every integer n.   ∎

Example 2   Solve $\sin x \tan x = \sin x$ for all values of x, where x is a real number.

Solution   Subtract $\sin x$ from both sides of the equation, factor out $\sin x$, set each factor equal to 0, and solve for x.

$$\sin x \tan x = \sin x$$

$$\sin x \tan x - \sin x = 0$$

$$\sin x(\tan x - 1) = 0$$

$$\sin x = 0 \quad \text{or} \quad \tan x - 1 = 0$$

$$\tan x = 1$$

The solutions of the equation $\sin x = 0$ are

$$0, \ \pm\pi, \ \pm 2\pi, \ \pm 3\pi, \ldots$$

Each solution of $\sin x = 0$ has the form

$$x = n\pi \qquad \text{for some integer } n$$

The only solution of the equation $\tan x = 1$ in the interval from 0 to $\frac{\pi}{2}$ is $\frac{\pi}{4}$. Because the period of the tangent function is π, all solutions can be obtained by adding multiples of π to $\frac{\pi}{4}$. Thus, each solution of $\tan x = 1$ has the form

$$x = \frac{\pi}{4} + n\pi \qquad \text{for some integer } n$$

The solutions of $\sin x = 0$ and $\tan x = 1$ are the only solutions of the equation $\sin x \tan x = \sin x$. Thus, its solutions are

$$n\pi \qquad \text{or} \qquad \frac{\pi}{4} + n\pi$$

for every integer n. ∎

Example 3 Solve $2\cos^2\theta + \cos\theta = 1$ for all values of θ, where $0° \le \theta < 360°$. Then give the general solutions to the equation.

Solution Subtract 1 from both sides of the equation, factor the left-hand side, set each factor equal to zero, and solve each equation for θ.

$$2\cos^2\theta + \cos\theta = 1$$
$$2\cos^2\theta + \cos\theta - 1 = 0$$
$$(2\cos\theta - 1)(\cos\theta + 1) = 0$$

$$2\cos\theta - 1 = 0 \qquad \text{or} \qquad \cos\theta + 1 = 0$$
$$2\cos\theta = 1 \qquad\qquad\qquad \cos\theta = -1$$
$$\cos\theta = \frac{1}{2}$$

The angles θ between $0°$ and $360°$ that have a cosine of $\frac{1}{2}$ or a cosine of -1 are

$$\theta = 60°, 300°, \text{ and } \theta = 180°$$

Thus, the three solutions of the given equation that lie between $0°$ and $360°$ are

$$\theta = 60°, 300°, \text{ and } 180°$$

Verify that each of these values satisfies the given equation.

Because values of trigonometric functions of coterminal angles are equal, the general solutions for this equation are angles of the form

$$\theta = 60° + n360°$$
$$\theta = 300° + n360°$$
$$\theta = 180° + n360°$$

for every integer n. ∎

Example 4 Solve $\sin 3x = \frac{1}{2}$ for all x, where $0 \le x < 2\pi$.

Solution 1 The only solutions of $\sin 3x = \frac{1}{2}$ where $3x$ is in the interval from 0 to 2π are

$$3x = \frac{\pi}{6} \qquad \text{or} \qquad 3x = \frac{5\pi}{6}$$

Because the period of the sine function is 2π, all solutions can be obtained by adding multiples of 2π to these basic solutions of $\frac{\pi}{6}$ and $\frac{5\pi}{6}$. Thus, the general solutions of $\sin 3x = \frac{1}{2}$ are

$$3x = \frac{\pi}{6} + 2\pi n \qquad \text{and} \qquad 3x = \frac{5\pi}{6} + 2\pi n$$

where n is every integer. Divide both sides of each of the previous equations by 3 to obtain

$$x = \frac{\pi}{18} + \frac{2\pi n}{3} \qquad \text{and} \qquad x = \frac{5\pi}{18} + \frac{2\pi n}{3}$$

Now write each fraction with a denominator of 18 to obtain

$$x = \frac{\pi}{18} + \frac{12\pi n}{18} \qquad \text{and} \qquad x = \frac{5\pi}{18} + \frac{12\pi n}{18}$$

To find the solutions x such that $0 \leq x < 2\pi$, substitute 0, 1, and 2 for n in each of the preceding equations to obtain

$$x = \frac{\pi}{18}, \frac{13\pi}{18}, \frac{25\pi}{18} \qquad \text{and} \qquad x = \frac{5\pi}{18}, \frac{17\pi}{18}, \frac{29\pi}{18}$$

Verify that if you substitute 3 for n, the resulting values of x are greater than 2π.

Solution 2 Because it is given that $0 \leq x < 2\pi$, you must find all values of $3x$ such that $\sin 3x = \frac{1}{2}$ and

$$3(0) \leq 3x < 3(2\pi)$$

or

$$0 \leq 3x < 6\pi$$

There are six such values. They are

$$3x = \frac{\pi}{6}, \frac{5\pi}{6}, \frac{13\pi}{6}, \frac{17\pi}{6}, \frac{25\pi}{6}, \frac{29\pi}{6}$$

To find x, divide both sides of the preceding equation by 3 to obtain

$$x = \frac{\pi}{18}, \frac{5\pi}{18}, \frac{13\pi}{18}, \frac{17\pi}{18}, \frac{25\pi}{18}, \frac{29\pi}{18}$$

Verify that all six roots satisfy the given equation. ■

Example 5 Solve $\sin \theta = \cos \theta$ for all θ between $0°$ and $360°$, including $0°$.

Solution If $\cos \theta \neq 0$, you can divide both sides of the equation $\sin \theta = \cos \theta$ by $\cos \theta$.

$$\sin \theta = \cos \theta$$

$$\frac{\sin \theta}{\cos \theta} = 1$$

$$\tan \theta = 1$$

The angles θ between $0°$ and $360°$ with a tangent of 1 are

$$\theta = 45°, 225°$$

Both values satisfy the given equation. Now if $\cos \theta = 0$, then

$$\theta = 90° \qquad \text{or} \qquad \theta = 270°$$

These values do not satisfy the given equation. Thus, the only solutions to the given equation are

$$\theta = 45°, 225°$$

∎

Example 6   Solve the equation $2 \cos^3 \theta = \cos \theta$ for θ, where $0° \le \theta < 360°$.

Solution   Subtract $\cos \theta$ from both sides of the equation, factor out $\cos \theta$, and proceed as follows:

$$2 \cos^3 \theta = \cos \theta$$
$$2 \cos^3 \theta - \cos \theta = 0$$
$$\cos \theta (2 \cos^2 \theta - 1) = 0$$

$$\cos \theta = 0 \qquad \text{or} \qquad 2 \cos^2 \theta - 1 = 0$$

$$\cos^2 \theta = \frac{1}{2}$$

$$\cos \theta = \pm \frac{1}{\sqrt{2}} = \pm \frac{\sqrt{2}}{2}$$

$$\theta = 90°, 270° \qquad\qquad \theta = 45°, 135°, 225°, 315°$$

Each of these six values satisfies the given equation. Note that if both sides of the equation had been divided by $\cos \theta$, the solutions of $90°$ and $270°$ would have been lost.

∎

Example 7   Solve $4 \sin^2 \dfrac{x}{2} = 1$ for x, where $0 \le x < 2\pi$.

Solution   Proceed as follows:

$$4 \sin^2 \frac{x}{2} = 1$$

$$4 \sin^2 \frac{x}{2} - 1 = 0 \qquad\qquad \text{Subtract 1 from both sides.}$$

$$\left(2 \sin \frac{x}{2} - 1 \right)\left(2 \sin \frac{x}{2} + 1 \right) = 0 \qquad\qquad \text{Factor the difference of two squares.}$$

$$2\sin\frac{x}{2} - 1 = 0 \qquad \text{or} \qquad 2\sin\frac{x}{2} + 1 = 0$$

$$\sin\frac{x}{2} = \frac{1}{2} \qquad\qquad\qquad \sin\frac{x}{2} = -\frac{1}{2}$$

$$\frac{x}{2} = \frac{\pi}{6}, \frac{5\pi}{6} \qquad\qquad \frac{x}{2} = \frac{7\pi}{6}, \frac{11\pi}{6}$$

$$x = \frac{\pi}{3}, \frac{5\pi}{3} \qquad\qquad x = \frac{7\pi}{3}, \frac{11\pi}{3}$$

Since $\frac{7\pi}{3}$ and $\frac{11\pi}{3}$ are greater than 2π, they must be excluded. The only solutions are

$$x = \frac{\pi}{3} \qquad \text{or} \qquad x = \frac{5\pi}{3}$$

Verify that each of these solutions satisfies the given equation. ∎

Example 8 Solve $4\sin^2\frac{x}{2} = 1$ for x $(0 \le x < 2\pi)$ by using a half-angle identity.

Solution From the half-angle identity for $\sin\frac{x}{2}$, it follows that

$$\sin^2\frac{x}{2} = \frac{1 - \cos x}{2}$$

Substitute $\dfrac{1 - \cos x}{2}$ for $\sin^2\dfrac{x}{2}$ in the original equation to obtain

$$4\sin^2\frac{x}{2} = 1$$

$$4\left(\frac{1 - \cos x}{2}\right) = 1$$

$$1 - \cos x = \frac{1}{2}$$

$$\cos x = \frac{1}{2}$$

$$x = \frac{\pi}{3}, \frac{5\pi}{3}$$

Verify that each value satisfies the given equation. ∎

Example 9 Solve $2\sin x \cos x + \cos x - 2\sin x - 1 = 0$, where $0 \le x < 2\pi$.

Solution Use the technique of *factoring by grouping*. The four terms on the left-hand side share no common factors. However, the first two terms share a common factor of $\cos x$, and the last two terms share a common factor of -1. Proceed as follows.

$$2 \sin x \cos x + \cos x - 2 \sin x - 1 = 0$$

$$\cos x(2 \sin x + 1) - 1(2 \sin x + 1) = 0$$

$$(2 \sin x + 1)(\cos x - 1) = 0 \qquad \text{Factor out the common factor of } 2 \sin x + 1.$$

$2 \sin x + 1 = 0$	or	$\cos x - 1 = 0$	Set each factor equal to 0.

$$2 \sin x = -1 \qquad\qquad \cos x = 1$$

$$\sin x = -\frac{1}{2}$$

$$x = \frac{7\pi}{6}, \frac{11\pi}{6} \qquad\qquad x = 0$$

Verify that each value satisfies the given equation. ∎

Example 10   Solve $\sin \theta + \cos \theta = 1$ for θ, where $0° \le \theta < 360°$.

Solution   Use the identity

$$A \sin \theta + B \cos \theta = \sqrt{A^2 + B^2} \, \sin(\theta + \alpha)$$

to write $\sin \theta + \cos \theta$ in the form $k \sin(\theta + \alpha)$.

$$\sin \theta + \cos \theta = 1$$

$$\sqrt{1^2 + 1^2} \, \sin(\theta + \alpha) = 1$$

$$\sqrt{2} \, \sin(\theta + \alpha) = 1$$

Remember that α is an angle for which $\sin \alpha = \cos \alpha = \dfrac{1}{\sqrt{2}} = \dfrac{\sqrt{2}}{2}$. Thus, $\alpha = 45°$

and you have

$$\sqrt{2} \, \sin(\theta + 45°) = 1$$

$$\sin(\theta + 45°) = \frac{1}{\sqrt{2}} = \frac{\sqrt{2}}{2}$$

$$\theta + 45° = 45° \qquad \text{or} \qquad \theta + 45° = 135°$$

$$\theta = 0° \qquad\qquad\qquad \theta = 90°$$

Verify that each value satisfies the given equation. ∎

Example 11   Solve $\sec^2 x \csc^2 x = \sec^2 x + \csc^2 x$ for x.

Solution   Transform the equation so that it involves only sines and cosines. Then find a common denominator for the fractions and add them. Finally, replace $\sin^2 x + \cos^2 x$ with 1.

$$\sec^2 x \csc^2 x = \sec^2 x + \csc^2 x$$

$$\frac{1}{\cos^2 x} \cdot \frac{1}{\sin^2 x} = \frac{1}{\cos^2 x} + \frac{1}{\sin^2 x}$$

$$\frac{1}{\cos^2 x \sin^2 x} = \frac{\sin^2 x + \cos^2 x}{\sin^2 x \cos^2 x}$$

$$\frac{1}{\cos^2 x \sin^2 x} = \frac{1}{\cos^2 x \sin^2 x}$$

Because this equation is true for all values in the domain of x, it is an identity.

∎

Example 12 Solve the equation $\sin \theta - 2 \csc \theta = 2$ for θ ($0° \leq \theta < 360°$).

Solution Transform the equation so that it involves only sines by multiplying both sides of the equation by $\sin \theta$ and simplifying to obtain

$$\sin \theta - 2 \csc \theta = 2$$

$$\mathbf{\sin \theta}(\sin \theta - 2 \csc \theta) = 2 \, \mathbf{\sin \theta}$$

$$\sin^2 \theta - 2 = 2 \sin \theta \qquad \text{Remember } \sin \theta \csc \theta = 1.$$

$$\sin^2 \theta - 2 \sin \theta - 2 = 0$$

This result is a quadratic equation in the variable $\sin \theta$, with $a = 1$, $b = -2$, and $c = -2$. Solve for $\sin \theta$ by using the quadratic formula.

$$\sin \theta = \frac{-b \pm \sqrt{b^2 - 4ac}}{2a}$$

$$\sin \theta = \frac{-(-2) \pm \sqrt{(-2)^2 - 4(1)(-2)}}{2(1)}$$

$$\sin \theta = \frac{2 \pm \sqrt{4 + 8}}{2}$$

$$\sin \theta = \frac{2 \pm \sqrt{12}}{2}$$

$$\sin \theta = 1 \pm \sqrt{3}$$

$$\sin \theta = 1 + \sqrt{3} \qquad \text{or} \qquad \sin \theta = 1 - \sqrt{3}$$
$$\approx 2.732 \qquad\qquad\qquad \approx -0.732$$

Because $\sin \theta$ cannot exceed 1, the equation $\sin \theta = 2.732$ has no solution.

To determine the solution of the equation $\sin \theta = -0.732$, use a calculator set in degree mode. Enter .732, press $\boxed{+/-}$ to change its sign, and press $\boxed{\text{INV}}$ $\boxed{\text{SIN}}$. Because the result $-47.0543°$ is not an angle between $0°$ and $360°$ as required, use its reference angle, $47.0543°$, to find QIII and QIV angles that have a sine of -0.732. The solutions of $\sin \theta = -0.732$ are

In QIII	In QIV
$\theta = 180° + 47.0543°$	$\theta = 360° - 47.0543°$
$= 227.0543°$	$= 312.9457°$

Verify that each value satisfies the equation.

∎

◼ EXERCISE 5.1 ◼

In Exercises 1–10, solve each equation for all values of the variable. Assume that x is a real number. **Do not use a calculator or tables.**

1. $\sin x = \dfrac{1}{2}$

2. $\cos x = -1$

3. $\tan x = \sqrt{3}$

4. $\tan x = -\sqrt{3}$

5. $\cos^2 x = \cos x$

6. $\sin^2 x + \sin x = 0$

7. $\tan^2 x = \tan x$

8. $\tan 2x = 1$

9. $4 \sin^2 x = 1$

10. $4 \cos^2 x = 1$

In Exercises 11–20, solve each equation for all values of the variable. Assume that θ is an angle in degrees. **Do not use a calculator or tables.**

11. $\sin \theta = \dfrac{\sqrt{3}}{2}$

12. $\cos \theta = \dfrac{\sqrt{2}}{2}$

13. $\tan \theta = -1$

14. $\sin \theta = -\dfrac{1}{2}$

15. $\sin 2\theta = 0$

16. $\cos 2\theta = 1$

17. $\sin \theta \tan \theta - \sin \theta = 0$

18. $2 \sin^2 \theta + \sin \theta - 1 = 0$

19. $\cos \theta \tan \theta - \cos \theta = 0$

20. $2 \cos^2 \theta - \cos \theta = 0$

In Exercises 21–70, solve each equation for all values of the variable, if any, between 0° and 360°, including 0°. **Do not use a calculator or tables.**

21. $\cos \dfrac{\theta}{2} = \dfrac{1}{2}$

22. $\sin \dfrac{\theta}{2} = \dfrac{\sqrt{3}}{2}$

23. $\sin \theta \cos \theta = 0$

24. $\sin \theta \cos \theta - \sin \theta = 0$

25. $\cos \theta \sin \theta - \cos \theta = 0$

26. $\cos \theta \sin \theta + \cos \theta = 0$

27. $\sin \theta \cos \theta + \sin \theta = 0$

28. $\cos^2 \theta = 1$

29. $\sin^2 \theta = \dfrac{1}{2}$

30. $2 \sin^2 \theta - \sin \theta - 1 = 0$

31. $2 \cos^2 \theta - \cos \theta - 1 = 0$

32. $\sin^2 \theta - 3 \sin \theta + 2 = 0$

33. $4 \sin^2 \theta + 4 \sin \theta + 1 = 0$

34. $\sec^2 \theta - 1 = 0$

35. $\tan^2 \theta - 3 = 0$

36. $2 \cos^2 \theta + 3 \cos \theta + 1 = 0$

37. $2 \sin^2 \theta + \sin \theta = 6$

38. $\cot^2 \theta = \cot \theta$

39. $\sqrt{3} \sin \theta \tan \theta = \sin \theta$

40. $\sin^2 \theta + \sin \theta = 6$

41. $2 \cos^2 B + \cos B = 1$

42. $\cos^2 A + 4 \cos A + 3 = 0$

43. $\sin 2A + \cos A = 0$

44. $\sin 2\theta - \cos \theta = 0$

45. $\cos B = \sin B$

46. $\cos B = \sqrt{3} \sin B$

47. $\cos \theta = \cos \dfrac{\theta}{2}$

48. $\cos \theta = \sin \dfrac{\theta}{2}$

49. $\cos 2\theta = \cos \theta$

50. $\cos 2\theta = \sin \theta$

51. $\tan A = -\sin A$

52. $\cot A = -\cos A$

53. $\cos^2 C - \sin^2 C = \dfrac{1}{2}$

54. $\cos^2 C + \sin^2 C = \dfrac{1}{2}$

55. $\sin \theta + \cos \theta = \sqrt{2}$
(*Hint:* Square both sides.)

56. $\sin \theta - \cos \theta = \sqrt{2}$

57. $\cos 2A = 1 - \sin A$

58. $\cos 2A = \cos A - 1$

59. $\cos 2B + \cos B + 1 = 0$

60. $\cos^2 B + \sin B - 1 = 0$

61. $\sin 4\theta = \sin 2\theta$

62. $\cos 4\theta = \cos 2\theta$

63. $\cos B = 1 + \sqrt{3} \sin B$

64. $\tan 2B + \sec 2B = 1$

65. $9 \cos^4 A = \sin^4 A$

66. $6 \cos^2 A = -9 \sin A$

67. $\sec C = \tan C + \cos C$

68. $1 - \tan C = \sqrt{2} \sec C$

69. $2 \cos A \sin A - \sqrt{2} \cos A = \sqrt{2} \sin A - 1$

70. $4 \sin B \cos B - 2 \sin B = 2 \cos B - 1$

In Exercises 71–76, solve each equation for θ, where $0° \leq \theta < 360°$. Recall that the expression $a \sin x + b \cos x$ can be written as $k \sin(x + \phi)$.

71. $\dfrac{\sqrt{3}}{2} \sin \theta + \dfrac{1}{2} \cos \theta = \dfrac{1}{2}$

72. $\sqrt{2} \sin \theta + \sqrt{2} \cos \theta = \sqrt{3}$

73. $\dfrac{1}{2} \sin \theta + \dfrac{\sqrt{3}}{2} \cos \theta = 1$

74. $\cos \theta - \sin \theta = \sqrt{2}$

75. $\cos \theta - \sqrt{3} \sin \theta = 1$

76. $\sin \theta + \cos \theta = -\sqrt{2}$

In Exercises 77–86, solve for x, where $0 \leq x < 2\pi$. Consider using a sum-to-product or a product-to-sum identity.

77. $\sin x \cos x = \dfrac{1}{2}$

78. $\cos x \sin x = \dfrac{\sqrt{3}}{4}$

79. $2 \sin \dfrac{3x}{2} \cos \dfrac{x}{2} = \sin x$

80. $2 \cos \dfrac{3x}{2} \cos \dfrac{x}{2} = \cos 2x$

81. $\sin 3x \cos x = \dfrac{1}{2} \sin 2x$

82. $\cos 3x \cos x = \cos^2 2x$

83. $\sin 4x = -\sin 2x$

84. $\cos 4x = -\cos 2x$

85. $\cos 9x = -\cos 3x$

86. $\sin 12x = -\sin 4x$

In Exercises 87–100, solve for x, where $0 \leq x < 2\pi$.

87. $2 \sin^4 x - 9 \sin^2 x + 4 = 0$

88. $2 \sin^3 x + \sin^2 x = \sin x$

89. $4 \cos x \sin^2 x - \cos x = 0$

90. $\tan^2 x = 1 + \sec x$

91. $\csc^4 x = 2 \csc^2 x - 1$

92. $\tan^2 x - 5 \tan x + 6 = 0$

93. $4 \sin^2 x + 4 \sin x + 1 = 0$

94. $2 \sin 5x = 1$

95. $2 \sin^2 5x = 1$

96. $2 \sin^2 2x + \sin 2x - 1 = 0$

97. $\tan^2 x - \tan x = 0$

98. $2 \sin x - \sqrt{2} \tan x = \sqrt{2} \sec x - 2$

99. $\cot^2 x \cos x + \cot^2 x - 3 \cos x - 3 = 0$

100. $\tan x + \cot x = -2$

In Exercises 101–108, solve each equation for all values of the variable in the interval from 0° to 360°. You may need to use a calculator. Give all answers to the nearest tenth of a degree. If no solution exists, so indicate.

101. $5\sin^2\theta = 4$ **102.** $9\cos^2\theta = 7$

103. $\sin^2\theta - 4\sin\theta - 1 = 0$ **104.** $\cos^2\theta - 2\cos\theta - 4 = 0$

105. $\sin\theta + 2\csc\theta = 6$ **106.** $1 + 2\sec\theta = \sec^2\theta$

107. $\sin^2\theta + 6\sin\theta\cos\theta = \cos^2\theta$ **108.** $\sin^2\theta + 2\sin 2\theta = 3\cos^2\theta$

109. Suppose that $\sin\sqrt{2}\,\pi t = 0$. Show that the values of $\cos\sqrt{2}\,\pi t$ are 1 or -1.

110. Suppose that $\cos(t/2) = 0$. Show that the values of $\sin(t/2)$ are 1 or -1.

111. Suppose that $\cos t - \sin t = 0$. Show that the values of $\cos t + \sin t$ are $\sqrt{2}$ and $-\sqrt{2}$.

112. Suppose that $\cos 2t = 1$. Show that the value of $2\sin t \cos t$ is 0.

5.2 INTRODUCTION TO THE INVERSE TRIGONOMETRIC RELATIONS

Solving trigonometric equations often leads to simple equations such as

$$\sin\theta = \frac{1}{2} \quad \text{or} \quad \sin\theta = \frac{\sqrt{3}}{2}$$

We can solve these equations for the variable θ because we know the values of the trigonometric functions of certain angles. For example, if $\sin\theta = \frac{1}{2}$, then θ can be 30° or 150° or any angle coterminal with 30° or 150°.

Solving more general equations such as $\sin\theta = y$ for θ requires a new concept called the **inverse sine of y**, denoted as $\mathbf{\sin^{-1}y}$. We can think of the expression $\sin^{-1}y$ as representing all angles whose sine is y. For example, $\sin^{-1}\frac{1}{2}$ represents all angles with a sine of $\frac{1}{2}$. Thus, $\sin^{-1}\frac{1}{2} = 30°$, 150°, and all angles that are coterminal with 30° and 150°:

$$\sin^{-1}\frac{1}{2} = 30° + n360° \quad \text{or} \quad 150° + n360°$$

where n is any integer.

The -1 in the notation $\sin^{-1}y$ refers to the inverse relation of the sine function. Because the -1 is not an exponent, the expression $\sin^{-1}y$ does not mean $(\sin y)^{-1}$.

$$\sin^{-1}y = \frac{1}{\sin y}$$

Rather,

$$\sin^{-1}y = \theta \quad \text{is equivalent to} \quad y = \sin\theta$$

Example 1 Find all values θ, where $\theta = \sin^{-1}\left(-\dfrac{\sqrt{3}}{2}\right)$.

Solution Translate the equation $\sin^{-1}(-\frac{\sqrt{3}}{2}) = \theta$ into the equivalent equation $\sin\theta = -\frac{\sqrt{3}}{2}$ and solve it as in Section 5.1.

$$\theta = \sin^{-1}\left(-\frac{\sqrt{3}}{2}\right)$$

$$\sin\theta = -\frac{\sqrt{3}}{2}$$

$$\theta = 240° + n360° \qquad \text{or} \qquad 300° + n360°$$

where n is any integer. ∎

The inverse relations for the remaining trigonometric functions are defined similarly. Whenever we see notation indicating the inverse relation of a trigonometric function, we can translate it into a more familiar form by using the following definition:

Definition.		
$\sin^{-1}y = \theta$	is equivalent to	$y = \sin\theta$
$\cos^{-1}y = \theta$	is equivalent to	$y = \cos\theta$
$\tan^{-1}y = \theta$	is equivalent to	$y = \tan\theta$
$\cot^{-1}y = \theta$	is equivalent to	$y = \cot\theta$
$\csc^{-1}y = \theta$	is equivalent to	$y = \csc\theta$
$\sec^{-1}y = \theta$	is equivalent to	$y = \sec\theta$

The inverse sine relation is also called the **arc sine** relation and the notation **arcsin** y is often used instead of $\sin^{-1}y$. Likewise, **arccos** y is equivalent to $\cos^{-1}y$, **arctan** y is equivalent to $\tan^{-1}y$, and so on.

Example 2 Find all values of θ, where $\theta = \arctan\sqrt{3}$.

Solution Translate the equation $\theta = \arctan\sqrt{3}$ into an equivalent equation and solve it for θ.

$$\theta = \arctan\sqrt{3}$$

$$\tan\theta = \sqrt{3}$$

$$\theta = 60° + n360° \qquad \text{or} \qquad 240° + n360°$$

where n is any integer. ∎

Example 3 Find θ if $\theta = \cos^{-1}(-1)$ and θ is in radians.

Solution Translate the equation $\theta = \cos^{-1}(-1)$ into the equation $\cos\theta = -1$. The angles (in radians) whose cosines are -1 are those coterminal with π. Thus,

$$\theta = \pi + 2\pi n$$

where n is any integer. ∎

Example 4   Find θ if $\theta = \sin^{-1}\dfrac{\pi}{2}$.

Solution   Don't confuse $\sin^{-1}\frac{\pi}{2}$ with $\sin\frac{\pi}{2}$; the answer is not 1. Instead, translate the equation $\theta = \sin^{-1}\frac{\pi}{2}$ into the equation $\sin\theta = \frac{\pi}{2}$. Thus, you are looking for those angles whose sines are $\frac{\pi}{2}$. Because $\frac{\pi}{2}$ is greater than 1 and the sine function can never be greater than 1, there are no values of θ possible.   ∎

Example 5   Find x in radians if $x = \arcsin 0.8330$.

Solution   Use your calculator, making sure that it is in the radian mode. Enter .833 and press the ⟨INV⟩ and ⟨SIN⟩ keys in succession. The answer .9845 is displayed. This is a first-quadrant angle. A second-quadrant angle whose sine is .833 is found by subtracting .9845 from π (.9845 is the reference angle in Figure 5-1). The result, 2.1571, is another possible answer. There are infinitely many more, found by adding multiples of 2π to either 0.9845 or 2.1571.

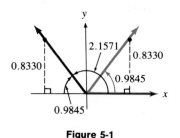

Figure 5-1   ∎

Example 6   If $x = \cos^{-1}\dfrac{1}{2}$, find all real numbers x that lie in the interval $0 \le x < 2\pi$.

Solution   Think of the equation "$x = \cos^{-1}\frac{1}{2}$" as "x is any real number (or angle in radians) whose cosine is $\frac{1}{2}$." Hence, $\cos x = \frac{1}{2}$. The real numbers (or angles in radians) in the interval $0 \le x < 2\pi$ whose cosines are $\frac{1}{2}$ are $\frac{\pi}{3}$ and $\frac{5\pi}{3}$.   ∎

Example 7   If $x = \sec^{-1}(-2)$, find all real numbers in the interval $0 \le x < 2\pi$.

Solution   Think of the equation "$x = \sec^{-1}(-2)$" as "x is any real number (or angle in radians) whose secant is -2." A real number whose secant is -2 is also a number whose cosine is $-\frac{1}{2}$. The numbers in the required interval that satisfy this condition are $\frac{2\pi}{3}$ and $\frac{4\pi}{3}$.   ∎

■ **EXERCISE 5.2** ━━━━━━━━━━━━━━━━━━━━━━━

In Exercises 1–16, find all values of θ, if any, that lie in the interval $0° \le \theta < 360°$. ***Do not use a calculator or tables.***

1. $\sin^{-1}\dfrac{1}{2} = \theta$ **2.** $\cos^{-1}\dfrac{1}{2} = \theta$ **3.** $\tan^{-1}(-1) = \theta$ **4.** $\cot^{-1}1 = \theta$

5. $\cos^{-1}\dfrac{-\sqrt{2}}{2} = \theta$ **6.** $\sin^{-1}(-2) = \theta$ **7.** $\cot^{-1}\sqrt{3} = \theta$ **8.** $\tan^{-1}\dfrac{-\sqrt{3}}{3} = \theta$

9. $\sin^{-1} 0 = \theta$ **10.** $\tan^{-1} 0 = \theta$ **11.** $\arccos\sqrt{3} = \theta$ **12.** $\text{arccot}(-\sqrt{3}) = \theta$

13. $\arcsin(-1) = \theta$ **14.** $\arctan\sqrt{3} = \theta$ **15.** $\sec^{-1} 2 = \theta$ **16.** $\csc^{-1}(-\sqrt{2}) = \theta$

*In Exercises 17–32, find all real numbers x, if any, that lie in the interval $0 \le x < 2\pi$. **Do not use a calculator or tables.***

17. $\cos^{-1}\dfrac{\sqrt{3}}{2} = x$ **18.** $\sin^{-1}\dfrac{\sqrt{2}}{2} = x$ **19.** $\sin^{-1}\dfrac{1}{2} = x$ **20.** $\cos^{-1} 0 = x$

21. $\tan^{-1}\sqrt{3} = x$ **22.** $\cot^{-1}\dfrac{\sqrt{3}}{3} = x$ **23.** $\cot^{-1}\dfrac{-\sqrt{3}}{3} = x$ **24.** $\tan^{-1}\dfrac{-\sqrt{3}}{3} = x$

25. $\arccos\sqrt{3} = x$ **26.** $\arccos\dfrac{\sqrt{2}}{2} = x$ **27.** $\arcsin 1 = x$ **28.** $\arcsin \pi = x$

29. $\arcsin 2 = x$ **30.** $\arccos\left(-\dfrac{\sqrt{2}}{2}\right) = x$ **31.** $\sec^{-1} 2 = x$ **32.** $\csc^{-1}\dfrac{2\sqrt{3}}{3} = x$

In Exercises 33–46, use a calculator to find all values of x in radians, if any, that lie in the interval $0 \le x < 2\pi$. Give all answers to the nearest thousandth.

33. $\sin^{-1} 0.8415 = x$ **34.** $\cos^{-1} 0.5403 = x$ **35.** $\tan^{-1} 1.557 = x$ **36.** $\cot^{-1} 0.6421 = x$

37. $\cos^{-1}(-0.4161) = x$ **38.** $\sin^{-1} 0.9193 = x$ **39.** $\cot^{-1}(-0.4577) = x$ **40.** $\tan^{-1}(-2.1850) = x$

41. $\arcsin 0.1411 = x$ **42.** $\arccos(-0.9900) = x$ **43.** $\text{arccot}(-7.0153) = x$ **44.** $\arctan(-0.1425) = x$

45. $\sec^{-1}(-0.9102) = x$ **46.** $\csc^{-1}(-1.1380) = x$

In Exercises 47–60, use a calculator to find all values of θ in degrees, if any, that lie in the interval $0° \le \theta < 360°$. Give all answers to the nearest degree.

47. $\sin^{-1} 0.8192 = \theta$ **48.** $\cos^{-1} 0.2588 = \theta$ **49.** $\tan^{-1}(-1.428) = \theta$ **50.** $\cot^{-1}(-1.3270) = \theta$

51. $\cos^{-1}(0.5736) = \theta$ **52.** $\sin^{-1}(-0.4226) = \theta$ **53.** $\cot^{-1}(-0.8391) = \theta$ **54.** $\tan^{-1} 0.0875 = \theta$

55. $\sec^{-1} 1.5557 = \theta$ **56.** $\csc^{-1} 1.2208 = \theta$ **57.** $\csc^{-1} 3.8637 = \theta$ **58.** $\sec^{-1}(-1.0353) = \theta$

59. $\cos^{-1}(1.3520) = \theta$ **60.** $\cos^{-1}(-0.3090) = \theta$

5.3   THE INVERSE SINE, COSINE, AND TANGENT FUNCTIONS

A **relation** in the set R of real numbers can be thought of as a nonempty set of ordered pairs of real numbers. Certain special relations are called **functions**. Recall that a function is a relation in which to each value of a first component there corresponds exactly one value of a second component.

The set of ordered pairs of real numbers (x, y) that satisfy the equation $y = \sin x$ is a function because every real number x gives a single value of y. Its graph appears in Figure 5-2.

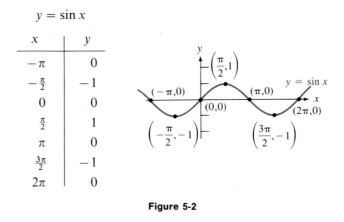

$$y = \sin x$$

x	y
$-\pi$	0
$-\frac{\pi}{2}$	-1
0	0
$\frac{\pi}{2}$	1
π	0
$\frac{3\pi}{2}$	-1
2π	0

Figure 5-2

Interchanging x and y in the function $y = \sin x$ yields the inverse relation $x = \sin y$, or by the notation used in Section 5.2, $y = \sin^{-1}x$. The graph in Figure 5-3 is the graph of the relation $y = \sin^{-1}x$.

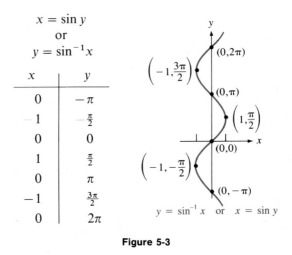

$$x = \sin y$$
$$\text{or}$$
$$y = \sin^{-1}x$$

x	y
0	$-\pi$
-1	$-\frac{\pi}{2}$
0	0
1	$\frac{\pi}{2}$
0	π
-1	$\frac{3\pi}{2}$
0	2π

Figure 5-3

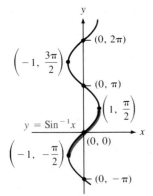

Figure 5-4

Because the graph of $y = \sin^{-1}x$ does not pass the vertical line test, it does not represent a function. However, the portion of the graph lying between $y = -\frac{\pi}{2}$ and $y = \frac{\pi}{2}$ as shown in Figure 5-4 does pass the vertical line test and does determine a function. The colored portion of Figure 5-4 is the graph of the *inverse sine function*, or the *arcsine function*. It is denoted either by $y = \text{Sin}^{-1}x$ or by $y = \text{Arcsin } x$. It is important to remember that uppercase letters "S" and "A" are used to denote the inverse sine *function*.

The graph of the inverse sine function appears in Figure 5-5.

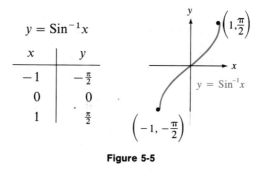

$y = \text{Sin}^{-1}x$

x	y
-1	$-\frac{\pi}{2}$
0	0
1	$\frac{\pi}{2}$

Figure 5-5

The formal definition of the inverse sine function is as follows.

Definition. The **inverse sine** (or **arcsine**) **function**, denoted by $y = \text{Sin}^{-1}x$ (or $y = \text{Arcsin } x$), has a domain of $\{x\colon -1 \le x \le 1\}$ and a range of $\{y\colon -\frac{\pi}{2} \le y \le \frac{\pi}{2}\}$.

$$y = \text{Sin}^{-1}x \quad \text{if and only if} \quad x = \sin y \text{ and } -\frac{\pi}{2} \le y \le \frac{\pi}{2}$$

Definition. The value of $\text{Sin}^{-1}x$ is called the **principal value** of the relation $y = \sin^{-1}x$.

The solid black curve in Figure 5-6 is that portion of $y = \sin x$ for which $-\frac{\pi}{2} \le x \le \frac{\pi}{2}$. That portion represents an increasing function, which is one-to-one, and therefore has an inverse. Its inverse is the colored curve of Figure 5-6, the function $y = \text{Sin}^{-1}x$. Note that the black and the colored curves are symmetric about the line $y = x$.

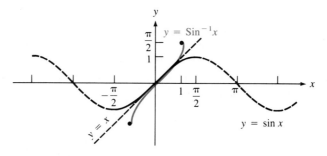

Figure 5-6

Although the domain and the range of the inverse sine function are defined to be sets of real numbers, it is often convenient to think of $\text{Sin}^{-1}x$ as the "angle whose sine is x." In this situation, the range of the inverse sine function may be thought of as a set of angles measured in radians. If $0 < x < 1$, then $\text{Sin}^{-1}x$ represents a QI angle. If $-1 < x < 0$, then $\text{Sin}^{-1}x$ is a QIV angle. If $x = -1, 0,$ or 1, then $\text{Sin}^{-1}x$ is $-\frac{\pi}{2}, 0,$ or $\frac{\pi}{2}$, respectively.

Example 1 Find **a.** $\text{Sin}^{-1}\frac{1}{2}$, **b.** $\text{Sin}^{-1}\left(-\frac{\sqrt{2}}{2}\right)$, **c.** $\text{Sin}^{-1}\pi$, **d.** $\text{Sin}^{-1}\frac{\pi}{4}$, and **e.** $\text{Arcsin}\,0.8330$.

Solution **a.** The expression $\text{Sin}^{-1}\frac{1}{2}$ represents the number between $-\frac{\pi}{2}$ and $\frac{\pi}{2}$, inclusive, whose sine is $\frac{1}{2}$. Because $\frac{\pi}{6}$ is the only number in this interval whose sine is $\frac{1}{2}$,

$$\text{Sin}^{-1}\frac{1}{2} = \frac{\pi}{6}$$

b. The expression $\text{Sin}^{-1}(-\frac{\sqrt{2}}{2})$ represents the number between $-\frac{\pi}{2}$ and $\frac{\pi}{2}$, inclusive, whose sine is $-\frac{\sqrt{2}}{2}$. Because $-\frac{\pi}{4}$ is the only number in this interval whose sine is $-\frac{\sqrt{2}}{2}$,

$$\text{Sin}^{-1}\left(-\frac{\sqrt{2}}{2}\right) = -\frac{\pi}{4}$$

c. The expression $\text{Sin}^{-1}\pi$ represents the number between $-\frac{\pi}{2}$ and $\frac{\pi}{2}$, inclusive, with a sine of π. Because no number has a sine greater than 1, $\text{Sin}^{-1}\pi$ is undefined.

d. It is easy to misread $\text{Sin}^{-1}\frac{\pi}{4}$ as $\text{Sin}\frac{\pi}{4}$ and incorrectly answer $\frac{\sqrt{2}}{2}$. Use a calculator to get the right answer. Divide π by 4. Make sure that your calculator is set for radian measure, and then press the $\boxed{\text{INV}}$ and $\boxed{\text{SIN}}$ keys in that order. The angle whose sine is $\frac{\pi}{4}$ is approximately 0.9033 radian as displayed on the calculator.

e. To find $\text{Arcsin}\,0.8330$, use a calculator that is set for radian measure. Enter .833 and press the $\boxed{\text{INV}}$ and $\boxed{\text{SIN}}$ keys in succession. The angle whose sine is 0.8330 is approximately 0.9845 radians as displayed on the calculator.

■

Because $y = \sin x$ (restricted to $-\frac{\pi}{2} \le x \le \frac{\pi}{2}$) and $y = \text{Sin}^{-1}x$ are inverse functions of each other, we consider the effect of performing the functions in succession. Because $\text{Sin}^{-1}x$ means "the angle whose sine is x," the notation $\sin(\text{Sin}^{-1}x)$ means "the sine of the angle whose sine is x." The obvious answer to this question is "x."

If $-1 \le x \le 1$, then $\sin(\text{Sin}^{-1}x) = x$

Similarly:

If $-\frac{\pi}{2} \le x \le \frac{\pi}{2}$, then $\text{Sin}^{-1}(\sin x) = x$

However, if x is not restricted to the proper interval, then $\text{Sin}^{-1}(\sin x)$ might *not* be x. For example, we have

$$\text{Sin}^{-1}\left(\sin\frac{5\pi}{6}\right) = \text{Sin}^{-1}\left(\frac{1}{2}\right) = \frac{\pi}{6}$$

Similar considerations produce inverse functions of the remaining five trigonometric functions.

Definition.   The **inverse cosine** (or **arccosine**) **function**, denoted by $y = \text{Cos}^{-1}x$ (or $y = \text{Arccos } x$), has a domain of $\{x: -1 \le x \le 1\}$ and a range of $\{y: 0 \le y \le \pi\}$.

$$y = \text{Cos}^{-1}x \quad \text{if and only if} \quad x = \cos y \quad \text{and} \quad 0 \le y \le \pi.$$

The graph of $y = \text{Cos}^{-1}x$, shown in Figure 5-7, is the reflection of a portion of the cosine curve with the line $y = x$ as an axis of symmetry.

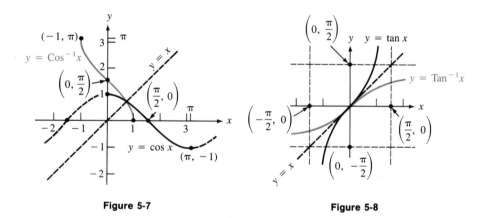

Figure 5-7   Figure 5-8

Definition.   The **inverse tangent** (or **arctangent**) **function**, denoted by $y = \text{Tan}^{-1}x$ (or $y = \text{Arctan } x$), has a domain of the real numbers and a range of $\{y: -\frac{\pi}{2} < y < \frac{\pi}{2}\}$.

$$y = \text{Tan}^{-1}x \quad \text{if and only if} \quad x = \tan y \quad \text{and} \quad -\frac{\pi}{2} < y < \frac{\pi}{2}.$$

The graph of $y = \text{Tan}^{-1}x$ shown in Figure 5-8 is the reflection of a portion of the tangent curve with the line $y = x$ as an axis of symmetry.

Example 2   Find **a.** $\text{Tan}^{-1}(-1)$, **b.** $\text{Cos}^{-1}\dfrac{\sqrt{3}}{2}$, and **c.** $\text{Tan}^{-1}\dfrac{\pi}{4}$.

Solution   **a.**   Because $-\dfrac{\pi}{4}$ is the only number in the interval between $-\dfrac{\pi}{2}$ and $\dfrac{\pi}{2}$ whose tangent is -1,

$$\text{Tan}^{-1}(-1) = -\frac{\pi}{4}$$

b.   Because $\dfrac{\pi}{6}$ is the only number in the interval from 0 to π whose cosine is $\dfrac{\sqrt{3}}{2}$,

$$\text{Cos}^{-1}\frac{\sqrt{3}}{2} = \frac{\pi}{6}$$

c.   To evaluate $\text{Tan}^{-1}\frac{\pi}{4}$, use a calculator set in radian mode. Divide π by 4. Then, press the $\boxed{\text{INV}}$ and $\boxed{\text{TAN}}$ keys in that order to get

$$\text{Tan}^{-1}\frac{\pi}{4} \approx 0.6658$$ ∎

Example 3   Find $\cos(\text{Sin}^{-1} 1)$.

Solution   Because $\text{Sin}^{-1} 1 = \frac{\pi}{2}$, it follows that

$$\cos(\text{Sin}^{-1} 1) = \cos\frac{\pi}{2} = 0$$ ∎

Example 4   Find the exact value of $\cos\left(\text{Sin}^{-1}\dfrac{5}{13}\right)$.

Solution   Because $\frac{5}{13} > 0$, $\text{Sin}^{-1}\frac{5}{13}$ represents a first-quadrant angle. Sketch a first-quadrant angle θ in standard position, as in Figure 5-9, with reference triangle OAP. Then determine a point $P(x, y)$ on its terminal side so that $\text{Sin}^{-1}\frac{5}{13} = \theta$, or equivalently, that $\sin\theta = \frac{y}{r} = \frac{5}{13}$. To do so, let $y = 5$ and $r = 13$ and use the Pythagorean theorem to determine x.

$$x^2 + y^2 = r^2$$
$$x^2 + 5^2 = 13^2$$
$$x^2 = 169 - 25$$
$$x = \sqrt{144}$$
$$x = 12$$

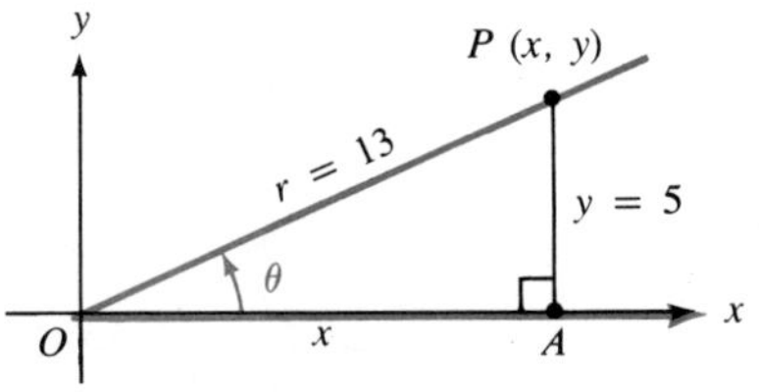

Now because $\text{Sin}^{-1}\frac{5}{13} = \theta$, you have

$$\cos\left(\text{Sin}^{-1}\frac{5}{13}\right) = \cos\theta$$

$$= \frac{x}{r} = \frac{12}{13}$$

Figure 5-19

∎

Example 5   Find $\cos(\text{Sin}^{-1}x)$, given that $0 < x \leq 1$.

Solution   Because the value of x is unknown, you don't know what number has a sine of x. However, you do know that $\text{Sin}^{-1}x$ is a first- or fourth-quadrant angle, because $-\frac{\pi}{2} \leq \text{Sin}^{-1}x \leq \frac{\pi}{2}$. In Figure 5-10, the angle denoted by $\text{Sin}^{-1}x$ has a sine of $x(x > 0)$, because the side opposite $\text{Sin}^{-1}x$ is labeled x and the hypotenuse is 1. Use the Pythagorean theorem to find that the remaining side of the triangle has length of $\sqrt{1 - x^2}$. You can then read the value of the cosine of $\text{Sin}^{-1}x$ from the figure.

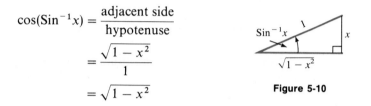

$$\cos(\text{Sin}^{-1}x) = \frac{\text{adjacent side}}{\text{hypotenuse}}$$

$$= \frac{\sqrt{1 - x^2}}{1}$$

$$= \sqrt{1 - x^2}$$

Figure 5-10

Because $-\frac{\pi}{2} \leq \text{Sin}^{-1}x \leq \frac{\pi}{2}$, the expression $\cos(\text{Sin}^{-1}x)$ is never negative and the positive value of the radical is correct.   ■

Example 6   Find $\tan(2\,\text{Cos}^{-1}x)$, where $-1 \leq x \leq 1$.

Solution   In this exercise you must consider two cases: when $0 \leq x \leq 1$ and when $-1 \leq x < 0$.

If x is positive or zero, then $\text{Cos}^{-1}x$ is an acute angle that can be drawn in standard position as in Figure 5-11. Because $\text{Cos}^{-1}x$ represents the angle whose cosine is x, the side adjacent to the angle can be labeled x and the hypotenuse labeled 1. The remaining side of the right triangle can be determined by the Pythagorean theorem.

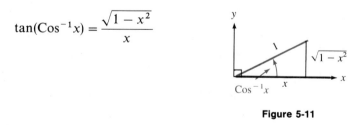

$$\tan(\text{Cos}^{-1}x) = \frac{\sqrt{1 - x^2}}{x}$$

Figure 5-11

The tangent of an acute angle is positive. Because x is also positive, the radical is chosen to be positive.

Now consider the case when x is negative. If x is negative, then $\text{Cos}^{-1}x$ is a second-quadrant angle that can be drawn in standard position as in Figure 5-12. Because $\text{Cos}^{-1}x$ is the angle whose cosine is x, the adjacent side may be labeled x and the hypotenuse labeled 1. Use the Pythagorean theorem to determine that the remaining side has length of $\sqrt{1 - x^2}$.

$$\tan(\mathrm{Cos}^{-1}x) = \frac{\sqrt{1-x^2}}{x}$$

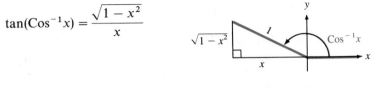

Figure 5-12

The tangent of a second-quadrant angle is negative. Because x is negative, the radical must again be chosen to be positive. Taken together, the results of these two cases determine that, for all x in the interval $-1 \le x \le 1$, we have

$$\tan(\mathrm{Cos}^{-1}x) = \frac{\sqrt{1-x^2}}{x}$$

To find $\tan(2\,\mathrm{Cos}^{-1}x)$, substitute the value of $\tan(\mathrm{Cos}^{-1}x)$ into the right-hand side of the double-angle identity for $\tan 2\theta$ and simplify.

$$\tan(2\,\mathrm{Cos}^{-1}x) = \frac{2\tan(\mathrm{Cos}^{-1}x)}{1 - \tan^2(\mathrm{Cos}^{-1}x)}$$

$$= \frac{2\dfrac{\sqrt{1-x^2}}{x}}{1 - \dfrac{1-x^2}{x^2}}$$

$$= \frac{\dfrac{2\sqrt{1-x^2}}{x}}{\dfrac{x^2 - 1 + x^2}{x^2}}$$

$$= \frac{2x\sqrt{1-x^2}}{2x^2 - 1}$$ ∎

Example 7   Write $\sin(\mathrm{Sin}^{-1}x + \mathrm{Sin}^{-1}y)$ as an algebraic expression in the variables x and y.

Solution   Note that both $\mathrm{Sin}^{-1}x$ and $\mathrm{Sin}^{-1}y$ are angles, and use the identity $\sin(A + B) = \sin A \cos B + \cos A \sin B$ to remove parentheses.

$$\sin(\mathrm{Sin}^{-1}x + \mathrm{Sin}^{-1}y) = \sin(\mathrm{Sin}^{-1}x)\cos(\mathrm{Sin}^{-1}y) + \cos(\mathrm{Sin}^{-1}x)\sin(\mathrm{Sin}^{-1}y)$$

Because $\sin(\mathrm{Sin}^{-1}z) = z$ and $\cos(\mathrm{Sin}^{-1}z) = \sqrt{1 - z^2}$ (see Example 5), it follows that

$$\sin(\mathrm{Sin}^{-1}x + \mathrm{Sin}^{-1}y) = x\sqrt{1 - y^2} + \sqrt{1 - x^2} \cdot y$$
$$= x\sqrt{1 - y^2} + y\sqrt{1 - x^2}$$ ∎

Example 8   Verify the identity $\mathrm{Sin}^{-1}x + \mathrm{Cos}^{-1}x = \dfrac{\pi}{2}$.

Solution   You are asked to verify that the sum of two angles is equal to a third angle. Normally, you would verify an identity by showing that one side may be transformed into the other. Here, however, verify the identity by showing that the sine of the left-hand side equals the sine of the right-hand side.

$$\text{Sin}^{-1}x + \text{Cos}^{-1}x = \frac{\pi}{2}$$

$$\sin(\text{Sin}^{-1}x + \text{Cos}^{-1}x) = \sin\frac{\pi}{2}$$

$$\sin(\text{Sin}^{-1}x)\cos(\text{Cos}^{-1}x) + \cos(\text{Sin}^{-1}x)\sin(\text{Cos}^{-1}x) = 1$$

$$x \cdot x + \sqrt{1-x^2} \cdot \sqrt{1-x^2} = 1$$

$$x^2 + 1 - x^2 = 1$$

$$1 = 1$$

Because $-\frac{\pi}{2} \le \text{Sin}^{-1}x \le \frac{\pi}{2}$ and $0 \le \text{Cos}^{-1}x \le \pi$, you know by adding these inequalities that

$$-\frac{\pi}{2} \le \text{Sin}^{-1}x + \text{Cos}^{-1}x \le \frac{3\pi}{2}$$

Because the left-hand side of the identity is an angle between $-\frac{\pi}{2}$ and $\frac{3\pi}{2}$, and because the sine of only one angle ($\frac{\pi}{2}$) in that interval attains the value of 1, the identity is verified. ■

■ EXERCISE 5.3

In Exercises 1–12, find the value of x, if any. **Do not use a calculator or tables.** *Note that each answer should be a real number.*

1. $\text{Sin}^{-1}\frac{1}{2} = x$
2. $\text{Cos}^{-1}\frac{\sqrt{3}}{2} = x$
3. $\text{Cos}^{-1}0 = x$
4. $\text{Sin}^{-1}\sqrt{3} = x$

5. $\text{Tan}^{-1}1 = x$
6. $\text{Tan}^{-1}0 = x$
7. $\text{Sin}^{-1}3 = x$
8. $\text{Tan}^{-1}(-\sqrt{3}) = x$

9. $\text{Cos}^{-1}\left(-\frac{\sqrt{2}}{2}\right) = x$
10. $\text{Sin}^{-1}\frac{\sqrt{2}}{2} = x$
11. $\text{Arcsin}\left(-\frac{\sqrt{3}}{2}\right) = x$
12. $\text{Arccos}\frac{1}{2} = x$

In Exercises 13–24, find the sine, cosine, and tangent of θ. **Do not use a calculator or tables.**

13. $\text{Sin}^{-1}\frac{1}{2} = \theta$
14. $\text{Cos}^{-1}\frac{\sqrt{3}}{2} = \theta$
15. $\text{Tan}^{-1}0 = \theta$
16. $\text{Tan}^{-1}1 = \theta$

17. $\text{Cos}^{-1}\left(-\frac{\sqrt{3}}{2}\right) = \theta$
18. $\text{Sin}^{-1}\left(-\frac{\sqrt{3}}{2}\right) = \theta$
19. $\text{Arcsin}\,1 = \theta$
20. $\text{Arctan}\,0 = \theta$

21. $\text{Cos}^{-1}(-1) = \theta$
22. $\text{Sin}^{-1}\frac{\sqrt{2}}{2} = \theta$
23. $\text{Arccos}\frac{\sqrt{2}}{2} = \theta$
24. $\text{Arcsin}\,0 = \theta$

In Exercises 25–36, find the value of x. **Do not use a calculator or tables.**

25. $\sin\left(\text{Sin}^{-1}\frac{1}{2}\right) = x$
26. $\cos\left(\text{Cos}^{-1}\frac{1}{2}\right) = x$
27. $\tan(\text{Tan}^{-1}1) = x$
28. $\sin(\text{Sin}^{-1}0) = x$

29. $\cos(\text{Cos}^{-1} 1) = x$ **30.** $\tan(\text{Tan}^{-1} 0) = x$ **31.** $\sin\left(\text{Cos}^{-1}\dfrac{\sqrt{3}}{2}\right) = x$ **32.** $\cos\left(\text{Sin}^{-1}\dfrac{1}{2}\right) = x$

33. $\tan\left[\text{Sin}^{-1}\left(-\dfrac{\sqrt{3}}{2}\right)\right] = x$ **34.** $\cot\left(\text{Cos}^{-1}\dfrac{\sqrt{2}}{2}\right) = x$ **35.** $\cos(\text{Arctan } 1) = x$ **36.** $\sin(\text{Arctan } 0) = x$

In Exercises 37–52, evaluate each expression. **Do not use a calculator or tables.**

37. $\cos\left(\text{Sin}^{-1}\dfrac{4}{5}\right)$

38. $\sin\left(\text{Cos}^{-1}\dfrac{3}{5}\right)$

39. $\sin\left(\text{Cos}^{-1}\dfrac{5}{13}\right)$

40. $\cos\left[\text{Sin}^{-1}\left(-\dfrac{5}{13}\right)\right]$

41. $\tan\left[\text{Sin}^{-1}\left(-\dfrac{4}{5}\right)\right]$

42. $\tan\left(\text{Cos}^{-1}\dfrac{3}{5}\right)$

43. $\tan\left(\text{Cos}^{-1}\dfrac{5}{13}\right)$

44. $\tan\left[\text{Sin}^{-1}\left(-\dfrac{12}{13}\right)\right]$

45. $\cos\left[\text{Tan}^{-1}\left(-\dfrac{5}{12}\right)\right]$

46. $\sin\left(\text{Tan}^{-1}\dfrac{12}{5}\right)$

47. $\sin\left(\text{Arccos}\dfrac{9}{41}\right)$

48. $\cos\left(\text{Arcsin}\dfrac{40}{41}\right)$

49. $\sin\left(\text{Cos}^{-1}\dfrac{1}{3}\right)$

50. $\cos\left(\text{Sin}^{-1}\dfrac{2}{3}\right)$

51. $\tan\left[\text{Sin}^{-1}\left(-\dfrac{3}{4}\right)\right]$

52. $\tan\left[\text{Cos}^{-1}\left(-\dfrac{2}{5}\right)\right]$

In Exercises 53–74, find each value. **Do not use a calculator or tables.**

53. $\sin\left(\text{Sin}^{-1}\dfrac{1}{2} + \text{Cos}^{-1}\dfrac{1}{2}\right)$

54. $\sin\left(\text{Sin}^{-1}\dfrac{1}{2} - \text{Cos}^{-1}\dfrac{1}{2}\right)$

55. $\sin\left(\text{Sin}^{-1}\dfrac{3}{5} + \text{Sin}^{-1}\dfrac{4}{5}\right)$

56. $\cos\left(\text{Cos}^{-1}\dfrac{5}{13} + \text{Cos}^{-1}\dfrac{12}{13}\right)$

57. $\cos\left(\text{Cos}^{-1}\dfrac{5}{13} - \text{Cos}^{-1}\dfrac{12}{13}\right)$

58. $\sin\left(\text{Sin}^{-1}\dfrac{4}{5} - \text{Sin}^{-1}\dfrac{3}{5}\right)$

59. $\cos\left(\text{Sin}^{-1}\dfrac{9}{41} - \text{Cos}^{-1}\dfrac{9}{41}\right)$

60. $\sin\left(\text{Cos}^{-1}\dfrac{40}{41} - \text{Sin}^{-1}\dfrac{40}{41}\right)$

61. $\sin\left[\pi + \text{Sin}^{-1}\left(-\dfrac{3}{5}\right)\right]$

62. $\cos\left[\dfrac{\pi}{2} - \text{Cos}^{-1}\left(-\dfrac{3}{5}\right)\right]$

63. $\sin\left(\text{Cos}^{-1}\dfrac{5}{13} + \dfrac{\pi}{2}\right)$

64. $\sin\left(\text{Sin}^{-1}\dfrac{4}{5} - \dfrac{3\pi}{2}\right)$

65. $\sin\left(2\,\text{Sin}^{-1}\dfrac{\sqrt{2}}{2}\right)$

66. $\cos\left(2\,\text{Sin}^{-1}\dfrac{1}{2}\right)$

67. $\sin\left(2\,\text{Sin}^{-1}\dfrac{3}{5}\right)$

68. $\cos\left(2\,\text{Cos}^{-1}\dfrac{4}{5}\right)$

69. $\cos\left(2\operatorname{Cos}^{-1}\dfrac{5}{13}\right)$

70. $\sin\left(2\operatorname{Sin}^{-1}\dfrac{12}{13}\right)$

71. $\tan\left(\operatorname{Sin}^{-1}\dfrac{4}{5}+\operatorname{Sin}^{-1}\dfrac{3}{5}\right)$

72. $\tan\left(\operatorname{Cos}^{-1}\dfrac{5}{13}+\operatorname{Cos}^{-1}\dfrac{12}{13}\right)$

73. $\tan\left(\operatorname{Tan}^{-1}\dfrac{3}{4}+\operatorname{Tan}^{-1}1\right)$

74. $\tan\left(\operatorname{Tan}^{-1}\dfrac{5}{12}+\pi\right)$

In Exercises 75–90, rewrite each value as an algebraic expression in the variable x.

75. $\sin(\operatorname{Tan}^{-1}x)$ **76.** $\cos(\operatorname{Tan}^{-1}x)$ **77.** $\tan(\operatorname{Sin}^{-1}x)$ **78.** $\tan(\operatorname{Cos}^{-1}x)$

79. $\sin(\operatorname{Cos}^{-1}x)$ **80.** $\cos(\operatorname{Sin}^{-1}x)$ **81.** $\sin(2\operatorname{Arcsin}x)$ **82.** $\cos(2\operatorname{Arccos}x)$

83. $\tan(2\operatorname{Arctan}x)$ **84.** $\sin(2\operatorname{Arccos}x)$ **85.** $\cos(2\operatorname{Sin}^{-1}x)$ **86.** $\sin\left(\dfrac{1}{2}\operatorname{Sin}^{-1}x\right)$

87. $\cos\left(\dfrac{1}{2}\operatorname{Cos}^{-1}x\right)$

88. $\tan\left(\dfrac{1}{2}\operatorname{Tan}^{-1}\dfrac{\sqrt{1-x^2}}{x}\right)$

89. $\sin\left(\dfrac{1}{2}\operatorname{Arccos}x\right)$

90. $\cos\left(\dfrac{1}{2}\operatorname{Arcsin}x\right)$

In Exercises 91–94, evaluate each of the following expressions.

91. $\operatorname{Sin}^{-1}\left(\sin\dfrac{11\pi}{6}\right)$ **92.** $\operatorname{Cos}^{-1}\left(\cos\dfrac{7\pi}{6}\right)$ **93.** $\operatorname{Cos}^{-1}\left(\sin\dfrac{3\pi}{4}\right)$ **94.** $\operatorname{Sin}^{-1}\left(\cos\dfrac{5\pi}{4}\right)$

In Exercises 95–102, verify each of the identities.

95. $\operatorname{Tan}^{-1}y+\operatorname{Tan}^{-1}\dfrac{1}{y}=\dfrac{\pi}{2}\quad(y>0)$

96. $\operatorname{Sin}^{-1}x=\dfrac{\pi}{2}-\operatorname{Cos}^{-1}x$

97. $\operatorname{Sin}^{-1}(-x)=-\operatorname{Sin}^{-1}x$

98. $\operatorname{Cos}^{-1}(-x)+\operatorname{Cos}^{-1}x=\pi$

99. $\operatorname{Arctan}x=\operatorname{Arccos}\dfrac{\sqrt{1+x^2}}{1+x^2}\,(x>0)$

100. $\operatorname{Arccos}2x=\operatorname{Arcsin}\sqrt{\dfrac{1-2x}{2}}$

101. $\operatorname{Tan}^{-1}(-x)=-\operatorname{Tan}^{-1}x$

102. $\operatorname{Tan}^{-1}x=\operatorname{Sin}^{-1}\dfrac{x}{\sqrt{1+x^2}}$

103. For what values of x is $\cos(\operatorname{Cos}^{-1}x)=x$?

104. For what values of x is $\operatorname{Cos}^{-1}(\cos x)=x$?

105. For what values of x is $\operatorname{Sin}^{-1}(\sin x)=x$?

106. For what values of x is $\sin(\operatorname{Sin}^{-1}x)=x$?

5.4 THE INVERSE COTANGENT, SECANT, AND COSECANT FUNCTIONS (OPTIONAL)

If we interchange the x- and y-coordinates of the ordered pairs of the relation defined by the equation $y=\cot x$, we obtain the inverse relation $y=\cot^{-1}x$. A graph of the equation $y=\cot^{-1}x$ is shown in Figure 5-13.

If we restrict the range of the relation defined by $y=\cot^{-1}x$ to the interval $0<y<\pi$, the equation $y=\cot^{-1}x$ defines a function. This function, graphed

$x = \cot y$ or $y = \cot^{-1}x$

x	y
0	$-\frac{\pi}{2}, \frac{\pi}{2}, \frac{3\pi}{2}$
1	$-\frac{3\pi}{4}, \frac{\pi}{4}, \frac{5\pi}{4}$
-1	$-\frac{\pi}{4}, \frac{3\pi}{4}, \frac{7\pi}{4}$

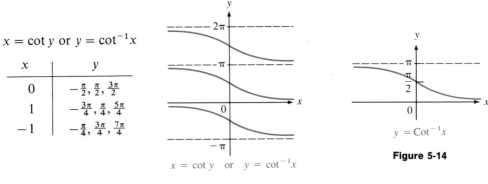

$x = \cot y$ or $y = \cot^{-1}x$

Figure 5-13

$y = \text{Cot}^{-1}x$

Figure 5-14

in Figure 5-14, is called the **inverse cotangent function**, or the **arccotangent function**, and is denoted by either $y = \text{Cot}^{-1}x$ or $y = \text{Arccot }x$.

> **Definition.** The **inverse cotangent** (or **arccotangent**) **function**, denoted by $y = \text{Cot}^{-1}x$ (or $y = \text{Arccot }x$), is a function with the real numbers for its domain and $\{y: 0 < y < \pi\}$ for its range.
>
> $y = \text{Cot}^{-1}x$ is equivalent to $x = \cot y$ and $0 < y < \pi$.

Example 1 Find **a.** $\text{Cot}^{-1}1$, **b.** $\text{Cot}^{-1}0$, **c.** $\text{Cot}^{-1}(-\sqrt{3})$, and **d.** $\text{Cot}^{-1}1.576$.

Solution **a.** $\text{Cot}^{-1}1$ is that number between 0 and π whose cotangent is 1. Hence, you have

$$\text{Cot}^{-1}1 = \frac{\pi}{4}$$

Note that $\cot\frac{\pi}{4} = 1$ and that $0 < \frac{\pi}{4} < \pi$.

b. $\text{Cot}^{-1}0$ is that number between 0 and π whose cotangent is 0. Hence, you have

$$\text{Cot}^{-1}0 = \frac{\pi}{2}$$

Note that $\cot\frac{\pi}{2} = 0$ and that $0 < \frac{\pi}{2} < \pi$.

c. $\text{Cot}^{-1}(-\sqrt{3})$ is that number between 0 and π whose cotangent is $-\sqrt{3}$. Hence, you have

$$\text{Cot}^{-1}(-\sqrt{3}) = \frac{5\pi}{6}$$

Note that $\cot\frac{5\pi}{6} = -\sqrt{3}$ and that $0 < \frac{5\pi}{6} < \pi$.

d.   To find $\text{Cot}^{-1}\,1.576$, use your calculator set in radian mode. Enter 1.576 and find its reciprocal by pressing the $\boxed{1/x}$ key. Then press the $\boxed{\text{INV}}$ and $\boxed{\text{TAN}}$ keys in that order to obtain the number 0.56541429. Thus, you have

$$\text{Cot}^{-1}\,1.576 \approx 0.5654$$

Note that $\cot 0.5654 \approx 1.576$ and that $0 < 0.5654 < \pi$.   ■

If we restrict the ranges of the relations defined by $y = \sec^{-1}x$ and $y = \csc^{-1}x$ to appropriate numbers, the equations $y = \sec^{-1}x$ and $y = \csc^{-1}x$ define functions also.

Definition.   The **inverse secant** (or **arcsecant**) **function**, denoted by $y = \text{Sec}^{-1}x$ (or $y = \text{Arcsec}\,x$), has a domain of $\{x: x \le -1 \text{ or } x \ge 1\}$ and a range of $\{y: 0 \le y \le \pi \text{ and } y \ne \frac{\pi}{2}\}$.*

$y = \text{Sec}^{-1}x$ is equivalent to $x = \sec y$ and $0 \le y \le \pi$ and $y \ne \dfrac{\pi}{2}$.

The graphs of $y = \sec^{-1}x$ and $y = \text{Sec}^{-1}x$ appear in Figure 5-15.

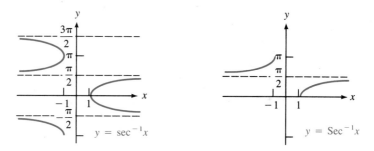

Figure 5-15

Definition.   The **inverse cosecant** (or **arccosecant**) **function**, denoted by $y = \text{Csc}^{-1}x$ (or $y = \text{Arccsc}\,x$), has a domain of $\{x: x \le -1 \text{ or } x \ge 1\}$ and a range of $\{y: -\frac{\pi}{2} \le y \le \frac{\pi}{2} \text{ and } y \ne 0\}$.†

$y = \text{Csc}^{-1}x$ is equivalent to $x = \csc y$ and $-\dfrac{\pi}{2} \le y \le \dfrac{\pi}{2}$ and $y \ne 0$.

The graphs of $y = \csc^{-1}x$ and $y = \text{Csc}^{-1}x$ appear in Figure 5-16.

---

* Some books restrict y to the interval $-\pi \le y < -\dfrac{\pi}{2}$ or $0 \le y < \dfrac{\pi}{2}$.

† Some books restrict y to the interval $-\pi < y \le -\dfrac{\pi}{2}$ or $0 < y \le \dfrac{\pi}{2}$.

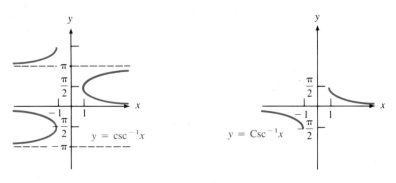

Figure 5-16

Example 2   Find **a.** $\text{Sec}^{-1} 2$, **b.** $\text{Sec}^{-1} 1.75$, **c.** $\text{Csc}^{-1}\left(-\dfrac{2\sqrt{3}}{3}\right)$, and **d.** $\text{Csc}^{-1}\dfrac{1}{2}$.

Solution   **a.** $\text{Sec}^{-1} 2$ is that number between 0 and π whose secant is 2 (or whose cosine is $\frac{1}{2}$). Hence, you have

$$\text{Sec}^{-1} 2 = \frac{\pi}{3}$$

Note that $\sec \frac{\pi}{3} = 2$ and that $0 \le \frac{\pi}{3} \le \pi$.

b.   To find $\text{Sec}^{-1} 1.75$, set your calculator to radian mode, enter 1.75, and find its reciprocal by pressing the $\boxed{1/x}$ key. Then press the $\boxed{\text{INV}}$ and $\boxed{\text{COS}}$ keys to obtain the number 0.96255075. Hence, you have

$$\text{Sec}^{-1} 1.75 \approx 0.9626$$

Note that $\sec 0.9626 \approx 1.75$ and that $0 \le 0.9626 \le \pi$.

c. $\text{Csc}^{-1}\left(-\frac{2\sqrt{3}}{3}\right)$ is that number between $-\frac{\pi}{2}$ and $\frac{\pi}{2}$ whose cosecant is $-\frac{2\sqrt{3}}{3}$ (or whose sine is $-\frac{3}{2\sqrt{3}}$, or $-\frac{\sqrt{3}}{2}$). Hence, you have

$$\text{Csc}^{-1}\left(-\frac{2\sqrt{3}}{3}\right) = -\frac{\pi}{3}$$

Note that $\csc(-\frac{\pi}{3}) = -\frac{2\sqrt{3}}{3}$ and that $-\frac{\pi}{2} \le -\frac{\pi}{3} < \frac{\pi}{2}$.

d. $\text{Csc}^{-1}\frac{1}{2}$ is that number whose cosecant is $\frac{1}{2}$ (or whose sine is 2). Because no such number exists, the expression $\text{Csc}^{-1}\frac{1}{2}$ is not defined.   ■

Example 3   Find $\sin\left(\text{Csc}^{-1}\dfrac{5}{4} + \text{Sec}^{-1}\dfrac{13}{5}\right)$.

Solution   Draw two right triangles with acute angles of $\text{Csc}^{-1}\frac{5}{4}$ and $\text{Sec}^{-1}\frac{13}{5}$. See Figure 5-17. These triangles will help you compute the sine and cosine of each angle. To evaluate $\sin(\text{Csc}^{-1}\frac{5}{4} + \text{Sec}^{-1}\frac{13}{5})$, use the identity for the sine of the sum of two angles and refer to Figure 5-17.

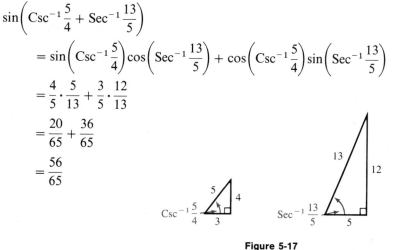

$$\sin\left(\text{Csc}^{-1}\frac{5}{4} + \text{Sec}^{-1}\frac{13}{5}\right)$$

$$= \sin\left(\text{Csc}^{-1}\frac{5}{4}\right)\cos\left(\text{Sec}^{-1}\frac{13}{5}\right) + \cos\left(\text{Csc}^{-1}\frac{5}{4}\right)\sin\left(\text{Sec}^{-1}\frac{13}{5}\right)$$

$$= \frac{4}{5}\cdot\frac{5}{13} + \frac{3}{5}\cdot\frac{12}{13}$$

$$= \frac{20}{65} + \frac{36}{65}$$

$$= \frac{56}{65}$$

Figure 5-17

Example 4 Express $\cot(\text{Csc}^{-1}x)$ as an algebraic expression in the variable x.

Solution The expression $\text{Csc}^{-1}x$ can represent an angle between $-\frac{\pi}{2}$ and $\frac{\pi}{2}$ whose cosecant is x. If $x \geq 1$, then $\text{Csc}^{-1}x$ is the first-quadrant angle shown in Figure 5-18. The ratio of the hypotenuse to the side opposite $\text{Csc}^{-1}x$ is $\frac{x}{1}$, or x, as required. The remaining side of the triangle is $\sqrt{x^2 - 1}$, as determined by the Pythagorean theorem.

The cotangent of $\text{Csc}^{-1}x$ is the ratio of the adjacent side to the opposite side. Thus, you have

$$\cot(\text{Csc}^{-1}x) = \sqrt{x^2 - 1} \qquad \text{for } x \geq 1$$

If $x \leq -1$, then the expression $\text{Csc}^{-1}x$ is a fourth-quadrant angle, and its cotangent is negative. Thus, you have

$$\cot(\text{Csc}^{-1}x) = -\sqrt{x^2 - 1} \qquad \text{for } x \leq -1$$

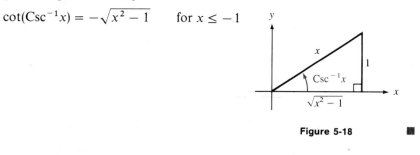

Figure 5-18

EXERCISE 5.4

In Exercises 1–6, evaluate each expression.

1. $\text{Cot}^{-1}\sqrt{3}$

2. $\text{Cot}^{-1}\left(-\frac{\sqrt{3}}{3}\right)$

3. $\text{Sec}^{-1}(-1)$

4. $\text{Sec}^{-1}(-2)$

5. $\text{Csc}^{-1}(-1)$

6. $\text{Csc}^{-1}2$

In Exercises 7–20, simplify each expression.

7. $\sec(\text{Sec}^{-1} 3)$

8. $\csc(\text{Csc}^{-1} 5)$

9. $\sec(\text{Csc}^{-1} 4)$

10. $\csc(\text{Sec}^{-1} 7)$

11. $\cot\left(\text{Sec}^{-1} \dfrac{41}{40}\right)$

12. $\tan\left(\text{Csc}^{-1} \dfrac{13}{5}\right)$

13. $\sec\left(\text{Csc}^{-1} \dfrac{5}{4}\right)$

14. $\csc\left(\text{Sec}^{-1} \dfrac{25}{20}\right)$

15. $\sin\left(2\,\text{Sec}^{-1} \dfrac{5}{3}\right)$

16. $\cos\left(2\,\text{Csc}^{-1} \dfrac{13}{5}\right)$

17. $\cos\left(\text{Cot}^{-1} \dfrac{3}{4} + \text{Sec}^{-1} \dfrac{13}{5}\right)$

18. $\sin(\text{Tan}^{-1} 1 + \text{Cot}^{-1} 1)$

19. $\sin\left(\text{Sec}^{-1} \dfrac{13}{12} - \text{Tan}^{-1} \dfrac{5}{12}\right)$

20. $\cos\left(\text{Cot}^{-1} \dfrac{4}{3} - \text{Csc}^{-1} \dfrac{5}{3}\right)$

In Exercises 21–26, rewrite each value as an algebraic expression in the variable x. Make sure that you consider all numbers in the domain of x.

21. $\sin(\text{Csc}^{-1} x)$

22. $\cos(\text{Cot}^{-1} x)$

23. $\sec(\text{Cot}^{-1} x)$

24. $\csc(\text{Sec}^{-1} x)$

25. $\tan(\text{Sec}^{-1} x)$

26. $\tan(\text{Cot}^{-1} x)$

5.5   APPLICATIONS OF TRIGONOMETRIC EQUATIONS AND INVERSE FUNCTIONS

Trigonometric equations and inverse trigonometric functions are used in several areas besides pure mathematics. Here are three applications from carpentry, optics, and mathematics.

1. Cutting Rafters

If the peak of a roof is h feet above the ceiling joists (see Figure 5-19) and the distance between the wall and the king post is r feet, then the **slope** of the roof is defined as $\frac{h}{r}$.

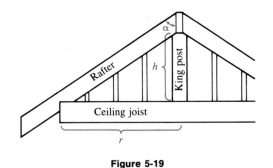

Figure 5-19

Example 1   At what angle α should the rafter in Figure 5-19 be cut?

Solution   The tangent of angle α is $\frac{r}{h}$. Hence, you have

$$\tan \alpha = \frac{r}{h}$$

$$\alpha = \text{Tan}^{-1}\frac{r}{h}$$

The rafter should be cut at an angle of $\text{Tan}^{-1}\frac{r}{h}$. Note that $\alpha = \text{Tan}^{-1}\left(\dfrac{1}{\text{slope}}\right)$.

■

2. Optics: Total Internal Reflection

Snell's law (see Section 2.8) indicates the relationship between the angle of incidence i and the angle of refraction r of a beam of light as it enters a substance such as glass, with an index of refraction n. The relationship is

$$n = \frac{\sin i}{\sin r}$$

Figure 5-20

See Figure 5-20. The ray of light is reversible: a beam passing right to left—from the glass into the air—will follow the same path. As angle r increases, so does angle i, until angle i becomes $90°$ and no light emerges from the glass. See Figure 5-21.

The angle r_c is called the **critical angle**. It is found by substituting $i = 90°$ into Snell's law and then solving for $r = r_c$.

$$\frac{\sin i}{\sin r} = n$$

$$\frac{\sin 90°}{\sin r_c} = n$$

$$\frac{1}{\sin r_c} = n$$

$$\sin r_c = \frac{1}{n}$$

$$r_c = \text{Sin}^{-1}\frac{1}{n}$$

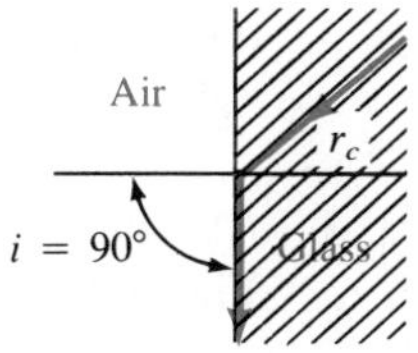

Figure 5-21

Although values of the inverse sine function are real numbers (or angles measured in radians), it is common in some applications to express answers in degrees.

Example 2   Given that the index of refraction for glass is 1.6, find its critical angle.

Solution   The critical angle is

$$r_c = \text{Sin}^{-1}\frac{1}{1.6} = \text{Sin}^{-1}0.625 \approx 38.7°$$

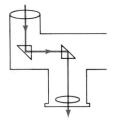

Figure 5-22

Any light striking the surface of the glass (from the inside) at an angle greater than 38.7° will not exit from the glass but will be internally reflected. ■

Glass prisms in binoculars use total internal reflection to increase the optical distance between the objective lens and the eyepiece. See Figure 5-22.

Rhinestones glitter because some of their surfaces are coated with a mirrorlike substance. With their higher index of refraction, real diamonds glitter because of internal reflections—they need no mirror coatings.

3. Polarized Light

Light energy is transmitted as an electromagnetic wave, just like a radio wave. Most light waves behave like the "wave" of a jump rope that is spun very rapidly. See Figure 5-23. A point on the rope moves in a circular pattern and has both an up/down and a side-to-side motion. If the rope is threaded through a picket fence, however, the side-to-side motion is filtered out and only the up/down motion continues. If two boards are nailed horizontally to the pickets, one above the rope and one below, the up/down motion would also stop, eliminating the jump rope wave completely.

Figure 5-23

The optical equivalent of a picket fence is a substance called **polaroid**, developed by Edwin H. Land in 1935. If two sheets of polaroid are oriented with their optical "pickets" at an angle θ with respect to each other, the intensity of the light transmitted by the second polaroid filter is $\cos^2\theta$ times the intensity of the light entering that filter.

Example 3   At what angle θ will half the light be transmitted through the second filter in Figure 5-24?

Solution   Solve the equation $I\cos^2\theta = \frac{1}{2}I$ for θ.

$$I\cos^2\theta = \frac{1}{2}I$$

$$\cos^2\theta = \frac{1}{2}$$

$$\cos\theta = \frac{1}{\sqrt{2}}$$

$$\cos\theta = \frac{\sqrt{2}}{2}$$

$$\theta = \mathrm{Cos}^{-1}\frac{\sqrt{2}}{2} = 45°$$

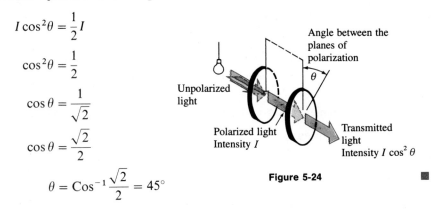

Figure 5-24

■

Light reflected from nonmetallic surfaces is partially polarized. Polaroid sunglasses reduce glare by partially blocking reflected light.

▬ EXERCISE 5.5

1. What is the critical angle for water with an index of refraction of 1.33?

2. What is the critical angle for a diamond with an index of refraction of 2.4?

3. At what angle α should the rafters for a house be cut if the slope of the roof is $\frac{5}{4}$?

4. If the rafters for a house were cut at an angle of 80°, what is the slope of the roof?

5. At what angle θ should two polarizing filters be oriented to transmit one-third of the intensity of the light striking the second filter?

6. One polarizing filter reduces the intensity of light by half. At what angle should the two filters be placed to reduce by 90% the intensity of the light passing through both filters?

CHAPTER SUMMARY

Key Words

critical angle (5.5)
inverse cosecant function (5.4)
inverse cosine function (5.3)
inverse cotangent function (5.4)
inverse secant function (5.4)

inverse sine function (5.3)
inverse tangent function (5.3)
polarization of light (5.5)
principal value (5.3)
Snell's law (5.5)

Key Ideas

(5.1) Solving trigonometric equations.

(5.2) The inverse trigonometric relations.

$$\sin^{-1}y = \theta \quad \text{is equivalent to} \quad y = \sin\theta.$$
$$\cos^{-1}y = \theta \quad \text{is equivalent to} \quad y = \cos\theta.$$
$$\tan^{-1}y = \theta \quad \text{is equivalent to} \quad y = \tan\theta.$$
$$\cot^{-1}y = \theta \quad \text{is equivalent to} \quad y = \cot\theta.$$
$$\csc^{-1}y = \theta \quad \text{is equivalent to} \quad y = \csc\theta.$$
$$\sec^{-1}y = \theta \quad \text{is equivalent to} \quad y = \sec\theta.$$

(5.3) If $y = \text{Sin}^{-1}x$, then $-\dfrac{\pi}{2} \le y \le \dfrac{\pi}{2}$, and $-1 \le x \le 1$.

If $y = \text{Cos}^{-1}x$, then $0 \le y \le \pi$, and $-1 \le x \le 1$.

If $y = \text{Tan}^{-1}x$, then $-\dfrac{\pi}{2} < y < \dfrac{\pi}{2}$, and x is a real number.

Figure 5-25 shows the graphs of the inverse trigonometric functions above.

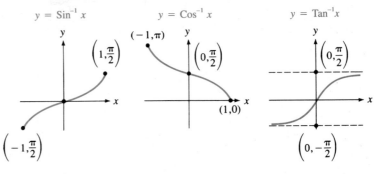

Figure 5-25

(5.4)   If $y = \text{Cot}^{-1}x$, then $0 < y < \pi$, and x is any real number.
(optional)   If $y = \text{Sec}^{-1}x$, then $0 \leq y \leq \pi$, $y \neq \dfrac{\pi}{2}$, and $x \leq -1$ or $x \geq 1$.

If $y = \text{Csc}^{-1}x$, then $-\dfrac{\pi}{2} \leq y \leq \dfrac{\pi}{2}$, $y \neq 0$, and $x \leq -1$ or $x \geq 1$.

Figure 5-26 shows the graphs of the inverse cotangent, inverse secant, and inverse cosecant functions.

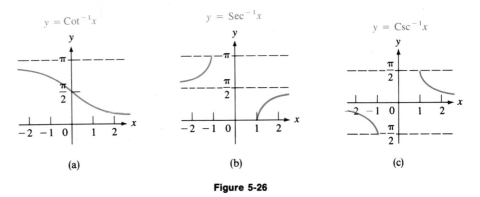

(a)   (b)   (c)

Figure 5-26

REVIEW EXERCISES

In Review Exercises 1–24, solve for all values of the variable between 0° and 360°, including 0°. **Do not use a calculator or tables.**

1. $\cos\theta = \dfrac{\sqrt{3}}{2}$

2. $\tan\theta = -\dfrac{\sqrt{3}}{3}$

3. $\sin 2\theta = 1$

4. $\cos 2\theta = -1$

5. $\sin\dfrac{\theta}{2} = -\dfrac{1}{2}$

6. $\cos\dfrac{\theta}{2} = -\dfrac{\sqrt{3}}{2}$

7. $\sin\theta\tan\theta + \tan\theta = 0$

8. $\cos\theta\tan\theta - \tan\theta = 0$

9. $\sin A \cos A - \cos A - \sin A + 1 = 0$

10. $2 \sin B \cos B + 2 \sin B + \cos B + 1 = 0$

11. $\sin \theta + 1 = \tan \theta + \cos \theta$

12. $\sin 2\theta = \sin \theta$

13. $2 \cos^2 A = -\sin A + 2$

14. $\cos 2B + \sin B = 0$

15. $\sin^2 \dfrac{C}{2} = \cos^2 C$

16. $\cos^2 \dfrac{A}{2} = 1 - \sin^2 A$

17. $4 \cos^2 \dfrac{\theta}{2} = 1$

18. $\tan \alpha - \cot \alpha = 0$

19. $\cos^2 \theta - \sin^2 \theta = 0$

20. $\csc \theta \sec \theta = \cot \theta + \tan \theta$

21. $\csc \beta \sec \beta = \sec \beta + \cot \beta$

22. $\sin^2 \theta - 2 = \cos \theta + \cos^2 \theta - 2$

23. $\sqrt{2} \sin \theta + \sqrt{2} \cos \theta = 2$

24. $\sqrt{3} \cos \theta - \sin \theta = 2$

In Review Exercises 25–36, find all values of x, if any, that lie in the interval $0 \le x < 2\pi$. Express all answers in radians. Do not use a calculator or tables.

25. $\sin^{-1} \dfrac{\sqrt{3}}{2} = x$

26. $\tan^{-1} \dfrac{\sqrt{3}}{3} = x$

27. $\cos^{-1}\left(-\dfrac{1}{2}\right) = x$

28. $\sin^{-1}\left(-\dfrac{\sqrt{3}}{2}\right) = x$

29. $\tan^{-1}(-\sqrt{3}) = x$

30. $\sin^{-1} \sqrt{2} = x$

31. $\cos^{-1} \sqrt{3} = x$

32. $\cot^{-1} \sqrt{3} = x$

33. $\cot^{-1} 0 = x$

34. $\csc^{-1} 2 = x$

35. $\csc^{-1} 0.5 = x$

36. $\sec^{-1}(-2) = x$

In Review Exercises 37–48, use a calculator to find all values of x, if any, that lie in the interval $0 \le x < 2\pi$. Give all answers in radians and to the nearest thousandth.

37. $\sin^{-1} 0.588 = x$

38. $\cos^{-1} 1.732 = x$

39. $\tan^{-1} 75 = x$

40. $\cot^{-1} 75 = x$

41. $\cos^{-1} 3.2 = x$

42. $\sin^{-1} 4.5 = x$

43. $\cot^{-1} 75.3 = x$

44. $\tan^{-1} 0.003 = x$

45. $\csc^{-1} 0.5 = x$

46. $\sec^{-1} 5 = x$

47. $\sec^{-1} 2 = x$

48. $\csc^{-1} 10 = x$

In Review Exercises 49–60, find the values of x, if any. Do not use a calculator or tables. Express all answers in radians.

49. $\mathrm{Sin}^{-1}\left(-\dfrac{1}{2}\right) = x$

50. $\mathrm{Cos}^{-1}\left(-\dfrac{1}{2}\right) = x$

51. $\mathrm{Tan}^{-1} \dfrac{\sqrt{3}}{3} = x$

52. $\mathrm{Tan}^{-1}\left(-\dfrac{\sqrt{3}}{3}\right) = x$

53. $\mathrm{Cos}^{-1} 0 = x$

54. $\mathrm{Sin}^{-1}(-1) = x$

55. $\mathrm{Arctan}(-1) = x$

56. $\mathrm{Arctan}\sqrt{3} = x$

57. $\mathrm{Arcsin}\, 0 = x$

58. $\mathrm{Arccos}\, 5 = x$

59. $\mathrm{Arccos}\, 1 = x$

60. $\mathrm{Arcsin}\, 1 = x$

In Review Exercises 61–74, find all real numbers x, if any. Do not use a calculator or tables.

61. $\sin\left(\mathrm{Sin}^{-1} \dfrac{4}{5}\right) = x$

62. $\cos\left(\mathrm{Cos}^{-1} \dfrac{1}{3}\right) = x$

63. $\sin\left(\mathrm{Cos}^{-1} \dfrac{1}{2}\right) = x$

64. $\cos\left(\mathrm{Sin}^{-1} \dfrac{1}{2}\right) = x$

65. $\tan\left(\mathrm{Cos}^{-1} \dfrac{\sqrt{2}}{2}\right) = x$

66. $\tan\left[\mathrm{Sin}^{-1}\left(-\dfrac{\sqrt{3}}{2}\right)\right] = x$

67. $\tan\left(\mathrm{Sin}^{-1} \dfrac{3}{5}\right) = x$

68. $\cot\left(\mathrm{Cos}^{-1} \dfrac{3}{5}\right) = x$

69. $\sec\left(\mathrm{Tan}^{-1} \dfrac{5}{12}\right) = x$

70. $\csc\left(\mathrm{Cos}^{-1} \dfrac{5}{13}\right) = x$

71. $\tan\left[\mathrm{Sin}^{-1}\left(-\dfrac{5}{13}\right)\right] = x$

72. $\cot\left[\mathrm{Cos}^{-1}\left(-\dfrac{5}{13}\right)\right] = x$

73. $\sin\left(\mathrm{Cos}^{-1} \dfrac{3}{7}\right) = x$

74. $\cos\left(\mathrm{Sin}^{-1} \dfrac{2}{7}\right) = x$

In Review Exercises 75–92, find the required value. **Do not use a calculator or tables.**

75. $\sin\left[\text{Sin}^{-1}\left(-\frac{1}{2}\right) + \text{Cos}^{-1}\left(-\frac{1}{2}\right)\right]$

76. $\sin\left(\text{Sin}^{-1}\frac{\sqrt{3}}{2} - \text{Cos}^{-1}\frac{\sqrt{3}}{2}\right)$

77. $\cos\left(\text{Sin}^{-1}\frac{3}{5} + \text{Tan}^{-1}\frac{3}{4}\right)$

78. $\sin\left(\text{Cos}^{-1}\frac{5}{13} + \text{Tan}^{-1}\frac{12}{5}\right)$

79. $\tan\left(\text{Tan}^{-1}1 + \text{Tan}^{-1}\frac{3}{4}\right)$

80. $\tan\left(\text{Tan}^{-1}\frac{4}{3} + \pi\right)$

81. $\sin\left(\text{Sin}^{-1}\frac{4}{5} + \text{Cos}^{-1}\frac{4}{5}\right)$

82. $\cos\left(\text{Cos}^{-1}\frac{3}{5} - \text{Sin}^{-1}\frac{3}{5}\right)$

83. $\sec\left(\text{Sin}^{-1}\frac{3}{5} - \text{Cos}^{-1}\frac{4}{5}\right)$

84. $\csc\left(\text{Cos}^{-1}\frac{12}{13} - \text{Sin}^{-1}\frac{12}{13}\right)$

85. $\sin\left(2\,\text{Sin}^{-1}\frac{1}{2}\right)$

86. $\cos\left(2\,\text{Sin}^{-1}\frac{1}{2}\right)$

87. $\tan\left(2\,\text{Sin}^{-1}\frac{1}{2}\right)$

88. $\cot\left(2\,\text{Cos}^{-1}\frac{1}{2}\right)$

89. $\sin\left(2\,\text{Sin}^{-1}\frac{3}{5}\right)$

90. $\cos\left(2\,\text{Cos}^{-1}\frac{12}{13}\right)$

91. $\cos\left(2\,\text{Tan}^{-1}\frac{4}{3}\right)$

92. $\sin\left(2\,\text{Tan}^{-1}\frac{12}{5}\right)$

In Review Exercises 93–104, rewrite each value as an algebraic expression in the variable u.

93. $\sin(\text{Cos}^{-1}u)$

94. $\cos(\text{Sin}^{-1}u)$

95. $\tan(\text{Sin}^{-1}u)$

96. $\sin(\text{Tan}^{-1}u)$

97. $\cos(\text{Tan}^{-1}u)$

98. $\tan(\text{Cos}^{-1}u)$

99. $\sin(2\,\text{Sin}^{-1}u)$

100. $\cos(2\,\text{Cos}^{-1}u)$

101. $\tan(2\,\text{Tan}^{-1}u)$

102. $\sin(2\,\text{Cos}^{-1}u)$

103. $\cos\left(\frac{1}{2}\text{Arcsin}\,u\right)$

104. $\cos\left(\frac{1}{2}\text{Arccos}\,u\right)$

In Review Exercises 105–108, evaluate each expression. These exercises are from an optional section.

105. $\text{Cot}^{-1}\left(\frac{\sqrt{3}}{3}\right)$

106. $\text{Sec}^{-1}\sqrt{2}$

107. $\text{Sec}^{-1}(-2)$

108. $\text{Cot}^{-1}(-\sqrt{3})$

In Review Exercises 109–114, simplify each expression. These exercises are from an optional section.

109. $\cos\left(\text{Csc}^{-1}\frac{5}{3}\right)$

110. $\sin\left(\text{Cot}^{-1}\frac{5}{12}\right)$

111. $\sin\left(\text{Sec}^{-1}\frac{13}{5}\right)$

112. $\cos\left(\text{Sec}^{-1}\frac{41}{40}\right)$

113. $\tan\left(\text{Sec}^{-1}\frac{7}{3}\right)$

114. $\cot\left(\text{Sec}^{-1}\frac{5}{2}\right)$

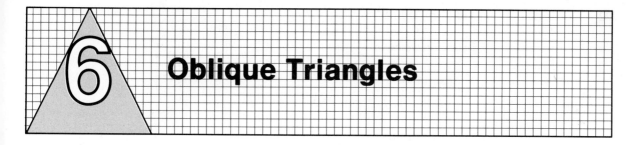
Oblique Triangles

Given two sides—or one acute angle and one side—of a *right* triangle, it is possible to compute the remaining parts. If the given triangle is *not* a right triangle, it is called an **oblique triangle** and the triangle-solving techniques that were discussed previously no longer apply. The solution of oblique triangles is the topic of this chapter.

In the discussion of oblique triangles, we shall use the following convention: capital letters will be used to name the vertices of a triangle, and the corresponding lowercase letters will name the sides opposite those vertices. Thus, side a is opposite angle A, side b is opposite angle B, and side c is opposite angle C.

In elementary geometry, you studied congruency of triangles and learned conditions that determine the size and shape of a triangle. For example, if two sides and the included angle of one triangle are equal to two sides and the included angle of a second triangle, then the two triangles are congruent. This fact (often abbreviated as SAS) indicates that if two sides and the angle between them are given, the triangle is uniquely determined. Because the triangle is determined, it is possible to compute the other side and the remaining two angles.

Similarly, if the three sides of a triangle are given, the triangle is determined (SSS) and the three angles of the triangle can be computed. If two angles and any side of a triangle are known (ASA or AAS), the triangle is determined and all remaining parts can be calculated.

The law of cosines is useful in solving oblique triangles when the given information is in SSS or SAS form.

6.1 THE LAW OF COSINES

Place triangle ABC in a coordinate system with vertex A at the origin, as indicated in Figure 6-1. Because angle A is in standard position and point B is on its terminal side, the cosine of A is the ratio of the x-coordinate of B to the distance that B is from the origin. Thus, we have

$$\cos A = \frac{x\text{-coordinate of } B}{c}$$

or

$$x\text{-coordinate of } B = c \cos A$$

Likewise, we have

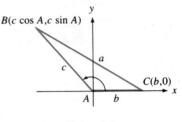

$$\sin A = \frac{y\text{-coordinate of } B}{c}$$

or

$$y\text{-coordinate of } B = c \sin A$$

Figure 6-1

Thus, the coordinates of point B are $(c \cos A, c \sin A)$, the coordinates of A are $(0,0)$ and the coordinates of C are $(b,0)$.

We can now use the distance formula to compute a^2.

$$a^2 = (c \cos A - b)^2 + (c \sin A - 0)^2$$
$$= c^2 \cos^2 A - 2bc \cos A + b^2 + c^2 \sin^2 A$$
$$= c^2 (\cos^2 A + \sin^2 A) + b^2 - 2bc \cos A$$
$$= c^2 + b^2 - 2bc \cos A$$

1. $\quad a^2 = b^2 + c^2 - 2bc \cos A$

Although the triangle in Figure 6-1 is obtuse, this derivation is valid for any triangle.

If point B were placed at the origin instead of point A, we would obtain a second formula:

2. $\quad b^2 = c^2 + a^2 - 2ca \cos B$

If point C were placed at the origin, we would obtain a third formula:

3. $\quad c^2 = a^2 + b^2 - 2ab \cos C$

These three formulas are called the **law of cosines**.

Note that a cyclical change of the letters in any one of these three formulas produces another: let a become b, let b become c, let c become a (and similarly for the capital letters).

The Law of Cosines. The square of any side of any triangle is equal to the sum of the squares of the remaining two sides, minus twice the product of these two sides and the cosine of the angle between them.

$$a^2 = b^2 + c^2 - 2bc \cos A$$
$$b^2 = c^2 + a^2 - 2ca \cos B$$
$$c^2 = a^2 + b^2 - 2ab \cos C$$

In the exercises you will be asked to prove that the Pythagorean theorem is a special case of the law of cosines.

The law of cosines is useful when we are given two sides and the included angle (SAS) or three sides (SSS) of a triangle. The first two examples illustrate these cases.

Example 1   In the oblique triangle ABC shown in Figure 6-2, $b = 27$, $c = 14$, and $A = 43°$. Find the length of side a.

Solution   Sides b and c and the included angle A are given (SAS). Use the law of cosines.

$$a^2 = b^2 + c^2 - 2bc \cos A$$
$$= 27^2 + 14^2 - 2(27)(14)\cos 43°$$
$$\approx 729 + 196 - 552.90$$
$$\approx 372.10$$
$$a \approx 19.29$$
$$\approx 19$$

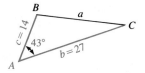

Figure 6-2

The value of a is approximately 19 units.   ■

Example 2   In the oblique triangle ABC shown in Figure 6-3, $a = 5.2$, $b = 3.7$, and $c = 7.1$ units. Find angle B.

Solution   Use the form of the law of cosines that involves angle B.

$$b^2 = a^2 + c^2 - 2ac \cos B$$

Solve this formula for $\cos B$, substitute the values for a, b, and c, and calculate angle B.

$$\cos B = \frac{a^2 + c^2 - b^2}{2ac}$$
$$\cos B = \frac{5.2^2 + 7.1^2 - 3.7^2}{2(5.2)(7.1)}$$
$$\cos B \approx 0.8635$$
$$B \approx 30.29°$$

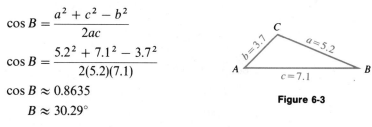

Figure 6-3

To the nearest degree, angle $B = 30°$.   ■

Example 3   A farmer uses two horses and two ropes to pull a tractor out of the mud. Each horse can pull with a force of 950 pounds and, when pulled tight, the ropes form an angle of 27°. What force is applied to the tractor?

Solution   By the parallelogram law for adding vectors, the combined force is length b in the diagram of Figure 6-4. Because opposite sides of a parallelogram are equal, each of the four sides is 950 units in length. Because consecutive angles of a parallelogram are supplementary, the obtuse angle at B is $180° - 27° = 153°$. Apply the law of cosines to triangle ABC.

$$b^2 = a^2 + c^2 - 2ac \cos B$$
$$b^2 = 950^2 + 950^2 - 2(950)(950) \cos 153°$$
$$\approx 3{,}413{,}266.8$$
$$b \approx 1847.5$$
$$\approx 1800$$

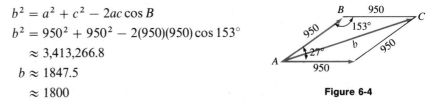

Figure 6-4

The combined pull of the horses is approximately 1800 pounds.   ■

Example 4   An airplane flies N 86.2° W for a distance of 143 kilometers. The pilot, experiencing engine problems, alters course and flies 79.5 kilometers in the direction S 32.7° E in an attempt to find a place to land. He crash-lands safely in a cornfield. How far, and in what direction, will the rescue team need to travel?

Solution   The course diagram appears in Figure 6-5. To find distance d, use the law of cosines.

$$d^2 = 79.5^2 + 143^2 - 2(79.5)(143) \cos 53.5°$$
$$\approx 13{,}244.8$$
$$d \approx 115.1$$
$$\approx 115$$

The pilot is about 115 kilometers from the airport.

As a first step toward finding the direction in which the rescue team must travel, use the law of cosines again to find angle β.

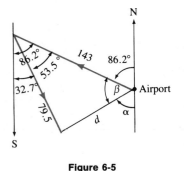

$$79.5^2 = 143^2 + 115^2 - 2(143)(115) \cos \beta$$
$$\cos \beta = \frac{143^2 + 115^2 - 79.5^2}{2(143)(115)}$$
$$\cos \beta \approx 0.8317$$
$$\beta \approx 33.7°$$

Figure 6-5

Angle α is $180° - 86.2° - 33.7°$, or $60.1°$. The rescue team must bear S 60.1° W.   ■

Example 5   A 100-pound weight is suspended by two cables as in Figure 6-6. The tension on the left-hand cable is 55 pounds and on the right-hand cable, 75 pounds. What angle does each cable make with the horizontal?

Solution   The forces in the two ropes are such that their resultant force is 100 pounds, directed upward to exactly counter the 100-pound downward pull. The force diagram appears in Figure 6-7. The angles that the cables make with the horizontal are the complements of angles α and β. Angles α and β can be found by using the law of cosines.

$$55^2 = 75^2 + 100^2 - 2(75)(100) \cos \alpha$$

$$\cos \alpha = \frac{75^2 + 100^2 - 55^2}{2(75)(100)}$$

$$\cos \alpha = 0.8400$$

$$\alpha \approx 32.9°$$

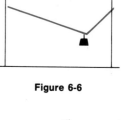

Figure 6-6

Similarly, you have

$$75^2 = 55^2 + 100^2 - 2(55)(100) \cos \beta$$

$$\cos \beta = \frac{55^2 + 100^2 - 75^2}{2(55)(100)}$$

$$\cos \beta \approx 0.6727$$

$$\beta \approx 47.7°$$

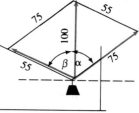

Figure 6-7

The cable on the left makes an angle of $90° - 47.7° \approx 42°$ with the horizontal. The right-hand cable is $90° - 32.9° \approx 57°$ from the horizontal.

■ EXERCISE 6.1

In Exercises 1–12, refer to Illustration 1 and find the value requested. **Use a calculator.**

1. $a = 42.9$ centimeters, $c = 37.2$ centimeters, $B = 99.0°$; find b.
2. $b = 192$ meters, $c = 86.9$ meters, $A = 21.2°$; find a.
3. $a = 2730$ kilometers, $b = 3520$ kilometers, $C = 21.7°$; find c.
4. $b = 2.1$ kilometers, $c = 1.3$ kilometers, $A = 14°$; find a.
5. $a = 91.1$ centimeters, $c = 87.6$ centimeters, $B = 43.2°$; find b.
6. $a = 107$ centimeters, $b = 205$ centimeters, $C = 86.5°$; find c.
7. $a = 19$ kilometers, $b = 23$ kilometers, $c = 18$ kilometers; find A.
8. $a = 14.3$ kilometers, $b = 29.7$ kilometers, $c = 21.3$ kilometers; find B.
9. $a = 30$ feet, $b = 40$ feet, $c = 50$ feet; find C.
10. $a = 130$ miles, $b = 50$ miles, $c = 120$ miles; find A.
11. $a = 1580$, $b = 2137$, $c = 3152$; find B.
12. $a = 0.0031$, $b = 0.0047$, $c = 0.0093$; find C.

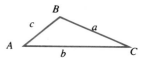

Illustration 1

13. Two men are pulling on ropes attached to the bumper of a car that is stuck in a snowdrift. If one man pulls with a force of 114 pounds and the other with a force of 97 pounds and the angle between the ropes is 13°, what is the force exerted on the car?

14. Two forces, one of 75 pounds and the other of 90 pounds, are exerted at an angle of 102° from each other. What is the magnitude of the resultant force?

15. In Exercise 13, what is the angle between the resultant force and the rope pulled by the stronger man?

16. In Exercise 14, what is the angle between the resultant force and the direction of the 90-pound force?

17. A donkey and a horse are tied to a large stone. The horse pulls with a force of 950 pounds; the donkey lazily tugs with a force of 150 pounds. The angle between their tethers is 19.5°. With what force do they pull on the stone?

18. A ship sails 21.2 nautical miles in a direction of N 42.0° W and then turns onto a course of S 15.0° E and sails 19.0 nautical miles. How far is the ship from its starting point?

19. A ship sails 14.3 nautical miles in a direction of S 28.0° W and then turns onto a course of S 52.0° W and sails 23.2 nautical miles. How far is the ship from its starting point?

20. To measure the length of a lake, a surveyor determines the measurements shown in Illustration 2. How long is the lake?

21. To estimate the cost of building a tunnel, a surveyor must find the distance through a hill. The surveyor determines the measurements shown in Illustration 3. How long must the tunnel be to pass through the hill?

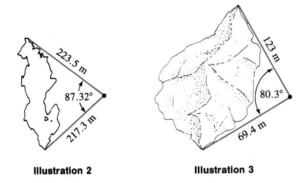

Illustration 2   **Illustration 3**

22. The three circles in Illustration 4 have radii of 4.0, 7.0, and 9.0 centimeters. If the circles are externally tangent to each other, what are the angles of the triangle that joins their centers?

23. The three circles in Illustration 4 have radii of 21.2, 19.3, and 31.2 centimeters. If the circles are externally tangent to each other, what are the angles of the triangle that joins their centers?

24. A tracking antenna is aimed at a satellite 42.32° above the horizon, and at a distance of 8752 miles. If the radius of the earth is 3960 miles, what is the distance, d, between the earth's surface and the satellite? See Illustration 5.

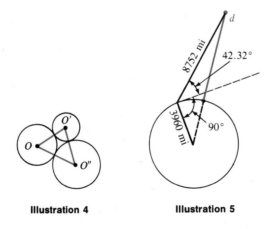

Illustration 4   **Illustration 5**

25. Circles with radii R and r have the same center, C. A chord of the larger circle is tangent to the smaller circle and subtends a central angle θ. See Illustration 6. Express the length, x, of the chord as a function of R and θ.

26. From one corner of a cube, diagonals are drawn along two of its faces. See Illustration 7. Find the angle, θ, between the diagonals.

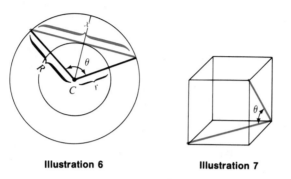

Illustration 6   **Illustration 7**

27. The diagonals of a parallelogram are 12 feet and 18 feet, and they intersect at an angle of 40°. Find the lengths of the sides of the parallelogram. (*Hint:* The diagonals of a parallelogram bisect each other.)

28. The pendulum of Illustration 8 is 1.73 meters long, and the horizontal distance between the extremities of its swing is 0.37 meter. Through what angle does the pendulum swing? Give the answer to the nearest tenth of a degree.

29. Show that the Pythagorean theorem is a special case of the law of cosines.

30. Consider triangle ABC in Figure 6-1. Rotate triangle ABC so that point B is at the origin. Prove that $b^2 = c^2 + a^2 - 2ac \cos B$.

31. Consider triangle ABC in Figure 6-1. Rotate triangle ABC so that point C is at the origin. Prove that $c^2 = a^2 + b^2 - 2ab \cos C$.

32. To determine whether two interior walls meet at a right angle, carpenters often mark a point 3 feet from the corner on one wall and a point (at the same height) on the other wall 4 feet from the corner. If the straight-line distance between those points is 5 feet, the walls are square. At what angle do the walls meet if the distance measures 4 feet, 10 inches?

33. To build a counter top for a kitchen, a cabinetmaker must determine the angle at which two walls meet. The method of Exercise 32 is used. What is the angle between the walls if the measured distance is 5 feet, 3 inches?

34. Triangle ABC is formed by points $A(3, 4)$, $B(1, 5)$, and $C(5, 9)$. Find angle A to the nearest tenth of a degree.

35. In Exercise 34, find angle B to the nearest tenth of a degree.

36. In Illustration 9, D is the midpoint of BC. Find angle 2 to the nearest tenth of a degree.

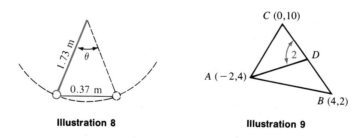

Illustration 8   **Illustration 9**

37. The three hydrogen atoms and one nitrogen atom in an ammonia molecule form a tetrahedron—a pyramid with four triangular faces. Three of these faces are isosceles triangles with the nitrogen atom at the vertex. If the distance between the hydrogen atoms is 1.64 angstroms, and the nitrogen atom is 1.02 angstroms from each hydrogen atom, determine the H–N–H bond angle.

38. A lighthouse is 15.0 nautical miles N 23.0° W of a dock. A ship leaves the dock heading due east at 26.3 knots (nautical miles per hour). How long will it take for the ship to reach a distance of 35.0 nautical miles from the lighthouse?

39. The sides of a parallelogram are 95 and 55 meters, and one angle is 110°. Find the length of the longest diagonal.

40. Circles of radii R and r, with centers at a distance d, have a common external tangent of length D. Refer to Illustration 10 and express D as a function of R, r, d, and θ.

41. Use the law of cosines to show that, in any triangle such as that in Illustration 11,

$$\frac{\cos A}{a} + \frac{\cos B}{b} + \frac{\cos C}{c} = \frac{a^2 + b^2 + c^2}{2abc}$$

42. Use the law of cosines to show that, in any triangle such as that in Illustration 11,

$$a = b \cos C + c \cos B$$

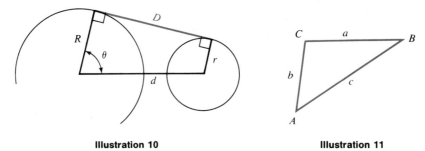

Illustration 10

Illustration 11

6.2 THE LAW OF SINES

To solve triangles when two angles and a side are given (ASA or AAS), another set of formulas called the **law of sines** is required. To develop the law of sines, we refer to the triangle of Figure 6-8 and the triangle of Figure 6-9 in which line segment CD is perpendicular to side AB (or to an extension of AB). If h is the length of segment CD in the right triangles of Figure 6-8, the following formulas are true.

$$h = b \sin A$$

and

$$h = a \sin B$$

Because both $b \sin A$ and $a \sin B$ are equal to h, they are equal to each other.

$$b \sin A = a \sin B$$

or

$$\frac{a}{\sin A} = \frac{b}{\sin B}$$

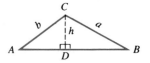

Figure 6-8

In the two right triangles of Figure 6-9,

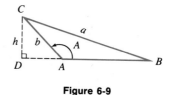

$$h = a \sin B$$

and

$$h = b \sin(180° - A)$$
$$= b \sin A$$

Figure 6-9

Because $a \sin B$ and $b \sin A$ both equal h, they are equal to each other.

$$b \sin A = a \sin B$$

or

$$\frac{a}{\sin A} = \frac{b}{\sin B}$$

In either of these triangles, the perpendicular need not be drawn from C to AB. If drawn from another vertex, similar reasoning yields

$$\frac{a}{\sin A} = \frac{c}{\sin C}$$

or

$$\frac{b}{\sin B} = \frac{c}{\sin C}$$

Because of the transitive law of equality, it follows that

$$\frac{a}{\sin A} = \frac{c}{\sin C} = \frac{b}{\sin B}$$

These results are summed up in a formula called the **law of sines**.

The Law of Sines. The sides in any triangle are proportional to the sines of the angles opposite those sides.

$$\frac{a}{\sin A} = \frac{b}{\sin B} = \frac{c}{\sin C}$$

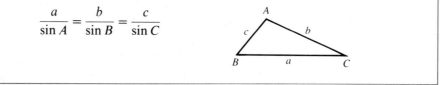

There are three equations implicit in the law of sines, each providing a relation between two angles of a triangle and the sides opposite those angles. If any three of these are known, the fourth can be calculated.

Example 1 In Figure 6-10, $a = 14$ kilometers, $A = 21°$, and $B = 35°$. Find side b.

Solution Note that the given information, in the pattern AAS, fills three of the four spots in the law of sines.

$$\frac{a}{\sin A} = \frac{b}{\sin B}$$

$$\frac{14}{\sin 21°} = \frac{b}{\sin 35°}$$

$$b \approx 22.4$$

$$\approx 22$$

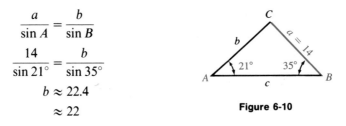

Figure 6-10

Thus, b is approximately 22 kilometers. ■

Example 2   In Figure 6-11, $a = 29$ meters, $B = 42°$, and $C = 31°$. Find side c.

Solution   Note that the given information is of the pattern ASA. To use the law of sines, you need to find the measure of angle A. Because the sum of angles A, B, and C is $180°$, you have

$$A = 180° - B - C$$

$$= 180° - 42° - 31°$$

$$= 107°$$

Now use the law of sines.

$$\frac{a}{\sin A} = \frac{c}{\sin C}$$

$$\frac{29}{\sin 107°} = \frac{c}{\sin 31°}$$

$$c \approx 15.62$$

$$\approx 16$$

Figure 6-11

Thus, c is approximately 16 meters. ■

Example 3   A ship is sailing due east. The skipper observes a lighthouse with a bearing of N 37.5° E. After the ship has sailed 4.70 nautical miles, the bearing to the lighthouse is N 9.0° W. How close to the lighthouse did the ship pass?

Solution   The information of the problem is contained in Figure 6-12. If you can find distance b first, then the required distance d can be obtained by solving right triangle ACD. The law of sines provides a way to compute b.

$$\frac{b}{\sin B} = \frac{c}{\sin C}$$

$$b = \sin 81° \left(\frac{4.70}{\sin 46.5°} \right)$$

$$b \approx 6.3996$$

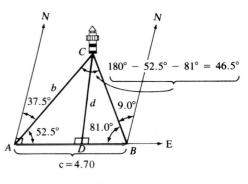

Figure 6-12

In right triangle ACD, it follows that

$$d = b \sin A$$
$$\approx 6.3996 \sin 52.5°$$
$$\approx 5.0771$$
$$\approx 5.08$$

The ship's closest approach to the lighthouse is 5.08 nautical miles. ■

Example 4 On a bright moonlit night, the moon is directly overhead in city A. In city B, 2500 miles to the north, the moonlight strikes a pole at an angle of 36.57°. If the radius of the earth is 4000 miles (to two significant digits), find the distance from the earth to the moon.

Solution Refer to Figure 6-13. Because 2500 miles is $\frac{1}{10}$ of the earth's circumference, angle BOA is $\frac{1}{10}$ of a complete revolution, or 36°. Angle OBM is supplementary to 36.57°. Hence, angle OBM is 143.43°. Because the sum of the angles in any triangle must be 180°, angle M is 0.57°. Use the law of sines to set up the following proportion and solve for x.

$$\frac{x}{\sin 143.43°} = \frac{4000}{\sin 0.57°}$$

$$x \approx \frac{0.5958(4000)}{0.00995}$$

$$x \approx 239{,}518$$

$$x \approx 240{,}000$$

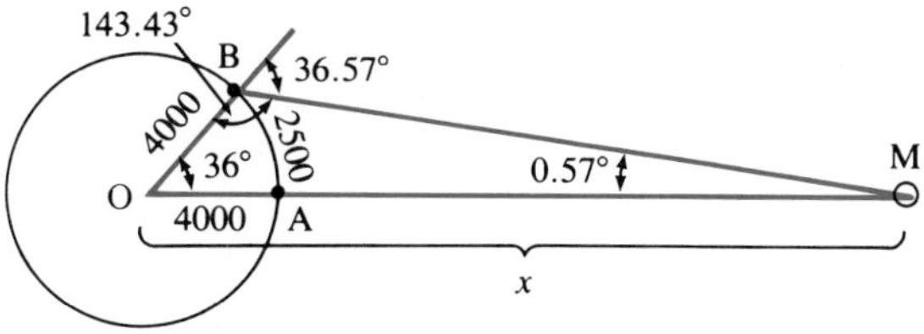

Figure 6-13

The estimate of the distance to the moon is $240{,}000 - 4000 = 236{,}000$ miles. To two significant digits, the distance is 240,000 miles. ■

Example 5 A pilot wishes to fly in the direction of 20.0° east of north, against a 45.0 mile per hour wind blowing from the east. The air speed of the plane is to be 185 miles per hour. What should be the pilot's heading, and what will be the ground speed of the plane?

Solution Refer to Figure 6-14. The pilot's intended direction of travel is represented by the vector OA. The direction of vector OA is 20.0° east of north. The length of vector OH represents the plane's air speed of 185 miles per hour. To find the pilot's heading, you must find angle θ, which is the direction of vector OH.

In triangle OAH, side AH has a length of 45.0 miles. Angle OAH is 90.0° + 20.0°, or 110.0°. Use the law of sines on triangle OAH to find $\theta - 20.0°$. You can then find θ.

$$\frac{185}{\sin 110.0°} = \frac{45.0}{\sin(\theta - 20.0°)}$$

$$\sin(\theta - 20.0°) = \frac{45.0 \sin 110.0°}{185}$$

$$\sin(\theta - 20.0°) \approx 0.2286$$

$$\theta - 20.0° \approx 13.2°$$

$$\theta \approx 33.2°$$

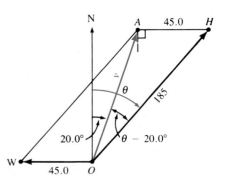

Figure 6-14

To attain the intended direction of travel, the pilot must set a heading of 33.2°.

In Figure 6-14, the ground speed of the plane is represented by the length, v, of vector OA. To determine this length, calculate angle H and use the law of cosines. Note that

$$\text{angle } H \approx 180.0° - 13.2° - 110.0°$$

$$\approx 56.8°$$

By the law of cosines,

$$v^2 \approx 45^2 + 185^2 - 2(45)(185) \cos 56.8°$$

$$\approx 27133$$

$$v \approx 164.7$$

$$\approx 165$$

The ground speed of the plane is approximately 165 miles per hour.   ■

■ EXERCISE 6.2 ■

In Exercises 1–16, refer to Illustration 1. Use the law of sines and your calculator to find the required value.

1. $A = 12°$, $B = 97°$, $a = 14$ kilometers; find b.

2. $A = 19°$, $C = 102°$, $c = 37$ feet; find a.

3. $A = 21.3°$, $B = 19.2°$, $a = 143$ meters; find c.

4. $A = 28.8°$, $C = 9.3°$, $c = 135$ meters; find b.

5. $B = 8.6°$, $C = 9.2°$, $c = 2.73$ meters; find b.

6. $A = 86.3°$, $C = 7.6°$, $a = 43.0$ kilometers; find c.

7. $A = 99.8°$, $C = 43.2°$, $b = 186$ meters; find a.

8. $A = 14.61°$, $B = 87.10°$, $c = 1437$ feet; find b.

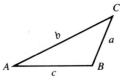

Illustration 1

9. $A = \dfrac{\pi}{7}$, $C = \dfrac{\pi}{2}$, $b = 44.3$ centimeters; find c.

10. $A = \dfrac{2\pi}{9}$, $B = \dfrac{\pi}{11}$, $c = 56.7$ centimeters; find a.

11. $B = 107°$, $C = 11°$, $a = 0.96$ miles, find b.

12. $A = 141°$, $B = 5°$, $c = 0.037$; find a.

13. $B = 32.0°$, $b = 120$, $a = c$; find a.

14. $C = 12°$, $c = 5.4$, $b = a$; find b.

15. $A = x°$, $C = 2x°$, $B = 3x°$, $b = 7.93$; find a.

16. $A = x°$, $C = 3x°$, $B = 5x°$, $c = 12.5$; find a.

17. To measure the distance up a steep hill, Mary determines the measurements shown in Illustration 2. What is the distance d?

18. A ship sails 3.2 nautical miles on a bearing of N 33° E. After reaching a lighthouse, the ship turns and sails 6.7 nautical miles to a position that is due east of the starting point. What is the bearing of the lighthouse from the ship's final position?

19. Points A and B are on opposite sides of a river. See Illustration 3. A tree at point C is 310 feet from point A. Angle A measures 125°, and angle C measures 32°. How wide is the river?

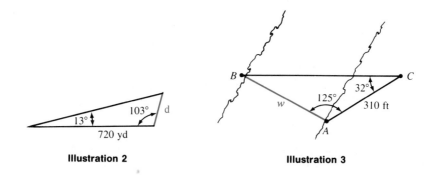

Illustration 2   Illustration 3

20. Observers at points A and B are directly in line with a hot-air balloon and are themselves 215 feet apart. See Illustration 4. The angles of elevation of the balloon from A and B are as shown in the illustration. How high is the balloon? (*Hint:* First use the law of sines to find b.)

21. A radio tower 175 feet high is located on top of a hill. At a point 800 feet down the hill, the angle of elevation to the top of the tower is 19.0°. What angle does the hill make with the horizontal? See Illustration 5.

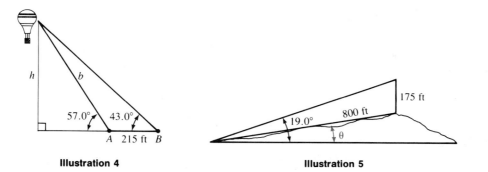

Illustration 4   Illustration 5

22. Two children are on one river bank, 120 feet apart. Each one sights the same tree on the opposite bank. See Illustration 6. Angle A measures 79°, and angle B measures 63°. How wide is the river?

23. A ship sails due north at 3.2 knots. At 2:00 P.M., the skipper sights a lighthouse in the direction of N 43° W. One hour later, the lighthouse bears S 78° W. How close to the lighthouse did the ship sail?

24. A ship sails on a course bearing N 21° E at a speed of 14 knots. At 12:00 noon, the first mate sights an island in the direction N 35° E and one hour later sights the same island due east. If the ship continues on its course, how close will it approach the island?

25. In Exercise 23, at what time was the ship closest to the lighthouse?

26. In Exercise 24, at what time will the ship be nearest the island?

27. A flagpole leans 10.5° from the vertical, toward an observer 17.0 feet from the flagpole's base. If the angle of elevation of the top of the flagpole is 75.5°, how long is the flagpole? See Illustration 7.

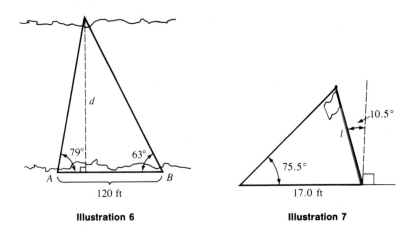

Illustration 6   **Illustration 7**

28. To determine the distance between points P and Q on opposite sides of a river, a surveyor measures the distances and angles shown in Illustration 8. What is the distance PQ?

29. A television tower is on the top of a building. See Illustration 9. From point A, the angle of elevation of the top of the tower is 40°. From B, a point 150 feet closer to the building, the angle of elevation of the top of the tower is 51°, and the angle of elevation of the base of the tower is 45°. Find the height of the tower.

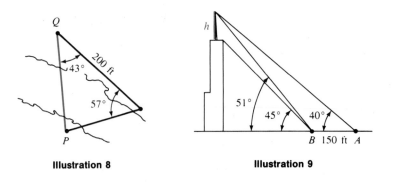

Illustration 8   **Illustration 9**

30. In Exercise 29, find the height of the building.

31. Show that in any triangle ABC, the following relation holds:

$$\frac{\sin A - \sin B}{\sin A + \sin B} = \frac{a - b}{a + b}$$

32. If R is the radius of a circle circumscribed about the triangle ABC in Illustration 10, and s is the **semiperimeter** (half the perimeter) of the triangle, then show that

$$R(\sin A + \sin B + \sin C) = s$$

(*Hint:* angle $BOC = 2A$, angle $BOA = 2C$, and angle $COA = 2B$.)

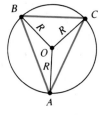

Illustration 10

6.3   THE AMBIGUOUS CASE: SSA

A single triangle is determined if two sides and the included angle of a triangle are given. In this case we can use the law of cosines to solve the triangle. A single triangle is also determined if two angles and a side of a triangle are given. In this case we can use the law of sines. However, a triangle may or may not be determined if two sides and a nonincluded angle are given. There may even be no triangle at all. The case SSA is called the **ambiguous case**.

Consider the problem of constructing a triangle with sides of 1 and 2 inches and a nonincluded angle of 20°, as indicated in Figure 6-15. For the 20° angle to not be the included angle, the 1-inch side must hang on to point A. There are two possible positions for it, and consequently, two possible triangles. The two possible positions are determined by the two intersections of a 1-inch radius circle with the third side.

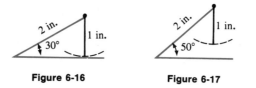

Figure 6-15

If the angle is given as 30°, only one triangle is formed because the 1-inch radius circle is tangent to the third side. The triangle is a 30°–60° right triangle. (In a 30°–60° right triangle, the side opposite the 30° angle is half the hypotenuse.) See Figure 6-16.

If the angle is greater than 30°, the 1-inch side is not long enough to make a triangle. See Figure 6-17.

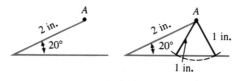

Figure 6-16   **Figure 6-17**

Here are some generalizations about the ambiguous case of SSA. If side b is given and angle A is acute, the length of side a determines the possible number of triangles. Side a could be too short as in Figure 6-18a, and no triangle would be

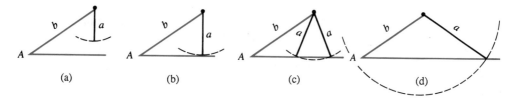

(a)   (b)   (c)   (d)

Figure 6-18

formed. Side a could be "just right," and one right triangle would be formed as in Figure 6-18b. As side a continues to get longer, the two possibilities of Figure 6-18c develop, but if a is longer than b, only one triangle is possible, as in Figure 6-18d.

If side b is given and angle A is obtuse, there are two more cases to consider. If a is less than or equal to b, no triangle is formed, as in Figure 6-19a. If a is greater than b, as in Figure 6-19b, one triangle is possible.

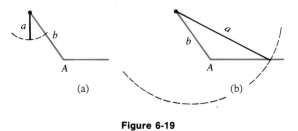

(a)   (b)

Figure 6-19

Memorizing all these possibilities is confusing and often unnecessary. If you are given information in the SSA form, sketch a careful scale drawing of the triangle. Then use common sense to determine whether two, one, or no triangles are expected. If inspection indicates a situation that is too close to call from the diagram alone, let the law of sines decide the outcome, as illustrated in Example 2.

Example 1   Three parts of a triangle are exactly $a = 1$, $b = 2$, and $A = 20°$. Find angles B and C to the nearest tenth of a degree and side c to the nearest hundredth.

Solution   A sketch indicates that there are two possible triangles. See Figure 6-20. By the law of sines, it follows that

$$\frac{a}{\sin A} = \frac{b}{\sin B}$$

$$\sin B = \frac{b \sin A}{a}$$

$$\sin B = \frac{2 \sin 20°}{1}$$

$$\sin B \approx 0.6840$$

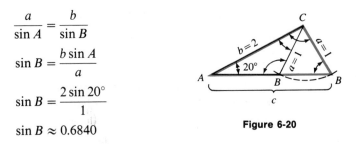

Figure 6-20

There are two possible values of B, one acute (a first-quadrant angle) and the other obtuse (a second-quadrant angle).

$$B \approx 43.2°$$

or

$$B \approx 180° - 43.2° = 136.8°$$

The third angle C has two possibilities also:

$$C = 180° - A - B$$
$$C \approx 180° - 20° - 43.2°$$
$$C \approx 116.8°$$

or

$$C \approx 180° - 20° - 136.8°$$
$$C \approx 23.2°$$

The third side c can also be found by using the law of sines. Side c has two possibilities:

$$\frac{c}{\sin C} = \frac{a}{\sin A}$$

$$c \approx \frac{1 \cdot \sin 116.8°}{\sin 20°}$$

$$\approx 2.61$$

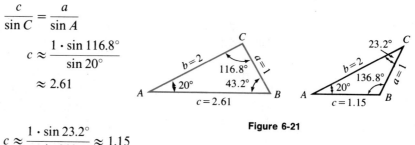

or

$$c \approx \frac{1 \cdot \sin 23.2°}{\sin 20°} \approx 1.15$$

Figure 6-21

The two triangles are shown in Figure 6-21.   ■

Example 2   In the triangle of Figure 6-22, $a = 1$, $b = 2$, and $A = 30°$. Solve the triangle.

Solution   A careful scale drawing indicates a situation too close to call from the diagram alone, so let the law of sines determine the number of possible triangles. By the law of sines, it follows that

$$\frac{a}{\sin A} = \frac{b}{\sin B}$$

$$\frac{1}{\sin 30°} = \frac{2}{\sin B}$$

$$\sin B = 2\left(\frac{1}{2}\right)$$

$$\sin B = 1$$

$$B = 90°$$

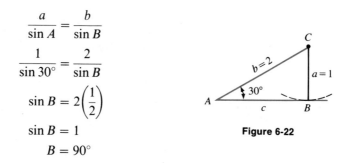

Figure 6-22

There is only one triangle possible—a right triangle. Because the right triangle has a 30° angle, the remaining angle must be 60° and the remaining side must be $\sqrt{3}$.   ■

Example 3   In triangle ABC, $a = 1$, $b = 2$, and $A = 45°$. If possible, solve the triangle.

Solution   By the law of sines, it follows that

$$\frac{a}{\sin A} = \frac{b}{\sin B}$$

$$\frac{1}{\sin 45°} = \frac{2}{\sin B}$$

$$\sin B = 2 \sin 45°$$

$$= 2\left(\frac{\sqrt{2}}{2}\right)$$

$$= \sqrt{2}$$

Because $\sqrt{2}$ is greater than 1, and $\sin B$ cannot be greater than 1, there is no triangle that satisfies the given conditions.   ■

Example 4   A vertical tower 255 feet tall stands on a hill. From a point 700 feet down the hill, the angle between the hill and an observer's line of sight to the top of the tower is 12.0°. What is the angle of inclination of the hill (the angle the ground makes with the horizontal)?

Solution   The given information is used to draw Figure 6-23. By the law of sines, it follows that

$$\frac{255}{\sin 12.0°} = \frac{700}{\sin B}$$

$$\sin B = \frac{700 \cdot \sin 12.0°}{255}$$

$$\approx 0.5707$$

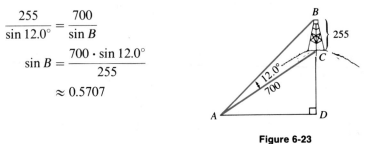

Figure 6-23

There are two possibilities for angle B—one acute and one obtuse. However, from the diagram, you want only the acute angle. Because $\sin B \approx 0.5707$, you have

$B \approx 34.8°$

Once you know the measure of angle B, it is easy to compute angle C (angle ACB).

$$C = 180° - A - B$$

$$\approx 180° - 12.0° - 34.8°$$

$$\approx 133.2°$$

The angle the hill makes with the horizontal is $C - 90°$, which is approximately 43.2°.   ■

■ EXERCISE 6.3

In Exercises 1–16, calculate all possibilities for the indicated value. If no triangle is possible, so indicate.

1. $A = 42.0°$, $a = 123$ feet, $b = 96.0$ feet; find B.

2. $B = 56.2°$, $b = 13.5$ yards, $c = 15.3$ yards; find C.

3. $C = 98.6°$, $a = 42.1$ centimeters, $c = 47.3$ centimeters; find A.

4. $B = 17.5°$, $a = 0.063$ meter, $b = 0.152$ meter; find A.

5. $A = 98.6°$, $a = 42.1$ inches, $c = 47.3$ inches; find C.

6. $C = 86°$, $b = 20$ feet, $c = 19$ feet; find B.

7. $C = 23°$, $b = 48$, $c = 52$; find A.   8. $B = 7.35°$, $b = 2.683$, $c = 4.752$; find A.

9. $A = 9.86°$, $a = 3761$, $b = 5293$; find C.   10. $B = 78.3°$, $a = 0.0570$, $b = 0.0930$; find C.

11. $A = 102.0°$, $a = 13.9$, $c = 15.0$; find B.   12. $C = 47.6°$, $a = 10.5$, $c = 7.35$; find B.

13. $A = 57°$, $b = 13$ meters, $a = 12$ meters; find c.

14. $C = 48°$, $b = 29$ kilometers, $c = 26$ kilometers; find a.

15. $B = 87°$, $a = 35$ centimeters, $b = 32$ centimeters; find c.

16. $B = 38°$, $a = 12$ centimeters, $b = 40$ centimeters; find c.

17.   The triangular piece of land owned by farmer Brown is bounded by three straight highways. The angle between two of them—U.S. 45 and county M—is 43°. Brown's property runs for 2500 feet along county M, and for 2000 feet along the third highway, scenic Silo Drive. How much land might Brown own fronting on U.S. 45?

18.   From the roof of farmer Brown's barn, the angle of elevation to the top of a ranger lookout tower is 17°. From the barn's ground level 43 feet below, the angle of elevation of the tower is 21°. How far above ground level is the top of the tower?

19.   A 210-foot television tower stands on the top of an office building. From a point on level ground, the angles of elevation to the top and base of the tower are 25.2° and 21.1°. How tall is the office building?

20.   A pilot leaves point A and flies 800 kilometers with a heading of 320° to point B. From B, she flies due south to a point C, which is 700 kilometers from point A. How long is the BC leg of the trip? Assume no wind.

21.   A 10-inch piston rod is connected to a 3-inch crank as in Illustration 1. When the piston rod makes an angle of 7° with its line of motion, what is the piston-to-crankshaft distance, d?

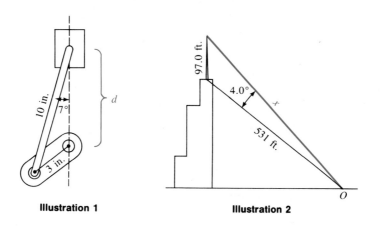

Illustration 1   **Illustration 2**

22. A 97.0-foot tower is on the roof of a building. See Illustration 2. Measured from position O, the difference between the angles of elevation of the top and bottom of the tower is 4.0°. If the line-of-sight distance from O to the bottom of the tower is 531 feet, how far is O from the top of the tower?

6.4 THE LAW OF TANGENTS

Another formula used to solve oblique triangles is called the **law of tangents**. To develop this formula, we consider triangle ABC shown in Figure 6-24. By the law of sines, we have

$$\frac{a}{c} = \frac{\sin A}{\sin C} \quad \text{and} \quad \frac{b}{c} = \frac{\sin B}{\sin C}$$

If these equations are added together, their sum is

$$\frac{a}{c} + \frac{b}{c} = \frac{\sin A}{\sin C} + \frac{\sin B}{\sin C}$$

1. $$\frac{a + b}{c} = \frac{\sin A + \sin B}{\sin C}$$

If the equations are subtracted, their difference is

$$\frac{a}{c} - \frac{b}{c} = \frac{\sin A}{\sin C} - \frac{\sin B}{\sin C}$$

2. $$\frac{a - b}{c} = \frac{\sin A - \sin B}{\sin C}$$

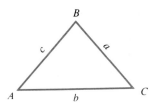

Figure 6-24

We can now divide Equation 2 by Equation 1 and simplify to obtain

$$\frac{\dfrac{a - b}{c}}{\dfrac{a + b}{c}} = \frac{\dfrac{\sin A - \sin B}{\sin C}}{\dfrac{\sin A + \sin B}{\sin C}}$$

$$\frac{a - b}{a + b} = \frac{\sin A - \sin B}{\sin A + \sin B}$$

After using the sum-to-product formulas for $\sin A - \sin B$ and $\sin A + \sin B$, we have

$$\frac{a - b}{a + b} = \frac{2 \cos\left(\dfrac{A + B}{2}\right) \sin\left(\dfrac{A - B}{2}\right)}{2 \sin\left(\dfrac{A + B}{2}\right) \cos\left(\dfrac{A - B}{2}\right)}$$

$$= \cot\left(\frac{A + B}{2}\right) \tan\left(\frac{A - B}{2}\right)$$

$$= \frac{1}{\tan\left(\dfrac{A+B}{2}\right)}\tan\left(\frac{A-B}{2}\right)$$

$$= \frac{\tan\left(\dfrac{A-B}{2}\right)}{\tan\left(\dfrac{A+B}{2}\right)}$$

$$= \frac{\tan\frac{1}{2}(A-B)}{\tan\frac{1}{2}(A+B)}$$

Thus, we have the following formula:

The Law of Tangents.   The formula

$$\frac{a-b}{a+b} = \frac{\tan\frac{1}{2}(A-B)}{\tan\frac{1}{2}(A+B)}$$

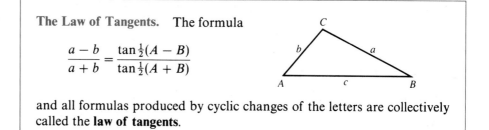

and all formulas produced by cyclic changes of the letters are collectively called the **law of tangents**.

The law of tangents can be used to solve triangles in which two angles and a side are given.

Example 1   Suppose that $A = 42°$, $B = 17°$, and $a = 13$ in a triangle ABC. Use the law of tangents to find b.

Solution   By the law of tangents, it follows that

$$\frac{a-b}{a+b} = \frac{\tan\frac{1}{2}(A-B)}{\tan\frac{1}{2}(A+B)}$$

$$\frac{13-b}{13+b} = \frac{\tan\frac{1}{2}(42° - 17°)}{\tan\frac{1}{2}(42° + 17°)}$$

$$\frac{13-b}{13+b} \approx 0.3918$$

Clear the fractions and solve for b.

$$13 - b \approx 0.3918(13 + b)$$

$$b \approx 5.68$$

$$b \approx 5.7 \text{ units}$$

■ EXERCISE 6.4 ■

In Exercises 1–12, use the law of tangents to solve for the value indicated.

1. $A = 14.3°$, $B = 46.8°$, $a = 25.1$; find b.
2. $A = 19.2°$, $B = 39.7°$, $b = 123$; find a.
3. $A = 27°$, $C = 29°$, $a = 47$; find b.
4. $A = 94.0°$, $C = 10.8°$, $b = 863$; find a.
5. $B = 7.83°$, $C = 56.37°$, $c = 271.3$; find b.
6. $B = 9.27°$, $C = 63.81°$, $b = 4.276$; find c.
7. $A = 47.9°$, $B = 32.2°$, $c = 13.7$; find a.
8. $A = 29.8°$, $B = 49.3°$, $c = 24.2$; find b.
9. $A = 56°$, $C = 72°$, $b = 18$; find a.
10. $A = 73.1°$, $C = 98.6°$, $b = 158$; find c.
11. $B = 99°$, $C = 12°$, $a = 800$; find b.
12. $B = 102°$, $C = 62°$, $a = 56$; find c.

13. A regular pentagon (a figure with five equal sides and five equal angles) is inscribed in a circle of radius 12.0 centimeters. Use the law of tangents to find the perimeter of the pentagon. See Illustration 1. (*Hint:* Angle $A = \frac{360°}{5} = 72°$ and angle $B = \frac{1}{2}(180° - A)$.)

14. A regular decagon (a figure with ten equal sides and ten equal angles) is inscribed in a circle of radius 23.0 centimeters. Use the law of tangents to find the perimeter of the decagon. See Illustration 2.

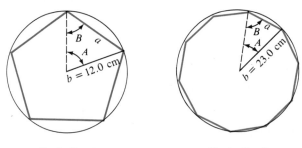

Illustration 1   Illustration 2

15. In an isosceles triangle with vertex angle B, show that

$$\tan\frac{1}{2}(A - B)\tan\frac{1}{2}A = \frac{a - b}{a + b}$$

16. Derive the following alternative form of the law of tangents,

$$\tan\frac{1}{2}(A - B)\tan\frac{1}{2}C = \frac{a - b}{a + b}$$

[*Hint:* Because $A + B + C = 180°$, $\frac{1}{2}(A + B) = \frac{1}{2}(180° - C)$.]

6.5 AREAS OF TRIANGLES

The area of any triangle is given by the formula

1. $$A = \frac{1}{2}bh$$

where b is the base and h is the height of the triangle. In any of the triangles of

Figure 6-25, the base is 6 and the height is 3. The area of each triangle is

$$A = \frac{1}{2}bh$$

$$= \frac{1}{2}(6)(3)$$

$$= 9 \text{ square units}$$

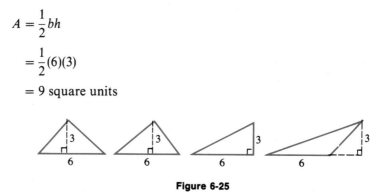

Figure 6-25

The height of the triangle must be known before we can use Equation 1 to find its area. We need other formulas when the height is unknown. To develop these formulas, we first consider a triangle for which two sides and their included angle are given (SAS).

In triangle ABC, shown in Figure 6-26, we assume that b, a, and angle C are given. Because $\sin C = \frac{h}{a}$, we can multiply both sides of this equation by a to obtain

$$h = a \sin C$$

The area K of triangle ABC is

$$K = \frac{1}{2}bh$$

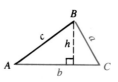

Figure 6-26

or, by substituting $a \sin C$ for h,

2. $\quad K = \frac{1}{2}ba \sin C$

If c, b, and angle A are given, a similar argument gives

3. $\quad K = \frac{1}{2}cb \sin A$

Finally, if a, c, and angle B are given, we have

4. $\quad K = \frac{1}{2}ac \sin B$

Equations 2–4 can be used to find the area of a triangle when two sides and their included angle are given.

Example 1   Find the area of the triangle shown in Figure 6-27.

Solution   Because you are given sides a and c and their included angle B, use Equation 4

and proceed as follows:

$$K = \frac{1}{2} ac \sin B$$

$$= \frac{1}{2}(15)(17) \sin 20°$$

$$\approx 43.6076$$

$$\approx 44$$

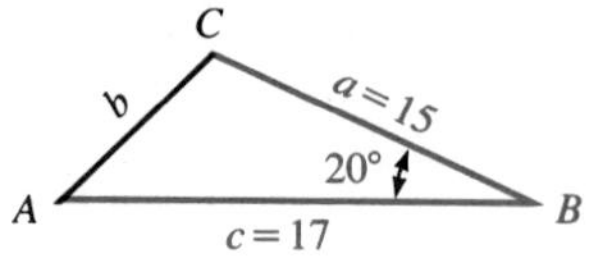

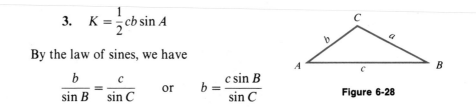

Figure 6-27

The area of the triangle is approximately 44 square units.   ■

Example 2   Find the area of an equilateral triangle with each side of length s.

Solution   Because each angle of an equilateral triangle is 60° and each side of the given triangle is of length s, you have

$$K = \frac{1}{2} ab \sin C$$

$$= \frac{1}{2} ss \sin 60°$$

$$= \frac{\sqrt{3}}{4} s^2$$

■

If two angles and a side are given (AAS or ASA), Equation 3 can be modified to provide the area. In Figure 6-28 the area K of triangle ABC is

3. $K = \frac{1}{2} cb \sin A$

By the law of sines, we have

$$\frac{b}{\sin B} = \frac{c}{\sin C} \quad \text{or} \quad b = \frac{c \sin B}{\sin C}$$

Figure 6-28

After substituting $(c \sin B)/\sin C$ for b in Equation 3, we have

$$K = \frac{1}{2} c \frac{c \sin B}{\sin C} \sin A$$

or

5. $K = \dfrac{c^2 \sin A \sin B}{2 \sin C}$

A similar argument gives the formulas

6. $K = \dfrac{a^2 \sin B \sin C}{2 \sin A}$

and

7. $K = \dfrac{b^2 \sin C \sin A}{2 \sin B}$

Equations 5–7 are used to find the area of a triangle when two angles and a side are given.

Example 3   Find the area of the triangle shown in Figure 6-29.

Solution   Angle $B = 180° - 20° - 15° = 145°$ and

$$K = \frac{c^2 \sin A \sin B}{2 \sin C}$$

$$= \frac{23^2 \sin 20° \sin 145°}{2 \sin 15°}$$

$$\approx 200.4806$$

$$\approx 200$$

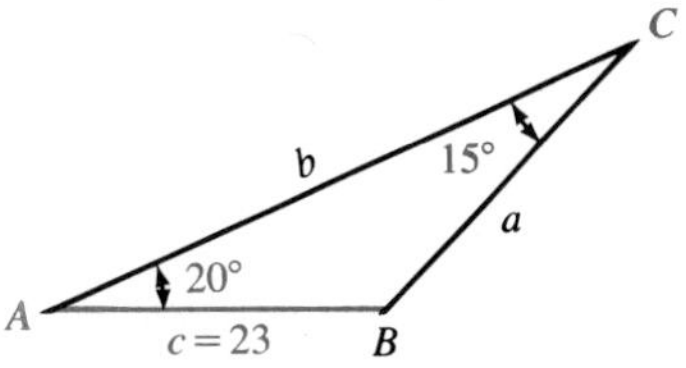

Figure 6-29

The area of the triangle is approximately 200 square units.   ∎

Example 4   Find the area of the isosceles triangle in Figure 6-30.

Solution   The vertex angle B is $180° - 2\theta$, and

$$K = \frac{b^2 \sin A \sin C}{2 \sin B}$$

$$= \frac{b^2 \sin \theta \sin \theta}{2 \sin(180° - 2\theta)}$$

$$= \frac{b^2 \sin^2 \theta}{2 \sin 2\theta}$$

$$= \frac{b^2 \sin^2 \theta}{2(2 \sin \theta \cos \theta)}$$

$$= \frac{b^2 \sin \theta}{4 \cos \theta}$$

$$= \frac{b^2}{4} \tan \theta$$

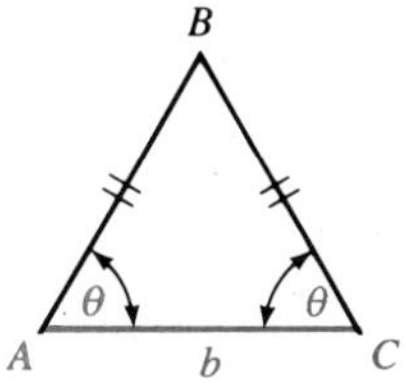

Figure 6-30

∎

Finally, if three sides of a triangle are given (SSS), the area can be calculated by a formula attributed to Heron (Hero) of Alexandria (circa 250 B.C.). To prove Heron's formula, we proceed as follows:

Proof   By the law of cosines, it follows that

8. $\cos A = \dfrac{b^2 + c^2 - a^2}{2bc}$

$$1 - \cos A = 1 - \frac{b^2 + c^2 - a^2}{2bc}$$ Subtract both sides of Equation 8 from 1.

$$= \frac{2bc - b^2 - c^2 + a^2}{2bc}$$ Find a common denominator and add.

$$= \frac{a^2 - (b - c)^2}{2bc}$$ Factor $2bc - b^2 - c^2$ as $-(b - c)^2$.

$$= \frac{(a + b - c)(a - b + c)}{2bc}$$ Factor the numerator, which is the difference of two squares.

Likewise, we have

$$1 + \cos A = 1 + \frac{b^2 + c^2 - a^2}{2bc}$$ Add 1 to both sides of Equation 8.

$$= \frac{2bc + b^2 + c^2 - a^2}{2bc}$$ Find a common denominator and add.

$$= \frac{(b + c)^2 - a^2}{2bc}$$ Factor $2bc + b^2 + c^2$ as $(b + c)^2$.

$$= \frac{(b + c - a)(b + c + a)}{2bc}$$ Factor the numerator, which is a difference of two squares.

Because $\sin^2 A + \cos^2 A = 1$, we have

$$\sin^2 A = 1 - \cos^2 A$$

$$= (1 - \cos A)(1 + \cos A)$$ Factor $1 - \cos^2 A$.

$$= \frac{(a + b - c)(a - b + c)(b + c - a)(b + c + a)}{4b^2 c^2}$$ Substitute values for $1 - \cos A$ and $1 + \cos A$.

We now substitute this result for $\sin^2 A$ in the formula $K^2 = \frac{1}{4} b^2 c^2 \sin^2 A$ to obtain

$$K^2 = \frac{1}{4} b^2 c^2 \sin^2 A$$

$$= \frac{1}{4} b^2 c^2 \frac{(a + b - c)(a - b + c)(b + c - a)(b + c + a)}{4b^2 c^2}$$

$$= \frac{(a + b - c)(a - b + c)(b + c - a)(b + c + a)}{16}$$

We now write each factor in the numerator in a slightly different form and give each factor its own divisor of 2 to obtain

$$K^2 = \left(\frac{a + b + c - 2c}{2} \right) \left(\frac{a + b + c - 2b}{2} \right) \left(\frac{a + b + c - 2a}{2} \right) \left(\frac{a + b + c}{2} \right)$$

$$= \left(\frac{a + b + c}{2} - c \right) \left(\frac{a + b + c}{2} - b \right) \left(\frac{a + b + c}{2} - a \right) \left(\frac{a + b + c}{2} \right)$$

The expression

$$\frac{a + b + c}{2}$$

is half the perimeter of the triangle. It is often called the **semiperimeter** and denoted by the letter s. Substituting s for each $(a + b + c)/2$ in the preceding equation gives

$$K^2 = (s - c)(s - b)(s - a)s$$

Finally, we take the square root of both sides of this equation and rearrange the terms to obtain

$$K = \sqrt{s(s - a)(s - b)(s - c)}$$

This formula is called **Heron's formula**. □

Heron's Formula. If a, b, and c are the three sides of a triangle and

$$s = \frac{a + b + c}{2}$$

then the area of the triangle is given by

$$K = \sqrt{s(s - a)(s - b)(s - c)}$$

Example 5 Find the area of a triangle that has sides that measure 5, 7, and 10 centimeters exactly.

Solution Let $a = 5$, $b = 7$, and $c = 10$. Then

$$s = \frac{5 + 7 + 10}{2} = 11$$

Use Heron's formula to find the area.

$$\begin{aligned}
K &= \sqrt{s(s - a)(s - b)(s - c)} \\
&= \sqrt{11(11 - 5)(11 - 7)(11 - 10)} \\
&= \sqrt{11 \cdot 6 \cdot 4 \cdot 1} \\
&= \sqrt{264} \\
&= 2\sqrt{66}
\end{aligned}$$

The area is $2\sqrt{66}$ square centimeters. ∎

EXERCISE 6.5

In Exercises 1–16, find the area of the triangle whose parts are given, if possible.

1. $b = 23$ feet, $a = 17$ feet, $C = 80°$

2. $c = 1.7$ yards, $b = 3.5$ yards, $A = 60°$

3. $a = 32.3$ centimeters, $c = 21.5$ centimeters, $B = 120.0°$

4. $B = 33.2°$, $a = 101$ kilometers, $c = 97.3$ kilometers

5. $a = 3.0, b = 5.0, c = 7.0$ **6.** $a = 2.1, b = 3.2, c = 5.7$

7. $a = 3, b = 4, c = 5$ **8.** $a = 1.2, b = 2.3, c = 3.4$

9. $A = 55°, B = 45°, c = 12$ **10.** $A = 102°, C = 47°, b = 82$

11. $B = 15°, A = 70°, b = 23$ **12.** $C = 41°, B = 62°, c = 17$

13. $a = 0.06$ millimeter, $b = 0.05$ millimeter, $c = 0.07$ millimeter

14. $a = 0.017$ centimeter, $b = 0.032$ centimeter, $c = 0.055$ centimeter

15. $a = 976$ kilometers, $b = 728$ kilometers, $c = 543$ kilometers

16. $a = 1860$ meters, $b = 2150$ meters, $c = 1590$ meters

Illustration 1

17. To find the area of a triangular lot, the owner starts at one corner and walks due west 205 feet to a second corner. After turning through an angle of 87.3°, he walks 307 feet to the third corner. What is the area of the lot in square feet?

18. A painter wishes to estimate the area of the gable end of a house. What is its area in square feet if the triangle has dimensions as shown in Illustration 1?

19. A printer wishes to make a sign in the form of an isosceles triangle with base angles of 70° and a side of 15 meters. Find the area of the triangle.

20. Point C has a bearing of N 20° E from point A and a bearing of N 10° E from point B. What is the area of triangle ABC if B is due east of A and 17 kilometers from C?

21. Three circles with radii of exactly 3, 5, and 9 centimeters are externally tangent. What is the area of the triangle joining their centers?

22. Three circles have diameters of 7.8, 5.0, and 11.4 centimeters. If they are tangent externally, what is the area of the triangle joining their centers?

23. A Boy Scout walks 520 feet, turns and walks 490 feet, turns again and walks 670 feet, returning to his starting point. What area did his walk encompass?

24. A Girl Scout hikes 523 meters, turns and jogs 412 meters, turns again, and runs 375 meters, returning to her starting point. What area did her trip encompass?

25. Prove that in any triangle,

$$\cos^2 \frac{A}{2} = \frac{s(s - a)}{bc}$$

where s is half the perimeter.

26. Prove that in any triangle,

$$\sin^2 \frac{A}{2} = \frac{(s - b)(s - c)}{bc}$$

where s is half the perimeter.

27. Prove that the area of a parallelogram is half the product of the diagonals and the sine of the angle between the diagonals.

28. Three externally tangent circles, as shown in Illustration 2, have radii of 5, 7, and 8. What is the area of the curve-sided "triangle" they enclose?

29. Find the area of an isosceles triangle with vertex angle α and base b.

30. Find the area of an isosceles triangle with vertex angle α and one of the equal sides of length a.

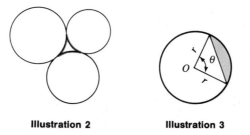

Illustration 2   **Illustration 3**

31. Find the area of an isosceles triangle if the base b is half the length of one of the equal sides.

32. Derive a formula for the area of the segment of the circle (the shaded area) in Illustration 3.

6.6 MORE ON VECTORS (OPTIONAL)

Some physical quantities, called **scalar quantities**, can be described completely by their numerical values, or magnitudes. Some scalar quantities are temperature, distance, area, volume, and elapsed time. Other quantities, called **vector quantities**, have both magnitude and direction. Some vector quantities are force, velocity, acceleration, and displacement. Although we have discussed vectors previously, we now examine some of their properties in more detail.

Recall that a **vector** is a directed line segment. A vector can be denoted by a boldface letter such as **V**. However, in handwritten work, we often express a vector as a letter with an arrow above it, such as $\vec{V}$. If a vector starts at point A and ends at point B, that vector can be denoted either as **AB** or as $\vec{AB}$.

> **Definition.** Two **vectors** are equal if and only if they have the same length and the same direction.

In Figure 6-31, note that vector **AB** could also be denoted as $\vec{AB}$ or as $\vec{V}$. Furthermore, because vectors **AB** and **CD** have the same length and direction, it is true that **AB** = **CD**. The length of a vector **V** is called its **magnitude** or its **norm**. Because the norm of a vector is a numerical value, it is a scalar quantity. This scalar is denoted by $|\mathbf{V}|$. By the distance formula, the vector **OA** in Figure 6-32 has a norm of 5, and $|\mathbf{OA}| = 5$. Similarly, $|\mathbf{OB}| = 5$ and $|\mathbf{OC}| = \sqrt{2}$. Note that $|\mathbf{OA}| = |\mathbf{OB}|$, but **OA** ≠ **OB**.

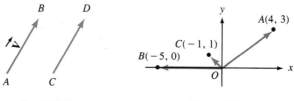

Figure 6-31   **Figure 6-32**

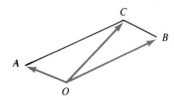

Figure 6-33

Recall that vectors can be added by using the **parallelogram law**. The sum of the two vectors **OA** and **OB** shown in Figure 6-33 is found by constructing the parallelogram $OACB$ with vectors **OA** and **OB** forming two of its adjacent sides. The **sum**, or **resultant**, of vectors **OA** and **OB** is the parallelogram's diagonal **OC**.

Vectors are easier to handle mathematically if they are placed on a coordinate system. The vector **V**, for example, in Figure 6-34, is placed on a coordinate system so that it starts at the origin and ends at the point $(3, 2)$. If we assume that *all* vectors start at the origin, then each is completely determined by its endpoint. Thus, we can denote the vector shown in Figure 6-34 by the ordered pair of numbers $\langle 3, 2 \rangle$. We use corner brackets $\langle\ \rangle$ to distinguish the vector from the point $(3, 2)$.

The distance formula can be used to find the norm of **V** as follows:

$$|\mathbf{V}| = \sqrt{3^2 + 2^2}$$
$$= \sqrt{13}$$

In general, we have the following theorem.

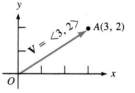

Figure 6-34

Theorem. If **V** is a vector placed on a coordinate system and **V** is represented by $\langle a, b \rangle$, then the norm of **V** is given by

$$|\mathbf{V}| = \sqrt{a^2 + b^2}$$

Example 1   If vector $\mathbf{V} = \langle 3, -5 \rangle$, find $|\mathbf{V}|$.

Solution
$$|\mathbf{V}| = \sqrt{3^2 + (-5)^2}$$
$$= \sqrt{9 + 25}$$
$$= \sqrt{34}$$
∎

It is easy to add vectors if we use ordered-pair notation. For example, the coordinates of point C in the parallelogram of Figure 6-35 are found by adding the corresponding coordinates of points A and B. Thus, if $\mathbf{OA} = \langle -3, 1 \rangle$ and $\mathbf{OB} = \langle 6, 1 \rangle$, then

$$\mathbf{OA} + \mathbf{OB} = \langle -3 + 6, 1 + 1 \rangle$$
$$= \langle 3, 2 \rangle$$
$$= \mathbf{OC}$$

The preceding discussion suggests the following definition.

Figure 6-35

> **Definition.**   If $V = \langle a, b \rangle$ and $W = \langle c, d \rangle$, then
>
> $$V + W = \langle a + c, b + d \rangle$$

Example 2   If $V = \langle 3, -5 \rangle$ and $W = \langle 1, 2 \rangle$, find $V + W$.

Solution $$\begin{aligned} V + W &= \langle 3, -5 \rangle + \langle 1, 2 \rangle \\ &= \langle 3 + 1, -5 + 2 \rangle \\ &= \langle 4, -3 \rangle \end{aligned}$$ ■

If k is a real number and V is a vector, we can define a type of multiplication called **scalar multiplication**.

> **Definition.**   If k is a scalar and V is the vector $\langle a, b \rangle$, then
>
> $$kV = k\langle a, b \rangle = \langle ka, kb \rangle$$

Example 3   If $V = \langle 3, -5 \rangle$, find **a.** $2V$ and **b.** $-8V$.

Solution   **a.** $$\begin{aligned} 2V &= 2\langle 3, -5 \rangle \\ &= \langle 6, -10 \rangle \end{aligned}$$

b. $$\begin{aligned} -8V &= -8\langle 3, -5 \rangle \\ &= \langle -24, 40 \rangle \end{aligned}$$ ■

If k is a real number and V is a vector, then the product kV is also a vector. Its norm is $|k|$ times the norm of V itself. If k is a positive real number, then kV has the same direction as V. If k is a negative real number, then kV has the opposite direction of V. See Figure 6-36. The vector kV is called a **scalar multiple** of V.

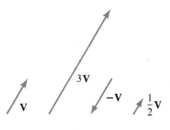

Figure 6-36

If any nonzero vector V is divided by its own norm, the result $\dfrac{V}{|V|}$ is a vector with the same direction as V and with length 1 unit. To show that this is true, we first note that $\dfrac{1}{|\mathbf{V}|}$ is a positive scalar. Thus, the vector $\dfrac{\mathbf{V}}{|\mathbf{V}|}$ has the same direction as $\mathbf{V}$. Then, to show that the norm of $\dfrac{\mathbf{V}}{|\mathbf{V}|}$ is 1, we let $\mathbf{V} = \langle a, b \rangle$ and proceed as follows:

$$\left| \frac{\mathbf{V}}{|\mathbf{V}|} \right| = \left| \frac{\langle a, b \rangle}{\sqrt{a^2 + b^2}} \right|$$

$$= \left| \left\langle \frac{a}{\sqrt{a^2 + b^2}}, \frac{b}{\sqrt{a^2 + b^2}} \right\rangle \right|$$

$$= \sqrt{\frac{a^2}{a^2 + b^2} + \frac{b^2}{a^2 + b^2}}$$

$$= \sqrt{\frac{a^2 + b^2}{a^2 + b^2}}$$

$$= 1$$

The vector $\dfrac{\mathbf{V}}{|\mathbf{V}|}$ is called the **unit vector in the direction of** $\mathbf{V}$.

If $\mathbf{i}$ is the vector $\langle 1, 0 \rangle$, then $\mathbf{i}$ is a vector of length 1 unit, pointing in the positive x-direction. Similarly, if $\mathbf{j}$ is the vector $\langle 0, 1 \rangle$, then $\mathbf{j}$ is a vector of length 1 unit, pointing in the positive y-direction. Any vector can be written as the sum of scalar multiples of the vectors $\mathbf{i}$ and $\mathbf{j}$. For example,

$$\langle 5, 2 \rangle = \langle 5, 0 \rangle + \langle 0, 2 \rangle$$
$$= 5 \langle 1, 0 \rangle + 2 \langle 0, 1 \rangle$$
$$= 5\mathbf{i} + 2\mathbf{j}$$

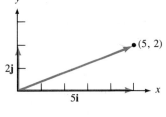

The two vectors $5\mathbf{i}$ and $2\mathbf{j}$ are called the x- and the y-**components** of the vector $\langle 5, 2 \rangle$. See Figure 6-37.

Figure 6-37

Definition. The vectors $\mathbf{i} = \langle 1, 0 \rangle$ and $\mathbf{j} = \langle 0, 1 \rangle$ are called **unit coordinate vectors.**

Any vector $\langle x, y \rangle$ is **resolved** into its x- or **horizontal component** $x\mathbf{i}$, and its y- or **vertical component** $y\mathbf{j}$ when it is written in the form $x\mathbf{i} + y\mathbf{j}$.

Example 4 Let $\mathbf{V} = 4\mathbf{i} + 3\mathbf{j}$ and $\mathbf{W} = 4\mathbf{i} - \mathbf{j}$. Calculate **a.** $\dfrac{\mathbf{V}}{|\mathbf{V}|}$, **b.** $5\mathbf{V} + 3\mathbf{W}$, and **c.** $\mathbf{V} - \mathbf{W}$.

Solution   **a.** $\dfrac{V}{|V|} = \dfrac{4i + 3j}{|4i + 3j|}$

$\qquad\qquad = \dfrac{4i + 3j}{\sqrt{4^2 + 3^2}}$

$\qquad\qquad = \dfrac{4}{5}i + \dfrac{3}{5}j$

b. $5V + 3W = 5(4i + 3j) + 3(4i - 1j)$

$\qquad\qquad\quad = (20i + 15j) + (12i - 3j)$

$\qquad\qquad\quad = 32i + 12j$

c. $V - W = V + (-1)W$

$\qquad\qquad = (4i + 3j) + (-1)(4i - 1j)$

$\qquad\qquad = (4i + 3j) + (-4i + 1j)$

$\qquad\qquad = 0i + 4j$

$\qquad\qquad = 4j$ ∎

Dot Product of Two Vectors

The definition of scalar multiplication provides the way to multiply a vector by a real number. We now define a way, called the **dot product**, to multiply one vector by another.

> **Definition.**   The **dot product** of vectors **V** and **W** is the *scalar*
>
> $$V \cdot W = |V||W|\cos\theta$$
>
> where θ is the angle between **V** and **W**.

It is not convenient to calculate the dot product of two vectors by using the preceding definition directly. It is easy, however, to calculate a dot product by using a theorem, which we state here without proof.

> **Theorem.**   Let $V = ai + bj$ and $W = ci + dj$. Then
>
> $$V \cdot W = ac + bd$$

Example 5   If $A = 2i + 3j$ and $B = 5i - 4j$, calculate $A \cdot B$.

Solution $A \cdot B = (2i + 3j) \cdot (5i - 4j)$

$\qquad\qquad = 2 \cdot 5 + 3 \cdot (-4)$

$\qquad\qquad = -2$ ∎

Example 6   Find the angle between $A = 3i + 4j$ and $B = 5i - 12j$.

Solution   See Figure 6-38. By definition, $A \cdot B = |A||B|\cos\theta$, where θ is the angle between the vectors. Solve for $\cos \theta$, and proceed as follows:

$$\mathbf{A} \cdot \mathbf{B} = |\mathbf{A}||\mathbf{B}| \cos \theta$$

$$\cos \theta = \frac{\mathbf{A} \cdot \mathbf{B}}{|\mathbf{A}||\mathbf{B}|}$$

$$= \frac{(3\mathbf{i} + 4\mathbf{j}) \cdot (5\mathbf{i} - 12\mathbf{j})}{\sqrt{3^2 + 4^2}\sqrt{5^2 + (-12)^2}}$$

$$= \frac{15 - 48}{5 \cdot 13}$$

$$\cos \theta = \frac{-33}{65}$$

$$\theta = \cos^{-1}\left(\frac{-33}{65}\right)$$

$$\theta \approx 120.5°$$

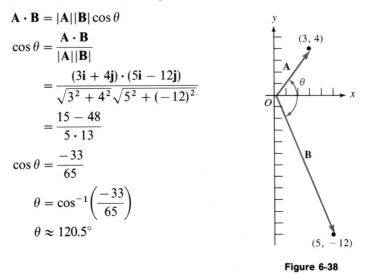

Figure 6-38   ∎

If the dot product, $|\mathbf{A}||\mathbf{B}| \cos \theta$, of two nonzero vectors $\mathbf{A}$ and $\mathbf{B}$ is 0, then $\cos \theta$ must be 0. If $\cos \theta = 0$, then $\theta = 90°$, and the two vectors must be perpendicular. Thus, the dot product provides a test for the perpendicularity of two vectors.

Theorem.   Two nonzero vectors are perpendicular if and only if their dot product is zero.

Example 7   Are the vectors $\mathbf{A} = 6\mathbf{i} - 2\mathbf{j}$ and $\mathbf{B} = \mathbf{i} + 3\mathbf{j}$ perpendicular?

Solution   Calculate $\mathbf{A} \cdot \mathbf{B}$. If $\mathbf{A} \cdot \mathbf{B} = 0$, the vectors are perpendicular. If $\mathbf{A} \cdot \mathbf{B} \neq 0$, the vectors are not perpendicular.

$$\mathbf{A} \cdot \mathbf{B} = (6\mathbf{i} - 2\mathbf{j}) \cdot (\mathbf{i} + 3\mathbf{j})$$

$$= 6 \cdot 1 + (-2)(+3)$$

$$= 6 - 6$$

$$= 0$$

Because the dot product is 0, the vectors $\mathbf{A}$ and $\mathbf{B}$ are perpendicular.   ∎

Another theorem relates the concepts of dot product and norm.

Theorem.   If $\mathbf{V}$ is any vector, then

$$\mathbf{V} \cdot \mathbf{V} = |\mathbf{V}|^2$$

You will be asked to prove this theorem in the exercises.

Example 8   Find the horizontal and vertical components of a 2.0-pound force that makes an angle of 30° with the x-axis. Express the 2.0-pound force in $a\mathbf{i} + b\mathbf{j}$ form.

Solution   The horizontal component of the given force is vector **OA**, which is one leg of the right triangle *OAC*. See Figure 6-39. Find the norm of **OA** as follows.

$$|\mathbf{OA}| = |\mathbf{OC}| \cos \theta$$
$$= 2.0(\cos 30°)$$
$$= 2.0\left(\frac{\sqrt{3}}{2}\right)$$
$$\approx 1.7$$

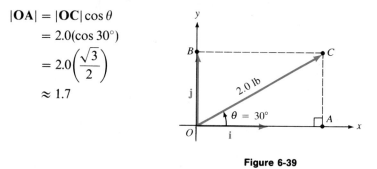

Figure 6-39

The horizontal component is approximately 1.7 pounds. Now find the vertical component **OB**.

$$|\mathbf{OB}| = |\mathbf{OC}| \sin \theta$$
$$= 2.0(\sin 30°) = 1.0$$

The vertical component is 1.0 pound. Thus, **OC** is approximately $1.7\mathbf{i} + 1.0\mathbf{j}$. If the measurements were exact, **OC** would be $\sqrt{3}\mathbf{i} + \mathbf{j}$.   ■

In Example 8, you resolved a given vector into its horizontal and vertical components. It is often necessary to determine a component of a given vector in a direction other than that of an axis. If **B** is not the zero vector, then the component of a vector **A** in the direction of **B** is called the **scalar projection of A on B** and is denoted as **comp$_\mathbf{B}$A**. Figure 6-40 shows that this component is the scalar $|\mathbf{A}| \cos \theta$, the length of vector **OP**. The vector **OP** itself is called the **vector projection of A on B**, and is denoted as **proj$_\mathbf{B}$A**.

The scalar projection of **A** on **B** can be expressed by using the dot product, as follows:

$$\text{comp}_\mathbf{B}\mathbf{A} = |\mathbf{A}| \cos \theta$$
$$= \frac{|\mathbf{A}||\mathbf{B}| \cos \theta}{|\mathbf{B}|} \qquad \text{Multiply numerator and denominator by } |\mathbf{B}|.$$
$$= \frac{\mathbf{A} \cdot \mathbf{B}}{|\mathbf{B}|} \qquad \text{Use the definition of } \mathbf{A} \cdot \mathbf{B}.$$

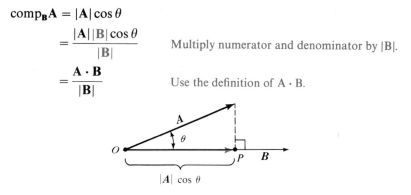

Figure 6-40

To derive a formula for the vector projection of **A** on **B**, we first form the vector $\frac{\mathbf{B}}{|\mathbf{B}|}$, which is the *unit* vector in the direction of **B**. We then multiply this unit vector by the scalar, comp$_\mathbf{B}$**A**, to stretch it to the length required for **proj$_\mathbf{B}$A**. Finally, we use properties of the dot product to write the result in compact form.

$$\mathbf{proj_B A} = (\text{comp}_\mathbf{B}\mathbf{A})(\text{unit vector in direction of } \mathbf{B})$$

$$= \frac{\mathbf{A} \cdot \mathbf{B}}{|\mathbf{B}|} \frac{\mathbf{B}}{|\mathbf{B}|}$$

$$= \frac{\mathbf{A} \cdot \mathbf{B}}{|\mathbf{B}|^2} \mathbf{B}$$

$$= \frac{\mathbf{A} \cdot \mathbf{B}}{\mathbf{B} \cdot \mathbf{B}} \mathbf{B} \qquad \text{For any vector, V, } |V|^2 = V \cdot V.$$

Thus, we have the following results:

If $\mathbf{B} \neq \mathbf{0}$, then

the scalar projection of **A** on **B** $= \text{comp}_\mathbf{B}\mathbf{A} = \dfrac{\mathbf{A} \cdot \mathbf{B}}{|\mathbf{B}|}$

the vector projection of **A** on **B** $= \mathbf{proj_B A} = \dfrac{\mathbf{A} \cdot \mathbf{B}}{\mathbf{B} \cdot \mathbf{B}} \mathbf{B}$

Example 9   A force **A** of exactly 2 pounds makes an angle of exactly 30° with the horizontal. **a.** Find the component of this force in the direction of $\mathbf{B} = 5\sqrt{3}\mathbf{i} + 3\mathbf{j}$, and **b.** find the vector projection of **A** on **B**.

Solution   **a.**   To find the component of the given force **A** in the direction of **B**, you must find comp$_\mathbf{B}$**A**. See Figure 6-41. From Example 8 and from the given information you have that

$$\mathbf{A} = \sqrt{3}\mathbf{i} + \mathbf{j} \quad \text{and} \quad \mathbf{B} = 5\sqrt{3}\mathbf{i} + 3\mathbf{j}$$

Thus, you have

$$\text{comp}_\mathbf{B}\mathbf{A} = \frac{\mathbf{A} \cdot \mathbf{B}}{|\mathbf{B}|}$$

$$= \frac{(\sqrt{3}\mathbf{i} + \mathbf{j}) \cdot (5\sqrt{3}\mathbf{i} + 3\mathbf{j})}{|5\sqrt{3}\mathbf{i} + 3\mathbf{j}|}$$

$$= \frac{15 + 3}{\sqrt{75 + 9}}$$

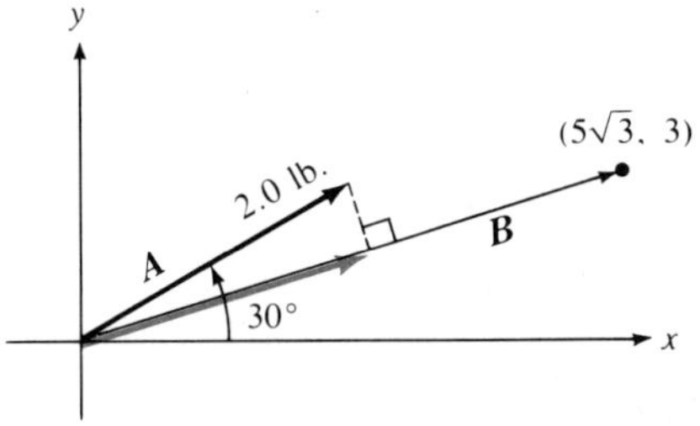

Figure 6-41

$$= \frac{18}{\sqrt{84}}$$

$$= \frac{9}{\sqrt{21}}$$ $\sqrt{84} = \sqrt{4 \cdot 21} = 2\sqrt{21}$

$$= \frac{3\sqrt{21}}{7}$$ Rationalize the denominator.

The component of the 2-pound force in the direction of **B** is $\dfrac{3\sqrt{21}}{7}$ pounds.

b.   To find the vector projection of **A** on **B**, proceed as follows:

$$\text{proj}_{\mathbf{B}}\mathbf{A} = \frac{\mathbf{A} \cdot \mathbf{B}}{\mathbf{B} \cdot \mathbf{B}}\mathbf{B}$$

$$= \frac{(\sqrt{3}\,\mathbf{i} + \mathbf{j}) \cdot (5\sqrt{3}\,\mathbf{i} + 3\mathbf{j})}{(5\sqrt{3}\,\mathbf{i} + 3\mathbf{j}) \cdot (5\sqrt{3}\,\mathbf{i} + 3\mathbf{j})}(5\sqrt{3}\,\mathbf{i} + 3\mathbf{j})$$

$$= \frac{18}{84}(5\sqrt{3}\,\mathbf{i} + 3\mathbf{j})$$

$$= \frac{18 \cdot 5\sqrt{3}}{84}\mathbf{i} + \frac{18 \cdot 3}{84}\mathbf{j}$$

$$= \frac{15\sqrt{3}}{14}\mathbf{i} + \frac{9}{14}\mathbf{j}$$ Simplify the fractions.

The vector projection of the 2-pound force **A** in the direction of **B** is $\dfrac{15\sqrt{3}}{14}\mathbf{i} + \dfrac{9}{14}\mathbf{j}$ ∎

EXERCISE 6.6

In Exercises 1–12, let $\mathbf{U} = \langle 2, -3 \rangle$, $\mathbf{V} = \langle 5, -2 \rangle$, *and* $\mathbf{W} = \langle -1, 1 \rangle$. *Calculate each quantity.*

1. $\mathbf{U} + \mathbf{V}$	**2.** $\mathbf{V} + \mathbf{W}$	**3.** $3\mathbf{U}$	**4.** $5\mathbf{V}$
5. $2\mathbf{U} + \mathbf{V}$	**6.** $3\mathbf{U} - \mathbf{W}$	**7.** $|\mathbf{U}|$	**8.** $|3\mathbf{U}|$
9. $|\mathbf{U} + \mathbf{W}|$	**10.** $|\mathbf{V} - \mathbf{W}|$	**11.** $|\mathbf{U}| + |\mathbf{W}|$	**12.** $|3\mathbf{V} - 5\mathbf{W}|$

In Exercises 13–16, find a unit vector in the direction of the given vector.

13. $3\mathbf{i} + 4\mathbf{j}$	**14.** $5\mathbf{i} - 12\mathbf{j}$	**15.** $\sqrt{3}\,\mathbf{i} - \mathbf{j}$	**16.** $2\mathbf{i} + \sqrt{5}\,\mathbf{j}$

In Exercises 17–22, resolve each vector into its horizontal and its vertical components by writing each vector in $a\mathbf{i} + b\mathbf{j}$ *form.*

17. $\langle 3, 5 \rangle + \langle 5, 3 \rangle$ **18.** $\langle -2, 7 \rangle + \langle 2, 3 \rangle$

19.   A vector of length 10, making an angle of 30° with the x-axis.

20.   A vector of length 10, making an angle of 45° with the x-axis.

21.   A vector of length 23.3, making an angle of 37.2° with the x-axis.

22.   A vector of length 19.1, making an angle of 183.7° with the x-axis.

In Exercises 23–28, find the dot product of the two given vectors.

23. $\langle 2, -3 \rangle$ and $\langle 3, -1 \rangle$

24. $\langle 1, -5 \rangle$ and $\langle 5, 1 \rangle$

25. $2\mathbf{i} + 5\mathbf{j}$ and $\mathbf{i} + \mathbf{j}$

26. $3\mathbf{i} - 3\mathbf{j}$ and $2\mathbf{i} + \mathbf{j}$

27. $\mathbf{i}$ and $\mathbf{j}$

28. $2\mathbf{i}$ and $3\mathbf{i}$

In Exercises 29–34, find the angle between the given vectors.

29. $\langle 2, 2 \rangle$ and $\langle 5, 0 \rangle$

30. $\langle \sqrt{3}, 1 \rangle$ and $\langle 3, 3 \rangle$

31. $\langle \sqrt{3}, -1 \rangle$ and $\langle -1, \sqrt{3} \rangle$

32. $2\mathbf{i} + 3\mathbf{j}$ and $-3\mathbf{i} + 2\mathbf{j}$

33. $3\mathbf{i} - \mathbf{j}$ and $3\mathbf{i} + \mathbf{j}$

34. $3\mathbf{i} - 4\mathbf{j}$ and $5\mathbf{i} + 12\mathbf{j}$

In Exercises 35–40, indicate whether the given vectors are perpendicular.

35. $\langle 2, 3 \rangle$ and $\langle -3, 2 \rangle$

36. $\langle 2, 3 \rangle$ and $\langle 3, 2 \rangle$

37. $5\mathbf{i} + \mathbf{j}$ and $\mathbf{i} + \mathbf{j}$

38. $6\mathbf{i} - 2\mathbf{j}$ and $\mathbf{i} + 3\mathbf{j}$

39. $\mathbf{i}$ and $\mathbf{j}$

40. $-\mathbf{i}$ and $3\mathbf{i}$

In Exercises 41–44, find the component of the first vector in the direction of the second vector.

41. $\langle 3, 4 \rangle, \langle 5, 12 \rangle$　　　**42.** $\langle 1, 1 \rangle, \langle 3, 2 \rangle$　　　**43.** $6\mathbf{i} + 8\mathbf{j}, 4\mathbf{i} - 3\mathbf{j}$　　　**44.** $\mathbf{i}, \mathbf{i} + \mathbf{j}$

In Exercises 45–48, find the vector projection of the first vector on the second.

45. $\langle 3, 4 \rangle, \langle 5, 12 \rangle$　　　**46.** $\langle 1, 1 \rangle, \langle 3, 2 \rangle$　　　**47.** $6\mathbf{i} + 8\mathbf{j}, 4\mathbf{i} - 3\mathbf{j}$　　　**48.** $\mathbf{i}, \mathbf{i} + \mathbf{j}$

49. A horizontal wind exerts a force of 17.0 pounds on the back of a child standing on roller skates, on a 7.0° incline. Find the vector that represents the force pushing him up the incline, and its norm.

50. A wind blowing from the southwest exerts a 213-pound force on a sailboat that is on a heading of 85.0°. Find the vector that represents the force propelling the boat, and its norm.

51. Find an example to illustrate that $|\mathbf{U} + \mathbf{V}| \neq |\mathbf{U}| + |\mathbf{V}|$.

52. Find an example to support the distributive law, $(a + b)\mathbf{V} = a\mathbf{V} + b\mathbf{V}$.

53. Find an example to support the distributive law, $a(\mathbf{V} + \mathbf{W}) = a\mathbf{V} + a\mathbf{W}$.

54. Find an example to support the associative law, $a(\mathbf{V} \cdot \mathbf{W}) = (a\mathbf{V}) \cdot \mathbf{W}$.

55. Prove the theorem: If $\mathbf{V}$ is any vector, then $\mathbf{V} \cdot \mathbf{V} = |\mathbf{V}|^2$.

56. Prove the theorem: If $|\mathbf{V}| = |\mathbf{W}|$, then $(\mathbf{V} + \mathbf{W}) \cdot (\mathbf{V} - \mathbf{W}) = 0$.

CHAPTER SUMMARY

Key Words

ambiguous case (6.3)

dot product (6.6)

Heron's formula (6.5)

law of cosines (6.1)

law of sines (6.2)

law of tangents (6.4)

norm of a vector (6.6)

oblique triangles (6.1)

scalars (6.6)

scalar multiplication (6.6)

scalar projection (6.6)   vector projection (6.6)
semiperimeter (6.5)   vectors (6.6)
unit vectors (6.6)

Key Ideas

(6.1)   **The law of cosines.** In triangle ABC with sides of a, b, and c,

$$a^2 = b^2 + c^2 - 2bc \cos A$$
$$b^2 = c^2 + a^2 - 2ca \cos B$$

and

$$c^2 = a^2 + b^2 - 2ab \cos C$$

The law of cosines can be used in the SAS and SSS cases.

(6.2)   **The law of sines.** In triangle ABC with sides of a, b, and c,

$$\frac{a}{\sin A} = \frac{b}{\sin B} = \frac{c}{\sin C}$$

The law of sines can be used in the AAS, ASA, and SSA cases.

(6.3)   When given information in the ambiguous SSA case, make a careful drawing to determine whether two, one, or no triangles exist.

(6.4)   **The law of tangents.** In triangle ABC with sides of a, b, and c,

$$\frac{a-b}{a+b} = \frac{\tan \frac{1}{2}(A-B)}{\tan \frac{1}{2}(A+B)}$$

(6.5)   Formulas for the area K of a triangle with vertices at A, B, and C and sides of a, b, and c.

$$K = \frac{1}{2} bh$$

Where b is the base and h is the altitude of the triangle

$$K = \frac{1}{2} ba \sin C$$

$$K = \frac{1}{2} cb \sin A$$

$$K = \frac{1}{2} ac \sin B$$

$$K = \frac{c^2 \sin A \sin B}{2 \sin C}$$

$$K = \frac{a^2 \sin B \sin C}{2 \sin A}$$

$$K = \frac{b^2 \sin C \sin A}{2 \sin B}$$

$$K = \sqrt{s(s-a)(s-b)(s-c)}$$ With s equal to half the triangle's perimeter (s is the semiperimeter)

(6.6)
(Optional)

Two vectors are equal if and only if they have the same length and the same direction.

If $\mathbf{V} = \langle a, b \rangle$, $\mathbf{W} = \langle c, d \rangle$, and k is a real number, then

$$|\mathbf{V}| = \sqrt{a^2 + b^2}$$
$$\mathbf{V} + \mathbf{W} = \langle a + c, b + d \rangle$$
$$k\mathbf{V} = \langle ka, kb \rangle$$
$$\mathbf{V} \cdot \mathbf{W} = |\mathbf{V}||\mathbf{W}| \cos \theta = ac + bd$$

$$\mathbf{i} = \langle 1, 0 \rangle$$
$$\mathbf{j} = \langle 0, 1 \rangle$$

Two nonzero vectors are perpendicular if and only if their dot product is zero.

$$\mathbf{V} \cdot \mathbf{V} = |\mathbf{V}|^2$$

$$\text{comp}_\mathbf{B}\mathbf{A} = \frac{\mathbf{A} \cdot \mathbf{B}}{|\mathbf{B}|}$$

$$\text{proj}_\mathbf{B}\mathbf{A} = \frac{\mathbf{A} \cdot \mathbf{B}}{\mathbf{B} \cdot \mathbf{B}} \mathbf{B}$$

REVIEW EXERCISES

In Review Exercises 1–3, state the indicated law in your own words.

1. The law of cosines.

2. The law of sines.

3. Heron's formula for the area of a triangle.

4. Explain why SSA is called the *ambiguous case.*

In Review Exercises 5–16, consider the given parts of triangle ABC. Use your calculator to solve for the required value, if possible. If more than one value is possible, give both.

5. $a = 12$, $c = 15$, $B = 30°$; find b.

6. $b = 23$, $a = 13$, $C = 125°$; find c.

7. $c = 0.5$, $b = 0.8$, $A = 50°$; find a.

8. $a = 28.7$, $b = 37.8$, $C = 11.2°$; find c.

9. $a = 12$, $c = 18$, $C = 40°$; find A.

10. $b = 17$, $a = 12$, $A = 25°$; find B.

11. $c = 31.5$, $b = 27.5$, $B = 16.2°$; find C.

12. $a = 315.2$, $b = 457.8$, $A = 32.51°$; find B.

13. $A = 24.3°$, $B = 56.8°$, $a = 32.3$; find b.

14. $B = 10.3°$, $C = 59.4°$, $c = 341$; find b.

15. $b = 17$, $c = 21$, $B = 42°$; find A.

16. $c = 189$, $a = 150$, $C = 85.3°$; find B.

In Review Exercises 17–24, find the area of the triangle with the given parts, if possible.

17. $a = 32, C = 47°, b = 55$

18. $b = 29, A = 96°, c = 85$

19. $B = 33°, C = 25°, a = 17$

20. $A = 85°, C = 80°, b = 7.5$

21. $A = 130°, B = 20°, a = 3.5$

22. $C = 15°, A = 110°, c = 91$

23. $a = 57, b = 85, c = 110$

24. $a = 8.50, b = 17.4, c = 22.3$

25. Two airplanes leave an airport at 2:00 P.M., one with a heading of 30.0° and ground speed of 425 miles per hour and the other with a heading of 85.0° and ground speed of 375 miles per hour. How far apart are they at 3:30 P.M.? (Assume no wind.)

26. From a point 542 feet away from the base of the Leaning Tower of Pisa, the angle of elevation to its top is 17.9°. If the tower makes an angle with the ground of 94.5° (see Illustration 1), how tall is the tower?

27. From a point 312 feet from the base of the great pyramid of Khufu (Cheops) at Giza, the angle of elevation to its top is 25.5°. If the pyramid makes an angle with the ground of 141.8° (see Illustration 2), find its slant height.

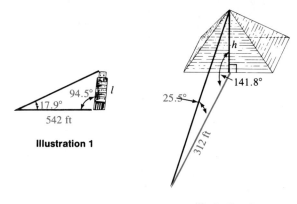

Illustration 1

Illustration 2

28. If R is the radius of a circle circumscribed about the triangle ABC in Illustration 3, then show that

$$R = \frac{a}{2 \sin A}$$

(*Hint:* Central angle α is twice inscribed angle A.)

29. An n-sided regular polygon is a geometric figure with n equal sides and n equal angles. If each side of an n-sided regular polygon is of length a, show that its area, A, is given by

$$A = \frac{na^2 \cot \dfrac{180°}{n}}{4}$$

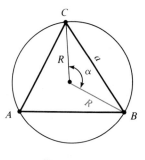

Illustration 3

(*Hint:* The area of a triangle is $\frac{1}{2} \cdot$ base $\cdot$ height)

30. Use the law of cosines to show that if $a^2 = b^2 + c^2$, then angle $A = 90°$.

31. The sides of a triangle are 13.0, 17.0, and 25.0 centimeters. What is the measure of the largest angle?

32. Find the angles of the triangle with vertices of $(0,0)$, $(5,0)$, and $(7,8)$.

33. Find the area of a triangular lot with sides of 21, 32, and 47 meters.

34. Find the area of the triangle joining points with coordinates of $(0, 0)$, $(0, 8)$, and $(6, 14)$.

In Review Exercises 35–38, assume that $\mathbf{V} = \langle 3, 7 \rangle$ *and* $\mathbf{W} = \langle -2, 5 \rangle$. *Calculate each quantity. These exercises come from an optional section.*

35. $2\mathbf{V} + 3\mathbf{W}$ **36.** $|3\mathbf{V}| - |\mathbf{W}|$ **37.** $5(\mathbf{V} - \mathbf{W})$ **38.** $|3\mathbf{V} - \mathbf{W}|$

In Review Exercises 39–42, find the angle between the two given vectors. These exercises come from an optional section.

39. $\langle 0, 5 \rangle$ and $\langle 2, 0 \rangle$ **40.** $\langle 8, 2 \rangle$ and $\langle 4, 1 \rangle$

41. $\sqrt{3}\,\mathbf{i} + \mathbf{j}$ and $\mathbf{i} - \mathbf{j}$ **42.** $2\mathbf{i} - 2\sqrt{3}\,\mathbf{j}$ and $\mathbf{i} + \sqrt{3}\,\mathbf{j}$

In Review Exercises 43–46, assume that $\mathbf{A} = \mathbf{i} + 2\sqrt{2}\,\mathbf{j}$ *and* $\mathbf{B} = -\sqrt{2}\,\mathbf{j}$. *Calculate each quantity. These exercises come from an optional section.*

43. a unit vector in the direction of $\mathbf{A}$ **44.** a unit vector in the direction of $\mathbf{B}$

45. $\text{proj}_\mathbf{B}\mathbf{A}$ **46.** $\text{comp}_\mathbf{B}\mathbf{A}$

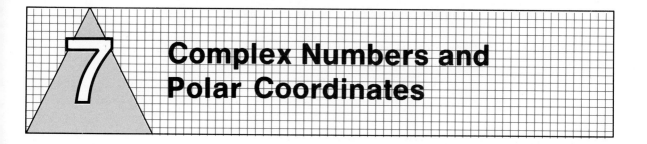

Complex Numbers and Polar Coordinates

All the numbers used thus far have been real numbers. Some situations in mathematics, such as solving certain quadratic equations, require a new set of numbers. This new set, called the set of **complex numbers**, is the main topic of this chapter.

7.1 COMPLEX NUMBERS

If we solve the equation $x^2 = 9$ for x, we find that the equation has two solutions.

$$x^2 = 9$$
$$x = \sqrt{9} \quad \text{or} \quad x = -\sqrt{9}$$
$$= 3 \qquad\qquad = -3$$

If we solve the equation $x^2 = -1$ in a similar way, we would expect it to have two solutions also.

$$x^2 = -1$$
$$x = \sqrt{-1} \quad \text{or} \quad x = -\sqrt{-1}$$

Each proposed solution of the equation $x^2 = -1$ involves the symbol $\sqrt{-1}$. The symbol $\sqrt{-1}$ cannot represent a real number, because no real number has a square that is equal to -1. For years, mathematicians believed that square roots of negative numbers, denoted by symbols such as $\sqrt{-3}$, $\sqrt{-4}$, and $\sqrt{-5}$ were nonsense. Even the great English mathematician Sir Isaac Newton (1642–1727) called them "impossible numbers." In the 17th century, these symbols were termed **imaginary numbers** by René Descartes. However, mathematicians no longer think of imaginary numbers as impossible. In fact, imaginary numbers have important uses, such as describing the behavior of alternating current in electronics.

The imaginary number $\sqrt{-1}$ occurs often enough to warrant a special symbol: the letter i is used to denote $\sqrt{-1}$. Because $i = \sqrt{-1}$, it follows that $i^2 = -1$. The powers of i with natural-number exponents produce an interesting pattern.

$$i = \sqrt{-1} = i \qquad\qquad i^5 = i^4 \cdot i = 1 \cdot i = i$$
$$i^2 = \sqrt{-1}\sqrt{-1} = -1 \qquad i^6 = i^4 \cdot i^2 = 1(-1) = -1$$
$$i^3 = i^2 \cdot i = -1 \cdot i = -i \qquad i^7 = i^4 \cdot i^3 = 1(-i) = -i$$
$$i^4 = i^2 \cdot i^2 = (-1)(-1) = 1 \qquad i^8 = i^4 \cdot i^4 = 1(1) = 1$$

The pattern continues $i, -1, -i, 1$.

We can simplify powers of i involving negative integral exponents by rationalizing denominators.

$$i^{-1} = \frac{1}{i} = \frac{1}{i} \cdot \frac{i}{i} = \frac{i}{-1} = -i$$

$$i^{-2} = \frac{1}{i^2} = \frac{1}{-1} = -1$$

$$i^{-3} = \frac{1}{i^3} = \frac{1}{i^3} \cdot \frac{i}{i} = \frac{i}{1} = i$$

$$i^{-4} = \frac{1}{i^4} = \frac{1}{1} = 1$$

If we define i^0 to be 1, then the familiar pattern for the powers of i carries over to integral exponents.

$$\vdots$$
$$i^{-3} = i$$
$$i^{-2} = -1$$
$$i^{-1} = -i$$
$$i^0 = 1$$
$$i^1 = i$$
$$i^2 = -1$$
$$i^3 = -i$$
$$i^4 = 1$$
$$\vdots$$

The sequence $i, -1, -i, 1$ repeats endlessly. Because the powers of i form a repeating sequence, it is easy to compute large powers of i.

Example 1   Simplify i^{365}.

Solution   Because $i^4 = 1$, each occurrence of i^4 is just a factor of 1. To determine how many factors of i^4 appear in i^{365}, divide 365 by 4: the quotient is 91 and the remainder is 1. Thus,

$$i^{365} = (i^4)^{91} \cdot i^1 = 1^{91} \cdot i = i$$

Note that the number 365 leaves a remainder of 1 when it is divided by 4.   ∎

Example 2   Suppose that i is raised to the nth power, where n is a natural number. Discuss the possible values for i^n.

Solution   If the exponent n of i^n is a multiple of 4, then $i^n = 1$.
If n leaves a remainder of 1 when divided by 4, then $i^n = i$.
If n leaves a remainder of 2 when divided by 4, then $i^n = -1$.
If n leaves a remainder of 3 when divided by 4, then $i^n = -i$.   ∎

Recall that, if at least one of two numbers x and y is not negative, then

$$\sqrt{xy} = \sqrt{x}\sqrt{y}$$

This allows us to write the square root of any negative number in the form bi, where b is a real number and $i = \sqrt{-1}$. For example, we have

$$\sqrt{-4} = \sqrt{4(-1)} = \sqrt{4}\sqrt{-1} = 2i$$
$$\sqrt{-9} = \sqrt{9(-1)} = \sqrt{9}\sqrt{-1} = 3i$$

and

$$\sqrt{-2} = \sqrt{2(-1)} = \sqrt{2}\sqrt{-1} = \sqrt{2}\,i$$

Because a square root of a number is one of two equal factors of that number, we know that $\sqrt{-3}\sqrt{-3} = -3$. It is incorrect to apply the rule $\sqrt{xy} = \sqrt{x}\sqrt{y}$ in this case:

$$\sqrt{-3}\sqrt{-3} = \sqrt{(-3)(-3)} = \sqrt{9} = 3$$

This example illustrates that, if x and y are both negative, then the rule $\sqrt{xy} = \sqrt{x}\sqrt{y}$ is not valid.

Expressions such as $3 + 4i$, $-5 + 7i$, and $-1 + 9i$ indicate the sum of a real number and an imaginary number. Such expressions form a new set of numbers called the set of **complex numbers**.

> **Definition.**   A **complex number** is any number that can be expressed in the form $a + bi$, where a and b are real numbers and $i = \sqrt{-1}$.
>   The number a is called the **real part** and the number b is called the **imaginary part** of the complex number $a + bi$.

If $b = 0$, the complex number $a + bi$ is the real number a. Thus, any real number is a complex number with a zero imaginary part. If $a = 0$ and $b \neq 0$, the complex number $a + bi$ is the imaginary number bi. Thus, any imaginary number is a complex number with a zero real part. It follows that the real-number set and the imaginary-number set are both subsets of the complex-number set.

We must make some definitions before doing any arithmetic with complex numbers.

> **Equality of Complex Numbers.**   Two complex numbers are equal if and only if their real parts are equal and their imaginary parts are equal. Thus, if $a + bi$ and $c + di$ are two complex numbers, then
>
> $$a + bi = c + di \quad \text{if and only if} \quad a = c \text{ and } b = d$$

Example 3   For what real-number values of x and y is $3x + 4i = (2y + x) + xi$?

Solution   Because the imaginary parts of these two complex numbers must be equal, $x = 4$. Because the real parts of these two complex numbers must be equal, $3x = 2y + x$. You can solve the system of equations

$$\begin{cases} x = 4 \\ 3x = 2y + x \end{cases}$$

by substituting 4 for x in the equation $3x = 2y + x$ and solving for y. You will find that $y = 4$. Hence, the solution is $x = 4$ and $y = 4$.   ∎

Complex numbers can be added and multiplied as if they were binomials.

> **Addition of Complex Numbers.**   Two complex numbers such as $a + bi$ and $c + di$ are added as if they were algebraic binomials:
>
> $$(a + bi) + (c + di) = (a + c) + (b + d)i$$

The definition for adding two complex numbers implies that the sum of two complex numbers is another complex number.

Example 4   Find the sum of $3 + 4i$ and $2 + 7i$.

Solution $(3 + 4i) + (2 + 7i) = 3 + 4i + 2 + 7i$
$$= 3 + 2 + 4i + 7i$$
$$= 5 + 11i$$ ∎

> **Multiplication of Complex Numbers.**   Two complex numbers such as $a + bi$ and $c + di$ are multiplied as if they were algebraic binomials, with $i^2 = -1$:
>
> $$(a + bi)(c + di) = (ac - bd) + (ad + bc)i$$

The definition for multiplying two complex numbers implies that the product of two complex numbers is another complex number.

Example 5 Find the product of $3 + 4i$ and $2 + 7i$.

Solution
$$(3 + 4i)(2 + 7i) = 6 + 21i + 8i + 28i^2$$
$$= 6 + 21i + 8i - 28$$
$$= -22 + 29i \qquad \blacksquare$$

When working with complex numbers, always express all numbers in $a + bi$ form before attempting any algebraic manipulations. This procedure will help you avoid making errors in determining the sign of the result.

Example 6 Find the product of $-2 + \sqrt{-16}$ and $4 - \sqrt{-9}$.

Solution First, change each complex number to $a + bi$ form.
$$-2 + \sqrt{-16} = -2 + \sqrt{16}\sqrt{-1}$$
$$= -2 + 4i$$
$$4 - \sqrt{-9} = 4 - \sqrt{9}\sqrt{-1}$$
$$= 4 - 3i$$

Then, find the product of $-2 + 4i$ and $4 - 3i$.
$$(-2 + 4i)(4 - 3i) = -8 + 6i + 16i - 12i^2$$
$$= -8 + 6i + 16i + 12$$
$$= 4 + 22i \qquad \blacksquare$$

The following definition is useful in finding the quotient of two complex numbers.

Definition. The complex numbers $a + bi$ and $a - bi$ are called **conjugates** of each other.

The conjugate of $3 + 4i$ is $3 - 4i$, and the conjugate of $-\frac{1}{2} - 4i$ is $-\frac{1}{2} + 4i$.

Example 7 Write the number $\dfrac{3}{2 + i}$ in $a + bi$ form.

Solution To write the given number in $a + bi$ form, multiply both the numerator and the denominator by the conjugate of the denominator, and simplify.

$$\frac{3}{2+i} = \frac{3}{2+i} \cdot \frac{2-i}{2-i}$$

$$= \frac{6-3i}{4-2i+2i-i^2}$$

$$= \frac{6-3i}{4+1}$$

$$= \frac{6}{5} - \frac{3}{5}i$$

$$= \frac{6}{5} + \left(-\frac{3}{5}\right)i$$

It is common practice to accept $\frac{6}{5} - \frac{3}{5}i$ as a substitute for $\frac{6}{5} + (-\frac{3}{5})i$. ■

Example 8   Write the number $\dfrac{2 - \sqrt{-64}}{3 + \sqrt{-1}}$ in $a + bi$ form.

Solution $\dfrac{2 - \sqrt{-64}}{3 + \sqrt{-1}} = \dfrac{2 - 8i}{3 + i}$

$$= \frac{2-8i}{3+i} \cdot \frac{3-i}{3-i}$$

$$= \frac{6 - 2i - 24i + 8i^2}{9 - 3i + 3i - i^2}$$

$$= \frac{-2 - 26i}{9 + 1}$$

$$= \frac{2(-1 - 13i)}{10}$$

$$= -\frac{1}{5} - \frac{13}{5}i$$ ■

The results of Examples 7 and 8 illustrate that the quotient of two **complex** numbers is another complex number.

The solutions of certain quadratic equations are complex numbers. The following example shows how the quadratic formula can be used to find these solutions.

Example 9   Use the quadratic formula to solve the quadratic equation $x^2 - 4x + 5 = 0$.

Solution   Substitute 1 for a, -4 for b, and 5 for c in the quadratic formula, and simplify.

$$x = \frac{-b \pm \sqrt{b^2 - 4ac}}{2a}$$

$$= \frac{-(-4) \pm \sqrt{(-4)^2 - 4(1)(5)}}{2(1)}$$

$$= \frac{4 \pm \sqrt{16 - 20}}{2}$$

$$= \frac{4 \pm \sqrt{-4}}{2}$$

$$= \frac{4 \pm 2i}{2}$$

$$= 2 \pm i$$

The two solutions of the equation $x^2 - 4x + 5 = 0$ are $x = 2 + i$ and $x = 2 - i$.

■

EXERCISE 7.1

In Exercises 1–12, simplify each expression.

1. i^9 **2.** i^{27} **3.** i^{38} **4.** i^{99} **5.** $\dfrac{1}{i^5}$ **6.** $\dfrac{1}{i^{10}}$

7. $\dfrac{1}{i^{99}}$ **8.** $\dfrac{1}{i^{1776}}$ **9.** i^{-1984} **10.** i^{-1492} **11.** i^{-1111} **12.** i^{-2001}

In Exercises 13–18, solve for the real numbers x and y. You may have to solve a system of equations after using the definition of equality of complex numbers.

13. $x + (x + y)i = y - i$ **14.** $x + 3y + 2xi + yi = 5x + 2 + 2yi$

15. $x + yi = y + 2xi$ **16.** $x + y + (x - y)i = x + yi$

17. $3x - 2yi = 2 + (x + y)i$ **18.** $\begin{cases} 2 + (x + y)i = 2 - i \\ \quad x + 3i = 2 + 3i \end{cases}$

In Exercises 19–46, perform any indicated operations, and express the final answer in a + bi form.

19. $(2 - 7i) + (3 + i)$ **20.** $(-7 + 2i) + (2 - 8i)$

21. $(5 - 6i) - (7 + 4i)$ **22.** $(11 + 2i) - (13 - 5i)$

23. $(14i + 2) + (2 - 4i)$ **24.** $(5 + 8i) - (23i - 32)$

25. $(2 + 3i)(-3 - 4i)$ **26.** $(-2 - 4i)(4 + 5i)$

27. $(-11 + 5i)(-2 - 6i)$ **28.** $(6 + 7i)(6 - 7i)$

29. $(2 + 3i)^2$ **30.** $(3 - 4i)^2$

31. $(\sqrt{-16} + 3)(2 + \sqrt{-9})$ **32.** $(12 - \sqrt{-4})(\sqrt{-25} + 7)$

33. $(2 - i)^3$ **34.** $(2 + 3i)^3$ **35.** $\dfrac{3}{4i^2}$ **36.** $\dfrac{-11}{3i^3}$ **37.** $\dfrac{1}{2 + i}$ **38.** $\dfrac{-2}{3 - i}$

39. $\dfrac{2i}{7 + i}$ **40.** $\dfrac{-3i}{2 + 5i}$ **41.** $\dfrac{2 + i}{3 - i}$ **42.** $\dfrac{4 - \sqrt{-25}}{2 + \sqrt{-9}}$

43. $\dfrac{-\sqrt{-16} + \sqrt{5}}{-8 + \sqrt{-4}}$ **44.** $\dfrac{34 + 2i}{\sqrt{2} - 4i}$ **45.** $\dfrac{2 + i\sqrt{3}}{3 + i}$ **46.** $\dfrac{3}{4 - i\sqrt{2}}$

47. Verify that $1 - i$ is a square root of $-2i$ by showing that $(1 - i)^2 = -2i$.

48. Verify that $2 - i$ is a square root of $3 - 4i$ by showing that $(2 - i)^2 = 3 - 4i$.

49. Verify that $1 + 2i$ is a square root of $-3 + 4i$ by showing that $(1 + 2i)^2 = -3 + 4i$.

50. Verify that $2 + i$ is a square root of $3 + 4i$ by showing that $(2 + i)^2 = 3 + 4i$.

51. Verify that $-\dfrac{1}{2} + \dfrac{\sqrt{3}}{2}i$ and $-\dfrac{1}{2} - \dfrac{\sqrt{3}}{2}i$ are roots of $x^2 + x + 1 = 0$.

52. Verify that $\dfrac{1}{2} + \dfrac{\sqrt{5}}{2}i$ and $\dfrac{1}{2} - \dfrac{\sqrt{5}}{2}i$ are not roots of $x^2 - x - 1 = 0$.

In Exercises 53–58, use the quadratic formula to solve each equation.

53. $x^2 + 2x + 2 = 0$ **54.** $x^2 + 4x + 8 = 0$ **55.** $x^2 + 4x + 5 = 0$ **56.** $x^2 + 2x + 5 = 0$

57. $9x^2 - 6x + 2 = 0$ **58.** $4x^2 - 4x + 5 = 0$

In Exercises 59–62, use a calculator and write each expression in $a + bi$ form.

59. $(2.3 + 4.5i)(8.9 - 3.2i)$ **60.** $(6.73 - 3.25i)^2 + (1.75 + 2.21i)$

61. $\dfrac{4.7 + 11.2i}{5.2 - 3.7i} - (6.5 - 7.4i)$ **62.** $\dfrac{29.8 - 45.3i}{-7.4 + 27.3i}$

63. Show that the product of the complex number $a + bi$ and its conjugate $a - bi$ is the real number $a^2 + b^2$.

64. Show that the sum of the complex number $a + bi$ and its conjugate $a - bi$ is a real number.

65. Show that the square of a complex number is a complex number.

66. Show that the quotient $\dfrac{a + bi}{c + di}$ is a complex number.

7.2   GRAPHING COMPLEX NUMBERS

Although complex numbers obey most of the properties of real numbers, there is one property that does not carry over. There is no way of ordering the complex numbers that is consistent with the ordering established for the real numbers. It makes no sense to say, for example, that $3i$ is larger than 5, that 0 is less than $3 + i$, or that $5 + 5i$ is greater than $-5 + i^2$. Because complex numbers are not linearly ordered, it is impossible to graph them on a number line in any way that preserves the ordering of the real numbers. This inability to graph complex numbers on the number line is not surprising because real numbers provide coordinates for all points on the number line. However, there *is* a method for graphing the complex numbers.

We construct two perpendicular axes as in Figure 7-1 and consider the horizontal axis to be the axis of real numbers and the vertical axis to be the axis of imaginary numbers. These axes determine what is called the **complex plane**.

Although these axes resemble the *x*-axis and *y*-axis encountered in earlier chapters, instead of plotting *x* and *y* values, we plot ordered pairs of real numbers (*a, b*), where *a* and *b* are the parts of the complex number *a* + *bi*. The axis of reals is used for plotting *a* and the axis of imaginaries is used for plotting *b*.

Example 1   Graph the complex number 3 + 2*i*.

Solution   To graph the complex number 3 + 2*i*, plot the point *P*(3, 2) as in Figure 7-1. The vector drawn from the origin to point *P* is considered to be the graph of the complex number 3 + 2*i*.

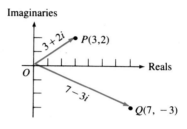

Figure 7-1

Example 2   Graph the complex number 7 − 3*i*.

Solution   To graph the complex number 7 − 3*i*, plot the point *Q*(7, −3) as in Figure 7-1. The vector *OQ* is the graph of the complex number 7 − 3*i*. ∎

The length of the vector that represents a complex number is called the **absolute value** of that complex number.

> **Absolute Value of a Complex Number.**   If *a* + *bi* is a complex number, then
> $$|a + bi| = \sqrt{a^2 + b^2}$$

See Figure 7-2 and note that $\sqrt{a^2 + b^2}$ is the length of the hypotenuse of a right triangle with sides of length |*a*| and |*b*|. On the complex plane, |*a* + *bi*| is the

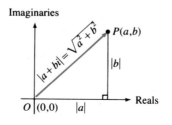

Figure 7-2

distance between the point (a, b) and the origin $(0, 0)$. This is consistent with the definition of the absolute value of a real number. The expression $|x|$ represents the distance on the number line between the point with coordinate x and the origin O.

Example 3   Find the value of **a.** $|3 + 4i|$ and **b.** $|2 - 5i|$.

Solution   **a.** $|3 + 4i| = \sqrt{3^2 + 4^2} = \sqrt{25} = 5$

b. $|2 - 5i| = \sqrt{2^2 + (-5)^2} = \sqrt{29}$ ∎

■ EXERCISE 7.2 ■

In Exercises 1–12, graph each complex number. You may have to write some numbers in $a + bi$ form first.

1. $2 + 3i$ **2.** $2 - 3i$ **3.** $-2 + 3i$ **4.** $-2 - 3i$ **5.** $(4 + i)^2$ **6.** $(4 - i)^3$

7. $\dfrac{2 + i}{3 - i}$ **8.** $\dfrac{-2 + 2i}{-3 - i}$ **9.** 6 **10.** -8 **11.** $7i$ **12.** $-5i$

In Exercises 13–24, compute each absolute value.

13. $|2 + 3i|$ **14.** $|-5 - i|$ **15.** $|-7 + 7i|$ **16.** $\left|\dfrac{1}{2} - \dfrac{1}{4}i\right|$

17. $|6|$ **18.** $|-6|$ **19.** $|5i|$ **20.** $|-4i|$

21. $\left|\dfrac{-3i}{2 + i}\right|$ **22.** $\left|\dfrac{5i}{i - 2}\right|$ **23.** $\left|\dfrac{4 - i}{4 + i}\right|$ **24.** $\left|\dfrac{2 + i}{2 - i}\right|$

In Exercises 25–30, let $z = a + bi$. Assume that $\bar{z}$ represents the conjugate of z, that is, $\bar{z} = a - bi$.

25. Show that $|z| = |\bar{z}|$.

26. Show that $|z| + |\bar{z}| = 2|z|$.

27. Show that $|z||\bar{z}| = |z|^2$

28. Show that $|z + \bar{z}| = 2|a|$.

29. Show that $\sqrt{z\bar{z}} = |z|$.

30. Show that $z = \bar{z}$ if and only if z is a real number.

31. Show that the addition of two complex numbers is commutative. Do this by adding the complex numbers $a + bi$ and $c + di$ in both orders and observing that the sums are equal.

32. Show that the multiplication of two complex numbers is commutative. Do this by multiplying the complex numbers $a + bi$ and $c + di$ in both orders and observing that the products are equal.

33. Show that the addition of complex numbers is associative.

34. Find three examples of complex numbers that are reciprocals of their own conjugates.

In Exercises 35–38, let $x = a + bi$ and $y = c + di$. Assume that the bar over each symbol is read as "the conjugate of." For example, $\overline{x + y}$ means "the conjugate of the sum of x and y."

35. Show that $\overline{x + y} = \bar{x} + \bar{y}$.

36. Show that $\overline{xy} = \bar{x}\bar{y}$.

37. Show that $\bar{\bar{x}} = x$

38. Show that $\overline{x^2} = \bar{x}^2$

7.3 TRIGONOMETRIC FORM OF A COMPLEX NUMBER

If a and b are real numbers, any complex number in the algebraic form $a + bi$ can be written in a form called **trigonometric form**. To write the complex number $a + bi$ in trigonometric form, we refer to Figure 7-3 in which

$$r = |a + bi|$$

$$\sin \theta = \frac{b}{r} \quad \text{or} \quad b = r \sin \theta$$

and

$$\cos \theta = \frac{a}{r} \quad \text{or} \quad a = r \cos \theta$$

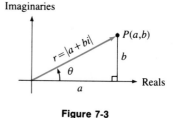

Figure 7-3

We then substitute $r \cos \theta$ for a and $r \sin \theta$ for b in the expression $a + bi$ to obtain

$$a + bi = r \cos \theta + r \sin \theta i$$
$$= r(\cos \theta + i \sin \theta)$$

The expression $r(\cos \theta + i \sin \theta)$ is the trigonometric form of the complex number $a + bi$.

Trigonometric Form of a Complex Number.   If $r = |a + bi|$ and θ is any angle for which $\sin \theta = \frac{b}{r}$ and $\cos \theta = \frac{a}{r}$, then

$$a + bi = r(\cos \theta + i \sin \theta)$$

If a complex number is written in $r(\cos \theta + i \sin \theta)$ form, the number r is called the **modulus**, and the angle θ is called the **argument**. The notation $r(\cos \theta + i \sin \theta)$ is often abbreviated as $r \operatorname{cis} \theta$. Thus, we have

---

$$r \operatorname{cis} \theta = r(\cos \theta + i \sin \theta)$$

Example 1   Write the complex number $3 + 4i$ in trigonometric form.

Solution   To write $3 + 4i$ in the form $r(\cos \theta + i \sin \theta)$, you must find the appropriate values for r and θ. To do this, graph the complex number as in Figure 7-4. Because

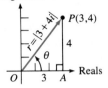

Imaginaries

$P(3,4)$

Reals

O 3 A

Figure 7-4

triangle OAP is a right triangle, you have $r^2 = 3^2 + 4^2$ or $r = 5$. From the figure, it follows that $\tan \theta = \frac{4}{3}$, so $\theta \approx 53.1°$. Substituting these values for r and θ gives

$$3 + 4i \approx 5(\cos 53.1° + i \sin 53.1°)$$

Note that $5(\cos 53.1° + i \sin 53.1°)$ can be written more compactly as $5 \operatorname{cis} 53.1°$. ■

Example 2   Write the complex number $-2 - 7i$ in trigonometric form.

Solution   To write $-2 - 7i$ in the form $r(\cos \theta + i \sin \theta)$, you must find the appropriate values of r and θ. To do this, graph the number $-2 - 7i$ as in Figure 7-5.

$$r = |-2 - 7i| = \sqrt{(-2)^2 + (-7)^2}$$
$$= \sqrt{4 + 49} = \sqrt{53}$$

Since $\tan \alpha = \frac{7}{2}$, it follows that $\alpha \approx 74.1°$. Find θ by noting that θ is a third-quadrant angle, so that

$$\theta = 180° + \alpha \approx 180° + 74.1° = 254.1°$$

Thus, you have

$$-2 - 7i \approx \sqrt{53}(\cos 254.1° + i \sin 254.1°)$$

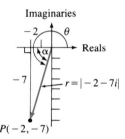

Figure 7-5

Example 3   Change the complex number $10(\cos 60° + i \sin 60°)$ to $a + bi$ form.

Solution $$10(\cos 60° + i \sin 60°) = 10\left(\frac{1}{2} + i\frac{\sqrt{3}}{2}\right)$$
$$= 5 + 5\sqrt{3}\,i$$ ■

Multiplying Numbers in Trigonometric Form

It is easy to add two complex numbers in $a + bi$ form—simply add their real and imaginary components. However, it is difficult to multiply and divide complex numbers if they are written in algebraic form. Fortunately, it is easy to multiply and divide two complex numbers written in trigonometric form. To show how, we suppose that $N_1 = r_1(\cos \theta_1 + i \sin \theta_1)$ and $N_2 = r_2(\cos \theta_2 + i \sin \theta_2)$. The product $N_1 N_2$ can be found as follows:

$$N_1 N_2 = r_1(\cos \theta_1 + i \sin \theta_1) \cdot r_2(\cos \theta_2 + i \sin \theta_2)$$
$$= r_1 r_2[\cos \theta_1 \cos \theta_2 + i \cos \theta_1 \sin \theta_2 + i \sin \theta_1 \cos \theta_2 - \sin \theta_1 \sin \theta_2]$$
$$= r_1 r_2[(\cos \theta_1 \cos \theta_2 - \sin \theta_1 \sin \theta_2) + i(\cos \theta_1 \sin \theta_2 + \sin \theta_1 \cos \theta_2)]$$
$$= r_1 r_2[\cos(\theta_1 + \theta_2) + i \sin(\theta_1 + \theta_2)]$$

This result can be stated more compactly by using $r \operatorname{cis} \theta$ notation:

$$r_1 \operatorname{cis} \theta_1 \cdot r_2 \operatorname{cis} \theta_2 = r_1 r_2 \operatorname{cis}(\theta_1 + \theta_2)$$

Thus, to form a product of two complex numbers written in trigonometric form, we multiply the moduli r_1 and r_2 and add the arguments θ_1 and θ_2. This property generalizes to products containing any number of factors.

$$r_1(\cos \theta_1 + i \sin \theta_1) \cdot r_2(\cos \theta_2 + i \sin \theta_2) \cdot \cdots \cdot r_n(\cos \theta_n + i \sin \theta_n)$$
$$= r_1 r_2 \cdot \cdots \cdot r_n[\cos(\theta_1 + \theta_2 + \cdots + \theta_n) + i \sin(\theta_1 + \theta_2 + \cdots + \theta_n)]$$

or

$$r_1 \operatorname{cis} \theta_1 \cdot r_2 \operatorname{cis} \theta_2 \cdot \cdots \cdot r_n \operatorname{cis} \theta_n$$
$$= r_1 r_2 \cdot \cdots \cdot r_n \operatorname{cis}(\theta_1 + \theta_2 + \cdots + \theta_n)$$

Example 4   Find the product of $2(\cos 40° + i \sin 40°)$ and $3(\cos 30° + i \sin 30°)$.

Solution   To find the product, multiply the moduli and add the arguments.

$$2(\cos 40° + i \sin 40°) \cdot 3(\cos 30° + i \sin 30°)$$
$$= 2 \cdot 3[\cos(40° + 30°) + i \sin(40° + 30°)]$$
$$= 6(\cos 70° + i \sin 70°) \qquad \blacksquare$$

Example 5   Find the product of $2 \operatorname{cis} 10°$, $4 \operatorname{cis} 20°$, and $9 \operatorname{cis} 50°$.

Solution   To find the product, multiply the moduli and add the arguments.

$$(2 \operatorname{cis} 10°)(4 \operatorname{cis} 20°)(9 \operatorname{cis} 50°) = 2 \cdot 4 \cdot 9 \operatorname{cis}(10° + 20° + 50°)$$
$$= 72 \operatorname{cis} 80° \qquad \blacksquare$$

Dividing Numbers in Trigonometric Form

If two complex numbers are written in trigonometric form, their quotient can be found by dividing their moduli and subtracting their arguments. To show that this is true, we let $N_1 = r_1(\cos \theta_1 + i \sin \theta_1)$ and $N_2 = r_2(\cos \theta_2 + i \sin \theta_2)$. Then we have

$$\frac{N_1}{N_2} = \frac{r_1(\cos \theta_1 + i \sin \theta_1)}{r_2(\cos \theta_2 + i \sin \theta_2)}$$
$$= \frac{r_1}{r_2} \cdot \frac{\cos \theta_1 + i \sin \theta_1}{\cos \theta_2 + i \sin \theta_2} \cdot \frac{\cos \theta_2 - i \sin \theta_2}{\cos \theta_2 - i \sin \theta_2}$$
$$= \frac{r_1}{r_2} \cdot \frac{\cos \theta_1 \cos \theta_2 - i \cos \theta_1 \sin \theta_2 + i \sin \theta_1 \cos \theta_2 - i^2 \sin \theta_1 \sin \theta_2}{\cos^2 \theta_2 - i^2 \sin^2 \theta_2}$$

$$= \frac{r_1}{r_2} \cdot \frac{\cos\theta_1 \cos\theta_2 + \sin\theta_1 \sin\theta_2 + i(\sin\theta_1 \cos\theta_2 - \cos\theta_1 \sin\theta_2)}{\cos^2\theta_2 + \sin^2\theta_2}$$

$$= \frac{r_1}{r_2} \cdot \frac{\cos(\theta_1 - \theta_2) + i\sin(\theta_1 - \theta_2)}{1}$$

$$= \frac{r_1}{r_2}[\cos(\theta_1 - \theta_2) + i\sin(\theta_1 - \theta_2)]$$

Thus, to divide complex numbers written in trigonometric form, we use the following rule:

$$\frac{r_1(\cos\theta_1 + i\sin\theta_1)}{r_2(\cos\theta_2 + i\sin\theta_2)} = \frac{r_1}{r_2}[\cos(\theta_1 - \theta_2) + i\sin(\theta_1 - \theta_2)]$$

or

$$\frac{r_1 \operatorname{cis}\theta_1}{r_2 \operatorname{cis}\theta_2} = \frac{r_1}{r_2}\operatorname{cis}(\theta_1 - \theta_2)$$

Example 6   Divide 8 cis 110° by 4 cis 50°.

Solution   To find the quotient, divide the moduli and subtract the arguments.

$$\frac{8 \operatorname{cis} 110°}{4 \operatorname{cis} 50°} = \frac{8}{4}\operatorname{cis}(110° - 50°)$$

$$= 2 \operatorname{cis} 60°$$

Note that 2 cis 60° can be written in the form $2(\cos 60° + i\sin 60°)$.   ■

■ EXERCISE 7.3

In Exercises 1–12, write each complex number in trigonometric form.

1. $6 + 0i$
2. $-7 + 0i$
3. $0 - 3i$
4. $0 + 4i$
5. $-1 - i$
6. $-1 + i$
7. $3 + 3i\sqrt{3}$
8. $7 - 7i$
9. $-1 - \sqrt{3}i$
10. $-3\sqrt{3} + 3i$
11. $-\sqrt{3} - i$
12. $1 + i\sqrt{3}$

In Exercises 13–24, write each complex number in a + bi form.

13. $2(\cos 30° + i\sin 30°)$
14. $5(\cos 45° + i\sin 45°)$
15. $7(\cos 90° + i\sin 90°)$
16. $12(\cos 0° + i\sin 0°)$
17. $-2\left(\cos\dfrac{2\pi}{3} + i\sin\dfrac{2\pi}{3}\right)$
18. $3\left(\cos\dfrac{4\pi}{3} + i\sin\dfrac{4\pi}{3}\right)$
19. $\dfrac{1}{2}(\cos\pi + i\sin\pi)$
20. $-\dfrac{2}{3}(\cos 2\pi + i\sin 2\pi)$

21. $-3 \operatorname{cis} 225°$ **22.** $3 \operatorname{cis} 300°$ **23.** $11 \operatorname{cis} \dfrac{11\pi}{6}$ **24.** $9 \operatorname{cis} 3$

In Exercises 25–36, find each product.

25. $[4(\cos 30° + i \sin 30°)][2(\cos 60° + i \sin 60°)]$

26. $[3(\cos 45° + i \sin 45°)][2(\cos 120° + i \sin 120°)]$

27. $(\cos 300° + i \sin 300°)(\cos 0° + i \sin 0°)$

28. $[5(\cos 85° + i \sin 85°)][2(\cos 65° + i \sin 65°)]$

29. $[2(\cos \pi + i \sin \pi)][3(\cos \pi + i \sin \pi)]$

30. $\left(\cos \dfrac{\pi}{2} + i \sin \dfrac{\pi}{2}\right)\left(\cos \dfrac{3\pi}{2} + i \sin \dfrac{3\pi}{2}\right)$

31. $\left[2\left(\cos \dfrac{\pi}{3} + i \sin \dfrac{\pi}{3}\right)\right]\left[3\left(\cos \dfrac{\pi}{6} + i \sin \dfrac{\pi}{6}\right)\right]$

32. $\left[3\left(\cos \dfrac{5\pi}{6} + i \sin \dfrac{5\pi}{6}\right)\right]\left[4\left(\cos \dfrac{7\pi}{6} + i \sin \dfrac{7\pi}{6}\right)\right]$

33. $(3 \operatorname{cis} 12°)(2 \operatorname{cis} 22°)(5 \operatorname{cis} 82°)$

34. $(2 \operatorname{cis} 50°)(3 \operatorname{cis} 100°)(6 \operatorname{cis} 2°)$

35. $\left(3 \operatorname{cis} \dfrac{\pi}{2}\right)\left(4 \operatorname{cis} \dfrac{\pi}{3}\right)\left(3 \operatorname{cis} \dfrac{\pi}{4}\right)$

36. $\left(4 \operatorname{cis} \dfrac{\pi}{6}\right)\left(2 \operatorname{cis} \dfrac{2\pi}{3}\right)\left(\operatorname{cis} \dfrac{\pi}{4}\right)$

In Exercises 37–44, find each quotient.

37. $\dfrac{12(\cos 60° + i \sin 60°)}{2(\cos 30° + i \sin 30°)}$

38. $\dfrac{24(\cos 150° + i \sin 150°)}{48(\cos 50° + i \sin 50°)}$

39. $\dfrac{18(\cos \pi + i \sin \pi)}{12\left(\cos \dfrac{\pi}{2} + i \sin \dfrac{\pi}{2}\right)}$

40. $\dfrac{15(\cos 2\pi + i \sin 2\pi)}{45(\cos \pi + i \sin \pi)}$

41. $\dfrac{12 \operatorname{cis} 250°}{5 \operatorname{cis} 120°}$

42. $\dfrac{365 \operatorname{cis} 370°}{20 \operatorname{cis} 255°}$

43. $\dfrac{\operatorname{cis} \dfrac{2\pi}{3}}{2 \operatorname{cis} \dfrac{\pi}{6}}$

44. $\dfrac{250 \operatorname{cis} \dfrac{7\pi}{16}}{50 \operatorname{cis} \dfrac{\pi}{3}}$

In Exercises 45–48, simplify each expression.

45. $\dfrac{(2 \operatorname{cis} 60°)(3 \operatorname{cis} 20°)}{6 \operatorname{cis} 40°}$

46. $\dfrac{36 \operatorname{cis} 200°}{(2 \operatorname{cis} 40°)(9 \operatorname{cis} 10°)}$

47. $\dfrac{48 \operatorname{cis} \dfrac{11\pi}{6}}{\left(3 \operatorname{cis} \dfrac{\pi}{3}\right)\left(4 \operatorname{cis} \dfrac{2\pi}{3}\right)}$

48. $\dfrac{(96 \operatorname{cis} \pi)(12 \operatorname{cis} 2\pi)}{\left(48 \operatorname{cis} \dfrac{\pi}{2}\right)\left(3 \operatorname{cis} \dfrac{3\pi}{2}\right)}$

7.4 DE MOIVRE'S THEOREM

We have shown that if two complex numbers are written in trigonometric form, their product is found by multiplying their moduli and adding their arguments. This property makes it easy to find powers of complex numbers that are expressed in trigonometric form. For example, we can find the cube of $3(\cos 40° + i \sin 40°)$ as follows:

$$[3(\cos 40° + i \sin 40°)]^3$$
$$= [3(\cos 40° + i \sin 40°)][3(\cos 40° + i \sin 40°)][3(\cos 40° + i \sin 40°)]$$
$$= 3^3[\cos(40° + 40° + 40°) + i \sin(40° + 40° + 40°)]$$
$$= 27[\cos 3(40°) + i \sin 3(40°)]$$
$$= 27(\cos 120° + i \sin 120°)$$

The generalization of the previous example is called **De Moivre's theorem.**

De Moivre's Theorem. If n is a real number and $r(\cos \theta + i \sin \theta)$ is a complex number in trigonometric form, then

$$[r(\cos \theta + i \sin \theta)]^n = r^n[\cos n\theta + i \sin n\theta]$$

or

$$(r \operatorname{cis} \theta)^n = r^n \operatorname{cis} n\theta$$

This theorem was first developed around 1730 by the French mathematician Abraham De Moivre. It can be proved for all natural numbers n by using mathematical induction. However, the theorem is true for negative and fractional powers as well.

Example 1 Find $[2(\cos 15° + i \sin 15°)]^4$.

Solution Use De Moivre's theorem.

$$[2(\cos 15° + i \sin 15°)]^4 = 2^4[\cos 4 \cdot 15° + i \sin 4 \cdot 15°]$$
$$= 16[\cos 60° + i \sin 60°]$$

This result can be changed to algebraic form if desired:

$$16(\cos 60° + i \sin 60°) = 16\left(\frac{1}{2} + i\frac{\sqrt{3}}{2}\right) = 8 + 8\sqrt{3}\,i$$

∎

Example 2 Find $[\sqrt{2}(\cos 10° + i \sin 10°)]^{10}$.

Solution Use De Moivre's theorem.

$$[\sqrt{2}(\cos 10° + i \sin 10°)]^{10} = (\sqrt{2})^{10}[(\cos 10 \cdot 10° + i \sin 10 \cdot 10°)]$$
$$= 32(\cos 100° + i \sin 100°)$$

∎

De Moivre's theorem can be used to find all the nth roots of any number. Because both real and complex numbers can be written in the form $a + bi$, they can be written in trigonometric form as well. Thus, one nth root of $a + bi$ is

$$\sqrt[n]{a + bi} = (a + bi)^{1/n}$$
$$= [r(\cos \theta + i \sin \theta)]^{1/n}$$
$$= \sqrt[n]{r}\left(\cos \frac{\theta}{n} + i \sin \frac{\theta}{n}\right)$$

Recall from algebra that the equation $x^2 = 9$ has two distinct roots, 3 and -3, and each qualifies as a square root of 9. Likewise, the equation $x^n = a + bi$ has n distinct roots, and each qualifies as an nth root of the complex number $a + bi$. It follows that there are n distinct nth roots of any complex number, and De Moivre's theorem can be used to find them all. Because $\sin \theta = \sin(\theta + k \cdot 360°)$ and $\cos \theta = \cos(\theta + k \cdot 360°)$ for all integers k, De Moivre's theorem implies that

$$[r(\cos \theta + i \sin \theta)]^{1/n} = \{r[\cos(\theta + k \cdot 360°) + i \sin(\theta + k \cdot 360°)]\}^{1/n}$$
$$= r^{1/n}\left(\cos \frac{\theta + k \cdot 360°}{n} + i \sin \frac{\theta + k \cdot 360°}{n}\right)$$

Substituting the numbers $0, 1, 2, \ldots, (n - 1)$ for k yields the n nth roots of the given complex number.

Example 3   Find the three cube roots of 8.

Solution   Because 8 can be expressed as $8 + 0i$, graph the complex number $8 + 0i$ as in Figure 7-6 to see that $r = 8$ and $\theta = 0°$. Write $8 + 0i$ in trigonometric form, and use the equation

$$[r(\cos \theta + i \sin \theta)]^{1/n} = \sqrt[n]{r}\left[\cos \frac{\theta + k \cdot 360°}{n} + i \sin \frac{\theta + k \cdot 360°}{n}\right]$$

Imaginaries

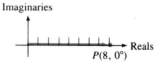

Reals

$P(8, 0°)$

Figure 7-6

Substituting the values for n, r, and θ gives

$$[8(\cos 0° + i \sin 0°)]^{1/3} = 8^{1/3}\left[\cos \frac{0° + k \cdot 360°}{3} + i \sin \frac{0° + k \cdot 360°}{3}\right]$$

Substituting 0 for k and replacing $8^{1/3}$ with 2 gives

$$2\left(\cos \frac{0°}{3} + i \sin \frac{0°}{3}\right) = 2(\cos 0° + i \sin 0°)$$
$$= 2(1 + 0i)$$
$$= 2$$

Now substitute 1 for k.

$$2\left(\cos\frac{0° + 360°}{3} + i\sin\frac{0° + 360°}{3}\right) = 2(\cos 120° + i\sin 120°)$$

$$= 2\left(-\frac{1}{2} + i\frac{\sqrt{3}}{2}\right)$$

$$= -1 + i\sqrt{3}$$

Finally, substitute 2 for k.

$$2\left(\cos\frac{0° + 720°}{3} + i\sin\frac{0° + 720°}{3}\right) = 2(\cos 240° + i\sin 240°)$$

$$= 2\left(-\frac{1}{2} + i\frac{-\sqrt{3}}{2}\right)$$

$$= -1 - i\sqrt{3}$$

The values 2, $-1 + i\sqrt{3}$, and $-1 - i\sqrt{3}$ are the three cube roots of 8. If these three cube roots of 8 are graphed in the complex plane, they are equally spaced around a circle of radius 2. If the endpoints of these vectors are joined by straight-line segments, an equilateral triangle is formed as in Figure 7-7.

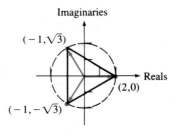

Figure 7-7

Example 4   Find the four fourth roots of $-16i$.

Solution   Express $-16i$ as $0 - 16i$ and graph it to determine that $r = 16$ and $\theta = 270°$. Hence,

$$\sqrt[4]{-16i} = [16(\cos 270° + i\sin 270°)]^{1/4}$$

$$= 16^{1/4}\left(\cos\frac{270° + k \cdot 360°}{4} + i\sin\frac{270° + k \cdot 360°}{4}\right)$$

Substitute 0, 1, 2, and 3 for k to get the four fourth roots of $-16i$.

$$2(\cos 67.5° + i\sin 67.5°) \approx 0.77 + 1.85i$$

$$2(\cos 157.5° + i\sin 157.5°) \approx -1.85 + 0.77i$$

$$2(\cos 247.5° + i\sin 247.5°) \approx -0.77 - 1.85i$$

$$2(\cos 337.5° + i\sin 337.5°) \approx 1.85 - 0.77i$$

Graphs of these four fourth roots are equally spaced about a circle of radius 2, and the endpoints of these vectors are the vertices of a square. See Figure 7-8.

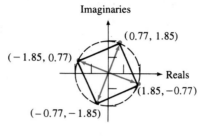

Figure 7-8

Example 5   Find the five fifth roots of $-4 - 4i$.

Solution   Express $-4 - 4i$ in trigonometric form by determining that $r = 4\sqrt{2}$ and $\theta = 225°$. Then, you have

$$[4\sqrt{2}(\cos 225° + i\sin 225°)]^{1/5}$$

$$= (4\sqrt{2})^{1/5}\left(\cos\frac{225° + k \cdot 360°}{5} + i\sin\frac{225° + k \cdot 360°}{5}\right)$$

$$= \sqrt{2}\left(\cos\frac{225° + k \cdot 360°}{5} + i\sin\frac{225° + k \cdot 360°}{5}\right)$$

Substituting 0, 1, 2, 3, and 4 for k generates the five fifth roots of $-4 - 4i$.

$$\sqrt{2}(\cos 45° + i\sin 45°)$$
$$\sqrt{2}(\cos 117° + i\sin 117°)$$
$$\sqrt{2}(\cos 189° + i\sin 189°)$$
$$\sqrt{2}(\cos 261° + i\sin 261°)$$
$$\sqrt{2}(\cos 333° + i\sin 333°)$$

If these five fifth roots are graphed, they are equally spaced about a circle with radius $\sqrt{2}$, and the endpoints of these vectors are the vertices of a regular pentagon. See Figure 7-9.

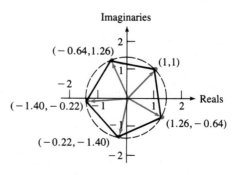

Figure 7-9

If $n > 2$ and the n nth roots of a complex number are graphed on the complex plane, the endpoints of the vectors that represent each root will always be at the vertices of a regular polygon.

■ EXERCISE 7.4

In Exercises 1–12, find each power. Leave all answers in trigonometric form.

1. $[3(\cos 30° + i \sin 30°)]^3$

2. $[4(\cos 15° + i \sin 15°)]^6$

3. $(\cos 15° + i \sin 15°)^{12}$

4. $[2(\cos 120° + i \sin 120°)]^9$

5. $[5 \operatorname{cis} 2°]^5$

6. $[0.5 \operatorname{cis} 100°]^3$

7. $\left[3\left(\cos \dfrac{\pi}{4} + i \sin \dfrac{\pi}{4}\right)\right]^4$

8. $\left[2\left(\cos \dfrac{3\pi}{2} + i \sin \dfrac{3\pi}{2}\right)\right]^6$

9. $[4(\cos 3 + i \sin 3)]^4$

10. $[2(\cos 5 + i \sin 5)]^{20}$

11. $\left[\dfrac{1}{3} \operatorname{cis} \dfrac{\pi}{2}\right]^3$

12. $\left[\dfrac{1}{2} \operatorname{cis} \dfrac{\pi}{6}\right]^5$

In Exercises 13–18, find the indicated nth root of each expression in a + bi form.

13. A cube root of $8(\cos 180° + i \sin 180°)$.

14. A fifth root of $32(\cos 150° + i \sin 150°)$.

15. A fifth root of $(\cos 300° + i \sin 300°)$.

16. A fourth root of $256(\cos \pi + i \sin \pi)$.

17. A sixth root of $64(\cos 2\pi + i \sin 2\pi)$.

18. A sixth root of $3^6(\cos \pi + i \sin \pi)$.

In Exercises 19–26, find and graph the indicated roots of each complex number. Change your answers to a + bi form only if that answer would be exact. Otherwise, leave the answers in trigonometric form.

19. The three cube roots of -8.

20. The four fourth roots of 16.

21. The two square roots of i.

22. The three cube roots of i.

23. The five fifth roots of $-i$.

24. The three cube roots of $\dfrac{\sqrt{2}}{2} + \dfrac{\sqrt{2}}{2} i$.

25. The four fourth roots of $-8 + 8\sqrt{3}\,i$.

26. The six sixth roots of $-i$.

In Exercises 27–30, substitute the given value of n into De Moivre's theorem with r = 1, and raise the binomial on the left to the nth power. Follow the additional directions.

27. $n = 2$. Set the real parts of the complex numbers equal to each other and thereby show that $\cos 2\theta = \cos^2\theta - \sin^2\theta$.

28. $n = 2$. Set the imaginary parts of the complex numbers equal to each other and thereby show that $\sin 2\theta = 2 \cos \theta \sin \theta$.

29. $n = 3$. Set the imaginary parts of the complex numbers equal to each other and thereby show that $\sin 3\theta = 3 \cos^2\theta \sin \theta - \sin^3\theta$.

30. $n = 3$. Set the real parts of the complex numbers equal to each other and thereby show that $\cos 3\theta = \cos^3\theta - 3 \cos \theta \sin^2\theta$.

31. Use the right side of the identity in Exercise 29 to show that $\sin 3\theta = 3 \sin \theta - 4 \sin^3\theta$.

32. Use the right side of the identity in Exercise 30 to show that $\cos 3\theta = 4 \cos^3\theta - 3 \cos \theta$.

7.5 POLAR COORDINATES

Some equations such as $(x^2 + y^2)^{3/2} = x$ are difficult to graph using the x and y coordinates of a rectangular coordinate system. However, these equations often can be written in a form using the variables r (a radius) and θ (an angle). These coordinates enable us to graph such equations in another coordinate system called the **polar coordinate system**.

The polar coordinate system is based on a ray called the **polar axis** and its source called the **pole**. In Figure 7-10, ray OA is the polar axis and point O is the pole. Any point $P(r, \theta)$ in the plane can be located if the length of a radius and an angle in standard position are known. For example, the polar coordinates $(10, 30°)$ determine the position of point R in Figure 7-10. If θ is in radians, the coordinates $(5, \frac{2\pi}{3})$ determine the point Q.

In Figure 7-11, point L is determined by the coordinates $(7, 225°)$ and point M by the coordinates $(6, \frac{5\pi}{3})$. Note that point M is also determined by many other pairs of polar coordinates such as $(6, 660°)$ or $(6, -60°)$. Any pair of polar coordinates locates a single point. However, any point has infinitely many pairs of polar coordinates.

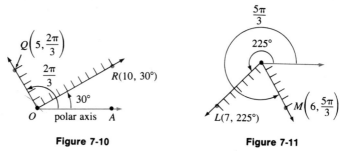

Figure 7-10   Figure 7-11

To plot point P with coordinates (r, θ) when r is positive, we draw angle θ in standard position and count r units along the terminal side of θ. This determines point P. See Figure 7-12.

To plot point P with coordinates (r, θ) when r is negative, we draw angle θ in standard position and count $|r|$ units along the extension of the terminal side of θ, but in the opposite direction. For example, to graph the point $P(-2, \frac{\pi}{6})$, we first draw an angle of $\frac{\pi}{6}$ in standard position as in Figure 7-13. We then draw the extension of ray OC in the opposite direction to obtain ray OB and count two units along ray OB to find point P. The graphs of the three points $R(5, \pi)$, $Q(-6, 100°)$, and $P(5, -30°)$ are shown in Figure 7-14.

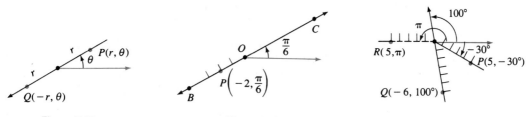

Figure 7-12   Figure 7-13   Figure 7-14

There is a relationship between the rectangular coordinates (x, y) and the polar coordinates (r, θ) of a point. Suppose point P in Figure 7-15 has rectangular coordinates (x, y) and polar coordinates (r, θ). Draw PA perpendicular to the x-axis to form right triangle OAP with $OA = x$, $AP = y$, and $OP = r$. Because angle θ is in standard position, the hypotenuse OP is the terminal side of angle θ. Thus, we have

$$\cos \theta = \frac{x}{r} \quad \text{and} \quad \sin \theta = \frac{y}{r}$$

or

$$x = r \cos \theta \quad \text{and} \quad y = r \sin \theta$$

To change from polar to rectangular coordinates, we use the following equations:

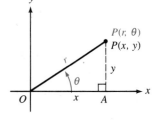

Figure 7-15

Formulas to Convert from Polar to Rectangular Coordinates.

$$x = r \cos \theta$$
$$y = r \sin \theta$$

Example 1   If the polar coordinates of point P are $(10, 150°)$, find the rectangular coordinates.

Solution
$$x = r \cos \theta = 10 \cos(150°) = 10\left(\frac{-\sqrt{3}}{2}\right) = -5\sqrt{3}$$

$$y = r \sin \theta = 10 \sin(150°) = 10\left(\frac{1}{2}\right) = 5$$

The rectangular coordinates of point P are $(-5\sqrt{3}, 5)$.   ∎

To find the polar coordinates from the rectangular coordinates of point P, we again refer to Figure 7-15. From the right triangle OAP, it follows that

$$r^2 = x^2 + y^2 \quad \text{and} \quad \tan \theta = \frac{y}{x}$$

To find possible polar coordinates for point P, we let $r = \sqrt{x^2 + y^2}$ and find an angle θ (equal to $\tan^{-1}\frac{y}{x}$) whose terminal side passes through the point (x, y). That is, if x is negative, for example, and y is positive, we choose θ to be a second-quadrant angle. If x and y are both negative, we choose θ to be a third-quadrant angle.

Formulas to Convert from Rectangular to Polar Coordinates.

$$r = \sqrt{x^2 + y^2}$$

$$\theta = \tan^{-1}\frac{y}{x}$$

where the terminal side of θ passes through the point (x, y). If $x = 0$, choose θ to be 90° (if $y > 0$) or 270° (if $y < 0$).

Example 2   If the rectangular coordinates of point P are $(4, 3)$, find a pair of polar coordinates for P.

Solution   Refer to Figure 7-16. First find the r-coordinate of point P as follows:

$$r = \sqrt{x^2 + y^2}$$
$$= \sqrt{4^2 + 3^2}$$
$$= 5$$

Second, determine the θ-coordinate of point P. Because x and y are both positive, θ is a first-quadrant angle.

$$\theta = \tan^{-1}\frac{y}{x}$$
$$= \tan^{-1}\frac{3}{4}$$
$$\approx 36.9°$$

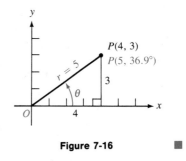

Figure 7-16

One possible choice of polar coordinates for point P is $(5, 36.9°)$, where θ is rounded to the nearest tenth of a degree.

Example 3   Change the rectangular coordinates $(-1, -4)$ to polar coordinates.

Solution   First find the r-coordinate of point P as follows:

$$r = \sqrt{x^2 + y^2}$$
$$= \sqrt{(-1)^2 + (-4)^2}$$
$$= \sqrt{17}$$

Second, determine the θ-coordinate of point P. Because x and y are both negative, θ is a third-quadrant angle.

$$\theta = \tan^{-1}\frac{y}{x}$$

$$= \tan^{-1} \frac{-4}{-1}$$

$$= \tan^{-1} 4$$

$$\approx 256.0° \qquad \text{Note that } 256.0° = 76.0° + 180°.$$

One possible choice of polar coordinates for point P is $(\sqrt{17}, 256.0°)$, where θ is rounded to the nearest tenth of a degree.

There are other possible choices for the polar coordinates of P. One is $(-\sqrt{17}, 76.0°)$, which has a *negative* value of r. See Figure 7-17.

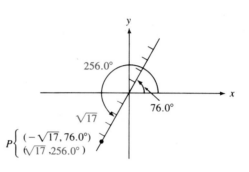

Figure 7-17

Equations involving the variables r and θ can be graphed in a polar coordinate system. The next several examples discuss the graphing of polar equations.

Example 4   Graph the polar equation $r = \theta$.

Solution   Make a table of values, plot the points, and join them with a smooth curve as in Figure 7-18. As θ increases, r increases, and the graph is a spiral called an **Archimedean spiral**.

$$r = \theta$$

θ	r
0	0
$\frac{\pi}{6}$	0.52
$\frac{\pi}{3}$	1.05
$\frac{\pi}{2}$	1.57
$\frac{2\pi}{3}$	2.09
$\frac{5\pi}{6}$	2.62
π	3.14
$\frac{3\pi}{2}$	4.71

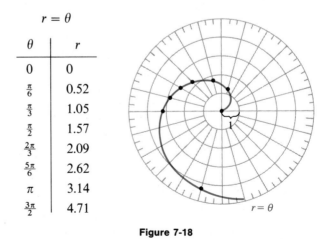

$r = \theta$

Figure 7-18

Example 5 Change the rectangular equation $x^2 + y^2 = y$ to polar coordinates and graph the curve.

Solution Substitute r^2 for $x^2 + y^2$ and $r \sin \theta$ for y in the original equation.

$$x^2 + y^2 = y$$
$$r^2 = r \sin \theta$$

If $r = 0$, the graph is the point at the pole for all θ. If $r \neq 0$, you can divide both sides of the equation by r to obtain the equation

$$r = \sin \theta$$

Make a table of values, plot the points, and graph as in Figure 7-19. The graph is a circle with center at $(\frac{1}{2}, \frac{\pi}{2})$ and with radius of $\frac{1}{2}$. You will be asked in an exercise to graph $x^2 + y^2 = y$ on a set of rectangular coordinate axes to verify that the graph is the described circle.

$r = \sin \theta$

θ	r
0	0
$\frac{\pi}{6}$	0.5
$\frac{\pi}{3}$	0.87
$\frac{\pi}{2}$	1
$\frac{2\pi}{3}$	0.87
$\frac{5\pi}{6}$	0.5
π	0
$\frac{7\pi}{6}$	−0.5
$\frac{3\pi}{2}$	−1

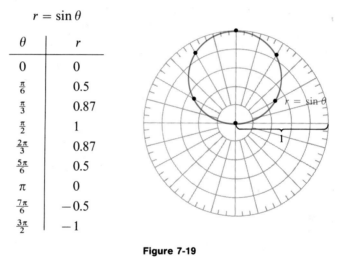

Figure 7-19

Example 6 Change the rectangular equation $(x^2 + y^2)^{3/2} = x$ to an equation having variables of r and θ. Then graph the curve using polar coordinates.

Solution Because $x^2 + y^2 = r^2$ and $x = r \cos \theta$, you have

$$(x^2 + y^2)^{3/2} = x$$
$$r^3 = r \cos \theta$$

If $r = 0$, the graph is the pole for all θ. If $r \neq 0$, you can divide both sides by r and obtain

$$r^2 = \cos \theta$$
$$r = \pm\sqrt{\cos \theta}$$

Make a table of values, and plot the points as in Figure 7-20. Because r^2 is positive, $\cos \theta$ must be positive. Therefore, θ is in quadrant I or IV.

$$r^3 = r \cos \theta \qquad \text{or} \qquad r = \pm\sqrt{\cos \theta}$$

θ	r
0	± 1
$\frac{\pi}{6}$	± 0.93
$\frac{\pi}{4}$	± 0.84
$\frac{\pi}{3}$	± 0.7
$\frac{\pi}{2}$	0
$\frac{5\pi}{3}$	± 0.7
$\frac{11\pi}{6}$	± 0.93

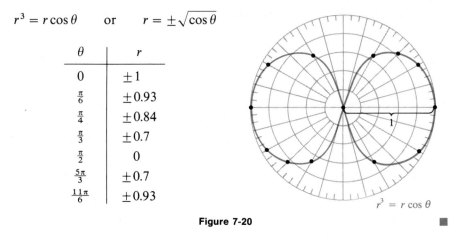

$$r^3 = r \cos \theta$$

Figure 7-20

Example 7 Change the polar equation $r(3 \cos \theta + 2 \sin \theta) = 7$ to an equation with rectangular coordinates.

Solution Use the distributive law to remove parentheses.

$$r(3 \cos \theta + 2 \sin \theta) = 7$$
$$3r \cos \theta + 2r \sin \theta = 7$$

Because $r \cos \theta = x$ and $r \sin \theta = y$, you can write this equation as

$$3x + 2y = 7$$

This is the equation of a line that can easily be graphed. ■

Example 8 Change the polar equation $r^2 = 9 \cos 2\theta$ to an equation with rectangular coordinates.

Solution Recall that $\cos 2\theta = \cos^2\theta - \sin^2\theta$ and substitute $\cos^2\theta - \sin^2\theta$ for $\cos 2\theta$ in the original equation.

$$r^2 = 9 \cos 2\theta$$
$$r^2 = 9(\cos^2\theta - \sin^2\theta)$$
$$r^2 = 9 \cos^2\theta - 9 \sin^2\theta$$

Multiply both sides by r^2 to obtain

$$r^4 = 9r^2 \cos^2\theta - 9r^2 \sin^2\theta$$
$$r^4 = 9(r \cos \theta)^2 - 9(r \sin \theta)^2$$
$$(x^2 + y^2)^2 = 9x^2 - 9y^2$$ ■

Example 9 Change the rectangular equation $(x^2 + y^2)^3 = 8x^2y^2$ to an equation with polar coordinates.

Solution Substitute $r \cos \theta$ for x, $r \sin \theta$ for y, and r^2 for $x^2 + y^2$ in the rectangular equation and simplify.

$$(x^2 + y^2)^3 = 8x^2 y^2$$
$$(r^2)^3 = 8r^2 \cos^2 \theta \, r^2 \sin^2 \theta$$
$$r^6 = 8r^4 \cos^2 \theta \sin^2 \theta$$

If $r = 0$, the graph is the point at the pole for all θ. If $r \neq 0$, you can divide both sides by r^4 to obtain

$$r^2 = 8 \cos^2 \theta \sin^2 \theta$$
$$r^2 = 2(2 \cos \theta \sin \theta)(2 \cos \theta \sin \theta)$$
$$r^2 = 2(\sin 2\theta)(\sin 2\theta)$$
$$r^2 = 2 \sin^2 2\theta$$

■

■ EXERCISE 7.5

In Exercises 1–20, the polar coordinates of point P are given. Find the rectangular coordinates of point P.

1. $(2, 30°)$ **2.** $(5, 135°)$ **3.** $(7, 300°)$ **4.** $(20, 225°)$

5. $(-3, 60°)$ **6.** $(-7, 210°)$ **7.** $\left(2, \dfrac{\pi}{2}\right)$ **8.** $\left(4, \dfrac{3\pi}{2}\right)$

9. $\left(-2, \dfrac{13\pi}{6}\right)$ **10.** $(-5, 3\pi)$ **11.** $\left(5, \dfrac{13\pi}{4}\right)$ **12.** $\left(3, \dfrac{17\pi}{6}\right)$

13. $(2, -30°)$ **14.** $(6, -225°)$ **15.** $(-10, -90°)$ **16.** $(-15, -45°)$

17. $(0, 39°)$ **18.** $(0, 0°)$ **19.** $(6, 1230°)$ **20.** $(35, 11.5\pi)$

In Exercises 21–36, the rectangular coordinates of point P are given. Find a pair of polar coordinates for point P.

21. $(1, 1)$ **22.** $(1, \sqrt{3})$ **23.** $(2\sqrt{3}, -2)$ **24.** $(-2, -2\sqrt{3})$

25. $(-\sqrt{3}, -1)$ **26.** $(-1, \sqrt{3})$ **27.** $(-\sqrt{3}, 1)$ **28.** $(0, 3)$

29. $(0, 0)$ **30.** $(7, 0)$ **31.** $(-5, 0)$ **32.** $(7, 7)$

33. $(3, -3)$ **34.** (π, π) **35.** $(7, 7\sqrt{3})$ **36.** $(-\sqrt{2}, -\sqrt{6})$

In Exercises 37–48, each equation contains rectangular coordinates. Change each equation to an equation containing polar coordinates.

37. $x = 3$ **38.** $y = -7$ **39.** $3x + 2y = 3$ **40.** $2x - y = 7$

41. $x^2 + y^2 = 9x$ **42.** $yx = 12$ **43.** $(x^2 + y^2)^3 = 4x^2 y^2$ **44.** $x^2 + y^2 = 9$

45. $y^2 = 2x - x^2$ **46.** $(x^2 + y^2)^2 = x^2 - y^2$ **47.** $x^2 = 2y + 1$ **48.** $y^2 = 2x + 1$

In Exercises 49–60, each equation contains polar coordinates. Change each equation to an equation containing rectangular coordinates.

49. $r = 3$ **50.** $r \sin \theta = 4$ **51.** $\cos \theta = \dfrac{5}{r}$ **52.** $3r \cos \theta + 2r \sin \theta = 2$

53. $r = \dfrac{1}{1 + \sin\theta}$ **54.** $r = \dfrac{1}{1 - \cos\theta}$ **55.** $r^2 = \sin 2\theta$ **56.** $r^2 = \cos 2\theta$

57. $\theta = \pi$ **58.** $\theta = 90°$ **59.** $r(2 - \cos\theta) = 2$ **60.** $r = 3\csc\theta + 2\sec\theta$

In Exercises 61–72, graph each equation.

61. $r = -\theta, \theta \geq 0$ **62.** $r\sin\theta = 3$ **63.** $r\cos\theta = -3$ **64.** $r\cos\theta + r\sin\theta = 1$

65. $r = \cos 2\theta$ **66.** $r = \sin 2\theta$ **67.** $r = \sin 3\theta$ **68.** $r = 3\cos 3\theta$

69. $r = 2(1 + \sin\theta)$ **70.** $r = \sqrt{2\cos\theta}$ **71.** $r = 2 + \cos\theta$ **72.** $r = 3(1 - \cos\theta)$

73. Graph $x^2 + y^2 = y$ on a Cartesian coordinate system and show that its graph is identical to the graph in Figure 7-19.

7.6 MORE ON POLAR COORDINATES

In Section 7.5, equations in polar coordinates were graphed by making an extensive table of values and plotting many points. A different approach is used in this section. Although some specific points are plotted, the behavior of functions of θ is considered as θ increases from 0 to 2π.

Example 1 Graph $r = \sin 2\theta$.

Solution Note that $\sin 2\theta$, hence r as well, is zero when $\theta = 0, \frac{\pi}{2}, \pi$, and $\frac{3\pi}{2}$. For these values, the curve passes through the pole.

When $\theta = \frac{\pi}{4}$ or $\frac{5\pi}{4}$, the value of $\sin 2\theta = 1$. When θ is $\frac{3\pi}{4}$ or $\frac{7\pi}{4}$, the value of $\sin 2\theta = -1$. Thus, the curve passes through the points $(1, \frac{\pi}{4}), (-1, \frac{3\pi}{4}), (1, \frac{5\pi}{4})$, and $(-1, \frac{7\pi}{4})$ and several times through the pole: $(0, 0), (0, \frac{\pi}{2}), (0, \pi)$, and $(0, \frac{3\pi}{2})$. Note that these last four ordered pairs represent four different pairs of polar coordinates for the pole. See Figure 7-21a.

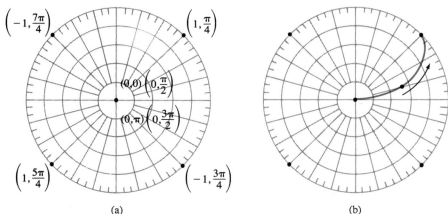

(a) (b)

Figure 7-21

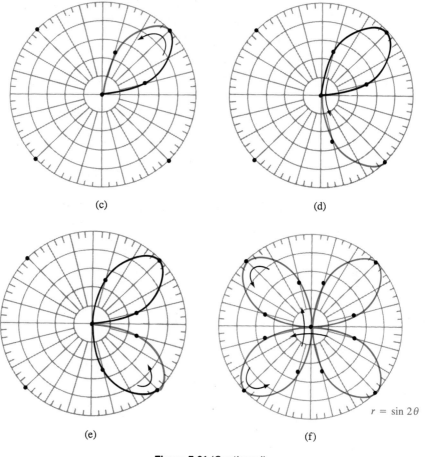

(c)

(d)

(e)

(f)

$r = \sin 2\theta$

Figure 7-21 (*Continued*)

As θ increases from 0 to $\frac{\pi}{4}$, the value of r increases from 0 to 1 and you can draw that portion of the curve shown in Figure 7-21b. As θ continues to increase from $\frac{\pi}{4}$ to $\frac{\pi}{2}$, r decreases, and the curve returns to the pole as in Figure 7-21c. As θ continues to increase from $\frac{\pi}{2}$ to $\frac{3\pi}{4}$, the value of r goes from 0 to -1 and the points (r, θ) trace the additional portion of the curve shown in Figure 7-21d. As θ increases from $\frac{3\pi}{4}$ to π, r goes from -1 back to 0 and the curve continues to develop as in Figure 7-21e. The complete curve, called a **four-leaved rose**, is shown in Figure 7-21f. ∎

Example 2 Graph $r = \cos 3\theta$.

Solution When $\theta = \frac{\pi}{6}, \frac{\pi}{2}, \frac{5\pi}{6}, \frac{7\pi}{6}, \frac{3\pi}{2}$, or $\frac{11\pi}{6}$, the corresponding values of $\cos 3\theta$ are zero. Thus, $r = 0$ and the curve passes through the pole at points

$$\left(0, \frac{\pi}{6}\right), \left(0, \frac{\pi}{2}\right), \left(0, \frac{5\pi}{6}\right), \left(0, \frac{7\pi}{6}\right), \left(0, \frac{3\pi}{2}\right), \text{ and } \left(0, \frac{11\pi}{6}\right)$$

When $\theta = 0$, $\frac{2\pi}{3}$, or $\frac{4\pi}{3}$, the radius r reaches its maximum value of 1. When $\theta = \frac{\pi}{3}$, π, or $\frac{5\pi}{3}$, the radius r reaches its minimum value of -1. Thus, the curve passes through the points

$$(1, 0), \left(1, \frac{2\pi}{3}\right), \left(1, \frac{4\pi}{3}\right), \left(-1, \frac{\pi}{3}\right), (-1, \pi), \text{ and } \left(-1, \frac{5\pi}{3}\right)$$

These pairs of coordinates represent only three distinct points as shown in Figure 7-22a.

As θ increases from 0 to $\frac{\pi}{6}$, $\cos 3\theta$ decreases from 1 to 0, and that portion of the curve is traced in Figure 7-22b. As θ continues to increase from $\frac{\pi}{6}$ to $\frac{\pi}{3}$, the value of $\cos 3\theta$ continues to decrease to its minimum value of -1, and the curve continues to develop as in Figure 7-22c. The complete graph of $r = \cos 3\theta$, called a **three-leaved rose**, appears in Figure 7-22d.

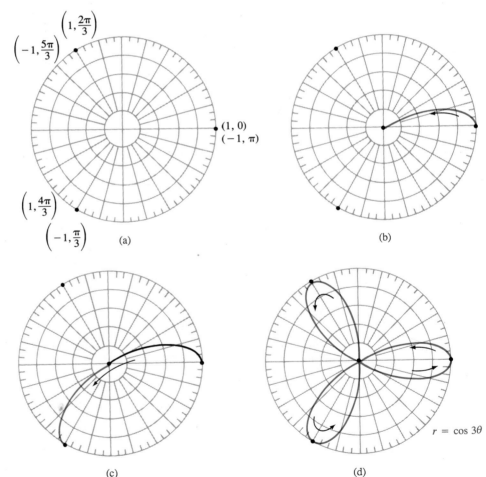

(a)

(b)

(c)

(d)

$r = \cos 3\theta$

Figure 7-22

All the rose curves fit into two categories.

> **Theorem.**   If n is an odd integer, then $r = \cos n\theta$ or $r = \sin n\theta$ represents an n-leaved rose.
>   If n is even, then $r = \cos n\theta$ or $r = \sin n\theta$ represents a $2n$-leaved rose.

Example 3   Graph the curve $r = 1 - \sin\theta$.

Solution   The easiest values of r to compute are those associated with the quadrantal angle values of θ. When $\theta = 0$ or π, then $r = 1$. When $\theta = \frac{\pi}{2}$, then $r = 0$. When $\theta = \frac{3\pi}{2}$,

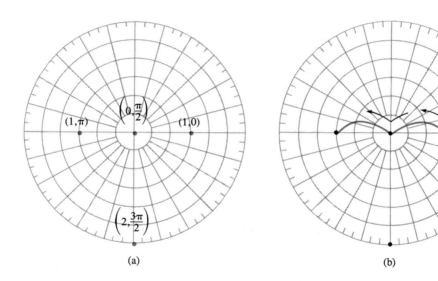

(a)

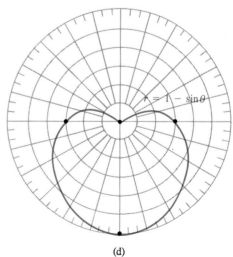

(b)

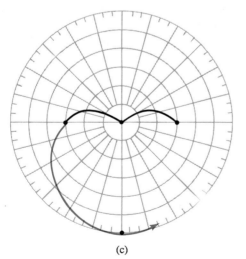

(c)

(d)

Figure 7-23

then $r = 2$. These four points

$$(1,0), \left(0, \frac{\pi}{2}\right), (1, \pi), \text{ and } \left(2, \frac{3\pi}{2}\right)$$

are the intercepts of the curve with the polar axis and the perpendicular line to the polar axis at the pole. See Figure 7-23a.

As θ increases from 0 to π, the value of $1 - \sin \theta$, hence the value of r, decreases from 1 to 0 and then increases back to 1. This accounts for the two bumps in the curve in Figure 7-23b. As θ increases from π to $\frac{3\pi}{2}$, the third-quadrant loop of Figure 7-23c is formed. The complete curve, called a **cardioid**, is shown in Figure 7-23d. ∎

Example 4 Graph the curve $r\theta = \pi$.

Solution Because the product of variables r and θ is a constant, r and θ are inversely proportional—as θ increases, then r decreases. Write the equation as $r = \frac{\pi}{\theta}$, and calculate the intercepts of the curve with the polar axis and a line perpendicular to the axis through the pole.

$$\text{at } \theta = \frac{\pi}{2}, \qquad r = \frac{\pi}{\frac{\pi}{2}} = 2$$

$$\text{at } \theta = \pi, \qquad r = 1$$

$$\text{at } \theta = \frac{3\pi}{2}, \qquad r = \frac{2}{3}$$

These points and a portion of the curve, called a **spiral**, are plotted in Figure 7-24a. The difficulty is in determining the shape of the curve as θ approaches 0 (and r approaches infinity). If θ is close to zero, and r consequently is very large, the distance PQ of Figure 7-24b is very close to the length of arc PQ', centered at the pole. Because this arc length is $r\theta$ or π, the distance PQ is approximately π. The horizontal line that is π units above the polar axis is an asymptote of this curve. The graph of $r\theta = \pi$ is a **hyperbolic spiral** and its graph appears in

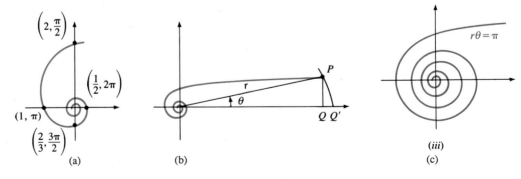

(a)   (b)   (c)

Figure 7-24

Figure 7-24c. The curve heading off to the right approaches the horizontal line that is parallel to the polar axis and π units above it.

Unlike the Archimedean spiral discussed in the preceding section, the hyperbolic spiral does not pass through the pole. ■

The following chart shows the graphs and the general equations for several important polar curves.

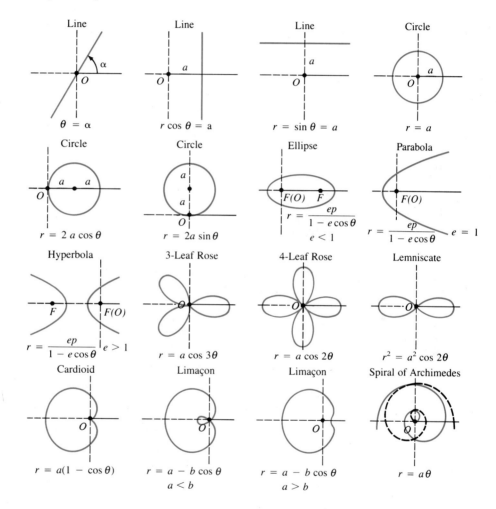

Line	Line	Line	Circle
$\theta = \alpha$	$r \cos \theta = a$	$r = \sin \theta = a$	$r = a$

Circle	Circle	Ellipse	Parabola
$r = 2a \cos \theta$	$r = 2a \sin \theta$	$r = \dfrac{ep}{1 - e \cos \theta}$ $e < 1$	$r = \dfrac{ep}{1 - e \cos \theta}$ $e = 1$

Hyperbola	3-Leaf Rose	4-Leaf Rose	Lemniscate
$r = \dfrac{ep}{1 - e \cos \theta}$ $e > 1$	$r = a \cos 3\theta$	$r = a \cos 2\theta$	$r^2 = a^2 \cos 2\theta$

Cardioid	Limaçon	Limaçon	Spiral of Archimedes
$r = a(1 - \cos \theta)$	$r = a - b \cos \theta$ $a < b$	$r = a - b \cos \theta$ $a > b$	$r = a\theta$

■ EXERCISE 7.6

Each of the following curves is important enough to receive a name. The equation and its title are given. Graph each polar equation.

1. $r = 2 \cos 2\theta$; four-leaved rose

2. $r = 3 \sin 2\theta$; four-leaved rose

3. $r = \sin 3\theta$; three-leaved rose

4. $r = 2 \cos 3\theta$; three-leaved rose

5. $r = 2(1 + \cos\theta)$;   cardioid

6. $r = 3(1 + \sin\theta)$;   cardioid

7. $r = 2 + \cos\theta$;   limaçon

8. $r = \dfrac{1}{2} + \cos\theta$;   limaçon

9. $r^2 = 2\sin 2\theta$;   lemniscate

10. $r = \sin\theta\cos^2\theta$;   bifolium

11. $r^2\theta = \pi, r > 0$;   lituus   (*Hint:* This curve has a horizontal asymptote.)

12. $r = \tan\theta$;   kappa curve $\Big\}$

13. $r = \sin\theta\tan\theta$;   cissoid $\Big\}$ (*Hint:* These curves have vertical asymptotes.)

14. $r = 2a\cos\theta$;   circle

15. $r = \dfrac{3}{2 - \cos\theta}$;   ellipse

16. $r = \dfrac{4}{2 - 3\cos\theta}$;   hyperbola

17. $r = \dfrac{4}{1 + \sin\theta}$;   parabola

18. $r = \dfrac{6}{4 - 3\cos\theta}$;   ellipse

7.7   APPLICATIONS OF COMPLEX NUMBERS AND POLAR COORDINATES

Because complex numbers of the form $a + bi$ can be represented as vectors from the origin to the point (a, b) in the complex plane, they can be used to solve problems involving vectors.

Example 1   Forces A and B, of 25 and 35 pounds, respectively, make an angle of $40°$ with each other. Find the magnitude of the resultant force, and the angle it makes with the smaller force.

Solution   Represent the forces as complex numbers written in trigonometric form as suggested by the diagram of Figure 7-25. Then, rewrite them in algebraic form.

$$\text{Vector } \mathbf{A} = 25(\cos 0° + i\sin 0°)$$
$$= 25 + 0i$$
$$\text{Vector } \mathbf{B} = 35(\cos 40° + i\sin 40°)$$
$$\approx 35(0.7660 + 0.6428i)$$
$$\approx 26.8 + 22.5i$$

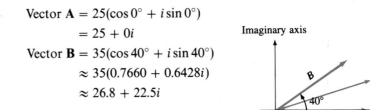

Figure 7-25

The sum of these two vectors is the sum of their complex representations.

$$\mathbf{A} + \mathbf{B} = (25 + 0i) + (26.8 + 22.5i)$$
$$= 51.8 + 22.5i$$

Converting this result to trigonometric form gives

$$\mathbf{A} + \mathbf{B} = k(\cos\phi + i\sin\phi)$$

where

$$k = \sqrt{51.8^2 + 22.5^2} \approx \sqrt{3189} \approx 56.5$$

and

$$\phi = \tan^{-1}\frac{22.5}{51.8} \approx 23.5°$$

Thus, you have

$$\mathbf{A} + \mathbf{B} = 56.5(\cos 23.5° + i\sin 23.5°)$$

Hence, the magnitude of the resultant force is 56.5 pounds, and the angle that it makes with the weaker force is 23.5°. ∎

The law of cosines can be used to derive a distance formula for points given in polar coordinates. If $P(r_1, \theta_1)$ and $Q(r_2, \theta_2)$ are two points in Figure 7-26, given in polar coordinates, then the distance d between them is found by applying the law of cosines to triangle OPQ.

$$d^2 = r_1{}^2 + r_2{}^2 - 2r_1r_2\cos(\theta_2 - \theta_1)$$

$$d = \sqrt{r_1{}^2 + r_2{}^2 - 2r_1r_2\cos(\theta_2 - \theta_1)}$$

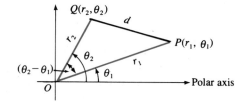

Figure 7-26

Example 2   What is the distance between $P\left(3, \dfrac{\pi}{6}\right)$ and $Q\left(2, \dfrac{\pi}{2}\right)$?

Solution   Let $r_1 = 3$, $\theta_1 = \frac{\pi}{6}$, $r_2 = 2$, and $\theta_2 = \frac{\pi}{2}$. Then, you have

$$d = \sqrt{r_1{}^2 + r_2{}^2 - 2r_1r_2\cos(\theta_2 - \theta_1)}$$

$$= \sqrt{3^2 + 2^2 - 2(3)(2)\cos\left(\frac{\pi}{2} - \frac{\pi}{6}\right)}$$

$$= \sqrt{9 + 4 - 12\cos\frac{\pi}{3}}$$

$$= \sqrt{13 - 12\left(\frac{1}{2}\right)}$$

$$= \sqrt{13 - 6}$$

$$= \sqrt{7}$$

The distance PQ is $\sqrt{7}$ units. ∎

■ EXERCISE 7.7

In Exercises 1 and 2, use complex numbers to solve each problem.

1. Forces of 15.2 and 29.7 pounds make an angle of 36.0° with each other. What is the magnitude of the resultant force, and what angle does it make with the weaker force?

2. Show that three forces of equal magnitude, and each acting at an angle of 120° from the other two, have a resultant force of zero.

3. Find the distance between points $P\left(5, \dfrac{5\pi}{6}\right)$ and $Q\left(3, \dfrac{-\pi}{6}\right)$.

4. Find the perimeter of the triangle formed by $P(2, 20°)$, $Q(5, 50°)$, and the pole.

CHAPTER SUMMARY

Key Words

absolute value of a complex
    number   (7.2)
Archimedean spiral   (7.5)
argument   (7.3)
cardioid   (7.6)
cis θ   (7.3)
complex conjugates   (7.1)
complex number   (7.1)
complex plane   (7.2)
De Moivre's theorem   (7.4)
four-leaved rose   (7.6)
hyperbolic spiral   (7.6)
i   (7.1)

imaginary number   (7.1)
imaginary part of a complex
    number   (7.1)
modulus   (7.3)
polar axis   (7.5)
polar coordinate system   (7.5)
pole   (7.5)
rationalizing the denominator   (7.1)
real part of a complex
    number   (7.1)
three-leaved rose   (7.6)
trigonometric form of a complex
    number   (7.3)

Key Ideas

(7.1) $i^2 = -1$

Two complex numbers are equal if and only if their real parts are equal and their imaginary parts are equal:

$$a + bi = c + di \quad \text{if and only if} \quad a = c \text{ and } b = d$$

Complex numbers are added as binomials:

$$(a + bi) + (c + di) = (a + c) + (b + d)i$$

Complex numbers are multiplied as binomials:

$$(a + bi)(c + di) = (ac - bd) + (ad + bc)i$$

(7.2)   To graph the complex number $a + bi$, plot the point (a, b) in the complex plane.

The vector drawn from the origin to the point (a, b) is the graph of the complex number $a + bi$.

$$|a + bi| = \sqrt{a^2 + b^2}$$

(7.3)   Complex numbers can be written either in algebraic $(a + bi)$ or trigonometric $[r(\cos \theta + i \sin \theta)]$ form, where $r = |a + bi|$, and θ is an angle such that $\sin \theta = \frac{b}{r}$ and $\cos \theta = \frac{a}{r}$.

$$r \operatorname{cis} \theta = r(\cos \theta + i \sin \theta)$$
$$(r_1 \operatorname{cis} \theta_1)(r_2 \operatorname{cis} \theta_2) \cdots (r_n \operatorname{cis} \theta_n) = r_1 r_2 \cdots r_n \operatorname{cis}(\theta_1 + \theta_2 + \cdots + \theta_n)$$
$$\frac{r_1 \operatorname{cis} \theta_1}{r_2 \operatorname{cis} \theta_2} = \frac{r_1}{r_2} \operatorname{cis}(\theta_1 - \theta_2)$$

(7.4)   De Moivre's theorem

$$[r(\cos \theta + i \sin \theta)]^n = r^n[\cos n\theta + i \sin n\theta]$$

or

$$[r \operatorname{cis} \theta]^n = r^n \operatorname{cis} n\theta$$

If $n > 2$ and the n nth roots of a complex number are graphed on the complex plane, the end points of the vectors that represent each root lie at the vertices of a regular polygon.

(7.5)   Formulas to convert from polar coordinates to rectangular coordinates.

$$\begin{cases} x = r \cos \theta \\ y = r \sin \theta \end{cases}$$

Formulas to convert from rectangular coordinates to polar coordinates.

$$\begin{cases} r = \sqrt{x^2 + y^2} \\ \theta = \tan^{-1} \frac{y}{x} \end{cases}$$

(7.6)   If n is an odd integer, then $r = \cos n\theta$ and $r = \sin n\theta$ represent roses with n leaves. If n is even, then $r = \cos n\theta$ and $r = \sin n\theta$ represent roses with $2n$ leaves.

REVIEW EXERCISES

1. Simplify i^{11}.
2. Simplify i^{5003}.
3. Simplify i^{-33}.
4. Simplify i^{-1812}.
5. Solve $x + (x + y)i = 2y + 2i$ for x and y.
6. Solve $3x + (x - y)i = 2y + 3 + 7i$ for x and y.

In Review Exercises 7–18, perform any indicated operations and express the final answer in $a + bi$ form.

7. $(3 + 2i) + (-7 - i)$
8. $(-2 - i) - (3 - 2i)$
9. $(2 + i)(-3 - i)$
10. $(3 + 2i)(5 - 3i)$
11. $\dfrac{1}{5i}$
12. $\dfrac{13}{-6i}$
13. $\dfrac{2}{4 + i}$
14. $\dfrac{-5}{3 - i}$

15. $\dfrac{1+\sqrt{-1}}{1-\sqrt{-1}}$ **16.** $\dfrac{2+\sqrt{-1}}{3+\sqrt{-16}}$ **17.** $\dfrac{2+3i}{1-\sqrt{2}\,i}$ **18.** $\dfrac{3-i}{1-\sqrt{3}\,i}$

In Review Exercises 19–22, graph each complex number.

19. $4-5i$ **20.** $-7+2i$ **21.** 6 **22.** $3i$

In Review Exercises 23–26, compute each absolute value.

23. $|8+3i|$ **24.** $|10-10i|$ **25.** $\left|\dfrac{3i}{i+3}\right|$ **26.** $\left|\dfrac{4-3i}{4+3i}\right|$

In Review Exercises 27–30, write each complex number in trigonometric form.

27. $-2+2i$ **28.** $5-5i$ **29.** $3+3i\sqrt{3}$ **30.** 4

In Review Exercises 31–34, write each complex number in a + bi form.

31. $3(\cos 60° + i\sin 60°)$ **32.** $2(\cos 330° + i\sin 330°)$

33. $3\left(\cos\dfrac{4\pi}{3} + i\sin\dfrac{4\pi}{3}\right)$ **34.** $7\left(\cos\dfrac{5}{6}\pi + i\sin\dfrac{5}{6}\pi\right)$

In Review Exercises 35–40, perform the indicated operation. Simplify, but leave your answer in trigonometric form.

35. $(\text{cis } 60°)(\text{cis } 50°)$ **36.** $(2\,\text{cis } 330°)(3\,\text{cis } 240°)$

37. $\left[3\left(\cos\dfrac{\pi}{12} + i\sin\dfrac{\pi}{12}\right)\right]\left[2\left(\cos\dfrac{\pi}{6} + i\sin\dfrac{\pi}{6}\right)\right]$ **38.** $\left[7\left(\cos\dfrac{\pi}{5} + i\sin\dfrac{\pi}{5}\right)\right]\left[3\left(\cos\dfrac{4\pi}{5} + i\sin\dfrac{4\pi}{5}\right)\right]$

39. $\dfrac{10\,\text{cis } 60°}{5\,\text{cis } 10°}$ **40.** $\dfrac{20(\cos 50° + i\sin 50°)}{30(\cos 40° + i\sin 40°)}$

41. Find one cube root of $\cos 60° + i\sin 60°$.   **42.** Find one fourth root of $7\sqrt{2} + 7\sqrt{2}\,i$.

43. Find the three cube roots of 125.   **44.** Find the four fourth roots of 81.

In Review Exercises 45–48, change the polar coordinates to rectangular coordinates.

45. $(5, 60°)$ **46.** $(-2, 390°)$ **47.** $\left(-1, \dfrac{7\pi}{6}\right)$ **48.** $\left(10, -\dfrac{5\pi}{4}\right)$

In Review Exercises 49–52, change the rectangular coordinates to a pair of polar coordinates.

49. $(-\sqrt{2}, \sqrt{2})$ **50.** $(-\sqrt{3}, 1)$ **51.** $(1, 0)$ **52.** $(1, -\sqrt{3})$

In Review Exercises 53–56, change the rectangular equation to a polar equation.

53. $4xy = 4$ **54.** $x + 2y = 2$ **55.** $x^2 = 3y$ **56.** $(x^2 + y^2)^2 = 4xy$

In Review Exercises 57–60, change the polar equation to a rectangular equation.

57. $r^2 = 9\cos 2\theta$ **58.** $r = 5\sin\theta$ **59.** $r = \dfrac{1}{4 + \sin\theta}$ **60.** $r = \dfrac{2}{1 - \cos\theta}$

In Review Exercises 61–64, graph each polar equation.

61. $r = \dfrac{6}{1 + \sin\theta}$ **62.** $r = \dfrac{2}{1 - \sin\theta}$ **63.** $r = 4(1 + \cos\theta)$ **64.** $r = 8 - 4\cos\theta$

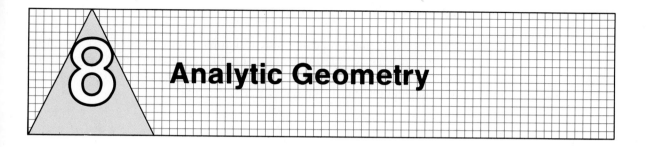

Analytic Geometry

The graphs of second-degree equations in x and y represent figures that have interested mathematicians since the time of the ancient Greeks. However, the equations of those graphs were not carefully studied until the 17th century, when René Descartes and another French mathematician, Blaise Pascal (1623–1662), began investigating them.

> **Definition.** If A, B, C, D, E, and F are real numbers, and if at least one of A, B, and C is not 0, then
>
> $$Ax^2 + Bxy + Cy^2 + Dx + Ey + F = 0$$
>
> is called the **general form of a second-degree equation in x and y**.

Descartes discovered that the graphs of second-degree equations always fall into one of seven categories: a single point, a pair of straight lines, a circle, a parabola, an ellipse, a hyperbola, or no graph at all. These graphs are called **conic sections** because each is the intersection of a plane and a right-circular cone. See Figure 8-1.

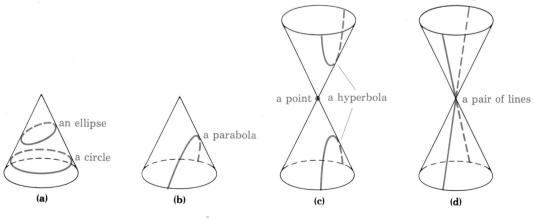

an ellipse

a circle

a parabola

a point · a hyperbola

a pair of lines

(a)　　　　(b)　　　　(c)　　　　(d)

Figure 8-1

281

The conic sections have many practical applications. For example, the properties of parabolas are used in building flashlights, satellite antennas, and solar furnaces. The orbits of the planets around the sun are ellipses. Hyperbolas are used in navigation and the design of gears.

8.1   THE STRAIGHT LINE

In Section 1.2 we discussed how to graph certain equations by plotting points. For example, to graph the equation $x + 2y = 5$ we determine ordered pairs (x, y) that satisfy the equation and plot the ordered pairs on a rectangular coordinate system. If $y = -1$, for example, we can find x as follows:

$$x + 2y = 5$$
$$x + 2(-1) = 5 \qquad \text{Substitute } -1 \text{ for } y.$$
$$x - 2 = 5$$
$$x = 7$$

Thus, one ordered pair that satisfies the equation is $(7, -1)$. This ordered pair and others that satisfy the equation are shown in the table of values of Figure 8-2. We then plot each of these ordered pairs and join them with a straight line. Later in the section we will show that these five points do lie on a straight line.

Because the line intersects the y-axis at the point $(0, \frac{5}{2})$, the number $\frac{5}{2}$ is called the **y-intercept** of the line. The number 5 is the **x-intercept**.

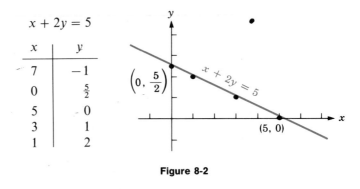

$x + 2y = 5$

x	y
7	-1
0	$\frac{5}{2}$
5	0
3	1
1	2

Figure 8-2

Slope of a Nonvertical Line

The slope of a line drawn in the Cartesian plane is a measure of its inclination. We consider the line l in Figure 8-3a, which passes through $P(x_1, y_1)$ and $Q(x_2, y_2)$. If line QR is perpendicular to the x-axis and PR is perpendicular to the y-axis, then triangle PRQ is a right triangle and point R has coordinates of (x_2, y_1). The distance from R to Q, called the **rise** of segment PQ, is denoted as Δy (delta y). Thus, Δy is the change in the y-coordinates of P and Q: $\Delta y = y_2 - y_1$. The horizontal distance from P to R, called the **run** and denoted as Δx, is the change in the x-coordinates of points P and Q: $\Delta x = x_2 - x_1$. If $x_1 \neq x_2$, then the slope of the line, denoted as m, in Figure 8-3a is the *rise* divided by the *run*:

$$m = \frac{\Delta y}{\Delta x}$$

Definition.   If $P(x_1, y_1)$ and $Q(x_2, y_2)$ are two points on a nonvertical line l in the Cartesian plane, the **slope of line l** is given by

$$m = \frac{\Delta y}{\Delta x} = \frac{y_2 - y_1}{x_2 - x_1}$$

If $x_1 = x_2$, then line l is a vertical line and has no defined slope.

In Figure 8-3b point S represents an arbitrary point on line l. Because triangles PRQ and PTS are similar, their corresponding sides are in proportion. Thus, the ratios of the rise to the run in the two triangles are equal, and we have

$$m = \frac{y_2 - y_1}{x_2 - x_1}$$

$$= \frac{y_3 - y_1}{x_3 - x_1}$$

This fact implies that the slope of a nonvertical line is a constant that can be calculated by using *any* two points on the line. Furthermore, if point P is on a line with slope m and the ratio of rise to run of a segment PQ is also m, then point Q is also on the line.

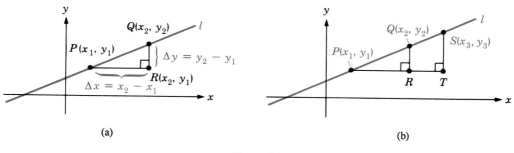

(a)   (b)

Figure 8-3

Example 1   Find the slope of the line passing through the points $P(-1, -2)$ and $Q(7, 8)$.

Solution   Let $x_1 = -1$, $y_1 = -2$, $x_2 = 7$, and $y_2 = 8$ in the formula that defines the slope, and simplify.

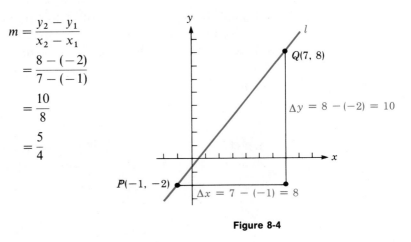

$$m = \frac{y_2 - y_1}{x_2 - x_1}$$

$$= \frac{8 - (-2)}{7 - (-1)}$$

$$= \frac{10}{8}$$

$$= \frac{5}{4}$$

Figure 8-4

Thus, the slope of the line is $\frac{5}{4}$. (See Figure 8-4.) ∎

If a line *rises* as it moves to the right—that is, if increasing x values result in increasing y values—the slope of the line is *positive* (see Figure 8-5a). If the line *drops* as it moves to the right—that is, if increasing x values result in decreasing y values—the slope of the line is *negative* (see Figure 8-5b). If a line is parallel to the x-axis, its slope is 0 (see Figure 8-5c). If a line is parallel to the y-axis, its slope is undefined (see Figure 8-5d).

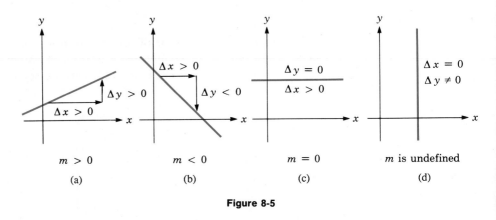

Figure 8-5

A theorem relates parallel lines to their slopes.

> **Theorem.**   Two nonvertical lines are parallel if and only if they have the same slope.

Proof   Suppose that the nonvertical lines l_1 and l_2 of Figure 8-6 have slopes of m_1 and m_2, respectively, and are parallel. Then the right triangles ABC and DEF are similar, and it follows that

$$m_1 = \frac{\text{rise of } l_1}{\text{run of } l_1}$$

$$= \frac{\text{rise of } l_2}{\text{run of } l_2}$$

$$= m_2$$

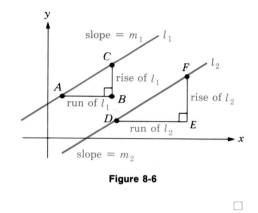

Figure 8-6

Thus, two nonvertical parallel lines have the same slope.

In the exercises you will be asked to prove that if two lines have the same slope, they are parallel.

It is also true that two lines with no defined slope are vertical lines, and vertical lines are parallel.

Example 2 The line passing through $P(-2, 3)$ and $Q(3, -1)$ is parallel to the line passing through $R(4, -5)$ and $S(x, 7)$. Find x.

Solution Because the lines PQ and RS are parallel, the slopes of PQ and RS are equal. Find the slope of each line, set them equal to each other, and solve the resulting equation.

$$\text{slope of } PQ = \text{slope of } RS$$

$$\frac{-1 - 3}{3 - (-2)} = \frac{7 - (-5)}{x - 4}$$

$$\frac{-4}{5} = \frac{12}{x - 4}$$

$$-4(x - 4) = 5 \cdot 12 \qquad \text{Multiply both sides by } 5(x - 4).$$

$$-4x + 16 = 60$$

$$-4x = 44$$

$$x = -11$$

Thus, $x = -11$. The line passing through $P(-2, 3)$ and $Q(3, -1)$ is parallel to the line passing through $R(4, -5)$ and $S(-11, 7)$. ■

Two numbers with a product of -1 are called **negative reciprocals** of each other. The following theorem describes the relationship between the slopes of perpendicular lines.

Theorem. Two nonvertical lines are perpendicular if and only if their slopes are negative reciprocals.

A line with a slope of 0 is perpendicular to a line with no defined slope.

Proof   Suppose that l_1 and l_2 are lines with slopes m_1 and m_2 intersecting at the origin. Let $P(a, b)$ be a point on l_1, let $Q(c, d)$ be a point on l_2, and let neither point P nor point Q be the origin. (See Figure 8-7.)

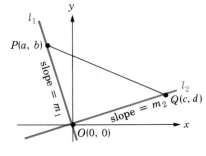

Figure 8-7

First, suppose lines l_1 and l_2 are perpendicular. Then triangle POQ is a right triangle with right angle at O, and by the Pythagorean theorem $d(OP)^2 + d(OQ)^2 = d(PQ)^2$. Thus,

$$d(OP)^2 + d(OQ)^2 = d(PQ)^2$$
$$(a - 0)^2 + (b - 0)^2 + (c - 0)^2 + (d - 0)^2 = (a - c)^2 + (b - d)^2$$
$$a^2 + b^2 + c^2 + d^2 = a^2 - 2ac + c^2 + b^2 - 2bd + d^2$$
$$0 = -2ac - 2bd$$
$$bd = -ac$$

1.
$$\frac{b}{a} = -\frac{c}{d}$$

The coordinates of P are (a, b) and the coordinates of O are $(0, 0)$. Using the definition of slope, we have

$$m_1 = \frac{b - 0}{a - 0} = \frac{b}{a}$$

Similarly, we have

$$m_2 = \frac{d}{c}$$

or

$$\frac{1}{m_2} = \frac{c}{d}$$

Substitute m_1 for $\dfrac{b}{a}$ and $\dfrac{1}{m_2}$ for $\dfrac{c}{d}$ in Equation 1 to get

$$m_1 = -\frac{1}{m_2}$$

Hence, if lines l_1 and l_2 are perpendicular, they have slopes that are negative reciprocals.

Conversely, suppose that the slopes of lines l_1 and l_2 are negative reciprocals of each other. Because the steps of the proof are reversible, $d(OP)^2 + d(OQ)^2 = d(PQ)^2$. By the Pythagorean theorem, triangle POQ is a right triangle. Thus, l_1 and l_2 are perpendicular.   ☐

It is also true that a line with a slope of 0 is horizontal and thus perpendicular to a vertical line, which has no defined slope.

Example 3   If two lines intersect at the point $P(-5, 3)$, one passes through the point $Q(-1, -3)$, and the other passes through $R(1, 7)$, are the lines perpendicular?

Solution   First find the slopes of lines PQ and PR.

$$\text{slope of } PQ = \frac{\Delta y}{\Delta x} = \frac{-3 - 3}{-1 - (-5)} = \frac{-6}{4} = -\frac{3}{2}$$

$$\text{slope of } PR = \frac{\Delta y}{\Delta x} = \frac{7 - 3}{1 - (-5)} = \frac{4}{6} = \frac{2}{3}$$

Because the slopes of these lines are negative reciprocals, the lines are perpendicular.   ∎

Point–Slope Form of the Equation of a Line

If two points on a line are known, its equation can be written by using the *point–slope* form of the equation of the line. Suppose that the nonvertical line l of Figure 8-8 has a slope of m and passes through the fixed point $P(x_1, y_1)$. If $Q(x, y)$ is another point on that line, then by the definition of *slope* we have

$$m = \frac{y - y_1}{x - x_1}$$

or

$$y - y_1 = m(x - x_1)$$

Because the equation $y - y_1 = m(x - x_1)$ displays the coordinates of a fixed point on the line and the line's slope, it is called the **point–slope form** of the equation of a line.

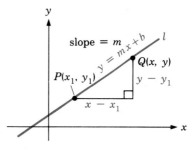

Figure 8-8

Point–Slope Form of the Equation of a Line.   The equation of the line passing through the point $P(x_1, y_1)$ and having a slope of m is

$$y - y_1 = m(x - x_1)$$

Proof   The proof that the equation $y - y_1 = m(x - x_1)$ is the equation of the line l has two parts. We must show two things:

1.   The coordinates of every point on the line satisfy the equation.
2.   Every point whose coordinates satisfy the equation lies on the line.

Part 1.   Let $Q(a, b)$ be any point on the line except point $P(x_1, y_1)$. Substitute a for x and b for y in the equation $y - y_1 = m(x - x_1)$ and solve for m.

$$y - y_1 = m(x - x_1)$$
$$b - y_1 = m(a - x_1)$$
$$m = \frac{b - y_1}{a - x_1}$$

Because the right-hand side of this equation is the slope m of the line, we have the identity

$$m = m$$

Thus, the coordinates of the point Q satisfy the equation of the line. The coordinates of point $P(x_1, y_1)$ itself satisfy the equation $y - y_1 = m(x - x_1)$, because $y_1 - y_1 = m(x_1 - x_1)$ reduces to $0 = 0$. Thus, *any point on line l has coordinates that satisfy the equation.*

Part 2.   Suppose that the coordinates of some point $R(a, b)$ satisfy the equation $y - y_1 = m(x - x_1)$ and that point R is not point P. Then $b - y_1 = m(a - x_1)$, and

$$m = \frac{b - y_1}{a - x_1}$$

Thus, the slope of the line RP is m. Because line l is the only line with slope m passing through P, point R must lie on line l. Thus, *any point with coordinates that satisfy the equation lies on the line l.*

The theorem is proved.   □

Example 4   Find the equation of the line passing through the point $P(3, -1)$ with a slope of $-\frac{5}{3}$. Then solve that equation for y.

Solution   Substitute 3 for x_1, -1 for y_1, and $-\frac{5}{3}$ for m into the point–slope form of the equation of a line.

$$y - y_1 = m(x - x_1)$$
$$y - (-1) = -\frac{5}{3}(x - 3)$$
$$y + 1 = -\frac{5}{3}x + 5$$
$$y = -\frac{5}{3}x + 4$$

■

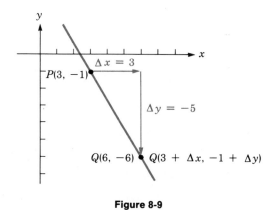

Figure 8-9

We can use the concept of *slope* to help graph the equation obtained in Example 4. We first plot the point $P(3, -1)$ as in Figure 8-9. Because the slope is $-\frac{5}{3}$, the line drops 5 units in a run of 3 units. Thus, we can locate another point Q by moving 3 units to the right of P and 5 units down. The graph of the equation is the line PQ.

Slope–Intercept Form of the Equation of a Line

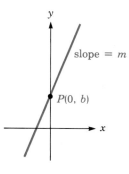

Figure 8-10

If the y-intercept of the line with slope m shown in Figure 8-10 is b, then the line intersects the y-axis at $P(0, b)$.

You can derive the equation of a line with slope m and y-intercept b from the point–slope form of the line:

$$y - y_1 = m(x - x_1)$$
$$y - b = m(x - 0) \qquad \text{Let } x_1 = 0 \text{ and } y_1 = b.$$
$$y = mx + b$$

Because the equation $y = mx + b$ displays both the slope and the y-intercept of a line, it is called the **slope–intercept form** of the equation of a line.

> **Slope–Intercept Form of the Equation of a Line.** The equation of the line with slope m and y-intercept b is
>
> $$y = mx + b$$

Example 5 Use the slope–intercept form to find the equation of the line that has slope $\frac{7}{3}$ and y-intercept -6.

Solution Substitute $\frac{7}{3}$ for m and -6 for b into the slope–intercept form of the equation of the line and simplify.

$$y = mx + b$$

$$y = \frac{7}{3}x + (-6)$$

$$y = \frac{7}{3}x - 6 \qquad \blacksquare$$

Example 6 Find the slope and the y-intercept of the line $3(y + 2) = 6x - 1$.

Solution Write the equation in the form $y = mx + b$ to determine the slope m and the y-intercept b.

$$3(y + 2) = 6x - 1$$
$$3y + 6 = 6x - 1$$
$$3y = 6x - 7$$
$$y = 2x - \frac{7}{3}$$

The slope of the line is 2, and the y-intercept is $-\dfrac{7}{3}$ $\qquad \blacksquare$

Example 7 Use the slope–intercept form to find the equation of the line perpendicular to the line

$$y = \frac{1}{3}x + 7$$

and passing through the point $(2, 5)$.

Solution The slope of the given line is $\frac{1}{3}$. The slope m of the required line must be -3, which is the negative reciprocal of $\frac{1}{3}$. Because the required line must pass through the point $(2, 5)$, the values $x = 2$ and $y = 5$ must satisfy the equation. Thus,

$$y = mx + b$$
$$5 = -3(2) + b$$
$$11 = b$$

The equation of the required line is

$$y = -3x + 11 \qquad \blacksquare$$

If a line is parallel to the x-axis, its slope is 0, and its equation is

$$y = 0x + b \qquad \text{or} \qquad y = b$$

If a line is parallel to the y-axis, it has no defined slope and cannot be the graph of any equation of the form $y = mx + b$. A vertical line does have an equation, however. If a vertical line passes through a point with an x-coordinate of a, then *every* point with an *x*-coordinate of *a* lies on the line and its equation is

$$x = a$$

Equations of Horizontal and Vertical Lines. The equation of the horizontal line passing through the point $P(a, b)$ is

$$y = b$$

The equation of the vertical line passing through the point $P(a, b)$ is

$$x = a$$

Example 8   Find the equation of the line passing through the points $P(-3, 4)$ and $Q(-3, -2)$.

Solution   Because the *x*-coordinates of points P and Q are equal, the line has no defined slope. It is a vertical line with equation $x = -3$ (see Figure 8-11).

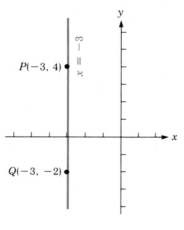

Figure 8-11

General Form of the Equation of a Line

We have shown that the graph of the equation

$$y - y_1 = m(x - x_1)$$

is a line. We can rewrite this equation as

$$-mx + y = y_1 - mx_1$$

which has the form $Ax + By = C$, where A, B, and C are constants: $A = -m$, $B = 1$, and $C = y_1 - mx_1$. In general, for any real numbers A, B, and C (where A and B are not *both* 0), the equation $Ax + By = C$ represents a line and is called the general form of the equation of a line. Any equation that can be written in general form is called a **linear equation in x and y.**

The General Form of the Equation of a Line.   If A, B, and C are real numbers and $B \neq 0$, the graph of the equation

$$Ax + By = C$$

is a nonvertical line with slope $-\dfrac{A}{B}$ and y-intercept $\dfrac{C}{B}$.

If $B = 0$ and $A \neq 0$, then the equation $Ax + By = C$ represents a vertical line with x-intercept $\frac{C}{A}$.

Example 9   Show that the two lines $3x - 2y = 5$ and $-6x + 4y = 7$ are parallel.

Solution   To show that two lines are parallel, show that they have the same slope. The first equation, $3x - 2y = 5$, is in general form, with $A = 3$, $B = -2$, and $C = 5$. By the previous theorem, the slope of the line is

$$m_1 = -\frac{A}{B} = -\left(\frac{3}{-2}\right) = \frac{3}{2}$$

Similarly, the second equation, $-6x + 4y = 7$, is in general form, with $A = -6$, $B = 4$, and $C = 7$. The slope of this line is

$$m_2 = -\frac{A}{B} = -\left(\frac{-6}{4}\right) = \frac{3}{2}$$

Because the slopes of the two lines are equal, the lines are parallel.   ■

EXERCISE 8.1

In Exercises 1–2, graph each equation.

1. $4x + 3y = 12$

2. $5x - 3y = 15$

In Exercises 3–8, find the slope of the line passing through each pair of points, if possible.

3. $P(2, 5)$; $Q(3, 10)$

4. $P(3, -1)$; $Q(5, 3)$

5. $P(8, -7)$; $Q(4, 1)$

6. $P(-4, 3)$; $Q(-4, -3)$

7. $P(a + b, c)$; $Q(b + c, a)$; $a \neq c$

8. $P(b, 0)$; $Q(a + b, 0)$

In Exercises 9–14, find two points on the line and determine the slope of the line.

9. $5x - 10y = 3$

10. $8y + 2x = 5$

11. $3(y + 2) = 2x - 3$

12. $4(x - 2) = 3y + 2$

13. $3(y + x) = 3(x - 2)$

14. $2x + 5 = 2(y + x)$

In Exercises 15–20, determine whether the lines with the given slopes are parallel, perpendicular, or neither.

15. $m_1 = 3$, $m_2 = -\dfrac{1}{3}$

16. $m_1 = \dfrac{2}{3}$, $m_2 = \dfrac{3}{2}$

17. $m_1 = \sqrt{8}, m_2 = 2\sqrt{2}$

18. $m_1 = 1, m_2 = -1$

19. $m_1 = -\sqrt{2}, m_2 = \dfrac{\sqrt{2}}{2}$

20. $m_1 = 2\sqrt{7}, m_2 = \sqrt{28}$

In Exercises 21–24, determine whether the line through the given points and the line through $R(-3, 5)$ and $S(2, 7)$ are parallel, perpendicular, or neither.

21. $P(2, 4); Q(7, 6)$

22. $P(-3, 8); Q(-13, 4)$

23. $P(-4, 6); Q(-2, 1)$

24. $P(0, -9); Q(4, 1)$

In Exercises 25–28, find the slopes of lines PQ and PR, and determine whether points P, Q, and R lie on the same line.

25. $P(-2, 8); Q(-6, 9); R(2, 5)$

26. $P(1, -1); Q(3, -2); R(-3, 0)$

27. $P(-3, 9); Q(-5, 5); R(-6, 3)$

28. $P(-1, 1); Q(13, -1); R(-8, 2)$

In Exercises 29–32, determine which, if any, of the three lines PQ, PR, and QR are perpendicular.

29. $P(5, 4), Q(2, -5), R(8, -3)$

30. $P(8, -2), Q(4, 6), R(6, 7)$

31. $P(0, 0), Q(a, b), R(-b, a)$

32. $P(1, 3), Q(1, 9), R(7, 3)$

In Exercises 33–36, use the point–slope form to write the equation of the line passing through the given point and having the given slope. Express the answer in slope–intercept form.

33. $P(2, 4); m = 2$

34. $P(3, 5); m = -3$

35. $P\left(-\dfrac{3}{2}, \dfrac{1}{2}\right); m = 2$

36. $P\left(\dfrac{1}{4}, \dfrac{4}{3}\right); m = 0$

In Exercises 37–40, use the slope–intercept form to write the equation of the line with the given slope and y-intercept. Give the answer in general form.

37. $m = 3, b = -2$

38. $m = -\dfrac{1}{3}, b = \dfrac{2}{3}$

39. $m = a, b = \dfrac{1}{a}$

40. $m = a, b = 2a$

In Exercises 41–44, find the slope and the y-intercept of the line defined by the given equation, if possible.

41. $3(14x + 12) = 5(y + 3)$

42. $-2(3x + 6) = -2y + 15$

43. $2(y - 3) + x = 2y - 7$

44. $3(y + 2) + 2x = 2(y + x)$

In Exercises 45–48, find the equation of the line parallel to the given line and passing through the given point. Give the answer in slope–intercept form.

45. $3x = 2y + 1; (-1, 1)$

46. $5y - 7x = 2; (5, -3)$

47. $3(x - y) = 5(y - 2x); (2, 3)$

48. $3x - 2(y + 2x) = x - 3; (1, -5)$

In Exercises 49–52, find the equation of the line perpendicular to the given line and passing through the given point. Give the answer in general form.

49. $3x = 2y + 1; (-1, 1)$

50. $5y - 7x = 2; (5, -3)$

51. $3(x - y) = 5(y - 2x); (2, 3)$

52. $3x - 2(y + 2x) = x - 3; (1, -5)$

In Exercises 53–56, write the equation of the line passing through the given point and parallel to the given axis.

53. $P(1, 0)$; y-axis

54. $P(-3, -7)$; x-axis

55. $P(-3, 5)$; x-axis

56. $P(-5, 9)$; y-axis

In Exercises 57–62, find the slope and the y-intercept of the line that has the given properties.

57.   The line passes through $P(-3, 5)$ and $Q(7, -5)$.

58.   The line passes through $P(15, 3)$ and $Q(-3, -6)$.

59.   The line is parallel to the line $y = 3(x - 7)$ and passes through $P(-2, 5)$.

60.   The line is parallel to the line $3y + 1 = 6(x - 2)$ and passes through $P(5, 0)$.

61.   The line is perpendicular to the line $y + 3x = 8$ and passes through $P(2, 3)$.

62.   The line is perpendicular to the line $x = 2y - 7$ and passes through $P(8, 0)$.

63.   Show that the three points $A(-1, -1)$, $B(-3, 4)$, and $C(4, 1)$ are the vertices of a right triangle.

64.   Show that the four points $A(1, -1)$, $B(3, 0)$, $C(2, 2)$, and $D(0, 1)$ are the vertices of a square.

65.   Show that the four points $A(-2, -2)$, $B(3, 3)$, $C(2, 6)$, and $D(-3, 1)$ are the vertices of a parallelogram. (Show that both pairs of opposite sides are parallel.)

66.   Show that the four points $E(1, -2)$, $F(5, 1)$, $G(3, 4)$, and $H(-3, 4)$ are the vertices of a trapezoid. (Show that only one pair of opposite sides are parallel.)

67.   Prove that the equation of the line with x-intercept a and y-intercept b may be written in the form

$$\frac{x}{a} + \frac{y}{b} = 1$$

68.   Prove that if $B \neq 0$ the graph of $Ax + By = C$ has a slope of $-\dfrac{A}{B}$ and a y-intercept of $\dfrac{C}{B}$.

69.   Prove that, if $B = 0$ and $A \neq 0$, the graph of $Ax + By = C$ is a vertical line with x-intercept $\dfrac{C}{A}$.

70.   Prove that the lines $Ax + By = C$ and $Bx - Ay = C$ are perpendicular.

71.   Prove that the equation of the line passing through the points (x_1, y_1) and (x_2, y_2) can be written in the form

$$y - y_1 = \frac{y_2 - y_1}{x_2 - x_1}(x - x_1)$$

This equation is called the **two-point form** of the equation of a line.

72.   Use the distance formula to find the equation of the perpendicular bisector of the line segment joining $A(3, 5)$ and $B(5, -3)$. [*Hint:* Let $P(x, y)$ be a point on the perpendicular bisector. Then $d(PA) = d(PB)$.]

73.   Prove that two lines with the same slope are parallel.

8.2   THE CIRCLE

> **Definition.**   A **circle** is the set of all points in a plane that are a fixed distance from a point called its **center**. The fixed distance is called the **radius of the circle**.

To find the general equation of a circle with radius r and center at the point $C(h, k)$, we must find all points $P(x, y)$ such that the length of the line segment PC

is r. See Figure 8-12. We can use the distance formula to find the length of CP, which is r:

$$r = \sqrt{(x - h)^2 + (y - k)^2}$$

We square both sides to get

$$r^2 = (x - h)^2 + (y - k)^2$$

This equation is called the **standard form of the equation of a circle.**

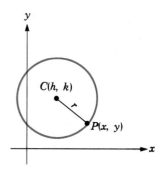

Figure 8-12

Theorem.   Any equation that can be written in the form

$$(x - h)^2 + (y - k)^2 = r^2, \qquad r > 0$$

has a graph that is a circle with radius r and center at the point (h, k).

If the center of the circle is the origin, then $(h, k) = (0, 0)$, and we have the following result.

Theorem.   Any equation that can be written in the form

$$x^2 + y^2 = r^2, \qquad r > 0$$

has a graph that is a circle with radius r and with center at the origin.

We can use the preceding theorems to write the equations of many circles.

Example 1   Find the equation of the circle with radius 5 and center $(3, 2)$. Express the equation in general form.

Solution   Substitute 5 for r, 3 for h, and 2 for k in the standard form of the equation of the circle and simplify:

$$(x - h)^2 + (y - k)^2 = r^2$$
$$(x - 3)^2 + (y - 2)^2 = 5^2$$
$$x^2 - 6x + 9 + y^2 - 4y + 4 = 25$$
$$x^2 + y^2 - 6x - 4y - 12 = 0$$

This final equation is a special case of the general form of a second-degree equation. The coefficient of the xy term is 0, and the coefficients of x^2 and y^2 are both 1. ∎

Example 2 Find the equation of the circle with endpoints of its diameter at $(8, -3)$ and $(-4, 13)$.

Solution First find the center (h, k) of the circle by finding the midpoint of its diameter. Use the midpoint formulas with $(x_1, y_1) = (8, -3)$ and $(x_2, y_2) = (-4, 13)$:

$$h = \frac{x_1 + x_2}{2} \qquad k = \frac{y_1 + y_2}{2}$$

$$h = \frac{8 + (-4)}{2} \qquad k = \frac{-3 + 13}{2}$$

$$= \frac{4}{2} \qquad\qquad = \frac{10}{2}$$

$$= 2 \qquad\qquad\quad = 5$$

Thus, the center of the circle is the point $(h, k) = (2, 5)$.

To find the radius of the circle, use the distance formula to find the distance between the center and one endpoint of the diameter. Because one endpoint is $(8, -3)$, substitute 8 for x_1, -3 for y_1, 2 for x_2, and 5 for y_2 in the distance formula and simplify:

$$r = \sqrt{(x_2 - x_1)^2 + (y_2 - y_1)^2}$$

$$r = \sqrt{(2 - 8)^2 + [5 - (-3)]^2}$$

$$= \sqrt{(-6)^2 + (8)^2}$$

$$= \sqrt{36 + 64}$$

$$= \sqrt{100}$$

$$= 10$$

Thus, the radius of the circle is 10.

To find the equation of the circle with radius 10 and center at the point $(2, 5)$, substitute 2 for h, 5 for k, and 10 for r in the standard form of the equation of the circle and simplify:

$$(x - h)^2 + (y - k)^2 = r^2$$

$$(x - 2)^2 + (y - 5)^2 = 10^2$$

$$x^2 - 4x + 4 + y^2 - 10y + 25 = 100$$

$$x^2 + y^2 - 4x - 10y - 71 = 0$$

Example 3 Graph the circle $x^2 + y^2 - 4x + 2y = 20$.

Solution To find the coordinates of the center and the radius, write the equation in standard form by completing the square on both x and y and then simplifying:

$$x^2 + y^2 - 4x + 2y = 20$$

$$x^2 - 4x + y^2 + 2y = 20$$

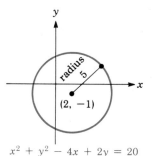

$$x^2 - 4x + 4 + y^2 + 2y + 1 = 20 + 4 + 1$$

Add 4 and 1 to both sides to complete the square.

$$(x - 2)^2 + (y + 1)^2 = 25$$

Factor $x^2 - 4x + 4$ and $y^2 + 2y + 1$.

$$(x - 2)^2 + [y - (-1)]^2 = 5^2$$

Note that the radius of the circle is 5 and the coordinates of its center are $h = 2$ and $k = -1$. Plot the center of the circle and construct the circle with a radius of 5 units, as shown in Figure 8-13. ∎

$x^2 + y^2 - 4x + 2y = 20$

Figure 8-13

■ EXERCISE 8.2

In Exercises 1–26, write an equation for the circle with the given properties.

1. Center at the origin; $r = 1$.
2. Center at the origin; $r = 4$.
3. Center at $(6, 8)$; $r = 4$.
4. Center at $(5, 3)$; $r = 2$.
5. Center at $(-5, 3)$ and tangent to the y-axis.
6. Center at $(-7, -2)$ and tangent to the x-axis.
7. Center at $(3, -4)$; $r = \sqrt{2}$.
8. Center at $(-9, 8)$; $r = 2\sqrt{3}$.
9. Ends of diameter at $(3, -2)$ and $(3, 8)$.
10. Ends of diameter at $(5, 9)$ and $(-5, -9)$.
11. Ends of diameter at $(-6, 9)$ and $(-4, -7)$.
12. Ends of diameter at $(17, 0)$ and $(-3, -3)$.
13. Center at $(-3, 4)$ and circle passing through the origin.
14. Center at $(4, 0)$ and circle passing through the origin.
15. Center at $(-2, -6)$ and circle passing through the origin.
16. Center at $(-19, -13)$ and circle passing through the origin.
17. Center at $(0, -3)$ and circle passing through $(6, 8)$.
18. Center at $(2, 4)$ and circle passing through $(1, 1)$.
19. Center at $(5, 8)$ and circle passing through $(-2, -9)$.
20. Center at $(7, -5)$ and circle passing through $(-3, -7)$.
21. Center at $(-4, -2)$ and circle passing through $(3, 5)$.
22. Center at $(0, -7)$ and circle passing through $(0, 7)$.
23. Radius of 6 and center at the intersection of $3x + y = 1$ and $-2x - 3y = 4$.
24. Radius of 8 and center at the intersection of $x + 2y = 8$ and $2x - 3y = -5$.
25. Radius of $\sqrt{10}$ and center at the intersection of $x - y = 12$ and $3x - y = 12$.
26. Radius of $2\sqrt{2}$ and center at the intersection of $6x - 4y = 8$ and $2x + 3y = 7$.

27. Can a circle with a radius of 10 have endpoints of its diameter at $(6, 8)$ and $(-2, -2)$?
28. Can a circle with radius 25 have endpoints of its diameter at $(0, 0)$ and $(6, 24)$?

In Exercises 29–38, graph each equation.

29. $x^2 + y^2 - 25 = 0$ **30.** $x^2 + y^2 - 8 = 0$

31. $(x - 1)^2 + (y + 2)^2 = 4$ **32.** $(x + 1)^2 + (y - 2)^2 = 9$

33. $x^2 + y^2 + 2x - 26 = 0$ **34.** $x^2 + y^2 - 4y = 12$

35. $9x^2 + 9y^2 - 12y = 5$ **36.** $4x^2 + 4y^2 + 4y = 15$

37. $4x^2 + 4y^2 - 4x + 8y + 1 = 0$ **38.** $9x^2 + 9y^2 - 6x + 18y + 1 = 0$

39. Write the equation of the circle passing through $(0, 8)$, $(5, 3)$, and $(4, 6)$.

40. Write the equation of the circle passing through $(-2, 0)$, $(2, 8)$, and $(5, -1)$.

41. Find the area of the circle $3x^2 + 3y^2 + 6x + 12y = 0$. (*Hint:* $A = \pi r^2$.)

42. Find the circumference of the circle $x^2 + y^2 + 4x - 10y - 20 = 0$. (*Hint:* $C = 2\pi r$.)

43. The rectangle in Illustration 1 is inscribed in the upper half of the circle $x^2 + y^2 = r^2$, with one vertex at $P(a, 0)$. Express the area of the rectangle as a function of a.

44. The triangle ABP in Illustration 2 is inscribed in the upper half of the circle $x^2 + y^2 = 1$. Find the area of the triangle as a function of a.

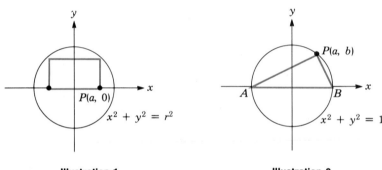

Illustration 1   Illustration 2

8.3 THE PARABOLA

> **Definition.** A **parabola** is the set of all points in a plane such that each point in the set is equidistant from a line, called the **directrix**, and a fixed point F, called the **focus**. The point on the parabola that is closest to the directrix is called the **vertex**. The line passing through the vertex and the focus is called the **axis**.

Consider the parabola in Figure 8-14, which opens to the right and has its vertex at the point $V(h, k)$. Let $P(x, y)$ be any point on the parabola. Because each point on the parabola is the same distance from the focus (point F) and from the

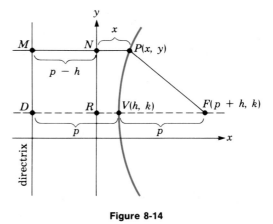

Figure 8-14

directrix, we can let $d(DV) = d(VF) = p$, where p is some positive constant. Because of the geometry of the figure,

$$d(MP) = p - h + x$$

Using the distance formula,

$$d(PF) = \sqrt{[x - (p + h)]^2 + (y - k)^2}$$

By the definition of the parabola, $d(MP) = d(PF)$. Thus,

$$p - h + x = \sqrt{[x - (p + h)]^2 + (y - k)^2}$$
$$(p - h + x)^2 = [x - (p + h)]^2 + (y - k)^2 \qquad \text{Square both sides.}$$

Finally, we expand the expression on each side of the equation and simplify:

$$p^2 - ph + px - ph + h^2 - hx + px - hx + x^2$$
$$= x^2 - 2px - 2hx + p^2 + 2ph + h^2 + (y - k)^2$$
$$-2ph + 2px = -2px + 2ph + (y - k)^2$$
$$4px - 4ph = (y - k)^2$$
$$4p(x - h) = (y - k)^2$$

The argument above proves the theorem that follows.

Theorem. The standard form of the equation of a parabola with vertex at point (h, k) and opening to the right is

$$(y - k)^2 = 4p(x - h)$$

where p is the distance from the vertex to the focus.

If the parabola has its vertex at the origin, both h and k are equal to zero, and we have the following theorem.

> **Theorem.**   The standard form of the equation of a parabola with vertex at the origin and opening to the right is
>
> $$y^2 = 4px$$
>
> where p is the distance from the vertex to the focus.

Equations of parabolas that open to the right, left, upward, and downward are summarized in Table 8-1. If $p > 0$, then

Table 8-1

Parabola opening	Vertex at origin	Vertex at $V(h, k)$
Right	$y^2 = 4px$	$(y - k)^2 = 4p(x - h)$
Left	$y^2 = -4px$	$(y - k)^2 = -4p(x - h)$
Upward	$x^2 = 4py$	$(x - h)^2 = 4p(y - k)$
Downward	$x^2 = -4py$	$(x - h)^2 = -4p(y - k)$

Example 1   Find the equation of the parabola with vertex at the origin and focus at $(3, 0)$.

Solution   Sketch the parabola as in Figure 8-15. Because the focus is to the right of the vertex, the parabola opens to the right. Because the vertex is the origin, the standard form of the equation is $y^2 = 4px$. The distance between the focus and the vertex is 3, which is p. Therefore, the equation of the parabola is $y^2 = 4(3)x$, or

$$y^2 = 12x$$

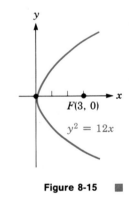

Figure 8-15   ■

Example 2   Find the equation of the parabola that opens upward, has vertex at the point $(4, 5)$, and passes through the point $(0, 7)$.

Solution   Use the standard form of a parabola that opens upward: $(x - h)^2 = 4p(y - k)$. Because the point $(0, 7)$ is on the curve, substitute 0 for x and 7 for y in the equation. Because the vertex (h, k) is $(4, 5)$, also substitute 4 for h and 5 for k. Then solve the equation to determine p:

$$(x - h)^2 = 4p(y - k)$$
$$(0 - 4)^2 = 4p(7 - 5)$$
$$16 = 8p$$
$$2 = p$$

To find the equation of the parabola, substitute 4 for h, 5 for k, and 2 for p in the standard form of the equation and simplify:

$$(x - h)^2 = 4p(y - k)$$
$$(x - 4)^2 = 4 \cdot 2(y - 5)$$
$$(x - 4)^2 = 8(y - 5)$$

The graph of this equation appears in Figure 8-16.

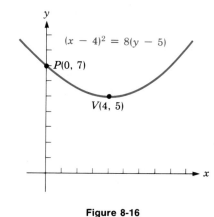

Figure 8-16

Example 3   Find the equations of the two parabolas each having its vertex at $(2, 4)$ and passing through the point $(0, 0)$.

Solution   Sketch the two possible parabolas as shown in Figure 8-17.

Part 1.   To find the parabola that opens to the left, use the standard form of the equation $(y - k)^2 = -4p(x - h)$. Because the curve passes through the point $(x, y) = (0, 0)$ and the vertex is $(h, k) = (2, 4)$, substitute 0 for x, 0 for y, 2 for h, and 4 for k in the equation $(y - k)^2 = -4p(x - h)$ and solve for p:

$$(y - k)^2 = -4p(x - h)$$
$$(0 - 4)^2 = -4p(0 - 2)$$
$$16 = 8p$$
$$2 = p$$

Since $h = 2$, $k = 4$, $p = 2$, and the parabola opens to the left, its equation is

$$(y - k)^2 = -4p(x - h)$$
$$(y - 4)^2 = -4(2)(x - 2)$$
$$(y - 4)^2 = -8(x - 2)$$

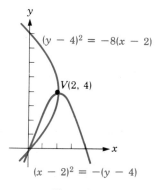

Figure 8-17

Part 2. To find the equation of the parabola that opens downward, use the standard form $(x - h)^2 = -4p(y - k)$. Substitute 2 for h, 4 for k, 0 for x, and 0 for y in the equation and solve for p:

$$(x - h)^2 = -4p(y - k)$$
$$(0 - 2)^2 = -4p(0 - 4)$$
$$4 = 16p$$
$$\frac{1}{4} = p$$

Since $h = 2, k = 4$, and $p = \frac{1}{4}$, and the parabola opens downward, its equation is

$$(x - h)^2 = -4p(y - k)$$
$$(x - 2)^2 = -4\left(\frac{1}{4}\right)(y - 4)$$
$$(x - 2)^2 = -(y - 4)$$

∎

Example 4 Find the vertex and y-intercepts of the parabola $y^2 + 8x - 4y = 28$. Then graph the parabola.

Solution Complete the square on y to write the equation in standard form:

$$y^2 + 8x - 4y = 28$$

$y^2 - 4y = -8x + 28$	Add $-8x$ to both sides.
$y^2 - 4y + 4 = -8x + 28 + 4$	Add 4 to both sides.
$(y - 2)^2 = -8(x - 4)$	Factor both sides.

Observe that this equation represents a parabola opening to the left with vertex at $(4, 2)$. To find the points where the graph intersects the y-axis, substitute 0 for x in the equation of the parabola.

$(y - 2)^2 = -8(x - 4)$	
$(y - 2)^2 = -8(0 - 4)$	Substitute 0 for x.
$y^2 - 4y + 4 = 32$	Remove parentheses.
$y^2 - 4y - 28 = 0$	

Use the quadratic formula to determine that the roots of this quadratic equation are $y \approx 7.7$ and $y \approx -3.7$.

The points with coordinates of approximately $(0, 7.7)$ and $(0, -3.7)$ are on the graph of the parabola. Using this information and the knowledge that the graph opens to the left and has a vertex at $(4, 2)$, draw the curve as shown in Figure 8-18.

∎

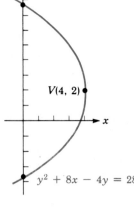

$y^2 + 8x - 4y = 28$

$V(4, 2)$

Figure 8-18

Example 5 A stone is thrown straight up. The equation $s = 128t - 16t^2$ expresses the height of the stone in feet t seconds after it was thrown. Find the maximum height reached by the stone.

Solution   The stone goes straight up and then straight down. The graph of $s = 128t - 16t^2$, expressing the height of the stone t seconds after it was thrown, is a parabola (see Figure 8-19). To find the maximum height reached by the stone, calculate the y-coordinate, k, of the vertex of the parabola. To find k, write the equation of the parabola, $s = 128t - 16t^2$, in standard form. To change this equation into standard form, complete the square on t:

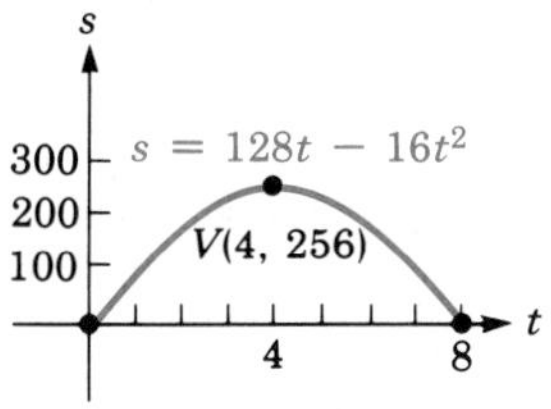

$$s = 128t - 16t^2$$

$$16t^2 - 128t = -s \qquad \text{Multiply both sides by } -1.$$

$$t^2 - 8t = \frac{-s}{16} \qquad \text{Divide both sides by 16.}$$

$$t^2 - 8t + 16 = \frac{-s}{16} + 16 \qquad \text{Add 16 to both sides.}$$

$$(t - 4)^2 = \frac{-s + 256}{16} \qquad \text{Factor } t^2 - 8t + 16 \text{ and combine terms.}$$

$$(t - 4)^2 = -\frac{1}{16}(s - 256) \qquad \text{Factor out } -\frac{1}{16}.$$

This equation indicates that the maximum height of 256 feet was reached in 4 seconds.   ■

s

300 — $s = 128t - 16t^2$
200
100 | $V(4, 256)$

4 8 t

Figure 8-19

EXERCISE 8.3

In Exercises 1–16, find the equation of each parabola.

1. Vertex at $(0, 0)$ and focus at $(0, 3)$.
2. Vertex at $(0, 0)$ and focus at $(0, -3)$.
3. Vertex at $(0, 0)$ and focus at $(3, 0)$.
4. Vertex at $(0, 0)$ and focus at $(-3, 0)$.
5. Vertex at $(3, 5)$ and focus at $(3, 2)$.
6. Vertex at $(3, 5)$ and focus at $(-3, 5)$.
7. Vertex at $(3, 5)$ and focus at $(3, -2)$.
8. Vertex at $(3, 5)$ and focus at $(6, 5)$.
9. Vertex at $(2, 2)$ and the parabola passing through $(0, 0)$.
10. Vertex at $(-2, -2)$ and the parabola passing through $(0, 0)$.
11. Vertex at $(-4, 6)$ and the parabola passing through $(0, 3)$.
12. Vertex at $(-2, 3)$ and the parabola passing through $(0, -3)$.
13. Vertex at $(6, 8)$ and the parabola passing through $(5, 10)$ and $(5, 6)$.
14. Vertex at $(2, 3)$ and the parabola passing through $(1, \frac{13}{4})$ and $(-1, \frac{21}{4})$.
15. Vertex at $(3, 1)$ and the parabola passing through $(4, 3)$ and $(2, 3)$.
16. Vertex at $(-4, -2)$ and the parabola passing through $(-3, 0)$ and $(\frac{9}{4}, 3)$.

In Exercises 17–26, change each equation to standard form and graph it.

17. $y = x^2 + 4x + 5$
18. $2x^2 - 12x - 7y = 10$
19. $y^2 + 4x - 6y = -1$
20. $x^2 - 2y - 2x = -7$

21. $y^2 + 2x - 2y = 5$

22. $y^2 - 4y = -8x + 20$

23. $x^2 - 6y + 22 = -4x$

24. $4y^2 - 4y + 16x = 7$

25. $4x^2 - 4x + 32y = 47$

26. $4y^2 - 16x + 17 = 20y$

27. A parabolic arch spans 30 meters and has a maximum height of 10 meters. Derive the equation of the arch using the vertex of the arch as the origin.

28. Find the maximum value of y in the parabola $x^2 + 8y - 8x = 8$.

29. The sum of two numbers is 20 and their product is maximum. What are the numbers?

30. The sum of two numbers is 10 and the sum of their squares is minimum. What are the numbers?

31. A resort owner plans to build and rent n cabins for d dollars per week. The price, d, that she can charge for each cabin depends on the number of cabins she builds, where $d = -45(\frac{n}{32} - \frac{1}{2})$. Find the number of cabins that the owner should build to maximize her weekly income.

32. A toy rocket is s feet above the earth at the end of t seconds, where $s = -16t^2 + 80\sqrt{3}\,t$. Find the maximum height of the rocket.

33. An engineer plans to build a tunnel whose arch is in the shape of a parabola. The tunnel will span a two-lane highway that is 8 meters wide. To allow safe passage for most vehicles, the tunnel must be 5 meters high at a distance of 1 meter from the tunnel's edge. What will be the maximum height of the tunnel?

34. The towers of a suspension bridge are 900 feet apart and rise 120 feet above the roadway. The cable between the towers has the shape of a parabola with a vertex 15 feet above the roadway. Find the equation of the parabola with respect to the indicated coordinate system. See Illustration 1.

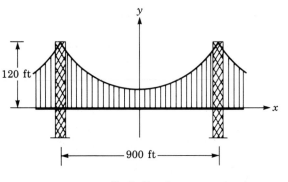

Illustration 1

35. A satellite antenna with a parabolic cross section is a dish 6 feet in diameter and 1 foot deep at its center. How far is the focus from the center of the dish?

36. A stone tossed upward is s feet above the earth after t seconds, where $s = -16t^2 + 128t$. Show that the stone's height x seconds *after* it is thrown is equal to its height x seconds *before* it returns to the ground.

37. Derive the standard form of the equation of a parabola that opens downward and has vertex at the origin.

38. Show that the result in Example 1 is a special case of the general form of the equation of second degree.

In Exercises 39 and 40, find the equation of the form $y = ax^2 + bx + c$ that determines a parabola passing through the three given points.

39. $(1, 8), (-2, -1),$ and $(2, 15)$

40. $(1, -3), (-2, 12),$ and $(-1, 3)$

8.4 THE ELLIPSE

> **Definition.** An **ellipse** is the set of all points P in a plane such that the sum of the distances from P to two other fixed points F and F' is a positive constant.

In the ellipse shown in Figure 8-20, the two fixed points F and F' are called **foci** of the ellipse, the midpoint of the chord FF' is called the **center**, the chord VV' is called the **major axis**, and each endpoint of the major axis is called a **vertex**. The chord BB', perpendicular to the major axis and passing through the center C, is called the **minor axis**.

To simplify the algebra, we will derive the equation of the ellipse shown in Figure 8-21, which has its center at $(0, 0)$. Because the origin is the midpoint of the chord FF', we can let $d(OF) = d(OF') = c$, where $c > 0$. Then the coordinates of point F are $(c, 0)$, and the coordinates of F' are $(-c, 0)$. We also let $P(x, y)$ be any point on the ellipse.

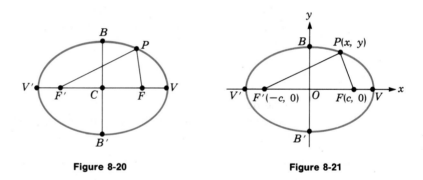

Figure 8-20   Figure 8-21

The definition of an ellipse requires that the sum of $d(F'P)$ and $d(PF)$ be a positive constant, which we will call $2a$. Thus,

1. $\quad d(F'P) + d(PF) = 2a$

We use the distance formula to compute the lengths of $F'P$ and PF:

$$d(F'P) = \sqrt{[x - (-c)]^2 + y^2}$$
$$d(PF) = \sqrt{(x - c)^2 + y^2}$$

and substitute these values into Equation 1 to obtain

$$\sqrt{[x - (-c)]^2 + y^2} + \sqrt{(x - c)^2 + y^2} = 2a$$

or

$$\sqrt{[x + c]^2 + y^2} = 2a - \sqrt{(x - c)^2 + y^2}$$

We square both sides of this equation and simplify to get

$$(x + c)^2 + y^2 = 4a^2 - 4a\sqrt{(x - c)^2 + y^2} + [(x - c)^2 + y^2]$$

$$x^2 + 2cx + c^2 + y^2 = 4a^2 - 4a\sqrt{(x - c)^2 + y^2} + x^2 - 2cx + c^2 + y^2$$

$$4cx = 4a^2 - 4a\sqrt{(x - c)^2 + y^2}$$

$$cx = a^2 - a\sqrt{(x - c)^2 + y^2}$$

$$cx - a^2 = -a\sqrt{(x - c)^2 + y^2}$$

We square both sides again and simplify to obtain

$$c^2x^2 - 2a^2cx + a^4 = a^2[(x - c)^2 + y^2]$$

$$c^2x^2 - 2a^2cx + a^4 = a^2(x^2 - 2cx + c^2 + y^2)$$

$$c^2x^2 - 2a^2cx + a^4 = a^2x^2 - 2a^2cx + a^2c^2 + a^2y^2$$

$$c^2x^2 + a^4 = a^2x^2 + a^2c^2 + a^2y^2$$

$$a^4 - a^2c^2 = a^2x^2 - c^2x^2 + a^2y^2$$

2. $$a^2(a^2 - c^2) = (a^2 - c^2)x^2 + a^2y^2$$

Because the shortest path between two points is a line segment, $d(F'P) + d(PF) > d(F'F)$. Therefore, $2a > 2c$. This implies that $a > c$ and that $a^2 - c^2$ is a positive number, which we will call b^2. Letting $b^2 = a^2 - c^2$ and substituting into Equation 2, we have

$$a^2b^2 = b^2x^2 + a^2y^2$$

Dividing both sides of this equation by a^2b^2 gives the standard form of the equation for an ellipse with center at the origin and major axis on the x-axis:

$$\frac{x^2}{a^2} + \frac{y^2}{b^2} = 1 \qquad \text{where } a > b > 0$$

To find the coordinates of the vertices V and V', we substitute 0 for y and solve for x:

$$\frac{x^2}{a^2} + \frac{y^2}{b^2} = 1$$

$$\frac{x^2}{a^2} + \frac{0^2}{b^2} = 1$$

$$\frac{x^2}{a^2} = 1$$

$$x^2 = a^2$$

$$x = a \qquad \text{or} \qquad x = -a$$

Thus, the coordinates of V are $(a, 0)$, and the coordinates of V' are $(-a, 0)$. In other words, a is the distance between the center of the ellipse, $(0, 0)$, and either of its vertices, and the center of the ellipse is the midpoint of the major axis.

To find the coordinates of B and B', we substitute 0 for x and solve for y:

$$\frac{x^2}{a^2} + \frac{y^2}{b^2} = 1$$

$$\frac{0^2}{a^2} + \frac{y^2}{b^2} = 1$$

$$y^2 = b^2$$

$$y = b \qquad \text{or} \qquad y = -b$$

Thus, the coordinates of B are $(0, b)$, and the coordinates of B' are $(0, -b)$. The distance between the center of the ellipse and either endpoint of the minor axis is b.

Theorem.   The standard form of the equation of an ellipse with center at the origin and major axis on the x-axis is

$$\frac{x^2}{a^2} + \frac{y^2}{b^2} = 1 \qquad \text{where } a > b > 0$$

If the major axis of an ellipse with center at $(0, 0)$ lies on the y-axis, the standard form of the equation of the ellipse is

$$\frac{y^2}{a^2} + \frac{x^2}{b^2} = 1 \qquad \text{where } a > b > 0$$

In either case, the length of the major axis is $2a$, and the length of the minor axis is $2b$.

If we develop the equation of the ellipse with center at (h, k), we obtain the following results.

Theorem.   The standard form of the equation of an ellipse with center at (h, k) and major axis parallel to the x-axis is

$$\frac{(x - h)^2}{a^2} + \frac{(y - k)^2}{b^2} = 1 \qquad \text{where } a > b > 0$$

If the major axis of an ellipse with center at (h, k) is parallel to the y-axis, the standard form of the equation of the ellipse is

$$\frac{(y - k)^2}{a^2} + \frac{(x - h)^2}{b^2} = 1 \qquad \text{where } a > b > 0$$

In either case, the length of the major axis is $2a$, and the length of the minor axis is $2b$.

Example 1   Find the equation of the ellipse with center at the origin, major axis of length 6 units located on the x-axis, and minor axis of length 4 units.

Solution   Because the center of the ellipse is the origin and the length of the major axis is 6, $a = 3$ and the coordinates of the vertices of the ellipse are $(3, 0)$ and $(-3, 0)$, as shown in Figure 8-22.

Because the length of the minor axis is 4, the value of b is 2 and the coordinates of B and B' are $(0, 2)$ and $(0, -2)$. To find the desired equation, substitute 3 for a and 2 for b in the standard form of the equation of an ellipse with center at the origin and major axis on the x-axis. Then simplify the equation:

$$\frac{x^2}{a^2} + \frac{y^2}{b^2} = 1$$

$$\frac{x^2}{3^2} + \frac{y^2}{2^2} = 1$$

$$\frac{x^2}{9} + \frac{y^2}{4} = 1$$

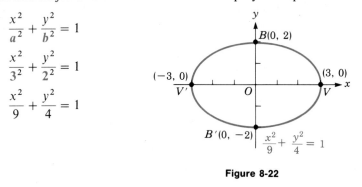

Figure 8-22

Example 2   Find the equation of the ellipse with focus $(0, 3)$ and vertices V and V' at $(3, 3)$ and $(-5, 3)$.

Solution   Because the midpoint of the major axis is the center of the ellipse, the coordinates of the center are $(-1, 3)$. Look at Figure 8-23 and note that the major axis is parallel to the x-axis. The standard form of the equation to use is

$$\frac{(x - h)^2}{a^2} + \frac{(y - k)^2}{b^2} = 1 \qquad \text{where } a > b > 0$$

The distance between the center of the ellipse and a vertex is $a = 4$; the distance between the focus and the center is $c = 1$. In the ellipse, $b^2 = a^2 - c^2$. From this equation, compute b^2:

$$b^2 = a^2 - c^2$$
$$= 4^2 - 1^2$$
$$= 15$$

To find the equation of the ellipse, substitute -1 for h, 3 for k, 16 for a^2, and 15 for b^2 in the standard form of the equation for an ellipse and simplify:

$$\frac{(x - h)^2}{a^2} + \frac{(y - k)^2}{b^2} = 1$$

$$\frac{[x - (-1)]^2}{16} + \frac{(y - 3)^2}{15} = 1$$

$$\frac{(x + 1)^2}{16} + \frac{(y - 3)^2}{15} = 1$$

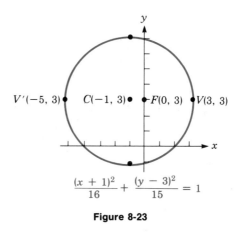

$$\frac{(x + 1)^2}{16} + \frac{(y - 3)^2}{15} = 1$$

Figure 8-23

Example 3   The orbit of the earth is approximately an ellipse, with the sun at one focus. The ratio of c to a (called the *eccentricity* of the ellipse) is about $\frac{1}{62}$, and the length of the major axis is approximately 186,000,000 miles. How close does the earth get to the sun?

Solution   Assume that this ellipse has its center at the origin and vertices V' and V at $(-93,000,000, 0)$ and $(93,000,000, 0)$, as shown in Figure 8-24. This implies that $a = 93,000,000$.

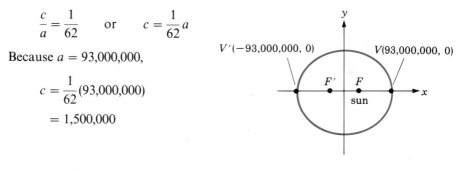

$$\frac{c}{a} = \frac{1}{62} \quad \text{or} \quad c = \frac{1}{62}a$$

Because $a = 93,000,000$,

$$c = \frac{1}{62}(93,000,000)$$
$$= 1,500,000$$

Figure 8-24

We use $d(FV)$ to represent the shortest possible distance between the earth and the sun. (You'll be asked to prove this in the exercises.) Thus,

$$d(FV) = a - c = 93,000,000 - 1,500,000 = 91,500,000 \text{ miles}$$

The earth's point of closest approach to the sun (called *perihelion*) is approximately 91.5 million miles.

Example 4   Graph the ellipse $\dfrac{(x + 2)^2}{4} + \dfrac{(y - 2)^2}{9} = 1$.

Solution   The center of the ellipse is at $(-2, 2)$, and the major axis is parallel to the y-axis. Because $a = 3$, the vertices are 3 units above and below the center at points $(-2, 5)$ and $(-2, -1)$. Because $b = 2$, the endpoints of the minor axis are 2 units to the right and left of the center at points $(0, 2)$ and $(-4, 2)$. Using these four points as guides, sketch the ellipse, as shown in Figure 8-25.

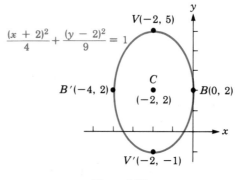

Figure 8-25

Example 5   Graph the equation $4x^2 + 9y^2 - 16x - 18y = 11$.

Solution   Write the equation in standard form by completing the square on x and y as follows:

$$4x^2 + 9y^2 - 16x - 18y = 11$$
$$4x^2 - 16x + 9y^2 - 18y = 11$$
$$4(x^2 - 4x) + 9(y^2 - 2y) = 11$$
$$4(x^2 - 4x + 4) + 9(y^2 - 2y + 1) = 11 + 16 + 9$$
$$4(x - 2)^2 + 9(y - 1)^2 = 36$$
$$\frac{(x - 2)^2}{9} + \frac{(y - 1)^2}{4} = 1$$

You can now see that the graph of the given equation is an ellipse with center at $(2, 1)$ and major axis parallel to the x-axis. Because $a = 3$, the vertices are at $(-1, 1)$ and $(5, 1)$. Because $b = 2$, the endpoints of the minor axis are at $(2, -1)$ and $(2, 3)$. Using these four points as guides, sketch the ellipse, as shown in Figure 8-26.

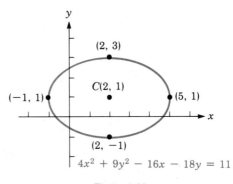

Figure 8-26

■ EXERCISE 8.4

In Exercises 1–6, write the equation of the ellipse that has its center at the origin.

1. Focus at $(3, 0)$ and a vertex at $(5, 0)$.

2. Focus at $(0, 4)$ and a vertex at $(0, 7)$.

3. Focus at $(0, 1)$; $\frac{4}{3}$ is half the length of the minor axis.

4. Focus at $(1, 0)$; $\frac{4}{3}$ is half the length of the minor axis.

5. Focus at $(0, 3)$ and major axis equal to 8.

6. Focus at $(5, 0)$ and major axis equal to 12.

In Exercises 7–16, write the equation of each ellipse.

7. Center at $(3, 4)$; $a = 3$, $b = 2$; the major axis is parallel to the y-axis.

8. Center at $(3, 4)$; the curve passes through $(3, 10)$ and $(3, -2)$; $b = 2$.

9. Center at $(3, 4)$; $a = 3$, $b = 2$; the major axis is parallel to the x-axis.

10. Center at $(3, 4)$; the curve passes through $(8, 4)$ and $(-2, 4)$; $b = 2$.

11. Foci at $(-2, 4)$ and $(8, 4)$; $b = 4$.

12. Foci at $(-8, 5)$ and $(4, 5)$; $b = 3$.

13. Vertex at $(6, 4)$ and foci at $(-4, 4)$ and $(4, 4)$.

14. Center at $(-4, 5)$; $\frac{c}{a} = \frac{1}{3}$; vertex at $(-4, -1)$.

15. Foci at $(6, 0)$ and $(-6, 0)$; $\frac{c}{a} = \frac{3}{5}$.

16. Vertices at $(2, 0)$ and $(-2, 0)$; $\frac{2b^2}{a} = 2$.

In Exercises 17–24, graph each equation.

17. $\dfrac{x^2}{25} + \dfrac{y^2}{49} = 1$

18. $4x^2 + y^2 = 4$

19. $\dfrac{x^2}{16} + \dfrac{(y + 2)^2}{36} = 1$

20. $(x - 1)^2 + \dfrac{4y^2}{25} = 4$

21. $x^2 + 4y^2 - 4x + 8y + 4 = 0$

22. $x^2 + 4y^2 - 2x - 16y = -13$

23. $16x^2 + 25y^2 - 160x - 200y + 400 = 0$

24. $3x^2 + 2y^2 + 7x - 6y = -1$

25. The moon has an orbit that is an ellipse with the earth at one focus. If the major axis of the orbit is 378,000 miles and the ratio of c to a is approximately $\frac{11}{200}$, how far does the moon get from earth? (This farthest point in an orbit is called *apogee*.)

26. An arch is a semiellipse 10 meters wide and 5 meters high. Write the equation of the ellipse if the ellipse is centered at the origin.

27. A track is built in the shape of an ellipse and has a maximum length of 100 meters and a maximum width of 60 meters. Write the equation of the ellipse and find its focal width; that is, find the length of a chord that is perpendicular to the major axis and that passes through either focus of the ellipse.

28. An arch has the shape of a semiellipse and has a maximum height of 5 meters. The foci are on the ground with a distance between them of 24 meters. Find the total distance from one focus to any point on the arch and back to the other focus.

29. Consider the ellipse in Illustration 1. If F is a focus of the ellipse and B is an endpoint of the minor axis, use the distance formula to prove that $d(FB)$ is a. (*Hint:* Remember that in an ellipse $a^2 - c^2 = b^2$.)

30. Consider the ellipse in Illustration 1. If F is a focus of the ellipse and P is any point on the ellipse, use the distance formula to show that $d(FP)$ is $a - \frac{c}{a}x$. (*Hint:* Remember that in an ellipse $a^2 - c^2 = b^2$.)

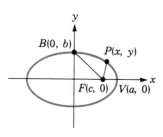

Illustration 1

31. Consider the ellipse in Illustration 2. Chord AA' passes through the focus F and is perpendicular to the major axis. Show that the length of AA' (called the **focal width** of the ellipse) is $\frac{2b^2}{a}$.

32. Prove that $d(FV)$ in Example 3 does represent the shortest distance between the earth and the sun. (*Hint*: You might find the result of Exercise 30 helpful.)

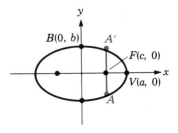

Illustration 2

33. The ends of a piece of string 6 meters long are attached to two thumbtacks that are 2 meters apart. A pencil catches the loop and draws it tight. As the pencil is moved about the thumbtacks (always keeping the tension), an ellipse is produced with the thumbtacks as foci. Write the equation of the ellipse. (*Hint*: You'll have to establish a coordinate system.)

34. Prove that a is greater than b in the development of the standard form of the equation of an ellipse.

35. Show that the expansion of the standard equation of an ellipse centered at (h, k) is a special case of the general second-degree equation.

36. The distance between point $P(x, y)$ and the point $(0, 2)$ is one-third the distance of point P from the line $y = 18$. Find the equation of the curve on which point P lies.

8.5   THE HYPERBOLA

The definition of the hyperbola is similar to the definition of the ellipse except that we demand a constant *difference* of $2a$ instead of a constant sum.

> **Definition.**   A **hyperbola** is the set of all points P in a plane such that the difference of the distances from point P to two other points in the plane, F and F', is a positive constant.

Points F and F' (see Figure 8-27) are called the **foci** of the hyperbola, and the midpoint of chord FF' is called the **center** of the hyperbola. The points V and V', where the hyperbola intersects the line segment FF', are called the **vertices** of the hyperbola, and the line segment VV' is called the **transverse axis**.

As with the ellipse, we will develop the equation of the hyperbola centered at the origin. Because the origin is the midpoint of chord FF', we can let $d(F'O) = d(OF) = c > 0$. Therefore, F is at $(c, 0)$ and F' is at $(-c, 0)$. The definition requires that $|d(F'P) - d(PF)| = 2a$, where $2a$ is a positive constant. Using the distance formula to compute the lengths of $F'P$ and PF gives

$$d(F'P) = \sqrt{[x - (-c)]^2 + y^2}$$
$$d(PF) = \sqrt{(x - c)^2 + y^2}$$

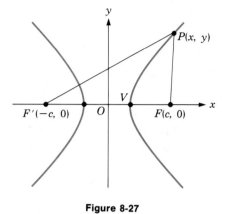

Figure 8-27

Substituting these values into the equation $d(F'P) - d(PF) = 2a$ gives

$$\sqrt{(x + c)^2 + y^2} - \sqrt{(x - c)^2 + y^2} = 2a$$

or

$$\sqrt{(x + c)^2 + y^2} = 2a + \sqrt{(x - c)^2 + y^2}$$

Squaring both sides of this equation and simplifying gives

$$(x + c)^2 + y^2 = 4a^2 + 4a\sqrt{(x - c)^2 + y^2} + (x - c)^2 + y^2$$
$$x^2 + 2cx + c^2 + y^2 = 4a^2 + 4a\sqrt{(x - c)^2 + y^2} + x^2 - 2cx + c^2 + y^2$$
$$4cx = 4a^2 + 4a\sqrt{(x - c)^2 + y^2}$$
$$cx - a^2 = a\sqrt{(x - c)^2 + y^2}$$

Squaring both sides again and simplifying gives

$$c^2x^2 - 2a^2cx + a^4 = a^2(x^2 - 2cx + c^2 + y^2)$$
$$c^2x^2 - 2a^2cx + a^4 = a^2x^2 - 2a^2cx + a^2c^2 + a^2y^2$$
$$c^2x^2 + a^4 = a^2x^2 + a^2c^2 + a^2y^2$$

1. $$(c^2 - a^2)x^2 - a^2y^2 = a^2(c^2 - a^2)$$

Because c exceeds a (you will be asked to prove this in the exercises), $c^2 - a^2$ is a positive number. Thus, we can let $b^2 = c^2 - a^2$ and substitute b^2 for $c^2 - a^2$ in Equation 1 to get

$$b^2x^2 - a^2y^2 = a^2b^2$$

Dividing both sides of the previous equation by a^2b^2 gives the standard form of the equation for a hyperbola with center at the origin and foci on the x-axis:

$$\frac{x^2}{a^2} - \frac{y^2}{b^2} = 1$$

If $y = 0$, the preceding equation becomes

$$\frac{x^2}{a^2} = 1 \qquad \text{or} \qquad x^2 = a^2$$

Solving this equation for x gives

$$x = a \qquad \text{or} \qquad x = -a$$

This implies that the coordinates of V and V' are $(a, 0)$ and $(-a, 0)$ and that the distance between the center of the hyperbola and either vertex is a. This, in turn, implies that the center of the hyperbola is the midpoint of the segment $V'V$ as well as of the segment FF'.

If $x = 0$, the equation becomes

$$\frac{-y^2}{b^2} = 1 \qquad \text{or} \qquad y^2 = -b^2$$

Because this equation has no real solutions, the hyperbola cannot intersect the y-axis. These results suggest the following theorem.

Theorem.   The standard form of the equation of a hyperbola with center at the origin and foci on the x-axis is

$$\frac{x^2}{a^2} - \frac{y^2}{b^2} = 1$$

The standard form of the equation of a hyperbola with center at the origin and foci on the y-axis is

$$\frac{y^2}{a^2} - \frac{x^2}{b^2} = 1$$

As with the ellipse, the standard equation of the hyperbola can be developed with center at (h, k). We state the results without proof.

Theorem.   The standard form of the equation of a hyperbola with center at (h, k) and foci on a line parallel to the x-axis is

$$\frac{(x - h)^2}{a^2} - \frac{(y - k)^2}{b^2} = 1$$

The standard form of the equation of a hyperbola with center at (h, k) and foci on a line parallel to the y-axis is

$$\frac{(y - k)^2}{a^2} - \frac{(x - h)^2}{b^2} = 1$$

Example 1   Write the equation of the hyperbola with vertices $(3, -3)$ and $(3, 3)$ and with a focus at $(3, 5)$.

Solution   First, plot the vertices and focus, as shown in Figure 8-28. Note that the foci lie on a vertical line. Therefore, the standard form to use is

$$\frac{(y-k)^2}{a^2} - \frac{(x-h)^2}{b^2} = 1$$

Because the center of the hyperbola is midway between the vertices V and V', the center is point $(3, 0)$, so $h = 3$, and $k = 0$. The distance between the vertex and the center of the hyperbola is $a = 3$, and the distance between the focus and the center is $c = 5$. Also, in a hyperbola, $b^2 = c^2 - a^2$. Therefore, $b^2 = 5^2 - 3^2 = 16$. Substituting the values for h, k, a^2, and b^2 into the standard form of the equation gives the desired result:

$$\frac{(y-0)^2}{9} - \frac{(x-3)^2}{16} = 1$$

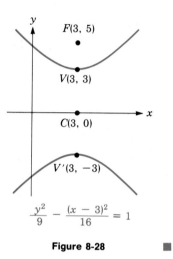

$$\frac{y^2}{9} - \frac{(x-3)^2}{16} = 1$$

Figure 8-28

Asymptotes of a Hyperbola

The values of a and b play an important role in graphing hyperbolas. To see their significance, we consider the hyperbola

$$\frac{x^2}{a^2} - \frac{y^2}{b^2} = 1$$

The center of this hyperbola is the origin, and the vertices are at $V(a, 0)$ and $V'(-a, 0)$. We plot points V, V', $B(0, b)$, and $B'(0, -b)$ and form rectangle $RSQP$, called the **fundamental rectangle**, as in Figure 8-29. The extended diagonals of this

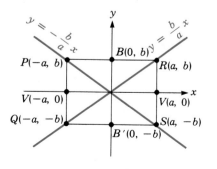

Figure 8-29

rectangle are asymptotes of the hyperbola. In the exercises, you will be asked to show that the equations of these two lines are

$$y = \frac{b}{a}x \quad \text{and} \quad y = -\frac{b}{a}x$$

To show that the extended diagonals are asymptotes of the hyperbola, we solve the equation

$$\frac{x^2}{a^2} - \frac{y^2}{b^2} = 1$$

for y and modify its form:

$$\frac{x^2}{a^2} - \frac{y^2}{b^2} = 1$$

$$b^2x^2 - a^2y^2 = a^2b^2 \qquad \text{Multiply both sides by } a^2b^2.$$

$$y^2 = \frac{b^2x^2 - a^2b^2}{a^2} \qquad \text{Add } -b^2x^2 \text{ to both sides and divide both sides by } -a^2.$$

$$y^2 = \frac{b^2x^2}{a^2}\left(1 - \frac{a^2}{x^2}\right) \qquad \text{Factor out a } b^2x^2 \text{ from the numerator.}$$

$$y = \pm\frac{bx}{a}\sqrt{1 - \frac{a^2}{x^2}} \qquad \text{Take the square root of both sides.}$$

If $|x|$ grows large without bound, the fraction $\frac{a^2}{x^2}$ in the preceding equation approaches 0 and $\sqrt{1 - \frac{a^2}{x^2}}$ approaches 1. Hence, the hyperbola approaches the lines

$$y = \frac{b}{a}x \quad \text{and} \quad y = -\frac{b}{a}x$$

This fact makes it easy to sketch a hyperbola. We convert the equation into standard form, find the coordinates of its vertices, and plot them. Then, we construct the fundamental rectangle and its extended diagonals. Using the vertices and the asymptotes as guides, we make a quick and relatively accurate sketch, as in Figure 8-30. The segment BB' is called the **conjugate axis** of the hyperbola.

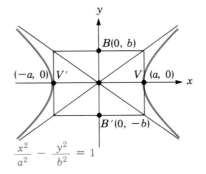

Figure 8-30

Example 2 Graph the hyperbola $x^2 - y^2 - 2x + 4y = 12$.

Solution First complete the square on x and y to convert the equation into standard form:

$$x^2 - 2x - y^2 + 4y = 12$$
$$x^2 - 2x - (y^2 - 4y) = 12$$
$$x^2 - 2x + 1 - (y^2 - 4y + 4) = 12 + 1 - 4$$
$$(x - 1)^2 - (y - 2)^2 = 9$$
$$\frac{(x - 1)^2}{9} - \frac{(y - 2)^2}{9} = 1$$

From the standard form of the equation of a hyperbola, observe that the center is $(1, 2)$, that $a = 3$ and $b = 3$, and that the vertices are on a line segment parallel to the x-axis, as shown in Figure 8-31. Therefore, the vertices V and V' are 3 units to the right and left of the center and have coordinates of $(4, 2)$ and $(-2, 2)$. Points B and B', 3 units above and below the center, have coordinates $(1, 5)$ and $(1, -1)$. After plotting points V, V', B, and B', construct the fundamental rectangle and its extended diagonals. Using the vertices as points on the hyperbola and the extended diagonals as asymptotes, sketch the graph.

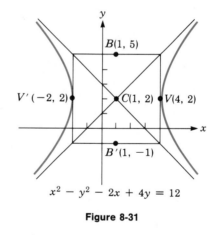

$$x^2 - y^2 - 2x + 4y = 12$$

Figure 8-31

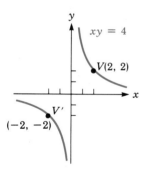

Figure 8-32

This discussion of the hyperbola has considered only cases where the segment that joins the foci is horizontal or vertical. However, there are hyperbolas where this is not true. For example, the graph of the equation $xy = 4$ is a hyperbola with vertices at $(2, 2)$ and $(-2, -2)$, as shown in Figure 8-32.

■ EXERCISE 8.5

In Exercises 1–12, write the equation of each hyperbola.

1. Vertices at $(5, 0)$ and $(-5, 0)$ and focus at $(7, 0)$.

2. Focus at $(3, 0)$, vertex at $(2, 0)$, and center at $(0, 0)$.

3. Center at $(2, 4)$; $a = 2$, $b = 3$; transverse axis is horizontal.

4.   Center at $(-1, 3)$, vertex at $(1, 3)$, and focus at $(2, 3)$.

5.   Center at $(5, 3)$, vertex at $(5, 6)$, hyperbola passes through $(1, 8)$.

6.   Foci at $(0, 10)$ and $(0, -10)$; $\dfrac{c}{a} = \dfrac{5}{4}$.

7.   Vertices at $(0, 3)$ and $(0, -3)$; $\dfrac{c}{a} = \dfrac{5}{3}$.

8.   Focus at $(4, 0)$, vertex at $(2, 0)$, and center at the origin.

9.   Center at $(1, -3)$; $a^2 = 4$, $b^2 = 16$.

10. Center at $(1, 4)$, focus at $(7, 4)$, and vertex at $(3, 4)$.

11. Center at the origin; hyperbola passes through points $(4, 2)$ and $(8, -6)$.

12. Center at $(3, -1)$, y-intercept of -1, x-intercept of $3 + \dfrac{3\sqrt{5}}{2}$.

In Exercises 13–16, find the area of the fundamental rectangle of each hyperbola.

13. $4(x - 1)^2 - 9(y + 2)^2 = 36$

14. $x^2 - y^2 - 4x - 6y = 6$

15. $x^2 + 6x - y^2 + 2y = -11$

16. $9x^2 - 4y^2 = 18x + 24y + 63$

In Exercises 17–20, write the equation of each hyperbola.

17. Center at $(-2, -4)$; $a = 2$; area of fundamental rectangle is 36 square units.

18. Center at $(3, -5)$; $b = 6$; area of fundamental rectangle is 24 square units.

19. One vertex at $(6, 0)$, one end of conjugate axis at $\left(0, \dfrac{5}{4}\right)$.

20. One vertex at $(3, 0)$, one focus at $(-5, 0)$, center at $(0, 0)$.

In Exercises 21–29, graph each equation.

21. $\dfrac{x^2}{9} - \dfrac{y^2}{4} = 1$

22. $\dfrac{y^2}{4} - \dfrac{x^2}{9} = 1$

23. $4x^2 - 3y^2 = 36$

24. $x^2 + 6x - y^2 + 2y = -11$

25. $y^2 - x^2 = 1$

26. $x^2 - y^2 - 4x - 6y = 6$

27. $4x^2 - 2y^2 + 8x - 8y = 8$

28. $9(y + 2)^2 - 4(x - 1)^2 = 36$

29. $y^2 - 4x^2 + 6y + 32x = 59$

In Exercises 30–32, graph each equation.

30. $xy = 9$

31. $-xy = 6$

32. $-xy = 20$

In Exercises 33–36, find the equation of each curve on which point P lies.

33. The difference of the distances between $P(x, y)$ and the points $(-2, 1)$ and $(8, 1)$ is 6.

34. The difference of the distances between $P(x, y)$ and the points $(3, -1)$ and $(3, 5)$ is 5.

35. The distance between point $P(x, y)$ and the point $(0, 3)$ is $\frac{3}{2}$ of the distance between P and the line $y = -2$.

36. The distance between point $P(x, y)$ and the point $(5, 4)$ is $\frac{5}{3}$ of the distance between P and the line $x = -3$.

37. Prove that c is greater than a for a hyperbola with center at the origin and line segment FF' on the x-axis.

38. Show that the equations of the extended diagonals of the fundamental rectangle of the hyperbola with equation $\frac{x^2}{a^2} - \frac{y^2}{b^2} = 1$ are $y = \frac{b}{a}x$ and $y = -\frac{b}{a}x$.

39. Show that the expansion of the standard form of the equation of a hyperbola centered at (h, k) is a special case of the general equation of second degree with $B = 0$.

40. The hyperbolas $\frac{x^2}{a^2} - \frac{y^2}{b^2} = 1$ and $-\frac{x^2}{a^2} + \frac{y^2}{b^2} = 1$ are **conjugate hyperbolas**. Show that they have the same asymptotes.

8.6   TRANSLATION AND ROTATION OF AXES

The graph of the equation

$$(x - 3)^2 + (y - 1)^2 = 4$$

is a circle with radius of 2 and with center at the point $(3, 1)$. See Figure 8-33. If we were to shift the black xy-coordinate system 3 units to the right and 1 unit up, we would establish the $x'y'$-coordinate system shown in color. With respect to this new $x'y'$-system, the center of the circle is the origin and its equation is

$$x'^2 + y'^2 = 4$$

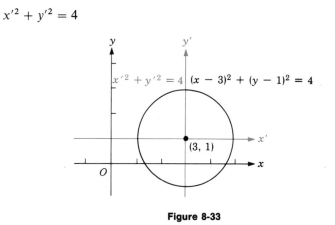

Figure 8-33

In this section, we will discuss how to simplify the equation of a graph by shifting the position of the x- and y-axes. A shift to the left, right, up, or down is called a **translation of the coordinate axes**.

Figure 8-34 shows both an xy- and an $x'y'$-coordinate system. The colored x'- and y'-axes are parallel to the black x- and y-axes, respectively, and the unit distance on each is the same. The origin of the $x'y'$-system is the point O' with $x'y'$-coordinates of $(0, 0)$ and with xy-coordinates of (h, k). The $x'y'$-system is called a **translated coordinate system**.

Relative to the xy-system in Figure 8-34, the coordinates of point P are (x, y). Relative to the $x'y'$-system, the coordinates of point P are (x', y'). By the geometry of the figure we can determine equations, called the **translation-of-axes formulas**, that enable us to find the coordinates of any point with respect to any translated coordinate system.

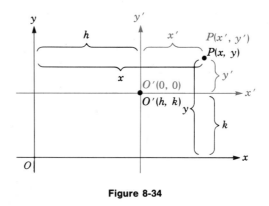

Figure 8-34

> **Translation-of-Axes Formulas.** If the origin of the $x'y'$-coordinate system is located at the point (h, k) of the xy-coordinate system, then
>
> $$x' = x - h$$
> $$y' = y - k$$

Often, a translation of axes can be used to simplify an equation, and thereby make it easier to graph the equation.

Example 1 Graph $y + 3 = (x - 2)^2$.

Solution Write the equation in the form

$$y - (-3) = (x - 2)^2$$

As suggested by the translation-of-axes formulas, make the substitutions

$$x' = x - h = x - 2$$
$$y' = y - k = y - (-3)$$

to obtain

$$y' = x'^2$$

This equation represents a parabola opening upward with its vertex at the origin of an $x'y'$-coordinate system. By the translation-of-axes formulas, you know that the origin of the $x'y'$-coordinate system is at the point $(h, k) = (2, -3)$ of the original xy-system. Thus, the graph of the original equation with respect to the xy-axis is a parabola with its vertex at the point $(2, -3)$. See Figure 8-35.

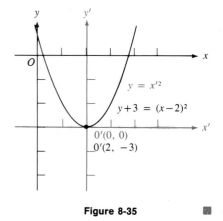

Figure 8-35

Example 2 Graph $y = \sin\left(x - \dfrac{\pi}{6}\right) + 1$.

Solution Write the equation in the form

$$y - 1 = \sin\left(x - \frac{\pi}{6}\right)$$

and make the substitutions

$$x' = x - h = x - \frac{\pi}{6}$$

$$y' = y - k = y - 1$$

to obtain

$$y' = \sin x'$$

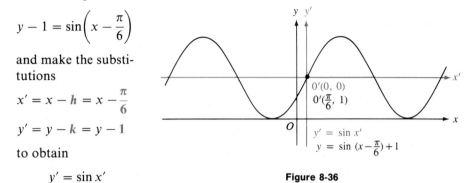

$0'(0, 0)$
$0'(\frac{\pi}{6}, 1)$

$y' = \sin x'$
$y = \sin\left(x - \frac{\pi}{6}\right) + 1$

Figure 8-36

This equation represents a sine curve, positioned in the $x'y'$-coordinate system as shown in Figure 8-36. By the translation-of-axes formulas, you know that the origin of the $x'y'$-coordinate system is at the point $(h, k) = (\frac{\pi}{6}, 1)$ of the original xy-system. Thus, the original equation represents the sine curve that passes through the point $(\frac{\pi}{6}, 1)$. Recall that the horizontal translation of the sine curve $\frac{\pi}{6}$ units to the right is also called a **phase shift**. ∎

Example 3 Graph $4x^2 + y^2 + 8x - 6y + 9 = 0$.

Solution To use the translation-of-axes formulas to simplify this equation, eliminate the first-degree terms by completing the square in both x and y and simplifying:

$4x^2 + y^2 + 8x - 6y + 9 = 0$	
$4x^2 + 8x + y^2 - 6y = -9$	Rearrange terms and add -9 to both sides.
$4(x^2 + 2x) + y^2 - 6y = -9$	Factor 4 from $4x^2 + 8x$.
$4(x^2 + 2x + 1) + (y^2 - 6y + 9) = -9 + 4 + 9$	Add 4 and 9 to both sides.
$4(x + 1)^2 + (y - 3)^2 = 4$	Factor both trinomials.
$\dfrac{(x + 1)^2}{1} + \dfrac{(y - 3)^2}{4} = 1$	Divide both sides by 4.

By the substitutions

$$x' = x - h = x - (-1)$$
$$y' = y - k = y - 3$$

you obtain the equation

$$\frac{x'^2}{1} + \frac{y'^2}{4} = 1$$

which represents an ellipse centered at the origin of an $x'y'$-coordinate system. By the translation-of-axes formulas, the ellipse is also centered at the point $(h, k) = (-1, 3)$ of the xy-coordinate system. The graph appears in Figure 8-37.

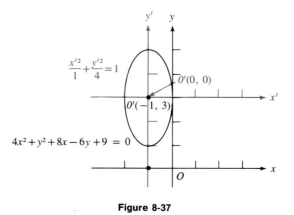

Figure 8-37

Rotation of Axes

The equation of a graph can often be simplified by rotating the xy-axes through some angle θ. This process is called **rotation of the coordinate axes**.

Figure 8-38 shows both an xy- and an $x'y'$-coordinate system with a common origin O, and the same unit of measure on each axis. The colored x'- and y'-axes are rotated through some angle θ with respect to the black x- and y-axes. The $x'y'$-system is called a **rotated coordinate system**.

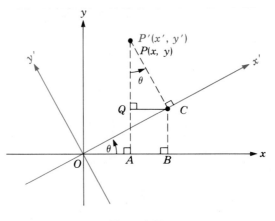

Figure 8-38

Suppose the xy-coordinates of point P are (x, y) and its $x'y'$-coordinates are (x', y'). From point P, we draw PC perpendicular to the x'-axis and PA perpendicular to the x-axis. Segments CQ and CB are perpendicular to PA and the x-axis, respectively, as indicated in Figure 8-38. This forms similar triangles

OBC and *PQC*, in which

$$\sin \theta = \frac{BC}{OC} = \frac{BC}{x'} \qquad \text{or} \qquad BC = x' \sin \theta$$

$$\cos \theta = \frac{OB}{OC} = \frac{OB}{x'} \qquad \text{or} \qquad OB = x' \cos \theta$$

$$\sin \theta = \frac{QC}{PC} = \frac{AB}{y'} \qquad \text{or} \qquad AB = y' \sin \theta$$

and

$$\cos \theta = \frac{QP}{PC} = \frac{QP}{y'} \qquad \text{or} \qquad QP = y' \cos \theta$$

The variables x and y can be expressed in terms of x', y', and θ:

$$
\begin{aligned}
x &= OA & y &= AP \\
&= OB - AB & &= BC + QP \\
&= x' \cos \theta - y' \sin \theta & &= x' \sin \theta + y' \cos \theta
\end{aligned}
$$

The equations relating the xy-coordinates of point P to its coordinates in the $x'y'$-system are as follows.

Equations of Rotation.

$$x = x' \cos \theta - y' \sin \theta$$
$$y = x' \sin \theta + y' \cos \theta$$

Example 4   The xy-coordinate system is rotated 30° counterclockwise to form the $x'y'$-coordinate system, as in Figure 8-39. The $x'y'$-coordinates of point P are $(\sqrt{3} + 1, \sqrt{3} - 1)$. What are the xy-coordinates of P?

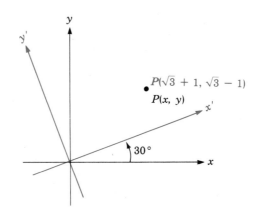

Figure 8-39

Solution Use the equations

$$x = x' \cos \theta - y' \sin \theta$$
$$y = x' \sin \theta + y' \cos \theta$$

with $x' = \sqrt{3} + 1$, $y' = \sqrt{3} - 1$, and $\theta = 30°$.

$$x = x' \cos \theta - y' \sin \theta$$

$$= (\sqrt{3} + 1)\frac{\sqrt{3}}{2} - (\sqrt{3} - 1)\frac{1}{2} \qquad$$ Substitute $\sqrt{3} + 1$ for x', $\sqrt{3} - 1$ for y', $\frac{\sqrt{3}}{2}$ for cos 30°, and $\frac{1}{2}$ for sin 30°.

$$= \frac{3}{2} + \frac{\sqrt{3}}{2} - \frac{\sqrt{3}}{2} + \frac{1}{2}$$

$$= 2$$

$$y = x' \sin \theta + y' \cos \theta$$

$$= (\sqrt{3} + 1)\frac{1}{2} + (\sqrt{3} - 1)\frac{\sqrt{3}}{2} \qquad$$ Substitute $\sqrt{3} + 1$ for x', $\sqrt{3} - 1$ for y', $\frac{1}{2}$ for sin 30°, and $\frac{\sqrt{3}}{2}$ for cos 30°.

$$= \frac{\sqrt{3}}{2} + \frac{1}{2} + \frac{3}{2} - \frac{\sqrt{3}}{2}$$

$$= 2$$

With respect to the xy-system, the coordinates of point P are $(2, 2)$. ∎

A rotation of the coordinate axes can transform the general second-degree equation

$$Ax^2 + Bxy + Cy^2 + Dx + Ey + F = 0$$

into another equation with no xy term. To determine the appropriate angle θ, use the following result. The proof is omitted.

The Angle-of-Rotation Formula. If the coordinate axes are rotated through an angle θ, $0° < \theta < 90°$, determined by

$$\cot 2\theta = \frac{A - C}{B}$$

then the equation

$$Ax^2 + Bxy + Cy^2 + Dx + Ey + F = 0$$

will be transformed into an equation with no xy term.

To determine the required rotation equations, we must use $\cot 2\theta$ to determine $\sin \theta$ and $\cos \theta$.

Example 5   Transform the equation

$$17x^2 - 48xy + 31y^2 + 49 = 0$$

into an equation with no xy term. Graph the resulting equation.

Solution   Rotate the coordinate system through an acute angle θ, where $\cot 2\theta = \frac{A-C}{B}$. To do so, first determine $\cot 2\theta$.

$$\cot 2\theta = \frac{A - C}{B}$$

$$= \frac{17 - 31}{-48} \qquad \text{Substitute 17 for } A, 31 \text{ for } C, \text{ and } -48 \text{ for } B.$$

$$= \frac{7}{24}$$

To determine the values of $\sin \theta$ and $\cos \theta$ required by the rotation equations, first determine the value of $\cos 2\theta$ and then use the identities

$$\sin \theta = \sqrt{\frac{1 - \cos 2\theta}{2}} \qquad \text{and} \qquad \cos \theta = \sqrt{\frac{1 + \cos 2\theta}{2}}$$

To evaluate $\cos 2\theta$, sketch a right triangle such as the one in Figure 8-40 and use the fact that $\cot 2\theta = \frac{7}{24}$ and the Pythagorean theorem to determine the sides of the triangle. You can then see that $\cos 2\theta = \frac{7}{25}$. Thus,

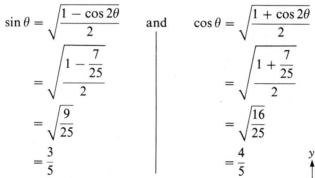

$$\sin \theta = \sqrt{\frac{1 - \cos 2\theta}{2}} \qquad \text{and} \qquad \cos \theta = \sqrt{\frac{1 + \cos 2\theta}{2}}$$

$$= \sqrt{\frac{1 - \frac{7}{25}}{2}} \qquad\qquad = \sqrt{\frac{1 + \frac{7}{25}}{2}}$$

$$= \sqrt{\frac{9}{25}} \qquad\qquad = \sqrt{\frac{16}{25}}$$

$$= \frac{3}{5} \qquad\qquad = \frac{4}{5}$$

The required rotation is given by the equations

$$x = x' \cos \theta - y' \sin \theta$$
$$y = x' \sin \theta + y' \cos \theta$$

where $\frac{3}{5} = \sin \theta$ and $\frac{4}{5} = \cos \theta$:

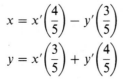

$$x = x'\left(\frac{4}{5}\right) - y'\left(\frac{3}{5}\right)$$

$$y = x'\left(\frac{3}{5}\right) + y'\left(\frac{4}{5}\right)$$

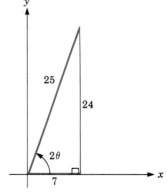

Figure 8-40

To transform the given equation $17x^2 - 48xy + 31y^2 + 49 = 0$ into an equation that has no xy term, substitute $\frac{4}{5}x' - \frac{3}{5}y'$ for x and $\frac{3}{5}x' + \frac{4}{5}y'$ for y in the given equation:

$$17x^2 - 48xy + 31y^2 + 49 = 0$$

$$17\left(\frac{4}{5}x' - \frac{3}{5}y'\right)^2 - 48\left(\frac{4}{5}x' - \frac{3}{5}y'\right)\left(\frac{3}{5}x' + \frac{4}{5}y'\right)$$

$$+ 31\left(\frac{3}{5}x' + \frac{4}{5}y'\right)^2 + 49 = 0$$

Perform the indicated operations to obtain

$$17\left[\frac{16}{25}x'^2 - 2\left(\frac{12}{25}\right)x'y' + \frac{9}{25}y'^2\right] - 48\left(\frac{12}{25}x'^2 + \frac{16}{25}x'y' - \frac{9}{25}x'y' - \frac{12}{25}y'^2\right)$$

$$+ 31\left[\frac{9}{25}x'^2 + 2\left(\frac{12}{25}\right)x'y' + \frac{16}{25}y'^2\right] + 49 = 0$$

Then remove parentheses and combine terms to obtain

$$-\frac{25}{25}x'^2 + 0x'y' + \frac{1225}{25}y'^2 + 49 = 0$$

or

$$\frac{x'^2}{49} - y'^2 = 1$$

The graph of this transformed equation is the hyperbola that appears in Figure 8-41.

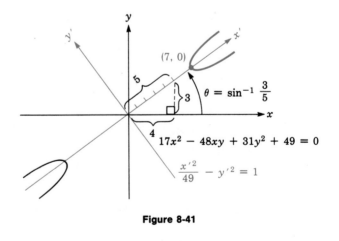

Figure 8-41

■ **EXERCISE 8.6**

In Exercises 1–6, use the translation-of-axes formulas to graph each equation. On each graph, include both the xy- and the x'y'-coordinate axes.

1. $y - 3 = (x - 2)^2$

2. $y - 3 = (x - 1)^2$

3. $y = 3(x + 1)^2$ **4.** $y = 2x^2 - 2$

5. $(x - 2)^2 + (y + 1)^2 = 4$ **6.** $(x + 2)^2 + 9(y - 1)^2 = 9$

In Exercises 7–12, use the translation-of-axes formulas to graph each equation. On each graph, include both the xy- and the x′y′-coordinate axes.

7. $y - 3 = \sin(x - \frac{\pi}{2})$ **8.** $y - 3 = \cos(x - \frac{\pi}{3})$

9. $y = \cos(x - \pi) + 1$ **10.** $y + \frac{1}{2} = \sin(x - \frac{\pi}{2})$

11. $y + \frac{1}{2} = \sec(x - \frac{\pi}{3})$ **12.** $y = \tan(x - \frac{\pi}{4}) - 2$

In Exercises 13–18, complete the square to determine the point to which the origin should be translated to eliminate the first-degree terms of the given equation. Then graph the equation, and show the relation of the graph to both coordinate systems.

13. $x^2 + y^2 + 4x - 10y - 6 = 0$ **14.** $x^2 + y^2 - 8x - 2y + 1 = 0$

15. $2x^2 + y^2 + 4x + 2y - 1 = 0$ **16.** $4x^2 + 9y^2 + 24x - 18y + 9 = 0$

17. $x^2 - y^2 - 6x - 4y + 4 = 0$ **18.** $4x^2 - 9y^2 + 8x - 36y = 68$

In Exercises 19–24, the xy-coordinate system is rotated counterclockwise through an angle θ to form the x′y′-coordinate system. Find the xy-coordinates of the point whose x′y′-coordinates are given.

19. $P(2\sqrt{3}, 0);\ \theta = 30°$ **20.** $Q(\sqrt{3} - 1, \sqrt{3} + 1);\ \theta = 30°$

21. $R(2, 6);\ \theta = 45°$ **22.** $S(\sqrt{2} - 1, \sqrt{2} + 1);\ \theta = 45°$

23. $T(7, 4);\ \theta = 60°$ **24.** $U(0, -8);\ \theta = 60°$

In Exercises 25–28, determine the angle θ through which the axes must be rotated to remove the xy term. **Do not transform the equation.**

25. $3x^2 - \sqrt{3}xy + 2y^2 - 3x + y - 10 = 0$ **26.** $5x^2 - 8xy + 5y^2 + y + 3 = 0$

27. $5\sqrt{3}x^2 + 8xy - 3\sqrt{3}y^2 - 9 = 0$ **28.** $x^2 - 6xy - 5y^2 - 27 = 0$

In Exercises 29–38, perform a rotation of axes to remove the xy term. Graph the equation.

29. $xy = 2$ **30.** $xy = 18$

31. $x^2 + 3xy + y^2 = 2$ **32.** $5x^2 + 26xy + 5y^2 - 72 = 0$

33. $x^2 + 2xy + y^2 - 2\sqrt{2}x + 2\sqrt{2}y = 0$ **34.** $x^2 - 2\sqrt{3}xy - y^2 - 2 = 0$

35. $2x^2 - 4xy + 5y^2 - 6 = 0$ **36.** $7x^2 + 2\sqrt{3}xy + 5y^2 = 8$

37. $6x^2 + 24xy - y^2 - 30 = 0$ **38.** $5x^2 - 4xy + 8y^2 - 36 = 0$

39. Show that the radius of a circle is not affected by a translation of axes.

40. Show that no translation of axes will remove the xy term of the equation $xy = 1$.

In Exercises 41 and 42, suppose that $Ax^2 + Bxy + Cy^2 + Dx + Ey + F = 0$ is changed into $A'x'^2 + B'x'y' + C'y'^2 + D'x' + E'y' + F' = 0$ by substituting $x' + h$ for x and $y' + k$ for y and then simplifying.

41. Show that $A + C = A' + C'$.   **42.** Show that $B^2 - 4AC = B'^2 - 4A'C'$.

In Exercises 43 and 44, suppose that $Ax^2 + Bxy + Cy^2 + Dx + Ey + F = 0$ is changed into $A'x'^2 + B'x'y' + C'y'^2 + D'x' + E'y' + F' = 0$ by rotating the axes through an angle θ.

43. Show that $A + C = A' + C'$.   **44.** Show that $B^2 - 4AC = B'^2 - 4A'C'$.

8.7 POLAR EQUATIONS OF THE CONICS

In this section we use polar coordinates to write equations for the parabola, ellipse, and hyperbola. The discussion is based on the fact that each of these conics is completely determined by a point, a directrix, and a positive real number called the **eccentricity** of the conic.

There are four standard forms for the polar equation of a conic. We derive one of these by letting point F be the pole and the directrix d be a line perpendicular to an extension of the polar axis, as in Figure 8-42. Furthermore, we let p ($p > 0$) be the distance from point F to the directrix d, and we let e be a positive constant representing the eccentricity. It can be shown that, if $P(r, \theta)$ lies on the plane in any location for which the ratio of $d(FP)$ to $d(QP)$ is equal to the constant e, then P lies on a conic with focus at F. Thus, for any point P on the conic,

$$e = \frac{d(FP)}{d(PQ)}$$

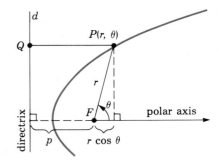

To derive the polar equation for the conic, we note that $d(FP) = r$ and that $d(QP) = p + r \cos \theta$. Then we have

$$e = \frac{r}{p + r \cos \theta}$$

or, by solving for r,

Figure 8-42

1. $$r = \frac{ep}{1 - e \cos \theta}$$

Equation 1 is a standard form of the equation of a conic with eccentricity of e, with focus at the pole, and with directrix perpendicular to an extension of the polar axis.

If $e = 1$, we can show that

$$r = \frac{ep}{1 - e \cos \theta}$$

is the equation of a parabola. To do so, we substitute 1 for e, convert the equation into rectangular form, and reduce it to the equation of a parabola:

$$r = \frac{1p}{1 - 1 \cos \theta}$$
$$r(1 - \cos \theta) = p$$
$$r - r \cos \theta = p$$
$$\sqrt{x^2 + y^2} - x = p$$
$$\sqrt{x^2 + y^2} = x + p$$
$$x^2 + y^2 = (x + p)^2 \qquad \text{Square both sides.}$$

$$x^2 + y^2 = x^2 + 2px + p^2$$
$$y^2 = 2px + p^2$$
$$y^2 = 2p\left(x + \frac{p}{2}\right) \qquad \text{Factor out } 2p \text{ from the right side.}$$

This result is the equation of a parabola with vertex (in the rectangular system) at the point $(-\frac{p}{2}, 0)$. Thus, if $e = 1$, Equation 1 is the equation of a parabola.

It can also be shown that, if $0 < e < 1$, Equation 1 is the equation of an ellipse, and, if $e > 1$, Equation 1 is the equation of a hyperbola.

There are three other standard forms for the equation of a conic. They are determined by rotating the directrix and graph in Figure 8-42 through angles of $90°$, $180°$, and $270°$. The four possibilities for the standard equations of a conic written in polar coordinates are summarized in Table 8-2.

Table 8-2

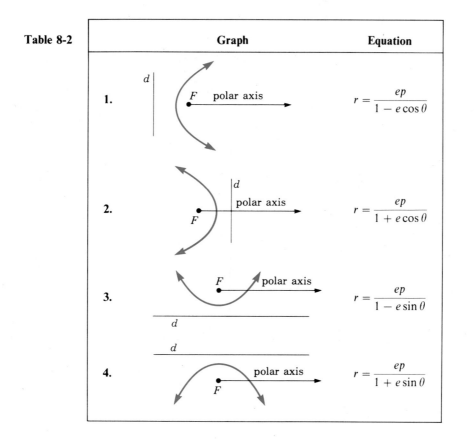

	Graph	Equation
1.	F polar axis	$r = \dfrac{ep}{1 - e\cos\theta}$
2.	F polar axis	$r = \dfrac{ep}{1 + e\cos\theta}$
3.	F polar axis	$r = \dfrac{ep}{1 - e\sin\theta}$
4.	F polar axis	$r = \dfrac{ep}{1 + e\sin\theta}$

Example 1 Graph the conic determined by $r = \dfrac{3}{4 - 2\cos\theta}$.

Solution By dividing numerator and denominator of the fraction by 4, you can write the equation in the form

$$r = \frac{\dfrac{3}{4}}{1 - \dfrac{1}{2}\cos\theta}$$

Compare this to Equation 1 and determine that $e = \frac{1}{2}$ and that $ep = \frac{3}{4}$. Then calculate p as follows:

$$ep = \frac{3}{4}$$

$$\frac{1}{2}p = \frac{3}{4} \qquad \text{Substitute } \tfrac{1}{2} \text{ for } e.$$

$$p = \frac{3}{2}$$

Because $e = \frac{1}{2} < 1$, the conic is an ellipse. Because $p = \frac{3}{2}$, the directrix is $\frac{3}{2}$ units to the left of the focus.

To graph the ellipse, first find its vertices by determining r for $\theta = 0$ and for $\theta = \pi$. Then find its intercepts with the vertical axis by letting $\theta = \frac{\pi}{2}$ and $\theta = \frac{3\pi}{2}$. Plot these points and sketch the graph, as in Figure 8-43.

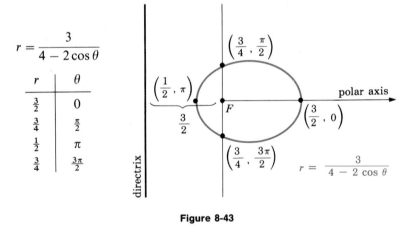

$$r = \frac{3}{4 - 2\cos\theta}$$

r	θ
$\frac{3}{2}$	0
$\frac{3}{4}$	$\frac{\pi}{2}$
$\frac{1}{2}$	π
$\frac{3}{4}$	$\frac{3\pi}{2}$

Figure 8-43

Example 2   Graph the conic determined by $r = \dfrac{5}{3 + 3\cos\theta}$.

Solution   Divide the numerator and the denominator of the fraction by 3, and write the equation in the form

$$r = \frac{\dfrac{5}{3}}{1 + 1\cos\theta}$$

Compare this result with Equation 2 in Table 8-2 and determine that $e = 1$ and that the graph of the equation is a parabola opening to the left. Because $p = \frac{5}{3}$, the directrix is $\frac{5}{3}$ units to the right of the pole.

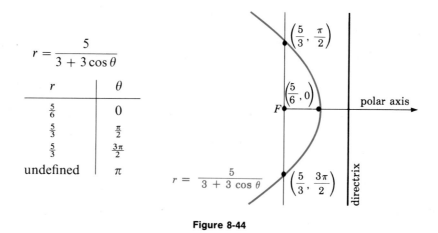

$$r = \frac{5}{3 + 3\cos\theta}$$

r	θ
$\frac{5}{6}$	0
$\frac{5}{3}$	$\frac{\pi}{2}$
$\frac{5}{3}$	$\frac{3\pi}{2}$
undefined	π

Figure 8-44

To graph the parabola, determine the coordinates of the vertex by letting $\theta = 0$: the vertex is the point $V(\frac{5}{6}, 0)$. Determine the intercepts with the vertical axis by letting $\theta = \frac{\pi}{2}$ and $\theta = \frac{3\pi}{2}$. From these three points it is easy to sketch the graph, as in Figure 8-44.

Note that the given function is not defined at $\theta = \pi$, because the denominator would be 0. Thus, no point on the parabola lies in the direction $\theta = \pi$. This is another indication that the parabola opens to the left. ■

Example 3 Classify and sketch the conic $r = \dfrac{\sqrt{3}}{\sqrt{3} - 2\sin\theta}$.

Solution Divide numerator and denominator of the fraction by $\sqrt{3}$ and write the equation in the form

$$r = \frac{1}{1 - \dfrac{2}{\sqrt{3}}\sin\theta}$$

Compare this result with Equation 3 in Table 8-2 and determine that $e = \frac{2}{\sqrt{3}}$ and that $ep = 1$. From this, determine that $p = \frac{\sqrt{3}}{2}$. Because $e > 1$, the conic is a hyperbola. Its directrix is parallel to the polar axis and $\frac{\sqrt{3}}{2}$ units below it. The hyperbola opens upward and downward.

To graph the hyperbola, first find its vertices by determining r for $\theta = \frac{\pi}{2}$ and for $\theta = \frac{3\pi}{2}$. Then find the intercepts with the polar axis by letting $\theta = 0$ and $\theta = \pi$. Plot these points and sketch the graph, as in Figure 8-45.

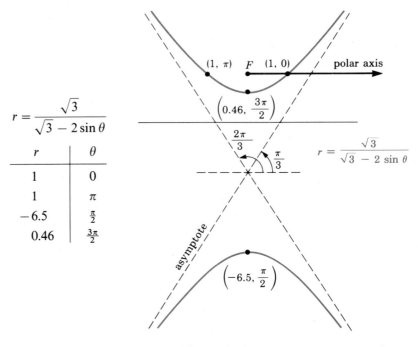

Figure 8-45

Note that the given function is not defined if its denominator is 0. Determine the values excluded from the domain of the function by setting the denominator equal to 0 and solving for θ:

$$1 - \frac{2}{\sqrt{3}} \sin \theta = 0$$

$$\frac{2}{\sqrt{3}} \sin \theta = 1$$

$$\sin \theta = \frac{\sqrt{3}}{2}$$

$$\theta = \frac{\pi}{3} \quad \text{or} \quad \frac{2\pi}{3}$$

No point on the hyperbola can be found in the directions $\theta = \frac{\pi}{3}$ or $\theta = \frac{2\pi}{3}$; these directions are parallel to the asymptotes of the hyperbola.

Note also that for $0 \leq \theta < 2\pi$, the *lower* branch of the hyperbola is traced as θ increases from $\frac{\pi}{3}$ to $\frac{2\pi}{3}$. Other values of θ determine the upper branch. ∎

Example 4 Determine the polar equation of the conic with eccentricity 2 and having a directrix with equation $r \sin \theta = 3$.

Solution   Because the eccentricity is given as 2, $e = 2$. Because $2 > 1$, the conic is a hyperbola. Convert the equation of the directrix to rectangular form to see that it is the line given by the equation $y = 3$. Thus, the directrix is a line parallel to, and 3 units above, the polar axis. Because p is the distance from the pole to the directrix, $p = 3$. Because the directrix is *above* the polar axis, substitute the values for e and p into Equation 4 of Table 8-2 and simplify. The result is the polar equation of the given conic:

$$r = \frac{ep}{1 + e \sin \theta}$$

$$r = \frac{2 \cdot 3}{1 + 2 \sin \theta}$$

$$r = \frac{6}{1 + 2 \sin \theta}$$

■

EXERCISE 8.7

In Exercises 1–6, classify and sketch the conic whose polar equation is given.

1. $r = \dfrac{3}{1 + \cos \theta}$ **2.** $r = \dfrac{5}{1 - \sin \theta}$ **3.** $r = \dfrac{4}{2 - \cos \theta}$ **4.** $r = \dfrac{3}{3 + 2 \sin \theta}$

5. $r = \dfrac{4}{2 - 3 \sin \theta}$ **6.** $r = \dfrac{2}{1 + \sqrt{2} \cos \theta}$

In Exercises 7–10, write the polar equation of the conic described.

7. Eccentricity $\frac{2}{3}$, equation of directrix $r \cos \theta = 5$.

8. Eccentricity 3, equation of directrix $r \sin \theta = -2$.

9. Eccentricity $\frac{2}{3}$, focal width $\frac{7}{3}$, and directrix to the right of the pole, perpendicular to the polar axis. (*Hint:* The focal width is the distance between the intercepts on the vertical axis.)

10. Eccentricity $\frac{4}{3}$, focal width 4, and directrix to the left of the pole, perpendicular to the polar axis.

11. Assume that $e < 1$. Express the length of the major and minor axes of the ellipse in terms of e and p.

12. Assume that $e > 1$. Express the length of the transverse axes of the hyperbola in terms of e and p.

13. In Equation 1 of Table 8-2, assume that $e = 1$. Find polar coordinates of the vertex of the parabola.

14. In Equation 1 of Table 8-2, assume that $e > 1$. Find polar coordinates of the center of the hyperbola.

15. Assuming that $0 < e < 1$, determine $2a$, the length of the major axis, and c, the distance between a focus and the center of the ellipse. Then prove that $e = \frac{c}{a}$.

16. Assuming that $e > 1$, determine $2a$, the length of the transverse axis, and c, the distance between a focus and the center of the hyperbola. Then prove that $e = \frac{c}{a}$.

8.8 PARAMETRIC EQUATIONS

Graphs of equations in two variables drawn on the Cartesian plane are called **plane curves**. If such a curve passes the vertical line test, its equation defines y to be a function of x. However, a plane curve such as an ellipse or hyperbola does not pass the vertical line test, and its equation does not define y to be a function of x. Thus, it is too restrictive to define a plane curve as the graph of an equation of the form $y = f(x)$. A more general approach defines x and y to be functions of yet another variable, say t. Such curves are defined by a system of **parametric equations**.

> **Definition.** If f and g are functions of the real variable t, then the set of ordered pairs (x, y) determined by the equations
>
> $$\begin{cases} x = f(t) \\ y = g(t) \end{cases}$$
>
> defines a plane curve. The curve is said to be **defined parametrically**; the equations are called **parametric equations**; and the variable t is called the **parameter**.

Example 1   Graph the curve defined parametrically by the equations

$$\begin{cases} x = t^2 - 1 \\ y = t + 3 \end{cases}$$

where the domain of the parameter t is the set of real numbers.

Solution   As t varies over the domain $\mathscr{R}$ (the set of real numbers), the point (x, y) traces the curve. When $t = 1$, for example, the values of x and y are

$$\begin{array}{ll} x = t^2 - 1 & y = t + 3 \\ x = 1^2 - 1 & y = 1 + 3 \\ \quad = 0 & \quad = 4 \end{array}$$

Thus, $(0, 4)$ is one point on the curve. Others appear in the table and graph shown in Figure 8-46.

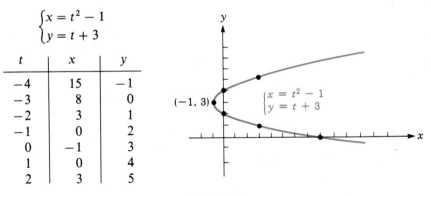

$$\begin{cases} x = t^2 - 1 \\ y = t + 3 \end{cases}$$

t	x	y
-4	15	-1
-3	8	0
-2	3	1
-1	0	2
0	-1	3
1	0	4
2	3	5

Figure 8-46

The two parametric equations of this curve can be combined into a single equation in the variables x and y by a method called **eliminating the parameter**. To do so, solve the second equation for t, and then substitute the result for t in the first equation:

$$\begin{cases} y = t + 3 \Rightarrow t = y - 3 \\ x = t^2 - 1 \end{cases}$$

$$x = (y - 3)^2 - 1$$

or

$$x + 1 = (y - 3)^2$$

The curve represented by the parametric equations is a parabola with vertex at the point $(-1, 3)$ and opening to the right. ∎

The next two examples illustrate that a curve defined parametrically might be only a portion of the graph of the rectangular equation obtained by eliminating the parameter.

Example 2   Graph the curve determined by the parametric equations

$$\begin{cases} x = t^2 \\ y = -2t^2 + 1 \end{cases}$$

where t is a real number.

Solution   Eliminate the parameter t by substitution:

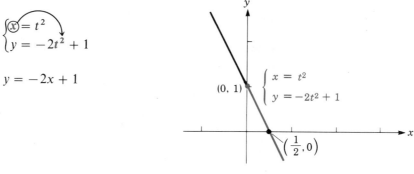

$$\begin{cases} x = t^2 \\ y = -2t^2 + 1 \end{cases}$$

$$y = -2x + 1$$

$$\begin{cases} x = t^2 \\ y = -2t^2 + 1 \end{cases}$$

$(0, 1)$

$\left(\dfrac{1}{2}, 0\right)$

Figure 8-47

The graph of this equation is the line with slope -2 and y-intercept 1. The graph of the parametric equations, however, is not the entire line. Since t^2 is nonnegative and $x = t^2$, it follows that $x \geq 0$. Thus the graph of the given parametric equations is only the colored portion of the line $y = -2x + 1$ shown in Figure 8-47. ∎

Example 3   Graph the curve given by the parametric equations

$$\begin{cases} x = \dfrac{1}{2}\sin^2 t \\ y = \cos^2 t \end{cases}$$

where t is a real number.

Solution   Eliminate the parameter t:

$$x = \frac{1}{2}\sin^2 t \qquad\qquad y = \cos^2 t$$

1. $2x = \sin^2 t$ **2.** $y = \cos^2 t$ Clear the first equation of fractions.

$2x + y = \sin^2 t + \cos^2 t$ Add Equations 1 and 2.

$2x + y = 1$ Use the identity $\sin^2 t + \cos^2 t = 1$.

or

$$y = -2x + 1$$

This is again the equation of a line with slope -2 and y-intercept 1. The graph of the given parametric equations is only a *portion* of this line, however, because, for any real number t,

$$0 \le \sin^2 t \le 1 \qquad\text{and}\qquad 0 \le \cos^2 t \le 1$$

$$0 \le \frac{1}{2}\sin^2 t \le \frac{1}{2}(1) \qquad\qquad 0 \le y \le 1$$

$$0 \le x \le \frac{1}{2}$$

Thus, the graph of the given parametric equations is only that portion of the line $y = -2x + 1$ determined by numbers x between 0 and $\frac{1}{2}$, inclusive. For these numbers x, the values of y lie between 0 and 1, inclusive. The graph is that portion of the line indicated in color in Figure 8-48.

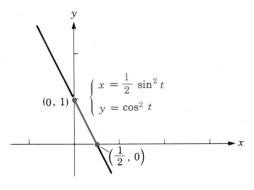

Figure 8-48

Example 4   Identify and graph the curve defined by the parametric equations

$$\begin{cases} x = a \sin t \\ y = b \cos t \end{cases}$$

Solution   Use the identity $\sin^2 t + \cos^2 t = 1$ to eliminate the parameter t:

$$x = a \sin t \qquad\qquad y = b \cos t$$

$$\frac{x}{a} = \sin t \qquad\qquad \frac{y}{b} = \cos t$$

1. $\dfrac{x^2}{a^2} = \sin^2 t$ **2.** $\dfrac{y^2}{b^2} = \cos^2 t$ Square both sides.

$$\frac{x^2}{a^2} + \frac{y^2}{b^2} = \sin^2 t + \cos^2 t \qquad\qquad \text{Add Equations 1 and 2.}$$

$$\frac{x^2}{a^2} + \frac{y^2}{b^2} = 1 \qquad\qquad \text{Use the identity } \sin^2 t + \cos^2 t = 1.$$

Thus, the given parametric equations determine an ellipse. Its graph appears in Figure 8-49.

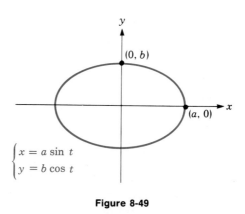

Figure 8-49   ■

Example 5   If a circle with radius a rolls on the x-axis without slipping, a point P on the circle traces a curve called a **cycloid**. Let C be the center of the circle, and assume that P is initially at the origin. Using as the parameter the angle t (in radians) through which the radius CP turns as the circle rolls, find parametric equations for the curve traced by P. See Figure 8-50.

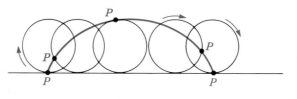

Figure 8-50

Solution   Let Q be the point at which the circle touches the x-axis. See Figure 8-51. Draw a vertical line through P meeting the x-axis at A and a horizontal line through P meeting CQ at B. Because the circle has rolled without slipping, $d(OQ)$ is equal to the arc length PQ. Because t is measured in radians,

$$d(OQ) = at$$

From the geometry of the figure,

$$
\begin{aligned}
x &= d(OA) \\
&= d(OQ) - d(QA) \\
&= at - a\sin t \\
&= a(t - \sin t)
\end{aligned}
\qquad \text{and} \qquad
\begin{aligned}
y &= d(AP) \\
&= d(QB) \\
&= d(QC) - d(CB) \\
&= a - a\cos t \\
&= a(1 - \cos t)
\end{aligned}
$$

Thus, the parametric equations for the cycloid are

$$
\begin{cases}
x = a(t - \sin t) \\
y = a(1 - \cos t)
\end{cases}
$$

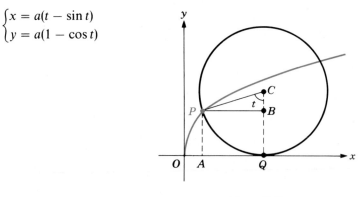

Figure 8-51   ■

The cycloid is also known as the *curve of quickest descent* because, of all possible curves joining P and Q in Figure 8-52, an object will slide (under the influence of gravity) from P to Q most quickly if the curve is a cycloid. The cycloid is also the *curve of equal descent* because, if R is any point on the cycloid, objects starting simultaneously from points P and R will slide to Q in equal times. These properties of the cycloid have been used in the design of clocks: a clock with a pendulum bob that traces a cycloid keeps excellent time.

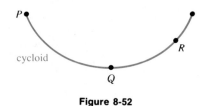

Figure 8-52

■ EXERCISE 8.8

In Exercises 1–8, sketch the curve determined by the given parametric equations.

1. $x = 2t, y = t - 5$

2. $x = 5t, y = 2 - t$

3. $x = 2(t - 3), y = 3(t - 2)$

4. $x = 3(t - 3), y = \frac{1}{2}(t - 2)$

5. $x = t^2, y = t - 2$

6. $x = 3t + 1, y = t^2$

7. $x = \sin t, y = \cos t$

8. $x = \sec t, y = \tan t$

In Exercises 9–16, eliminate the parameter and identify the curve defined by the given parametric equations.

9. $x = 3t + 2, y = 2t$

10. $x = t^2 - 1, y = t + 1$

11. $x = 3t, y = 5$

12. $x = t^3, y = t^3 + 3$

13. $x = 3\cos t, y = 5\sin t$

14. $x = 2\tan t, y = 6\cot t$

15. $x = \cos t, y = \sec t$

16. $x = \sin t, y = \csc t$

In Exercises 17–28, the graph of the parametric equations is only a portion of the graph of the rectangular equation obtained by eliminating the parameter. Determine the graph of the given parametric equations.

17. $x = t^2, y = 2t^2 + 1$

18. $x = 3t^2, y = 5t^2$

19. $x = 2t^2 - 1, y = 1 - t^2$

20. $x = 2(t^2 + 1), y = 1 - t^2$

21. $x = \sqrt{t}, y = t$

22. $x = \sqrt{t - 1}, y = \sqrt{t}$

23. $x = |t|, y = |t|$

24. $x = |t| - 2, y = |t| - 1$

25. $x = 2\sin t, y = 3\sin t$

26. $x = \sin t, y = \cos^2 t$

27. $x = 3\sin^2 t, y = 4\cos^2 t$

28. $x = \sec^2 t, y = \tan^2 t$

In Exercises 29 and 30, calculate pairs (x, y) determined by the given parametric equations, and sketch the curve. Do not attempt to eliminate the parameter.

29. $x = \sin t, y = \sin 2t$ Lissajous curve

30. $x = \cos^3 t, y = \sin^3 t$ hypocycloid

31. Find parametric equations for a circle with radius a and center at the origin. Let P be a point on the circle, and let the parameter t be the angle that OP makes with the x-axis. See Illustration 1.

32. Find parametric equations of the parabola $x^2 = 4py$. Let the parameter t be the slope of the line OP. See Illustration 2.

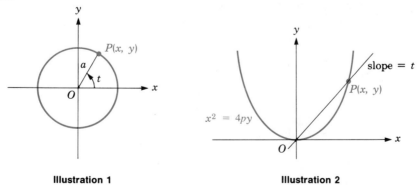

Illustration 1   Illustration 2

33. Circles of radius a and b are centered at the origin. See Illustration 3. Line OBA makes an angle t with the x-axis. Line PB is horizontal, and line PA is vertical. Using t as the parameter, determine parametric equations for the curve traced by P.

34. The line segment AB in Illustration 4 is $2k$ units long. Point A remains on the positive x-axis, and B lies on the positive y-axis. Determine parametric equations for the midpoint of segment AB. Use as the parameter the angle OM makes with the x-axis.

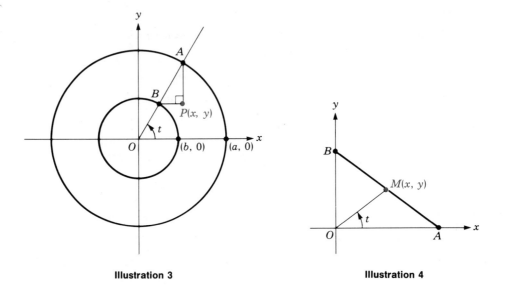

Illustration 3   **Illustration 4**

35. A projectile fired at an angle θ with the horizontal and with an initial velocity v_0 has position $P(x, y)$ at any time t given by

$$x = v_0 t \cos \theta, \qquad y = v_0 t \sin \theta - 16t^2$$

Eliminate t, and identify the curve.

36. In Exercise 35, determine the range of the projectile (the distance it travels before it strikes the ground) and its maximum height.

37. Find parametric equations for the line through $P_1(x_1, y_1)$ and $P_2(x_2, y_2)$. See Illustration 5.

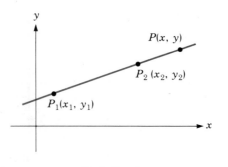

Illustration 5

38. A string is wound around a circle of radius a centered at the origin. The string is held taut, and, as it is unwound, point P traces a curve known as an **involute**. See Illustration 6. Let t be the **angle** (in radians) that OQ makes with the x-axis. Show that parametric equations for the curve are

$$x = a(\cos t + t \sin t), \qquad y = a(\sin t - t \cos t)$$

(*Hint: OQ* and QP are perpendicular, and QP = arc QA.)

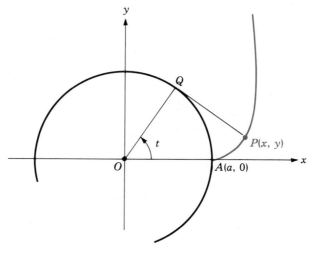

Illustration 6

CHAPTER SUMMARY

Key Words

center of a circle (8.2)
center of an ellipse (8.4)
center of a hyperbola (8.5)
circle (8.2)
conjugate axis of a hyperbola (8.5)
directrix of a parabola (8.3)
eccentricity of a conic (8.7)
ellipse (8.4)
foci of an ellipse (8.4)
foci of a hyperbola (8.5)
focus of a parabola (8.3)
fundamental rectangle (8.5)
hyperbola (8.5)

major axis of an ellipse (8.4)
minor axis of an ellipse (8.4)
parabola (8.3)
parametric equations (8.8)
plane curves (8.8)
radius of a circle (8.2)
rotation of axes (8.6)
slope of a nonvertical line (8.1)
translation of axes (8.6)
transverse axis of a hyperbola (8.5)
vertex of a parabola (8.3)
vertices of an ellipse (8.4)
vertices of a hyperbola (8.5)

Key Ideas

(8.1) **Slope of a nonvertical line:** $m = \dfrac{\Delta y}{\Delta x} = \dfrac{y_2 - y_1}{x_2 - x_1}$

Nonvertical parallel lines have the same slope.

Nonvertical perpendicular lines have slopes that are negative reciprocals.

Point–slope form of the equation of a line: $y - y_1 = m(x - x_1)$

Slope–intercept form of the equation of a line: $y = mx + b$

General form of the equation of a line: $Ax + By = C$

(8.2)   **Equation of a circle with center at (h, k) and radius r:** $(x - h)^2 + (y - k)^2 = r^2$

Equation of a circle with center at the origin and radius r: $x^2 + y^2 = r^2$

(8.3)   The standard forms of the equations for parabolas that open to the right, left, upward, and downward are as follows. (Consider $p > 0$.)

Parabola opening	Vertex at origin	Vertex at $V(h, k)$
Right	$y^2 = 4px$	$(y - k)^2 = 4p(x - h)$
Left	$y^2 = -4px$	$(y - k)^2 = -4p(x - h)$
Upward	$x^2 = 4py$	$(x - h)^2 = 4p(y - k)$
Downward	$x^2 = -4py$	$(x - h)^2 = -4p(y - k)$

(8.4)   The standard form of the equation of an ellipse with center at the origin and major axis on the x-axis is

$$\frac{x^2}{a^2} + \frac{y^2}{b^2} = 1 \qquad \text{where } a > b > 0$$

If the major axis of the ellipse with center at the origin lies on the y-axis, the standard form of the equation of the ellipse is

$$\frac{y^2}{a^2} + \frac{x^2}{b^2} = 1 \qquad \text{where } a > b > 0$$

In either case, the length of the major axis is $2a$; the length of the minor axis is $2b$.

The standard form of the equation of an ellipse with center at (h, k) and major axis parallel to the x-axis is

$$\frac{(x - h)^2}{a^2} + \frac{(y - k)^2}{b^2} = 1 \qquad \text{where } a > b > 0$$

If the major axis of the ellipse with center (h, k) is parallel to the y-axis, the standard form of the equation of the ellipse is

$$\frac{(y - k)^2}{a^2} + \frac{(x - h)^2}{b^2} = 1 \qquad \text{where } a > b > 0$$

In either case, the length of the major axis is $2a$; the length of the minor axis is $2b$.

(8.5) The standard form of the equation of a hyperbola with center at the origin and foci on the x-axis is

$$\frac{x^2}{a^2} - \frac{y^2}{b^2} = 1$$

The standard form of the equation of a hyperbola with center at the origin and foci on the y-axis is

$$\frac{y^2}{a^2} - \frac{x^2}{b^2} = 1$$

The standard form of the equation of a hyperbola with center at (h, k) and foci on a line parallel to the x-axis is

$$\frac{(x - h)^2}{a} - \frac{(y - k)^2}{b^2} = 1$$

The standard form of the equation of a hyperbola with center at (h, k) and foci on a line parallel to the y-axis is

$$\frac{(y - k)^2}{a^2} - \frac{(x - h)^2}{b^2} = 1$$

The extended diagonals of the fundamental rectangle are asymptotes of the graph of a hyperbola.

(8.6) **The translation-of-axes formulas:** $x' = x - h$
$$y' = y - k$$

The rotation-of-axes formulas: $x = x' \cos\theta - y' \sin\theta$
$$y = x' \sin\theta + y' \cos\theta$$

The angle-of-rotation formula: $\cot 2\theta = \dfrac{A - C}{B}$ $(0° < \theta < 90°)$

(8.7) Standard equations of a conic written in polar coordinates:

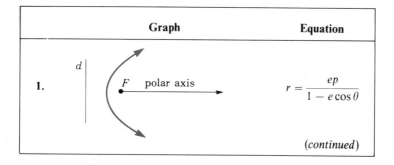

Graph		Equation
1.	F polar axis	$r = \dfrac{ep}{1 - e\cos\theta}$

(continued)

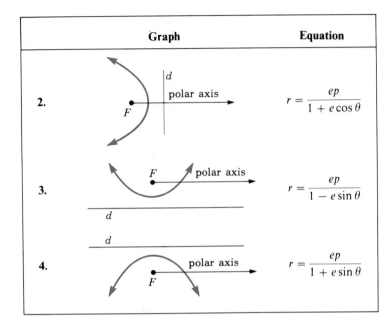

Graph	Equation
2.	$r = \dfrac{ep}{1 + e\cos\theta}$
3.	$r = \dfrac{ep}{1 - e\sin\theta}$
4.	$r = \dfrac{ep}{1 + e\sin\theta}$

(8.8)   Plane curves can be defined by a system of parametric equations.

REVIEW EXERCISES

In Review Exercises 1–4, graph each equation.

1. $2x - 5y = 10$ **2.** $9x + 2y = 18$ **3.** $3x + 2y = 1$ **4.** $3x - 7y = 5$

In Review Exercises 5–8, find the slope of the line PQ, if possible.

5. $P(3, -5); Q(1, 7)$ **6.** $P(2, 7); Q(-5, -7)$

7. $P(b, a); Q(a, b)$ **8.** $P(a + b, b); Q(b, b - a)$

In Review Exercises 9–18, write the equation of the line with the given properties. Express the answer in general form.

9. The line passes through the origin and the point $(-5, 7)$.

10. The line passes through $(-2, 1)$ and has a slope of -4.

11. The line passes through $(7, -5)$ and $(4, 1)$.

12. The line has a slope of $\frac{2}{3}$ and a y-intercept of 3.

13. The line has a slope of 0 and passes through $(-5, 17)$.

14. The line has no defined slope and passes through $(-5, 17)$.

15. The line passes through $(8, -2)$ and is parallel to the line segment joining $(2, 4)$ and $(4, -10)$.

16. The line passes through $(8, -2)$ and is perpendicular to the line segment joining $(2, 4)$ and $(4, -10)$.

17. The line is parallel to $3x - 4y = 7$ and passes through $(2, 0)$.

18. The line passes through $(7, 0)$ and is perpendicular to the line $3y + x - 4 = 0$.

19. Write the equation of the circle with center at the origin and curve passing through point $(5, 5)$.

20. Write the equation of the circle with center at the origin and curve passing through point $(6, 8)$.

21. Write the equation of a circle with endpoints of its diameter at $(-2, 4)$ and $(12, 16)$.

22. Write the equation of a circle with endpoints of its diameter at $(-3, -6)$ and $(7, 10)$.

23. Write in standard form the equation of the circle $x^2 + y^2 - 6x + 4y = 3$ and graph the circle.

24. Write in standard form the equation of the circle $x^2 + 4x + y^2 - 10y = -13$ and graph the circle.

25. Write the equation of the parabola with vertex at the origin and curve passing through $(-8, 4)$ and $(-8, -4)$.

26. Write the equation of the parabola with vertex at the origin and curve passing through $(-8, 4)$ and $(8, 4)$.

27. Write the equation of the parabola $y = ax^2 + bx + c$ in standard form to show that the x-coordinate of the vertex of the parabola is $-\frac{b}{2a}$.

28. Find the equation of the parabola with vertex at $(-2, 3)$, curve passing through point $(-4, -8)$, and opening downward.

29. Graph the parabola $x^2 - 4y - 2x + 9 = 0$.

30. Graph the parabola $y^2 - 6y = 4x - 13$.

31. Graph $y^2 - 4x - 2y + 13 = 0$.

32. Write the equation of the ellipse with center at the origin, major axis that is horizontal and 12 units long, and minor axis 8 units long.

33. Write the equation of the ellipse with center at point $(-2, 3)$ and curve passing through points $(-2, 0)$ and $(2, 3)$.

34. Graph the ellipse $4x^2 + y^2 - 16x + 2y = -13$.

35. Graph the curve $x^2 + 9y^2 - 6x - 18y + 9 = 0$.

36. Write the equation of the hyperbola with vertices at points $(-3, 3)$ and $(3, 3)$ and a focus at point $(5, 3)$.

37. Graph the hyperbola $9x^2 - 4y^2 - 16y - 18x = 43$.

38. Graph the hyperbola $-2xy = 9$.

In Review Exercises 39–42, the origin of the $x'y'$-system is at the point $(2, -3)$ of the xy-system. Express each equation in terms of the variables x' and y'. Draw a sketch that shows the relation of the graphs to both coordinate systems.

39. $x^2 + y^2 = 25$ **40.** $x^2 - y^2 = 4$ **41.** $4x^2 + 9y^2 = 36$ **42.** $x^2 - 4x - 3y = 5$

In Review Exercises 43 and 44, determine the point at which the origin of the $x'y'$-system should be located to eliminate the first-degree terms of the given equation. Draw a sketch that shows the relation of the graphs of the equations to both coordinate systems.

43. $x^2 + y^2 - 6x - 4y + 12 = 0$ **44.** $4x^2 - y^2 - 8x + 4y = 4$

In Review Exercises 45–46, determine the rotation of axes needed to remove the xy term of the given equation. Draw a sketch that shows the relation of the graphs of the equations to both coordinate systems.

45. $13x^2 + 6\sqrt{3}xy + 7y^2 - 16 = 0$ **46.** $5x^2 + 2\sqrt{3}xy + 7y^2 - 8 = 0$

In Review Exercises 47–50, graph each polar equation.

47. $r = 4(1 + \cos \theta)$ **48.** $r = 8 - 4 \cos \theta$ **49.** $r = \dfrac{6}{1 + \sin \theta}$ **50.** $r = \dfrac{2}{1 - 2 \sin \theta}$

In Review Exercises 51–56, sketch the curve determined by the given parametric equations, and eliminate the parameter to determine an equation for the curve in x and y alone.

51. $x = 3t + 1, \; y = t - 1$ **52.** $x = 3t, \; y = t + 3$

53. $x = \sin t, \; y = 2 \cos t$ **54.** $x = 2 \sin t, \; y = 2 \csc t$

55. $x = 2 \sin t, \; y = \sin t$ **56.** $x = \cos t, \; y = 2 + \cos t$

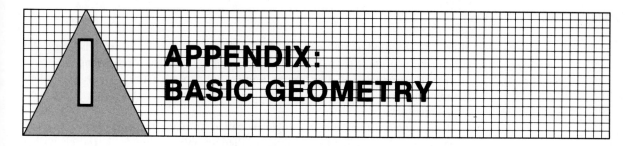

APPENDIX: BASIC GEOMETRY

I.1 ANGLES, TRIANGLES, AND THE PYTHAGOREAN THEOREM

> **Definition.** An **angle** is a figure determined by two rays (not necessarily different) originating from a common point called the **vertex**. A line segment and a ray, or two line segments, with a common endpoint also determine an angle.

In Figure I-1, the two rays BA and BC originate at point B, the vertex of the angle. There are various ways of naming this angle. One is to name the angle after its vertex: angle B. A second way is to use three letters: angle ABC or angle CBA. A third way is to use a Greek letter such as β: angle β.

In Figure I-2, angle ABC is the same angle as angle β. Angle CBD and angle α are two names for the same angle.

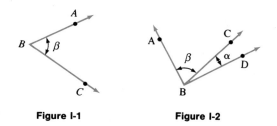

Figure I-1   Figure I-2

In geometry, angles are usually measured in **degrees**. One degree, denoted as $1°$, is $1/360$ of one complete revolution.

An **acute angle** is an angle greater than $0°$, but less than $90°$. See Figure I-3. An angle of $90°$ is called a **right angle**. An angle that is greater than $90°$ but less

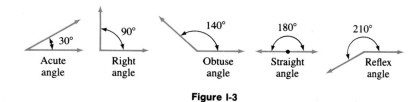

| Acute angle | Right angle | Obtuse angle | Straight angle | Reflex angle |

Figure I-3

than 180° is called an **obtuse angle**. An angle of 180° is called a **straight angle**. An angle that is greater than 180° but less than 360° is called a **reflex angle**.

If the measures of two acute angles add to 90°, the angles are called **complementary**. If the measures of two angles add to 180°, they are called **supplementary**.

A triangle with three acute angles is called an **acute triangle**. See Figure I-4. A triangle with a single obtuse angle is called an **obtuse triangle**. A triangle with a 90° angle (a right angle) is called a **right triangle**, and the side opposite the right angle is called the **hypotenuse**. The other two sides are often called **legs** of the triangle.

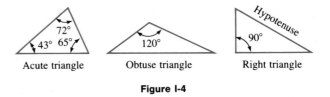

| Acute triangle | Obtuse triangle | Right triangle |

Figure I-4

If a triangle is not a right triangle, it is called an **oblique triangle**. The acute and obtuse triangles in Figure I-4 are oblique triangles. If a triangle has at least two sides of equal length, the triangle is called **isosceles**. In an isosceles triangle, the angles opposite the sides with equal lengths have equal measures also. A triangle with all three sides of equal length is called an **equilateral triangle**. Because an equilateral triangle has three equal angles, it is called **equiangular** also. See Figure I-5.

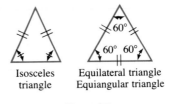

Isosceles triangle   Equilateral triangle Equiangular triangle

Figure I-5

There are two theorems from geometry that cannot be overemphasized. Their proofs can be found in any geometry book.

Theorem. The sum of the three angles in any triangle is equal to 180°.

Example 1   If the vertex angle of an isosceles triangle is equal to 112°, find the measure of each of the base angles.

Solution   You know that the base angles of an isosceles triangle are equal and that the vertex angle is 112°. Let x represent the measure of each base angle as in Figure I-6. Because the sum of the three angles in a triangle is 180°, it follows that

$$x + x + 112 = 180$$
$$2x + 112 = 180$$
$$2x = 68$$
$$x = 34$$

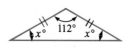

Figure I-6

Each base angle measures 34°. ∎

The Pythagorean Theorem. A triangle is a right triangle if and only if the square of the length of one side is equal to the sum of the squares of the lengths of the other two sides.

Example 2   A triangle has sides of 5, 12, and 13 meters. Is this triangle a right triangle?

Solution   By the Pythagorean theorem, the triangle is a right triangle if the square of the length of one side of the triangle equals the sum of the squares of the other two sides. Because

$$13^2 = 169 \quad \text{and} \quad 5^2 + 12^2 = 169$$

the triangle is a right triangle. ∎

Example 3   If two legs of a right triangle measure 9 and 11 centimeters, find the length of the triangle's hypotenuse.

Solution   Let the length of the hypotenuse be h and use the Pythagorean theorem.

$$h^2 = 9^2 + 11^2$$
$$h^2 = 81 + 121$$
$$h^2 = 202$$
$$h = \sqrt{202}$$

The hypotenuse measures $\sqrt{202}$ centimeters. ∎

There are infinitely many natural numbers x, y, and z that have the property that $x^2 + y^2 = z^2$. Such triples are called **Pythagorean triples**. Some common Pythagorean triples are

3, 4, 5	5, 12, 13	6, 8, 10
7, 24, 25	8, 15, 17	9, 40, 41

Recall that $\sqrt{202}$ represents the positive, or **principal square root**, of 202. The negative square root of 202 is denoted as $-\sqrt{202}$. Although the equation $h^2 = 202$ has two solutions, $h = \sqrt{202}$ and $h = -\sqrt{202}$, only the principal square root is considered in Example 3 because the hypotenuse of a right triangle cannot be negative.

The following theorems are based on the Pythagorean theorem.

> **Theorem.**   If a right triangle is isosceles, the length of the hypotenuse is $\sqrt{2}$ times one of the equal legs.

Proof   We consider the isosceles right triangle in Figure I-7 with equal legs of x units and hypotenuse of h units. By the Pythagorean theorem,

$$h^2 = x^2 + x^2$$
$$h^2 = 2x^2$$
$$h = x\sqrt{2}$$

Figure I-7

Only the principal square root is considered because the sides of a triangle cannot be negative. The theorem is proved.

Example 4   Find the length of the hypotenuse of an isosceles right triangle with one leg of length 5 centimeters.

Solution   Because the length of the hypotenuse is $\sqrt{2}$ times the length of one of its equal legs, the hypotenuse measures $5\sqrt{2}$ centimeters.

Example 5   If the hypotenuse of an isosceles right triangle measures 9 meters, how long is each leg of the triangle?

Solution   Let each of the equal legs measure x meters. Because of the preceding theorem it follows that

$$\sqrt{2}\,x = 9$$
$$x = \frac{9}{\sqrt{2}}$$
$$x = \frac{9\sqrt{2}}{2}$$

Each leg of the triangle measures $\dfrac{9\sqrt{2}}{2}$ meters.

The following theorem is presented without proof.

> **Theorem.**   If a right triangle has acute angles that measure 30° and 60°, the hypotenuse is twice as long as the leg opposite the 30° angle. The leg opposite the 60° angle is $\sqrt{3}$ times as long as the leg opposite the 30° angle.

Example 6   The leg opposite the 30° angle in a 30°, 60° right triangle is 10 centimeters. Find the other two sides.

Solution   The hypotenuse measures twice 10 centimeters, or 20 centimeters. The leg opposite the 60° angle is $\sqrt{3}$ times 10 centimeters, or $10\sqrt{3}$ centimeters.   ∎

Example 7   The leg opposite the 60° angle in a 30°, 60° right triangle is 12 decimeters. Find the length of the other two sides.

Solution   Refer to Figure I-8. If DB measures x decimeters, side CD measures $\sqrt{3}\,x$ decimeters. Hence, you have

$$\sqrt{3}\,x = 12$$
$$x = \frac{12}{\sqrt{3}}$$
$$x = 4\sqrt{3}$$

Because the hypotenuse CB is twice as long as the leg opposite the 30° angle, it follows that

$$CB = 2(4\sqrt{3})$$
$$= 8\sqrt{3}$$

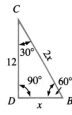

Figure I-8

DB measures $4\sqrt{3}$ decimeters and CB measures $8\sqrt{3}$ decimeters.   ∎

▬ EXERCISE I.1 ▬▬▬▬▬▬▬▬▬▬▬▬▬▬▬▬▬▬▬▬▬▬▬

Exercises 1–6 refer to Illustration 1. Give another name for each angle.

1.   angle α

2.   angle θ

3.   angle BAC

4.   angle FBG

5.   angle D

6.   angle F

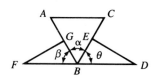

Illustration 1

In Exercises 7–16, classify each angle as acute, right, obtuse, straight, reflex, or as none of these.

7.   30°   **8.**   89°   **9.**   135°   **10.**   180°

11.   90°   **12.**   146°   **13.**   0°   **14.**   212°

15.   An angle whose measure equals the sum of the measures of two right angles.

16.   An angle whose measure equals the sum of the measures of one acute and one right angle.

Exercises 17–26 refer to Illustration 2. Find the measure of angle BAC based on the given information.

17.   angle $B = 20°$ and angle $C = 30°$ **18.**   angle $B = 89°$ and angle $C = 17°$

19. $AB = AC$ and angle $B = 42°$ **20.** $BC = AC$ and angle $B = 42°$

21. $AB = BC$ and angle $C = 63°$

22. $CB = CA$ and angle $C = 63°$

23. Triangle BCA is equilateral.

24. Triangle CAB is equiangular.

25. Angle CAD measures $57°$ and points B, A, and D are on the same line.

26. angle B + angle $C = 43°$

Illustration 2

In Exercises 27–32, tell whether the triangles with the given sides are right triangles.

27. 3, 4, and 5 centimeters

28. 6, 8, and 10 feet

29. 4, 5, and 6 centimeters

30. 7, 9, and 11 centimeters

31. 15, 20, and 25 feet

32. 10, 26, and 24 centimeters

In Exercises 33–38, find the measure of the hypotenuse of a right triangle with the given legs.

33. 30 and 40 meters

34. 20 and 48 centimeters

35. 1 and 2 centimeters

36. 11 and 17 inches

37. a and b feet

38. b and $(b + d)$ centimeters

Exercises 39–44 refer to Illustration 3. Give the measure of each of the remaining sides.

39. $AC = 2$ centimeters

40. $BC = 3$ feet

41. $AB = 10$ inches

42. $AB = 8$ centimeters

43. $AC = 3\sqrt{3}$ kilometers

44. $BC = 4\sqrt{5}$ centimeters

Exercises 45–56 refer to Illustration 4. Give the measure of each of the remaining sides.

45. $AC = 3$ centimeters

46. $BC = 3$ centimeters

47. $BC = 5$ miles

48. $AC = 5$ centimeters

49. $AB = 10$ centimeters

50. $AB = 15$ yards

51. $AC = 2\sqrt{3}$ centimeters

52. $AC = 2\sqrt{5}$ centimeters

53. $BC = 2\sqrt{3}$ feet

54. $BC = 2\sqrt{5}$ inches

55. $AB = 2\sqrt{3}$ centimeters

56. $AB = 2\sqrt{5}$ centimeters

Illustration 3

Illustration 4

Recall that $\sqrt{x^2}$ denotes the principal square root of x^2. If x can be negative, then absolute value notation is required to guarantee that $\sqrt{x^2}$ is positive. For this reason, $\sqrt{x^2} = |x|$. Absolute values are not required for cube roots, because all real numbers have a single real cube root; the cube root of a negative number is negative, the cube root of zero is zero, and the cube root of a positive number is positive.

In Exercises 57–64, simplify each radical using absolute value notation when necessary.

57. $\sqrt{16x^2}$ **58.** $\sqrt{25x^4}$ **59.** $\sqrt{625x^8}$ **60.** $\sqrt{64x^6}$ **61.** $\sqrt[3]{x^3}$ **62.** $\sqrt[3]{x^6}$

63. $\sqrt[n]{x^n}$ and n is even   **64.** $\sqrt[n]{x^n}$ and n is odd

I.2 SIMILAR TRIANGLES

Historically, much of the work of applied trigonometry has involved indirect measurement. For example, it would be inconvenient to measure the length of a flagpole by climbing up the pole with a tape measure. It is better to stay safely on the ground and measure the flagpole indirectly. We can do this by using similar triangles.

> **Definition.** Two triangles are called **similar** if and only if the three angles of one triangle are equal, respectively, to the three angles of the second triangle, and all pairs of corresponding sides are in proportion.

This definition implies that similar triangles have the same shape. If similar triangles also have the same size (area), they are called **congruent**. We will accept the following theorem without proof.

> **Theorem.** Two triangles are similar if and only if two angles of one triangle are equal to two angles of the other triangle.

Example 1 Find the values of x and y in Figure I-9 if angle A equals angle D and angle C equals angle F.

Solution Because two angles of the larger triangle are equal to two angles of the smaller triangle, the triangles are similar. Because they are similar, all corresponding sides are in proportion. So, you can set up the proportions

$$\frac{5}{2} = \frac{10}{y} \quad \text{and} \quad \frac{5}{2} = \frac{7}{x}$$

and solve for x and y.

$$5y = 20 \quad \text{and} \quad 5x = 14$$

$$y = 4 \qquad\qquad x = \frac{14}{5}$$

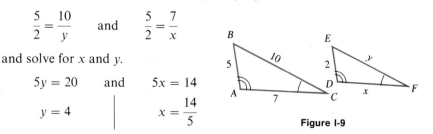

Figure I-9

Example 2   A flagpole casts a shadow of 8 meters at the same time a meter stick held perpendicular to the ground casts a shadow of 50 centimeters. Find the height of the flagpole.

Solution   Because the triangles in Figure I-10 are similar, all corresponding sides are in proportion. Note that 50 centimeters = 0.5 meters. Form the proportion

$$\frac{h}{1} = \frac{8}{0.5}$$

and solve for h.

$$0.5h = 8$$
$$h = 16$$

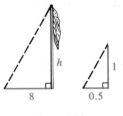

Figure I-10

The height of the flagpole is 16 meters.   ∎

Example 3   Two Boy Scouts wish to know the width of the river represented in Figure I-11. How can they find the width of the river without crossing to the other side?

Solution   The boys see a tree and a large rock on the bank across the river. By using their line of sight, they put stakes at points A, B, and C.

They can measure sides AB and BC directly and can estimate the measure of side DE indirectly by measuring the segment CF (F is on the bank straight across from E, and C is on the bank straight across from D). The boys make some measurements and find that $AB = 3$ meters, $CB = 4$ meters, and $CF = DE = 30$ meters. Note that angles ACB and ECD are vertical angles and, therefore, must be equal. Also note that angles B and D are right angles and, therefore, must be equal. It follows that triangles ABC and EDC are similar triangles. Thus, the boys can form the following proportion and solve for w.

$$\frac{3}{30} = \frac{4}{w}$$
$$120 = 3w$$
$$40 = w$$

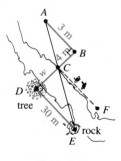

The river is 40 meters wide.

Figure I-11   ∎

In Example 3, the triangles ABC and EDC were shown to be similar. Because all corresponding pairs of angles in similar triangles are equal, it follows that angles A and E are equal. Note that line AE, called a **transversal**, intersects lines AB and DE to form angles A and E, which lie on opposite sides of the transversal AE and on the interior of lines AB and DE. Such angles are called **alternate**

interior angles. Because transversal AE intersects lines AB and DE to form *equal alternate interior angles*, the lines AB and DE can never intersect. Such lines are called **parallel lines**. This important idea, along with its converse, is stated formally in the next theorem.

> **Theorem.**   If two lines are cut by a transversal so that alternate interior angles are equal, then the lines are parallel.
>
>   If two parallel lines are cut by a transversal, then alternate interior angles are equal.

▬ EXERCISE I.2 ▬

Exercises 1–4 refer to the similar triangles in Illustration 1.

1. If $AC = 7$, $DF = 4$, and $AB = 15$ centimeters, find DE.
2. If $CB = 12$, $AB = 15$, and $DE = 7$ meters, find FE.
3. If $AC = 21$, $BC = 25$, and $FE = 8$ centimeters, find DF.
4. If $DE = 6$, $AC = 12$, and $DF = 4$ centimeters, find AB.

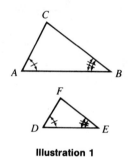

Illustration 1

5. A building casts a shadow of 50 feet at the same time a person casts a shadow of 8 feet. If the person is 6 feet tall, how tall is the building?

6. A person 5.5 feet tall casts a 3-foot shadow. How long is the shadow of a 60-foot flagpole?

7. A straight road going up a constant grade rises 3 feet in the first 50 feet of roadway. If the car travels 1000 feet from the bottom of the hill to the top, how high is the hill?

8. A ski hill has a run 0.5 mile long. If the hill drops 5 feet as the skier slides 9 feet, how high is the hill?

9. Sally is 5 feet, 6 inches tall. To measure the height of a smokestack, she asks her 6-foot, 6-inch boyfriend to help her. They stand as in Illustration 2 with her line of sight to the top of the stack being tangent to his head. If the boyfriend stands 126 feet from the smokestack and the girl is 3 feet from her friend, how tall is the smokestack?

10. Find the width of the river in Illustration 3. You may use the measurements in the figure, and you may assume that M and P are on the riverbank, directly across from N and R, respectively.

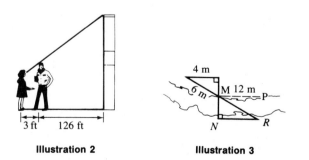

Illustration 2   **Illustration 3**

11. An airplane ascends 100 feet as it travels 1000 feet. How much altitude will it gain if it flies one mile?

12. In a landing approach an airplane descends 45 feet as it travels 500 feet. How far does the plane travel as it drops 600 feet?

13. The two trees in Illustration 4 are 54 feet apart. To find the height of the large tree, a 5-foot-tall Girl Scout backs 6 feet away from the small tree and sights to both the top and bottom of the taller tree. If her lines of sight are separated by 4 feet on the smaller tree, how tall is the larger tree?

14. A 12-foot pole is placed on each side of a river as in Illustration 5. A trigonometry student steps back from one pole a distance of 10 feet. She sights to the top and bottom of the second pole. If her lines of sight are separated by 5 feet on the first pole, how wide is the river?

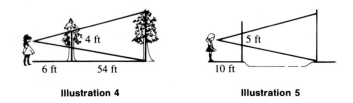

Illustration 4   **Illustration 5**

Exercises 15–20 refer to Illustration 6. Assume angle ACB = 90° and angle CDB = 90°.

15. If angle $A = 40°$, find angle 1.   16. If angle $2 = 60°$, find angle B.

17. Show that angle A is always equal to angle 1 and that angle B is always equal to angle 2.

18. Show that triangles ADC and ACB are similar.   19. Show that triangles BDC and BCA are similar.

20. Show that triangles ADC and CDB are similar.

Exercises 21 and 22 refer to Illustration 7.

21. If $AE = 4, DE = 6, BE = 3$, and AB is parallel to CD, find the length of CE.

22. If $AE = 4, DE = 6, BC = 8$, and AB is parallel to CD, find the length of BE.

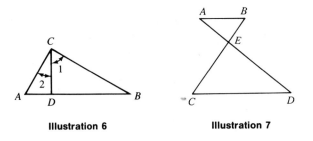

Illustration 6   **Illustration 7**

I.3 CIRCLES

A knowledge of certain properties of circles is essential to understanding the principles of trigonometry. As a matter of fact, the trigonometric functions are often called the **circular functions**.

> **Definition.** A **circle** is a set of points all in the same plane that are equidistant from a fixed point. The fixed point is called the **center** of the circle.

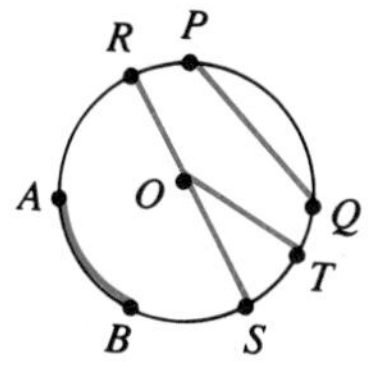

Consider the circle with center at point O in Figure I-12. A segment, such as PQ, that joins two points on a circle is called a **chord**. Any chord that passes through the center of the circle is called a **diameter**. For example, RS is a diameter of circle O. A portion of the circle (such as $\overset{\frown}{AB}$) is called an **arc**. Any segment joining the center of the circle to a point on the circle (such as OT) is called a **radius**. Any angle determined by two radii (such as angle TOS) is called a **central angle**. The distance around the entire circle is called the **circumference** of the circle.

Figure I-12

If the circumference of any circle is divided by the length of its diameter, the quotient is a constant that is independent of the size of the circle. This constant is the number pi ($\pi = 3.14159265\ldots$).

$$\frac{C}{D} = \pi$$

Multiplying both sides by D gives the formula for the circumference of a circle.

Formula for Circumference of a Circle. $C = \pi D$

Because $D = 2r$, where r is the length of a radius of the circle, this formula can be rewritten as

---

Formula for Circumference of a Circle. $C = 2\pi r$

A rigorous proof for the formula for the area of a circle requires techniques from calculus. For this reason, we state without proof that the area of a circle is given by the formula

---

Formula for Area of a Circle. $A = \pi r^2$

Example 1   Find the circumference and area of a circle with a diameter of 26 centimeters.

Solution   Because the diameter is 26 centimeters, the radius is 13 centimeters. Use the formulas for the circumference and area of a circle.

$$C = \pi(26\,\text{cm}) = 26\pi \text{ centimeters}$$
$$A = \pi(13\,\text{cm})^2 = 169\pi \text{ square centimeters}$$

Figure I-13

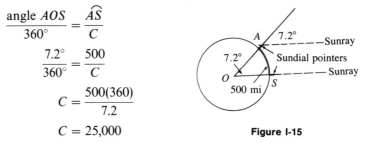

Figure I-14

> **Definition.** A **sector** of a circle is the set of points bounded by two radii and the arc intercepted by those radii.

The area shaded in Figure I-13 represents the area of sector AOB of circle O. Note from Figure I-14 that the central angle AOB is the same fractional part of one complete revolution that its intercepted arc is of the circle's circumference.

$$\frac{\text{angle } AOB}{360°} = \frac{\overset{\frown}{AB}}{C}$$

The Greek mathematician Eratosthenes (275–195 B.C.) was aware of this relationship and used it to estimate the circumference of the earth. He knew that at noontime during the summer solstice, the sun was directly overhead in Syene, Egypt (now Aswan). At this time, the sundials in Syene produced no shadows. However, the sundials did cast a shadow in Alexandria, a city about 5000 stades (approximately 500 miles) north of Syene. See Figure I-15. In the figure, line OA passes from the center of the earth through the tip of a sundial in Alexandria. Line OS passes from the center of the earth through the tip of a sundial in Syene. Eratosthenes was able to determine that in Alexandria, the sun crossed the pointer of a sundial at an angle that was one-fiftieth of a complete revolution, or 7.2°. Because the rays of the sun are nearly parallel and because two parallel lines cut by a transversal determine equal corresponding angles, angle AOS is also 7.2°. Although Eratosthenes reasoned geometrically, elementary algebra can be used to solve a proportion for C.

$$\frac{\text{angle } AOS}{360°} = \frac{\overset{\frown}{AS}}{C}$$

$$\frac{7.2°}{360°} = \frac{500}{C}$$

$$C = \frac{500(360)}{7.2}$$

$$C = 25,000$$

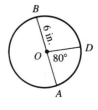

Figure I-15

Eratosthenes' estimate of the circumference of the earth was 25,000 miles. The actual circumference is closer to 24,900 miles.

■ EXERCISE I.3

Exercises 1 and 2 refer to Illustration 1. Circle O has AB as one of its diameters.

1. Find the circumference of circle O.

2. Find the area of circle O.

3. Find the radius of the circle with an area of 81π square units.

4. Find the radius of the circle with an area of 121π square units.

5. Find the radius of the circle with a circumference of 81π units.

6. Find the radius of the circle with a circumference of 121π units.

Illustration 1

7. Find the radius of the circle with an area of 63 square units.

8. Find the radius of the circle with an area of 47 square units.

9. Find the radius of the circle whose area has the same number of units as its circumference.

10. Find the radius of the circle whose area has twice the number of units as its circumference.

11. Find the diameter of the circle whose circumference has one-third the number of units as its area.

12. Find the diameter of the circle whose circumference has three times the number of units as its area.

Exercises 13–16 refer to Illustration 2.

13. If $AB = 10$ centimeters, find the circumference of circle O.

14. If $AB = 12$ meters, find the area of circle O.

15. If the area of circle O is 12 square inches, find AB.

16. If the circumference of circle O is 18 feet, find AB.

Illustration 2

17. John wants to plant marigolds to border his circular garden. If he plants the marigolds 8 inches apart and the garden has a diameter of 20 feet, how many marigold plants does he need?

18. If canvas sells for $4 a square yard, how much will it cost for material to make a cover for a round swimming pool 25 feet in diameter? Ignore any waste or overlap.

19. If the cost of a belt is $4 per yard, how much would it cost to buy a belt to encircle the earth at its equator? Consider the radius of the earth to be 4000 miles.

20. If paint costs $10 a gallon and if one gallon covers 500 square feet, how much would it cost to paint a circular helicopter landing pad 60 feet in diameter?

21. A bicycle with 27-inch diameter wheels is ridden for 1 mile. How many times will each wheel rotate during the trip?

22. A vehicle with 30-inch diameter wheels travels down the road. How many miles does the vehicle go if the tires rotate 1000 times?

23. If a college dining room has a circular floor plan with a diameter of 40 feet, how much will it cost for material to carpet the floor? Assume no waste and that carpet can be purchased for $15 per square yard.

24. Find the area of the shaded portion of Illustration 3. The large circle has a radius of R and the small circle has a radius of r.

Illustration 3

25. If a central angle of a circle measures 15° and its intercepted arc measures 1 meter, what is the circumference of the circle?

26. If the intercepted arc of a central angle in a circle measures 12 centimeters and the circumference of the circle is 31 centimeters, what is the measure of the central angle to the nearest tenth of a degree?

27. A central angle of a circle measures 125.5° and intercepts an arc of x centimeters. Find x if the diameter of the circle is 50 centimeters.

28. If the radius of a circle is 10 meters and an arc of the circle is 60 centimeters, find, to the nearest tenth of a degree, the measure of the central angle that intercepts the given arc.

29. Use Eratosthenes' estimate of the earth's circumference to estimate the length of the earth's radius.

30. Use Eratosthenes' estimate of the earth's circumference to estimate the earth's volume. $(V = \frac{4}{3}\pi r^3)$

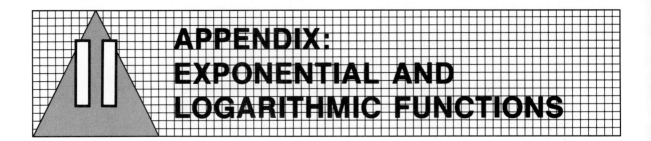

APPENDIX: EXPONENTIAL AND LOGARITHMIC FUNCTIONS

In the late 16th century, logarithms were invented to simplify computations in astronomy. They proved to be so successful that they were soon used to simplify calculations in many other fields. Today, the widespread use of calculators has eliminated the need for logarithms as a computational aid. However, logarithms have retained their importance in mathematics, science, engineering, and economics in other ways such as expressing the acidity of a solution, expressing the voltage gain of an amplifier, or computing the intensity of an earthquake. Logarithms are related to functions that involve exponential expressions. For this reason we begin by considering exponential functions.

II.1 EXPONENTIAL FUNCTIONS

We have previously defined exponential expressions such as b^x, where x is a rational number. We now give meaning to the expression b^x, where x is an irrational number. To do so we consider the expression $3^{\sqrt{2}}$, where $\sqrt{2}$ is an irrational number with a decimal value of 1.414213562... and, consequently, $1 < \sqrt{2} < 2$. Because 3 is a number greater than 1, it can be shown that

$$3^1 < 3^{\sqrt{2}} < 3^2$$

Similarly, because $1.4 < \sqrt{2} < 1.5$,

$$3^{1.4} < 3^{\sqrt{2}} < 3^{1.5}$$

Because the numbers 1.4 and 1.5 are rational numbers, the exponential expressions $3^{1.4}$ and $3^{1.5}$ are defined. If we continue in this manner, $3^{\sqrt{2}}$ becomes fenced in by two values, one that is greater than $3^{\sqrt{2}}$ and one that is less than $3^{\sqrt{2}}$. As the list continues we can see that the value of $3^{\sqrt{2}}$ gets squeezed into a smaller and smaller interval:

$$
\begin{aligned}
3^1 &= 3 & &< 3^{\sqrt{2}} < 9 & &= 3^2 \\
3^{1.4} &\approx 4.656 & &< 3^{\sqrt{2}} < 5.196 & &\approx 3^{1.5} \\
3^{1.41} &\approx 4.7070 & &< 3^{\sqrt{2}} < 4.7590 & &\approx 3^{1.42} \\
3^{1.414} &\approx 4.727695 & &< 3^{\sqrt{2}} < 4.732892 & &\approx 3^{1.415}
\end{aligned}
$$

In calculus classes it is shown that $3^{\sqrt{2}}$ has a fixed value greater than any of the increasing values on the left-hand side of the preceding list, but less than any of the decreasing values on the right. To find an approximation for $3^{\sqrt{2}}$, we can press these keys on a calculator:

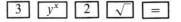

The display will show 4.728804388. Thus,

$$3^{\sqrt{2}} \approx 4.728804388\dots$$

The preceding discussion suggests this important fact: *If b is a positive real number and x is a real number (either rational or irrational), the exponential expression b^x represents a single real number.*

The function defined by the equation $y = f(x) = b^x$ is called an **exponential function**.

> **Definition.**   The **exponential function with base *b*** is defined by the equation
>
> $$y = b^x$$
>
> where $b > 0$, $b \neq 1$, and x is a real number.
>   The **domain** of the exponential function is the set of real numbers, and the **range** is the set of positive real numbers.

Because the exponential function defined by $y = f(x) = b^x$ has the real number set for its domain and the positive real number set for its range, we can graph the function on a Cartesian coordinate system.

Example 1   Graph the exponential functions defined by **a.** $y = 2^x$ and **b.** $y = 4^x$.

Solution   **a.**   To graph the function defined by the equation $y = 2^x$, calculate several pairs of points (x, y) that satisfy the equation, plot the points, and join them with a smooth curve. The graph appears in Figure II-1a.

b.   To graph the function defined by the equation $y = 4^x$, proceed as in part a. The graph appears in Figure II-1b.

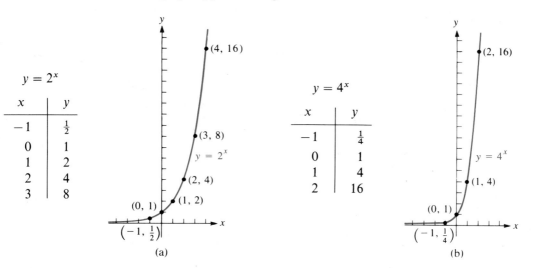

| $y = 2^x$ | |
x	y
-1	$\frac{1}{2}$
0	1
1	2
2	4
3	8

| $y = 4^x$ | |
x	y
-1	$\frac{1}{4}$
0	1
1	4
2	16

(a)

(b)

Figure II-1

In each graph of Example 1, the values of y increase as the values of x increase; thus, the functions defined by the equations $y = 2^x$ and $y = 4^x$ are called **increasing functions**. Note that each graph approaches the x-axis as x gets smaller and that each graph passes through the point $(0, 1)$. Note also that the graph of $y = 2^x$ passes through the point $(1, 2)$ and that the graph of $y = 4^x$ passes through the point $(1, 4)$. In general, the graph of $y = b^x$ passes through the points $(0, 1)$ and $(1, b)$.

Example 2 Graph the exponential functions $y = \left(\dfrac{1}{2}\right)^x$ and $y = \left(\dfrac{1}{4}\right)^x$.

Solution Calculate and plot several pairs (x, y) that satisfy each equation. The graph of $y = (\frac{1}{2})^x$ appears in Figure II-2a, and the graph of $y = (\frac{1}{4})^x$ appears in Figure II-2b.

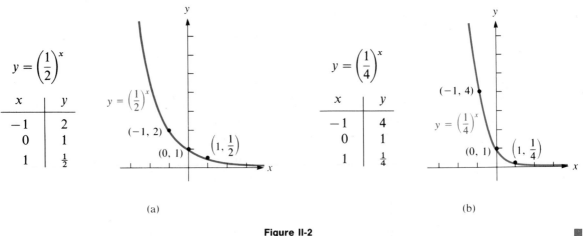

$y = \left(\dfrac{1}{2}\right)^x$

x	y
-1	2
0	1
1	$\frac{1}{2}$

$y = \left(\dfrac{1}{4}\right)^x$

x	y
-1	4
0	1
1	$\frac{1}{4}$

(a)   (b)

Figure II-2

In each graph of Example 2, the values of y decrease as the values of x increase; thus, the functions defined by the equations $y = (\frac{1}{2})^x$ and $y = (\frac{1}{4})^x$ are called **decreasing functions**. Each graph approaches the x-axis as x gets larger and each graph passes through the point $(0, 1)$. Note that the graph of $y = (\frac{1}{2})^x$ passes through the point $(1, \frac{1}{2})$ and that the graph of $y = (\frac{1}{4})^x$ passes through the point $(1, \frac{1}{4})$. In general, the graph of $y = (1/b)^x$ passes through the points $(0, 1)$ and $(1, 1/b)$.

Example 3 On the same set of coordinate axes, graph the exponential functions defined by
a. $y = (\frac{3}{2})^x$ and **b.** $y = (\frac{2}{3})^x$.

Solution   Calculate and plot several pairs (x, y) that satisfy each equation. The graphs appear in Figure II-3. Note that the equation $y = (\frac{3}{2})^x$ represents an increasing function and that the equation $y = (\frac{2}{3})^x$ represents a decreasing function. Note also that the bases of the functions are reciprocals of each other and that the graphs are reflections of each other in the y-axis.

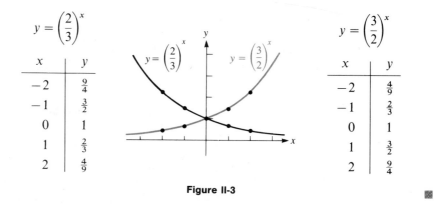

$$y = \left(\frac{2}{3}\right)^x$$

x	y
-2	$\frac{9}{4}$
-1	$\frac{3}{2}$
0	1
1	$\frac{2}{3}$
2	$\frac{4}{9}$

$$y = \left(\frac{3}{2}\right)^x$$

x	y
-2	$\frac{4}{9}$
-1	$\frac{2}{3}$
0	1
1	$\frac{3}{2}$
2	$\frac{9}{4}$

Figure II-3

Examples 1 through 3 suggest that an exponential function with base b is either increasing (when $b > 1$) or decreasing (when $0 < b < 1$). Thus, distinct real numbers x determine distinct values b^x. The exponential function, therefore, is one-to-one. This fact is the basis of an important fact involving exponential expressions.

> If $b > 0$, $b \neq 1$, and $b^m = b^n$, then $m = n$.

Because exponential functions are one-to-one, they all have inverse functions. The next example shows how to find the inverse of an exponential function.

Example 4   Find the inverse function of the exponential function defined by $y = 3^x$. Graph both functions on the same set of coordinate axes.

Solution   The graph of $y = 3^x$ appears in Figure II-4. The equation that defines the inverse function of $y = 3^x$ is found by interchanging x and y. Thus, the inverse of $y = 3^x$ is

$$x = 3^y$$

The graph of $x = 3^y$ also appears in the figure. As expected, the two graphs are symmetric with respect to the line $y = x$.

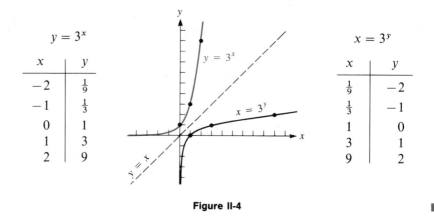

Figure II-4

Applications of Exponential Functions

A mathematical description of an observed event is called a **model** of that event. Many observed events can be modeled by functions defined by equations of the form $y = ab^{kx}$. For example, the amount of money in a savings account that earns *compound interest* is modeled by the formula

Formula for
Compound Interest

$$A = A_0\left(1 + \frac{r}{k}\right)^{kt}$$

where A represents the amount in the account after t years, with interest paid k times a year at an annual rate of r percent on an initial deposit of A_0.

Example 5   If $1500 is deposited in a bank account that pays 10 percent annual interest, compounded quarterly, how much will be in the account after 20 years?

Solution   Calculate A using the formula

$$A = A_0\left(1 + \frac{r}{k}\right)^{kt}$$

with $A_0 = 1500$, $r = 0.10$, and $t = 20$. Because with quarterly compounding interest is paid four times a year, $k = 4$.

$$A = A_0\left(1 + \frac{r}{k}\right)^{kt}$$

$$A = 1500\left(1 + \frac{0.10}{4}\right)^{4(20)}$$

$$= 1500(1.025)^{80}$$

$$= 10,814.35$$

Use a calculator and press these keys:
1.025 $\boxed{y^x}$ 80 $\boxed{\times}$ 1500 $\boxed{=}$.

After 20 years, the account will contain $10,814.35.

Base-e Exponential Functions

To make their services more attractive, banks often compound interest more frequently than quarterly. Monthly and daily compounding is common, but some banks compound interest *continuously.*

If interest is compounded continuously, the amount A contained in a savings account is given by the formula

Formula for
Continuous
Compounding
of Interest

$$A = A_0 e^{rt}$$

where A_0 is the initial deposit, r is the annual interest rate, t is the length of time in years, and e is an irrational number with a value of

$$e \approx 2.718281828$$

To compute the amount to which $1500 will grow if invested for 10 years at 10 percent annual interest compounded continuously, we substitute 1500 for A_0, 0.10 for r, and 10 for t in the formula $A = A_0 e^{rt}$ and simplify:

$$A = A_0 e^{rt}$$
$$A = 1500(e)^{0.10(10)}$$
$$= 1500(e)^1$$
$$\approx 4077.42$$

Use a calculator and press these keys:

$1 \boxed{e^x} \boxed{\times} 1500 \boxed{=}$.

After 10 years, the account will contain $4077.42.

If your calculator does not have an $\boxed{e^x}$ key, try pressing $\boxed{\text{INV}}$ $\boxed{\ln x}$ in place of $\boxed{e^x}$ or consult your owner's manual.

In mathematical models of natural events, the number e appears often as the base of an exponential function. For example, in the **Malthusian model for population growth**, the current population of a colony is given by the formula

Formula for
Population Growth

$$P = P_0 e^{kt}$$

where P_0 is the population at $t = 0$, and k is the difference between the birth and death rates. If t is measured in years, k is called the **annual growth rate**.

Example 6 The annual growth rate within a city is 2 percent. If this growth rate remains constant and if the current population of the city is 15,000, what will the population be in 100 years?

Solution Use the Malthusian model for population growth.

$$P = P_0 e^{kt}$$

with $P_0 = 15,000$, $k = 0.02$, and $t = 100$.

$$P = P_0 e^{kt}$$
$$P = 15{,}000e^{0.02(100)}$$
$$= 15{,}000e^2$$
$$\approx 110{,}835.84 \qquad \text{Use a calculator and press these keys:}$$

2 $\boxed{e^x}$ $\boxed{\times}$ 15000 $\boxed{=}$

In 100 years the population will be approximately 111,000.   ■

EXERCISE II.1

In Exercises 1–12, graph the function defined by each equation.

1. $y = 3^x$ **2.** $y = 5^x$ **3.** $y = \left(\dfrac{1}{5}\right)^x$ **4.** $y = \left(\dfrac{1}{3}\right)^x$

5. $y = \left(\dfrac{3}{4}\right)^x$ **6.** $y = \left(\dfrac{5}{3}\right)^x$ **7.** $y = 2 + 2^x$ **8.** $y = -2 + 3^x$

9. $y = 3(2^x)$ **10.** $y = 2(3^x)$ **11.** $y = 2^{-x}$ **12.** $y = 3^{-x}$

In Exercises 13–16, graph the exponential function defined by each equation and graph its inverse on the same set of coordinate axes.

13. $y = 2^x$ **14.** $y = 4^x$ **15.** $y = \left(\dfrac{1}{4}\right)^x$ **16.** $y = \left(\dfrac{1}{2}\right)^x$

In Exercises 17–22, find the value of b, if any, that would cause the graph of $y = b^x$ to look like the graph indicated.

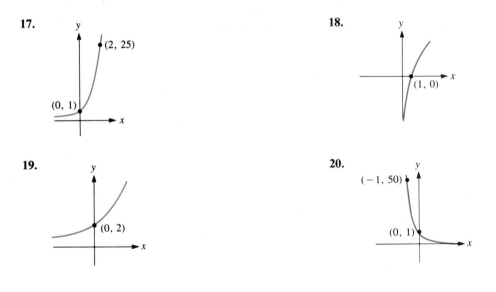

17.   **18.**

19.   **20.**

21.

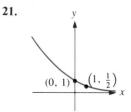

$(0, 1)$ $\left(1, \frac{1}{2}\right)$

22.

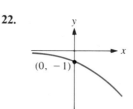

$(0, -1)$

23. An initial deposit of $10,000 earns 8 percent annual interest compounded quarterly. How much will be in the account after 10 years?

24. An initial deposit of $10,000 earns 8 percent annual interest compounded monthly. How much will be in the account after 10 years?

25. An initial deposit of $20,000 earns 7 percent annual interest compounded monthly. How much will be in the account after 25 years?

26. An initial deposit of $20,000 earns 7 percent annual interest compounded daily. How much will be in the account after 25 years?

27. An initial deposit of $8000 earns 6 percent annual interest compounded continuously. How much will be in the account after 10 years?

28. An initial deposit of $22,000 earns 9 percent annual interest compounded continuously. How much will be in the account after 20 years?

29. A population of fish is growing according to the Malthusian model. How many fish will there be in 10 years if the annual growth rate is 3 percent and the initial population is 2700 fish?

30. The population of a small town is 1350. The town is expected to grow according to the Malthusian model with an annual growth rate of 6 percent. What will be the population of the town in 20 years?

31. The population of the world is approximately 5 billion. If the world population is growing according to the Malthusian model with an annual growth rate of 1.8 percent, what will be the population of the world in 30 years?

32. The population of the world is approximately 5 billion. If the world population is growing according to the Malthusian model with an annual growth rate of 1.8 percent, what will be the population of the world in 60 years?

33. A colony of 6 million bacteria is growing in a culture medium. The population P after t hours is given by the formula $P = (6 \times 10^6)(2.3)^t$. What is the population after 4 hours?

34. The population of North Rivers is growing exponentially according to the formula $P = 375(1.3)^t$, where t is measured in years from the present date. What will be the population of North Rivers in 6 years and 9 months?

35. The charge remaining in a battery decreases as the battery slowly discharges. The charge C after t days is given by the formula $C = (3 \times 10^{-4})(0.7)^t$. What is the charge after 5 days?

36. A radioactive material decays according to the formula $A = A_0(\frac{2}{3})^t$, where A_0 is the initial amount present and t is measured in years. What amount will be present in 5 years?

II.2 LOGARITHMIC FUNCTIONS

The exponential function defined by the equation $y = b^x$ is one-to-one and, therefore, possesses an inverse function. The equation that defines its one-to-one inverse is obtained by interchanging x and y. Thus, the inverse of the function $y = b^x$ is defined by the equation $x = b^y$. To express the equation $x = b^y$ in a form that is solved for y, we need the following definition:

> **Definition.** The **logarithmic function with base** b is a one-to-one function defined by the equation
>
> $$y = \log_b x$$
>
> where $b > 0$ and $b \neq 1$. This equation is read as "y equals log base b of x," and is equivalent to the exponential equation $x = b^y$.
>
> The **domain** of the logarithmic function is the set of positive real numbers. The **range** is the set of real numbers.

The preceding definition points out that any pair of values (x, y) that satisfies the equation $y = \log_b x$ also satisfies the equation $x = b^y$. From this definition, it follows that

$$\log_{10} x = y \qquad \text{means} \qquad 10^y = x$$
$$\log_7 1 = 0 \qquad \text{means} \qquad 7^0 = 1$$
$$\log_3 m = n \qquad \text{means} \qquad 3^n = m$$
$$\log_4 16 = 2 \qquad \text{means} \qquad 4^2 = 16$$
$$\log_2 32 = 5 \qquad \text{means} \qquad 2^5 = 32$$
$$\log_5 \tfrac{1}{25} = -2 \qquad \text{means} \qquad 5^{-2} = \tfrac{1}{25}$$

In each case the logarithm of a number appears as the exponent in the corresponding exponential expression. This is no coincidence, because the logarithm of any number is always an exponent. In fact, $\log_b x$ is the exponent to which a base b must be raised in order to obtain x:

$$b^{\log_b x} = x$$

Because there is no real number x such that $2^x = -8$, for example, there is no meaning to the expression $\log_2(-8)$. In general, *there are no logarithms for nonpositive numbers.*

Because base-10 logarithms appear so often in mathematics and science, it is a good idea to learn the following base-10 logarithms:

$$\log_{10} \tfrac{1}{100} = -2 \qquad \text{because} \qquad 10^{-2} = \tfrac{1}{100}$$
$$\log_{10} \tfrac{1}{10} = -1 \qquad \text{because} \qquad 10^{-1} = \tfrac{1}{10}$$
$$\log_{10} 1 = 0 \qquad \text{because} \qquad 10^0 = 1$$
$$\log_{10} 10 = 1 \qquad \text{because} \qquad 10^1 = 10$$
$$\log_{10} 100 = 2 \qquad \text{because} \qquad 10^2 = 100$$
$$\log_{10} 1000 = 3 \qquad \text{because} \qquad 10^3 = 1000$$

In general, we have

$$\log_{10} 10^x = x$$

Example 1   Find the value of x in each equation: **a.** $\log_3 81 = x$, **b.** $\log_x 125 = 3$, and **c.** $\log_4 x = 3$

Solution   **a.** $\log_3 81 = x$ is equivalent to the equation $3^x = 81$. Because $3^4 = 81$, it follows that $x = 4$.

b. $\log_x 125 = 3$ is equivalent to the equation $x^3 = 125$. Because $5^3 = 125$, it follows that $x = 5$.

c. $\log_4 x = 3$ is equivalent to the equation $4^3 = x$. Because $4^3 = 64$, it follows that $x = 64$.   ∎

Example 2   Find the value of x in each equation: **a.** $\log_{1/3} x = 2$, **b.** $\log_{1/3} x = -2$, and **c.** $\log_{1/3} \frac{1}{27} = x$.

Solution   **a.** $\log_{1/3} x = 2$ is equivalent to $\left(\frac{1}{3}\right)^2 = x$. Thus, $x = \frac{1}{9}$.

b. $\log_{1/3} x = -2$ is equivalent to $\left(\frac{1}{3}\right)^{-2} = x$. Thus,

$$x = \left(\frac{1}{3}\right)^{-2} = 3^2 = 9$$

c. $\log_{1/3} \frac{1}{27} = x$ is equivalent to $\left(\frac{1}{3}\right)^x = \frac{1}{27}$. Because $\left(\frac{1}{3}\right)^3 = \frac{1}{27}$, it follows that $x = 3$.   ∎

Because both the domain and the range of the logarithmic function are subsets of the real numbers, it is possible to graph logarithmic functions on the rectangular coordinate system.

Example 3   Graph the logarithmic function defined by $y = \log_2 x$.

Solution   The equation $y = \log_2 x$ is equivalent to the equation $x = 2^y$. Calculate a table of values (x, y) that satisfy the equation $x = 2^y$, plot the corresponding points, and connect them with a smooth curve. The graph appears in Figure II-5.

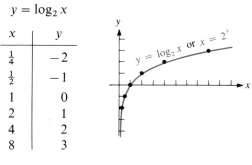

$$y = \log_2 x$$

x	y
$\frac{1}{4}$	-2
$\frac{1}{2}$	-1
1	0
2	1
4	2
8	3

Figure II-5   ∎

Example 4   Graph the logarithmic function defined by $y = \log_{1/2} x$.

Solution   The equation $y = \log_{1/2} x$ is equivalent to the equation $x = \left(\frac{1}{2}\right)^y$. Calculate a table of values (x, y) that satisfy the equation $x = (\frac{1}{2})^y$, plot the corresponding points, and connect them with a smooth curve. The graph appears in Figure II-6.

$$y = \log_{1/2} x$$

x	y
8	-3
4	-2
2	-1
1	0
$\frac{1}{2}$	1
$\frac{1}{4}$	2

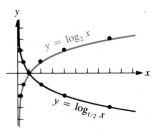

Figure II-6

If the graphs of Examples 3 and 4 are merged onto a single set of coordinate axes as in Figure II-7, it is apparent that they are reflections of each other in the x-axis. Note that both graphs pass through the point with coordinates of $(1, 0)$, and that both curves approach but never touch the y-axis.

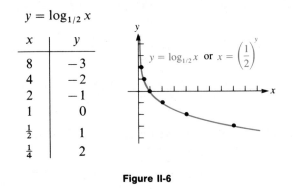

Figure II-7

Properties of Logarithms

Because logarithmic equations can be written as exponential equations, it is not surprising that the laws of exponents have counterparts in logarithmic notation. We will use this fact to develop many of the following important properties of logarithms.

Properties of Logarithms. If M, N, p, and b are positive numbers and $b \neq 1$, then

1. $\log_b 1 = 0$ 2. $\log_b b = 1$
3. $\log_b b^x = x$ 4. $b^{\log_b x} = x$

5. $\log_b MN = \log_b M + \log_b N$ 6. $\log_b \dfrac{M}{N} = \log_b M - \log_b N$

7. $\log_b M^p = p \log_b M$ 8. If $\log_b x = \log_b y$, then $x = y$.

Properties 1 through 4 follow directly from the definition of logarithm.

To prove Property 5, we let $x = \log_b M$ and $y = \log_b N$. Because of the definition of logarithm, these equations can be written in the form

$$M = b^x \quad \text{and} \quad N = b^y$$

Then,

$$MN = b^x b^y$$

and the properties of exponents give

$$MN = b^{x+y}$$

Using the definition of logarithms in reverse gives

$$\log_b MN = x + y$$

Substituting the values of x and y completes the proof:

$$\log_b MN = \log_b M + \log_b N \qquad \square$$

To prove Property 6, we again let $x = \log_b M$ and $y = \log_b N$. These equations can be written as

$$M = b^x \quad \text{and} \quad N = b^y$$

Then,

$$\frac{M}{N} = \frac{b^x}{b^y}$$

and the properties of exponents give

$$\frac{M}{N} = b^{x-y}$$

Using the definition of logarithms in reverse gives

$$\log_b \frac{M}{N} = x - y$$

Substituting the values of x and y completes the proof:

$$\log_b \frac{M}{N} = \log_b M - \log_b N \qquad \square$$

To prove Property 7, we let $x = \log_b M$, write the expression in exponential form, and raise both sides to the pth power:

$$M = b^x$$
$$(M)^p = (b^x)^p$$
$$M^p = b^{px}$$

Using the definition of logarithms in reverse gives

$$\log_b M^p = px$$

Substituting the value for x completes the proof:

$$\log_b M^p = p \log_b M \qquad \square$$

Property 8 follows from the fact that the logarithmic function is a one-to-one function.

Property 5 of logarithms asserts that the logarithm of the *product* of two numbers is equal to the *sum* of their logarithms. The logarithm of a *sum* or a *difference* usually does not simplify. In general,

$$\log_b(M + N) \neq \log_b M + \log_b N$$

Likewise,

$$\log_b(M - N) \neq \log_b M - \log_b N$$

Property 6 of logarithms asserts that the logarithm of the *quotient* of two numbers is equal to the *difference* of their logarithms. The logarithm of a quotient is not the quotient of the logarithms:

$$\log_b \frac{M}{N} \neq \frac{\log_b M}{\log_b N}$$

Example 5 Simplify each expression: **a.** $\log_3 1$, **b.** $\log_4 4$, **c.** $\log_7 7^3$, and **d.** $b^{\log_b 3}$.

Solution **a.** By Property 1, $\log_3 1 = 0$.

b. By Property 2, $\log_4 4 = 1$.

c. By Property 3, $\log_7 7^3 = 3$.

d. By Property 4, $b^{\log_b 3} = 3$. ∎

The properties of logarithms are often used to expand or condense a logarithmic expression, as in Examples 6 and 7.

Example 6 Assume x, y, z, and b are positive numbers ($b \neq 1$). Use properties of logarithms to write each expression in terms of the logarithms of x, y, and z.

a. $\log_b \dfrac{xy}{z}$, **b.** $\log_b(x^3 y^2 z)$, and **c.** $\log_b \dfrac{y^2 \sqrt{z}}{x}$

Solution **a.** $\log_b \dfrac{xy}{z} = \log_b(xy) - \log_b z$ Use Property 6.

$$= \log_b x + \log_b y - \log_b z \qquad \text{Use Property 5.}$$

b. $\log_b(x^3 y^2 z) = \log_b x^3 + \log_b y^2 + \log_b z$ Use Property 5 twice.

$= 3\log_b x + 2\log_b y + \log_b z$ Use Property 7 twice.

c. $\log_b \dfrac{y^2\sqrt{z}}{x} = \log_b(y^2\sqrt{z}) - \log_b x$ Use Property 6.

$= \log_b y^2 + \log_b\sqrt{z} - \log_b x$ Use Property 5.

$= \log_b y^2 + \log_b z^{1/2} - \log_b x$ Write $\sqrt{z}$ as $z^{1/2}$.

$= 2\log_b y + \tfrac{1}{2}\log_b z - \log_b x$ Use Property 7 twice.   ∎

Example 7 Assume x, y, z, and b are positive numbers ($b \neq 1$). Use properties of logarithms to write each expression as the logarithm of a single quantity.

a. $2\log_b x + \dfrac{1}{3}\log_b y$ and   **b.** $\dfrac{1}{2}\log_b(x - 2) - \log_b y + 3\log_b z$

Solution **a.** $2\log_b x + \dfrac{1}{3}\log_b y = \log_b x^2 + \log_b y^{1/3}$ Use Property 7 twice.

$= \log_b(x^2 y^{1/3})$ Use Property 5.

$= \log_b(x^2 \sqrt[3]{y})$ Write $y^{1/3}$ as $\sqrt[3]{y}$.

b. $\dfrac{1}{2}\log_b(x - 2) - \log_b y + 3\log_b z$

$= \log_b(x - 2)^{1/2} - \log_b y + \log_b z^3$ Use Property 7 twice.

$= \log_b \dfrac{(x - 2)^{1/2}}{y} + \log_b z^3$ Use Property 6.

$= \log_b \dfrac{z^3\sqrt{x - 2}}{y}$ Use Property 5 and write $(x - 2)^{1/2}$ as $\sqrt{x - 2}$.   ∎

Example 8 Given that $\log_{10} 2 \approx 0.3010$ and $\log_{10} 3 \approx 0.4771$, find the approximate values of **a.** $\log_{10} 6$, **b.** $\log_{10} 9$, **c.** $\log_{10} 5$, **d.** $\log_{10}\sqrt{5}$, and **e.** $\log_{10} 1.5$.

Solution **a.** $\log_{10} 6 = \log_{10}(2 \cdot 3)$ Factor 6 as $2 \cdot 3$.

$= \log_{10} 2 + \log_{10} 3$ Use Property 5 of logarithms.

$\approx 0.3010 + 0.4771$ Substitute the value of each logarithm.

≈ 0.7781

b. $\log_{10} 9 = \log_{10} 3^2$ Write 9 as 3^2.

$= 2\log_{10} 3$ Use Property 7 of logarithms.

$\approx 2(0.4771)$ Substitute the value of the logarithm of 3.

≈ 0.9542

c. $\log_{10} 5 = \log_{10} \dfrac{10}{2}$ $\qquad$ Write 5 as $\frac{10}{2}$.

$\qquad\qquad = \log_{10} 10 - \log_{10} 2$ $\qquad$ Use Property 6 of logarithms.

$\qquad\qquad = 1 - \log_{10} 2$ $\qquad$ Use Property 2 of logarithms $(\log_{10} 10 = 1)$.

$\qquad\qquad \approx 1 - 0.3010$ $\qquad$ Substitute the value of the logarithm of 2.

$\qquad\qquad \approx 0.6990$

d. $\log_{10} \sqrt{5} = \log_{10} 5^{1/2}$ $\qquad$ Write $\sqrt{5}$ as $5^{1/2}$.

$\qquad\qquad = \dfrac{1}{2} \log_{10} 5$ $\qquad$ Use Property 7 of logarithms.

$\qquad\qquad \approx \dfrac{1}{2}(0.6990)$ $\qquad$ Use the answer from part c.

$\qquad\qquad \approx 0.3495$

e. $\log_{10} 1.5 = \log_{10} \dfrac{3}{2}$ $\qquad$ Write 1.5 as $\frac{3}{2}$.

$\qquad\qquad = \log_{10} 3 - \log_{10} 2$ $\qquad$ Use Property 6 of logarithms.

$\qquad\qquad \approx 0.4771 - 0.3010$ $\qquad$ Substitute the value of each logarithm.

$\qquad\qquad \approx 0.1761$ $\qquad\qquad\qquad\qquad\qquad\qquad\qquad\qquad$ ■

Example 9 Verify that $\log_{10}(\tan x \sec x) = \log_{10}(\sin x) - 2\log(\cos x)$.

Solution Work with the left-hand side, using properties of logarithms and the trigonometric functions.

$$\log_{10}(\tan x \sec x) = \log_{10}\left(\dfrac{\sin x}{\cos x} \cdot \dfrac{1}{\cos x}\right)$$

$$= \log_{10}\left(\dfrac{\sin x}{\cos^2 x}\right)$$

$$= \log_{10}(\sin x) - \log_{10}(\cos^2 x) \qquad \log\dfrac{M}{N} = \log M - \log N$$

$$= \log_{10}(\sin x) - 2\log_{10}(\cos x) \qquad \log M^2 = 2\log M \qquad ■$$

▬ EXERCISE II.2 ▬

In Exercises 1–8, write each equation in exponential form.

1. $\log_3 81 = 4$ $\qquad$ **2.** $\log_7 7 = 1$ $\qquad$ **3.** $\log_{1/2} \dfrac{1}{8} = 3$ $\qquad$ **4.** $\log_{1/5} 1 = 0$

5. $\log_4 \dfrac{1}{64} = -3$ $\qquad$ **6.** $\log_6 \dfrac{1}{36} = -2$ $\qquad$ **7.** $\log_x y = z$ $\qquad$ **8.** $\log_m n = \dfrac{1}{2}$

In Exercises 9–16, write each equation in logarithmic form.

9. $8^2 = 64$ $\qquad$ **10.** $10^3 = 1000$ $\qquad$ **11.** $4^{-2} = \dfrac{1}{16}$ $\qquad$ **12.** $3^{-4} = \dfrac{1}{81}$

13. $\left(\dfrac{1}{2}\right)^{-5} = 32$ **14.** $\left(\dfrac{1}{3}\right)^{-3} = 27$ **15.** $x^y = z$ **16.** $m^n = p$

In Exercises 17–24, find each value.

17. $\log_7 1$ **18.** $\log_2 4$ **19.** $\log_5 25$ **20.** $\log_9 81$

21. $\log_{1/8}\dfrac{1}{8}$ **22.** $\log_{1/6}\dfrac{1}{36}$ **23.** $\log_{2/5}\dfrac{8}{125}$ **24.** $\log_{3/4}\dfrac{81}{256}$

In Exercises 25–50, find each value of x.

25. $\log_2 8 = x$ **26.** $\log_3 27 = x$ **27.** $\log_5 125 = x$ **28.** $\log_3 81 = x$

29. $\log_4 x = 2$ **30.** $\log_5 x = 3$ **31.** $\log_3 x = 4$ **32.** $\log_2 x = 6$

33. $\log_x 16 = 4$ **34.** $\log_x 9 = 2$ **35.** $\log_x 32 = 5$ **36.** $\log_x 125 = 3$

37. $\log_{1/5}\dfrac{1}{25} = x$ **38.** $\log_{1/5} x = -3$ **39.** $\log_x\dfrac{1}{9} = -2$ **40.** $\log_{1/4} x = -4$

41. $\log_x 9 = \dfrac{1}{2}$ **42.** $\log_x 8 = \dfrac{1}{3}$ **43.** $\log_{125} x = \dfrac{2}{3}$ **44.** $\log_{100}\dfrac{1}{1000} = x$

45. $\log_{5/2}\dfrac{4}{25} = x$ **46.** $\log_x\dfrac{9}{4} = 2$ **47.** $\log_x\dfrac{\sqrt{3}}{3} = \dfrac{1}{2}$ **48.** $\log_{1/2} x = 0$

49. $\log_{\sqrt{2}} 4 = x$ **50.** $\log_{\sqrt{3}} 9 = x$

In Exercises 51–58, graph each logarithmic function.

51. $y = \log_3 x$ **52.** $y = \log_4 x$ **53.** $y = \log_{1/4} x$ **54.** $y = \log_{1/3} x$

55. $y = 1 + \log_2 x$ **56.** $y = -2 + \log_5 x$ **57.** $y = 3\log_3 x$ **58.** $y = \log_2(3x)$

In Exercises 59 and 60, graph each pair of functions on a single set of coordinate axes. Note the line of symmetry.

59. $y = 4^x$ and $y = \log_4 x$ **60.** $y = 3^x$ and $y = \log_3 x$

In Exercises 61–66, find the value of b, if any, that would cause the graph of $y = \log_b x$ to look like the graph shown.

61.

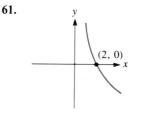

62.

63.

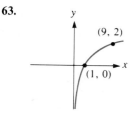

64.

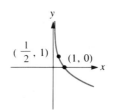

65.

66.

In Exercises 67–74, assume that x, y, z, and b are positive numbers. Use the properties of logarithms to write each expression in terms of the logarithms of x, y, and z.

67. $\log_b xyz$

68. $\log_b \dfrac{x}{yz}$

69. $\log_b \left(\dfrac{x}{y}\right)^2$

70. $\log_b (xz)^{1/3}$

71. $\log_b x\sqrt{xz}$

72. $\log_b \dfrac{\sqrt[3]{x}}{\sqrt[3]{yz}}$

73. $\log_b \sqrt[4]{\dfrac{x^3 y^2}{z^4}}$

74. $\log_b x\sqrt{\dfrac{\sqrt{y}}{z}}$

In Exercises 75–82, assume that x, y, z, and b are positive numbers. Use the properties of logarithms to write each expression as the logarithm of a single quantity.

75. $\log_b(x + 1) - \log_b x$

76. $\log_b x + \log_b(x + 2) - \log_b 8$

77. $2\log_b x + \dfrac{1}{3}\log_b y$

78. $-2\log_b x - 3\log_b y + \log_b z$

79. $-3\log_b x - 2\log_b y + \dfrac{1}{2}\log_b z$

80. $3\log_b(x + 1) - 2\log_b(x + 2) + \log_b x$

81. $\log_b\left(\dfrac{x}{z} + x\right) - \log_b\left(\dfrac{y}{z} + y\right)$

82. $\log_b(xy + y^2) - \log_b(xz + yz) + \log_b x$

In Exercises 83–98, tell whether each statement is true. If not, so indicate.

83. $\log_b 0 = 1$

84. $7^{\log_7 7} = 7$

85. $\log_b 1 = 0$

86. $\log_b 1 = 1$

87. $\log_b 1 = b$

88. $\log_b b = 1$

89. $\log_7 7^7 = 7$

90. $\dfrac{\log_b A}{\log_b B} = \log_b A - \log_b B$

91. $\log_b ab = 1 + \log_b a$

92. $\log_b \dfrac{1}{a} = -\log_b a$

93. $\log_b 2 = \log_2 b$

94. A logarithm cannot be negative.

95. $\log_b xy = (\log_b x)(\log_b y)$

96. $\dfrac{1}{3}\log_b a^3 = \log_b a$

97. If $\log_a b = c$, then $\log_b a = c$.

98. If $\log_b a = c$, then $\log_b a^p = pc$.

In Exercises 99–114, assume that $\log_{10} 4 \approx 0.6021$, $\log_{10} 7 \approx 0.8451$, and $\log_{10} 9 \approx 0.9542$. Use these values and the properties of logarithms to find the approximate value of each quantity.

99. $\log_{10} 28$

100. $\log_{10} \dfrac{7}{4}$

101. $\log_{10} 2.25$

102. $\log_{10} 36$

103. $\log_{10} \dfrac{63}{4}$ **104.** $\log_{10} \dfrac{4}{63}$ **105.** $\log_{10} 252$ **106.** $\log_{10} 49$

107. $\log_{10} 2$ **108.** $\log_{10} 3$ **109.** $\log_{10} 6$ **110.** $\log_{10} 81$

111. $\log_{10} 50$ **112.** $\log_{10} \sqrt{14}$ **113.** $\log_{10} 0.5$ **114.** $\log_{10} 27$

In Exercises 115–120, verify each identity. Assume that all logarithms are base-10, and that the values of the trigonometric functions are such that the logarithmic expressions are defined.

115. $\log(\sec x) = -\log(\cos x)$

116. $\log(\tan x) + \log(\cot x) = 0$

117. $\log[(1 - \sin^2 x)\csc x \sec^2 x] = -\log(\sin x)$

118. $\log(\sec x) + \log(\csc x) = -\log\left(\dfrac{1}{2}\sin 2x\right)$

119. $\log(\sec x + \tan x) = -\log(\sec x - \tan x)$

120. $\log(\csc x - \cot x) = -\log(\csc x + \cot x)$

II.3   APPLICATIONS OF BASE-10 AND BASE-e LOGARITHMS

Logarithms are necessary to solve many applied problems. In the past mathematicians had to use tables such as Tables C and D in Appendix IV to find logarithms of numbers, but today logarithms are easy to find with a calculator.

For computational purposes, base-10 logarithms, often called **common logarithms**, are the most convenient. In this book, if the base b is not given in the notation $\log_b x$, we will always assume that b is 10; thus,

$$\log x \quad \text{means} \quad \log_{10} x$$

To use a calculator to find $\log 2.34$, for example, we enter the number 2.34 into a calculator and press the ⎡ log ⎤ key. (*Caution*: Some calculators require pressing a ⎡ 2nd ⎤ function key first.) The display on the calculator will read ⎡ .3692158574 ⎤. Thus, $\log 2.34 \approx 0.3692$. Note that this statement is equivalent to the statement $10^{0.3692} \approx 2.34$.

Example 1 Use a calculator to find the value of x in each equation: **a.** $\log 8.75 = x$, **b.** $\log 379 = x$, and **c.** $\log x = -2.1180$.

Solution **a.** To find $\log 8.75$, enter the number 8.75 and press the ⎡ log ⎤ key. The display will read ⎡ 0.942008053 ⎤. Thus, $\log 8.75 \approx 0.9420$.

b. To find $\log 379$, enter the number 379 and press the ⎡ log ⎤ key. The display will read ⎡ 2.57863921 ⎤. Thus, $\log 379 \approx 2.5786$.

c. To find x in the equation $\log x = -2.1180$, change the equation to exponential form:

$$10^{-2.1180} = x$$

To evaluate $10^{-2.1180}$, enter the number 10, press the ⎡ y^x ⎤ key, and enter the number 2.1180. Then press the ⎡ +/− ⎤ key to change 2.1180 to -2.1180, and press the $\boxed{=}$ key. The display will read $\boxed{.0076207901}$. Thus, $x = 10^{-2.1180} \approx 0.0076$.

If your calculator has a $\boxed{10^x}$ key, you can enter 2.1180, press the $\boxed{+/-}$ key, and then press the $\boxed{10^x}$ key. ∎

Applications of Base-10 Logarithms

Common logarithms are used in chemistry to express the acidity of solutions. The more acidic a solution, the greater the concentration of hydrogen ions. This hydrogen ion concentration is indicated indirectly by the **pH scale**, defined by the equation

Formula for
pH Scale

$$pH = -\log[H^+]$$

where $[H^+]$ is the hydrogen ion concentration in gram-ions per liter. Pure water, for example, has some hydrogen ions—approximately 10^{-7} gram-ion per liter. Thus, the pH of pure water is

$$pH = -\log 10^{-7}$$
$$= -(-7)\log 10 \qquad \text{Use Property 7 of logarithms.}$$
$$= 7 \qquad \text{Recall that } \log 10 = 1.$$

Example 2   Seawater has a pH of approximately 8.5. Find its hydrogen ion concentration.

Solution   To find the hydrogen ion concentration of seawater, solve the equation $8.5 = -\log[H^+]$ for $[H^+]$:

$$8.5 = -\log[H^+]$$
$$-8.5 = \log[H^+] \qquad \text{Multiply both sides by } -1.$$
$$[H^+] = 10^{-8.5} \qquad \text{Change the equation to exponential form.}$$
$$[H^+] \approx 3.2 \times 10^{-9} \qquad \text{Use a calculator.}$$

Thus, seawater has a hydrogen ion concentration of approximately 3.2×10^{-9} gram-ions per liter. ∎

Common logarithms are used in electrical engineering to express the voltage gain (or loss) of an electronic device such as an amplifier. The unit of gain (or loss), called the **decibel**, is defined by a logarithmic relation. If E_O is the output voltage of a device and E_I is the input voltage, the decibel voltage gain is given by the formula

Formula for
Decibel Voltage Gain

$$\text{Decibel voltage gain} = 20\log\frac{E_O}{E_I}$$

Example 3   If the input to an amplifier is 0.5 volt and the output is 40 volts, find the decibel voltage gain of the amplifier.

Solution   The decibel voltage gain is calculated by substituting 0.5 for E_I and 40 for E_O in the formula and simplifying:

$$\text{Decibel voltage gain} = 20\log\frac{E_O}{E_I}$$

$$= 20\log\frac{40}{0.5}$$

$$= 20\log 80$$

$$\approx 38 \qquad \text{Use a calculator.}$$

The decibel voltage gain is approximately 38.   ■

The intensity of earthquakes is measured on the **Richter scale**, which is based on the formula

Richter Scale $R = \log\dfrac{A}{P}$

where A is the amplitude of the tremor (measured in micrometers) and P is the period of the tremor (measured in seconds).

Example 4   Find the measure on the Richter scale of an earthquake with an amplitude of 10,000 micrometers and a period of 0.1 second.

Solution   Substitute $10,000$ for A and 0.1 for P in the formula defining the Richter scale and proceed as follows:

$$R = \log\frac{A}{P}$$

$$R = \log\frac{10,000}{0.1}$$

$$= \log 100,000$$

$$= 5$$

The earthquake measures 5 on the Richter scale.   ■

Natural Logarithms

A second important base for logarithms is the number e. Logarithms with a base of e are called **natural logarithms** and are denoted by the symbol $\ln x$ (read as "the natural logarithm of x") rather than the symbol $\log_e x$. Thus, in this book

$$\ln x \quad \text{means} \quad \log_e x$$

Natural logarithms are found by using the $\boxed{\ln x}$ key on a calculator. For example, to find the value of $\ln 25.3$, enter 25.3 and press the $\boxed{\ln x}$ key. The display will read $\boxed{3.230804396}$. Thus, $\ln 25.3 \approx 3.2308$.

Example 5   Use a calculator to find the value of x in each equation: **a.** $\ln 8.75 = x$, **b.** $\ln 379 = x$, and **c.** $\ln x = -2.1180$.

Solution   **a.**   To find $\ln 8.75$, enter the number 8.75 and press the $\boxed{\ln x}$ key. The display will read $\boxed{2.1690537}$. Thus, $x = \ln 8.75 \approx 2.1691$.

b.   To find $\ln 379$, enter the number 379 and press the $\boxed{\ln x}$ key. The display will read $\boxed{5.937536205}$. Thus, $x = \ln 379 \approx 5.9375$.

c.   To find x in the equation $\ln x = -2.1180$, write the equation in exponential form:

$$e^{-2.1180} = x$$

To evaluate $e^{-2.1180}$, enter 2.1180, press the $\boxed{+/-}$ key, and then press the $\boxed{e^x}$ key. The display will read $\boxed{0.120271932}$. Thus, $x = e^{-2.1180} \approx 0.1203$.

∎

Applications of Natural Logarithms

The length of time t that it takes for money to double if left in an account where interest is compounded continuously is given by the formula

$$t = \frac{\ln 2}{r}$$

where r is the annual rate of interest.

Example 6   How long will it take $1000 to double if it is invested at an annual interest rate of 8 percent compounded continuously?

Solution   Substitute **0.08** for r and simplify:

$$t = \frac{\ln 2}{r}$$

$$t = \frac{\ln 2}{0.08}$$

$$\approx 8.6643 \qquad \text{Use a calculator.}$$

It will take approximately $8\frac{2}{3}$ years for the money to double. ∎

The loudness of a sound is not directly proportional to the intensity of the sound. Experiments in physiology suggest that the relationship between loudness and intensity is a logarithmic one known as the **Weber–Fechner Law**. This law states that the apparent loudness L of a sound is directly proportional to the natural logarithm of its intensity I. In symbols,

Weber–Fechner Law $L = k \ln I$

where k is the constant of proportionality.

Example 7 Find the increase in intensity that will cause the apparent loudness of a sound to double.

Solution If the original loudness L_0 is caused by an actual intensity I_0, then

$$\textbf{1.} \quad L_0 = k \ln I_0$$

To double the apparent loudness, multiply both sides of Equation 1 by 2 and apply Property 7 of logarithms:

$$2L_0 = 2k \ln I_0$$
$$= k \ln(I_0)^2$$

Thus, to double the apparent loudness of the sound, the intensity must be squared. ∎

▬ EXERCISE II.3

In Exercises 1–18, use a calculator to find the value of the variable. Express all answers to four decimal places.

1. $\log 3.25 = x$ **2.** $\log 32.1 = x$ **3.** $\log \dfrac{17}{7} = x$ **4.** $\log \dfrac{3}{31} = x$

5. $\log N = 3.29$ **6.** $\log M = -9.17$ **7.** $\log M = -\dfrac{21}{5}$ **8.** $\log N = \dfrac{4}{13}$

9. $\ln 3.9 = y$ **10.** $\ln 0.087 = z$ **11.** $\ln \dfrac{1}{37} = t$ **12.** $\ln 4300 = r$

13. $\ln M = -8.3$ **14.** $\ln N = 0.763$ **15.** $\ln N = 2.83$ **16.** $\ln M = -23.2$

17. $\ln A = \log 7$ **18.** $\log B = \ln 8$

In Exercises 19–32, solve each word problem.

19. Find the pH of a solution with a hydrogen ion concentration of 1.7×10^{-5} gram-ions per liter.

20. The hydrogen ion concentration of sour pickles is 6.31×10^{-4}. What is the pH?

21. What is the hydrogen ion concentration of a saturated solution of calcium hydroxide with pH = 13.2?

22. Find the hydrogen ion concentration of the water in a pond if the pH of the water is 6.5.

23. The safe pH for a freshwater tank containing tropical fish can range from 6.8 to 7.6. What is the corresponding range in hydrogen ion concentration?

24. The pH of apples ranges from 2.9 to 3.3. What is the corresponding range in hydrogen ion concentration?

25. The decibel voltage gain of an amplifier is 29. If the output is 20 volts, what is the input?

26. The decibel voltage gain of an amplifier is 35. If the input voltage is 0.05 volt, what is the output?

27. An amplifier produces an output of 30 volts when driven by an input of 0.1 volt. What is the decibel voltage gain of the amplifier?

28. An amplifier produces an output of 80 volts when driven by an input of 0.12 volt. What is the amplifier's decibel voltage gain?

29. An earthquake has an amplitude of 5000 micrometers and a period of 0.2 second. What does the earthquake measure on the Richter scale?

30. An earthquake with an amplitude of 8000 micrometers measures 6 on the Richter scale. What is its period?

31. An earthquake with a period of 0.25 second measures 4 on the Richter scale. What is the earthquake's amplitude?

32. By what factor must the period of an earthquake change to increase its severity by 1 point on the Richter scale? Assume that the amplitude remains constant.

In Exercises 33 and 34, use the formula

$$t = \frac{\ln 3}{r}$$

which gives the length of time t for an amount of money to triple if left in an account paying an annual rate r, compounded continuously.

33. Find the length of time it will take $2500 to triple if invested in an account paying 9 percent annual interest compounded continuously.

34. Find the length of time it will take $15,500 to triple if invested in an account paying 15 percent annual interest compounded continuously.

In Exercises 35 and 36, use the formula

$$t = \frac{\ln 4}{r}$$

which gives the length of time t for an amount of money to quadruple if left in an account paying an annual rate r, compounded continuously.

35. Find the length of time it will take $1700 to quadruple if it is invested in an account that pays 6 percent annual interest compounded continuously.

36. Find the length of time it will take $50,500 to quadruple if it is invested in an account paying 9 percent annual interest compounded continuously.

In Exercises 37–40, solve each word problem.

37. If the intensity of a sound is doubled, what will be the apparent change in loudness?

38. If the intensity of a sound is tripled, what will be the apparent change in loudness?

39. What increase in intensity will cause an apparent tripling of the loudness of a sound?

40. What increase in the intensity of a sound will cause the apparent loudness to be multiplied by 4?

II.4 EXPONENTIAL AND LOGARITHMIC EQUATIONS

An **exponential equation** is an equation in which the variable appears in an exponent, and a **logarithmic equation** is an equation that involves the logarithm of an expression containing the variable. Some exponential and logarithmic equations are difficult to solve, but others can be solved by using the properties of exponents and logarithms and the usual equation-solving techniques.

Because logarithms of equal positive numbers are equal, we can solve many equations by taking the logarithm of both sides and then applying Property 7 of logarithms to remove the variable from its position as an exponent. To solve the equation $3^x = 5$, for example, we take the common logarithm of both sides of the equation and use Property 7 of logarithms:

$$3^x = 5$$
$$\log 3^x = \log 5$$
$$x \log 3 = \log 5$$

We then divide both sides of the equation by $\log 3$ to obtain

1. $x = \dfrac{\log 5}{\log 3}$

To obtain a decimal approximation, we use a calculator to determine that $\log 5 \approx 0.6990$ and $\log 3 \approx 0.4771$ and then substitute these values into Equation 1 and simplify:

$$x \approx \frac{0.6990}{0.4771}$$
$$\approx 1.465$$

A careless reading of Equation 1 leads to a common error. Because $\log \frac{A}{B} = \log A - \log B$, many students believe that the expression $\frac{\log 5}{\log 3}$ also involves a subtraction. It does not. The expression $\frac{\log 5}{\log 3}$ indicates the quotient that is obtained when $\log 5$ is divided by $\log 3$.

Example 1   Solve the equation $6^{x-3} = 2$.

Solution   Take the common logarithm of both sides of the equation and proceed as follows:

$$6^{x-3} = 2$$
$$\log 6^{x-3} = \log 2$$
$$(x-3)\log 6 = \log 2 \qquad \text{Use Property 7 of logarithms.}$$
$$x - 3 = \frac{\log 2}{\log 6} \qquad \text{Divide both sides by } \log 6.$$
$$x = \frac{\log 2}{\log 6} + 3 \qquad \text{Add 3 to both sides.}$$

To obtain a decimal approximation, use a calculator and press 2 $\boxed{\log}$ $\boxed{\div}$ 6 $\boxed{\log}$ $\boxed{+}$ 3 $\boxed{=}$ to get the following result rounded to four decimal places:

$$x \approx 3.3869$$

If pressing these keys gives you a different result, consult the owner's manual for your calculator. ∎

Example 2   Solve the equation $\log x + \log(x - 3) = 1$.

Solution   Use Property 5 of logarithms to write the left-hand side of the equation as $\log x(x - 3)$ to obtain

$$\log[x(x - 3)] = 1$$

This equation is equivalent to the exponential equation

$$x(x - 3) = 10^1 \qquad \log_{10}[x(x - 3)] = 1 \text{ is equivalent to } 10^1 = x(x - 3).$$

which is a quadratic equation that can be solved by factoring:

$$x(x - 3) = 10^1$$
$$x^2 - 3x - 10 = 0$$
$$(x + 2)(x - 5) = 0$$
$$x = -2 \quad \text{ or } \quad x = 5$$

Because logarithms of negative numbers are not defined, the number -2 is not a solution. However, 5 is a solution because it satisfies the original equation.

Check:

$$\log x + \log(x - 3) = 1$$
$$\log 5 + \log(5 - 3) \stackrel{?}{=} 1$$
$$\log 5 + \log 2 \stackrel{?}{=} 1$$
$$\log 10 \stackrel{?}{=} 1 \qquad \log 5 + \log 2 = \log(5 \cdot 2) = \log 10.$$
$$1 = 1 \qquad \text{Use Property 5 of logarithms.} \qquad \blacksquare$$

Example 3   Solve the equation $\log_b(3x + 2) - \log_b(2x - 3) = 0$.

Solution   Proceed as follows:

$$\log_b(3x + 2) - \log_b(2x - 3) = 0$$
$$\log_b(3x + 2) = \log_b(2x - 3) \qquad \text{Add } \log_b(2x - 3) \text{ to both sides.}$$
$$3x + 2 = 2x - 3 \qquad \text{Use Property 8 of logarithms.}$$
$$x = -5$$

Check:

$$\log_b(3x + 2) - \log_b(2x - 3) = 0$$
$$\log_b[3(-5) + 2] - \log_b[2(-5) - 3] = 0$$
$$\log_b(-13) - \log_b(-13) = 0$$

Because there are no logarithms of negative numbers, the previous equation is meaningless. The original equation has no solutions. $\qquad \blacksquare$

The Change-of-Base Formula

A formula for finding logarithms with different bases can be found by solving the equation $b^x = y$ for the variable x. To produce the desired result, we begin by taking the base-*a* logarithm of both sides:

$$b^x = y$$

$$\log_a b^x = \log_a y$$

$$x \log_a b = \log_a y \qquad \text{Use Property 7 of logarithms.}$$

$$\mathbf{2.} \qquad x = \frac{\log_a y}{\log_a b}$$

Because the original equation $b^x = y$ is equivalent to the equation $x = \log_b y$, we can substitute $\log_b y$ for x in Equation 2 to obtain the **change-of-base formula**:

The Change-of-Base Formula.

$$\log_b y = \frac{\log_a y}{\log_a b}$$

If we know the logarithm of some number y to some base a (for example, $a = 10$), we can then use the change-of-base formula to find the logarithm of y to a different base b. To do so, we divide the base-*a* logarithm of y by the base-*a* logarithm of the new base b.

Example 4 Use the change-of-base formula to find $\log_3 5$.

Solution Substitute 3 for b, 5 for y, and 10 for a in the change-of-base formula and simplify:

$$\log_b y = \frac{\log_a y}{\log_a b}$$

$$\log_3 5 = \frac{\log_{10} 5}{\log_{10} 3}$$

$$\approx \frac{0.6990}{0.4771} \qquad \log_{10} 5 \approx 0.6990 \text{ and } \log_{10} 3 \approx 0.4771.$$

$$\approx 1.465 \qquad \qquad \blacksquare$$

Applications of Exponential and Logarithmic Equations

The atomic structure of radioactive material changes as the material emits radiation. Uranium, for example, decays into thorium, then into radium, and eventually into lead. Experiments have determined the time for half of a given amount of radioactive material to decompose. This time, called the **half-life**, is constant for any given substance. The amount A of radioactive material present decays exponentially according to the model

Formula for Radioactive Decay
$$A = A_0 2^{t/h}$$

where A_0 is the amount present at time $t = 0$ and h is the half-life of the material.

Example 5   How long will it take 1 gram of radium, with a half-life of 1600 years, to decompose to 0.75 gram?

Solution   Substitute **1** for A_0, **0.75** for A, and **1600** for h in the formula for radioactive decay, and solve for t:

$$A = A_0 2^{-t/h}$$
$$0.75 = 1 \cdot 2^{-t/1600}$$

Take the common logarithm of both sides to obtain

$$\log 0.75 = \log 2^{-t/1600}$$

$$\log 0.75 = -\frac{t}{1600} \log 2 \qquad \text{Use Property 7 of logarithms.}$$

$$t = -1600 \frac{\log 0.75}{\log 2}$$

$$t \approx 664$$

It will take approximately 664 years for 1 gram of radium to decompose to 0.75 gram.   ∎

When a living organism dies, the oxygen/carbon dioxide cycle common to all living things stops, and carbon-14 (^{14}C), a radioactive isotope of carbon that is present in the atmosphere, is no longer absorbed. Because the half-life of ^{14}C is about 5700 years, a wooden artifact with half of the radioactivity of wood from a living tree would be 5700 years old. Carbon dating is a common technique used by archaeologists to date objects as old as 25,000 years.

Example 6   How old is a wooden statue if it contains one-third as much carbon-14 as a living tree?

Solution   Substitute $\frac{1}{3}A_0$ for A and **5700** for h in the formula for radioactive decay, and solve for t:

$$A = A_0 2^{-t/h}$$
$$\frac{1}{3}A_0 = A_0 2^{-t/5700}$$

$$\frac{1}{3} = 2^{-t/5700} \qquad \text{Divide both sides by } A_0.$$

Take the common logarithm of both sides of the equation, and solve for t:

$$\log \frac{1}{3} = \log 2^{-t/5700}$$

$$\log 1 - \log 3 = -\frac{t}{5700} \log 2 \qquad \text{Use Properties 6 and 7 of logarithms.}$$

$$-\log 3 = -\frac{t}{5700} \log 2 \qquad \log 1 = 0$$

$$t = 5700 \frac{\log 3}{\log 2}$$

$$\approx 9034.29$$

To the nearest thousand years, the statue is approximately 9000 years old. ∎

While there is a sufficient food supply and space to grow, populations of living things tend to increase according to the Malthusian model of population growth:

Formula for
Population Growth

$$P = P_0 e^{kt}$$

where P_0 is the initial population at time $t = 0$ and k depends on the rate of growth.

Example 7   A bacteria culture increased from an initial population of 500 to a population of 1500 in 3 hours. How long will it take the population to reach 10,000?

Solution   Substitute 1500 for P, 500 for P_0, and 3 for t in the formula for population growth and solve for k.

$$P = P_0 e^{kt}$$

$$1500 = 500 e^{k \cdot 3}$$

$$3 = e^{3k}$$

Because the base of e^{3k} is e, take the natural logarithm of both sides and simplify to obtain

$$\ln 3 = \ln e^{3k}$$

$$\ln 3 = 3k \ln e \qquad \text{Use Property 7 of logarithms.}$$

$$\ln 3 = 3k \qquad\quad \ln e = 1$$

$$k = \frac{\ln 3}{3}$$

To find when the population will reach 10,000, substitute 10,000 for P, 500 for P_0, and $\frac{\ln 3}{3}$ for k in the equation for population growth, and solve for t:

$$P = P_0 e^{kt}$$

$$10,000 = 500 e^{[(\ln 3)/3]t}$$

$$20 = e^{[(\ln 3)/3]t} \qquad\qquad \text{Divide both sides by 500.}$$

$$\ln 20 = \ln e^{[(\ln 3)/3]t} \qquad\quad \text{Take the natural logarithm of both sides.}$$

$$\ln 20 = \frac{\ln 3}{3} t \ln e \qquad\qquad \text{Use Property 7 of logarithms.}$$

$$t = \frac{3 \ln 20}{\ln 3} \qquad\qquad\quad \text{Remember that } \ln e = 1.$$

$$\approx 8.1805$$

In approximately 8 hours the culture will have a population of 10,000. ∎

■ **EXERCISE II.4** ■

In Exercises 1–16, solve each equation for x.

1. $4^x = 5$ **2.** $7^x = 12$ **3.** $2^{x-1} = 10$ **4.** $3^{x+1} = 100$

5. $2^{x+1} = 2^{2x+3}$ **6.** $3^{3x+2} = 3^{2x+7}$ **7.** $10^{x^2} = 100$ **8.** $10^{x^2} = 1000$

9. $3^x = 4^x$ **10.** $3^{2x} = 4^x$ **11.** $3^{x+1} = 4^x$ **12.** $3^x - 4^{x+1} = 0$

13. $10^{\sqrt{x}} = 1000$ **14.** $100^{\sqrt{x}} = 10$ **15.** $10^{\sqrt[3]{x}} - 100 = 0$ **16.** $1000^{\sqrt[3]{x}} - 10 = 0$

In Exercises 17–38, solve each equation for x.

17. $\log 2x = \log 4$ **18.** $\log 3x = \log 9$

19. $\log(3x + 1) = \log(x + 7)$ **20.** $\log(x^2 + 4x) = \log(x^2 + 16)$

21. $\log(2x - 3) - \log(x + 4) = 0$ **22.** $\log(3x + 5) - \log(2x + 6) = 0$

23. $\log\dfrac{4x + 1}{2x + 9} = 0$ **24.** $\log\dfrac{5x + 2}{2(x + 7)} = 0$

25. $\log x^2 = 2$ **26.** $\log x^3 = 3$

27. $\log x + \log(x - 48) = 2$ **28.** $\log x + \log(x + 9) = 1$

29. $\log x + \log(x - 15) = 2$ **30.** $\log x + \log(x + 21) = 2$

31. $\log(x + 90) = 3 - \log x$ **32.** $\log(x - 90) = 3 - \log x$

33. $\log(x - 6) - \log(x - 2) = \log\dfrac{5}{x}$ **34.** $\log(x - 1) - \log 6 = \log(x - 2) - \log x$

35. $\log_7(2x - 3) - \log_7(x - 1) = 0$ **36.** $\log x^2 = (\log x)^2$

37. $\log(\log x) = 1$ **38.** $\log_3 x = \log_3\dfrac{1}{x} + 4$

In Exercises 39–46, use the change-of-base formula to find each logarithm.

39. $\log_3 7$ **40.** $\log_7 3$ **41.** $\log_{1/3} 3$ **42.** $\log_{1/2} 6$

43. $\log_8 \sqrt{2}$ **44.** $\log_6 \sqrt{6}$ **45.** $\log_\pi \sqrt{2}$ **46.** $\log_{\sqrt{2}} \pi$

47. Show that $\log_{b^2} x = \dfrac{1}{2}\log_b x$.   **48.** Show that $\log_b b^x = x$.

49. Show that $b^{\log_b x} = x$.   **50.** Show that $e^{x \ln a} = a^x$.

In Exercises 51–62, solve each word problem.

51. The half-life of tritium is 12.4 years. How long will it take for 25 percent of a sample of tritium to decompose?

52. Twenty percent of a newly discovered radioactive element decays in 2 years. What is its half-life?

53. An isotope of thorium, ^{227}Th, has a half-life of 18.4 days. How long will it take 80 percent of a sample to decompose?

54. An isotope of lead, ^{201}Pb, has a half-life of 8.4 hours. How many hours ago was there 30 percent more of the substance?

55. A parchment fragment is found in a newly discovered ancient tomb. It contains 60 percent of the ^{14}C it is assumed to have had initially. Approximately how old is the fragment?

56. Only 10 percent of the ^{14}C in a small wooden bowl remains. How old is the bowl?

57. A bacteria culture doubles in size every 24 hours. By how much will it have increased in 36 hours?

58. A bacteria culture triples in size every 3 days. By what factor will it have increased in 1 day?

59. The population of Waving Grass, Kansas, is presently 140 and is expected to triple every 15 years. When can the city planners expect the population to be double the present census?

60. The rodent population in a city is currently estimated at 20,000. It is expected to double every 5 years. When will the population reach 1 million?

61. An initial deposit of p dollars in a bank that pays 12 percent annual interest, compounded annually, will grow to an amount A in t years, where

$$A = p(1.12)^t$$

How long will it take $1000 to increase to $3000?

62. An initial deposit of p dollars in a bank that pays 12 percent annual interest, compounded monthly, will grow to an amount A in t years, where

$$A = p(1.01)^{12t}$$

How long will it take $1000 to increase to $3000?

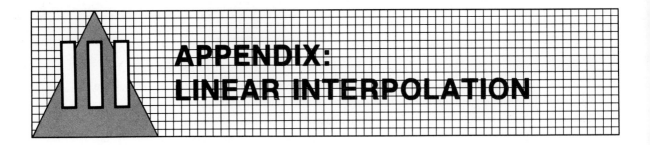
Before the widespread use of pocket calculators, people had to rely on published tables such as Table A in Appendix IV to find the values of the trigonometric functions. However, Table A includes angles that are measured to the nearest tenth of a degree only. Fortunately, there is a method, called **linear interpolation**, that lets us calculate good estimates of the values of the functions for angles measured to the nearest hundredth.

The method of linear interpolation is based on the fact that the graph of a trigonometric curve appears straight when we look at only a very small part of it. For example, the sine curve appears linear between points A and B in Figure III-1. If AB is assumed to *be* straight, we can set up proportions involving points A and B. We illustrate with an example.

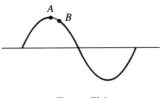

Figure III-1

Example 1   Use Table A in Appendix IV to find $\sin 33.23°$.

Solution   As expected, Table A does not give the value of $\sin 33.23°$. It does, however, list values for $\sin 33.20°$ and $\sin 33.30°$.

$$10\left(3\left(\begin{matrix}\sin 33.20° & \dots\dots\dots\dots & 0.5476 \\ \sin 33.23° & \dots\dots\dots\dots & ? \end{matrix}\right)x \\ \sin 33.30° \quad \dots\dots\dots\dots \quad 0.5490 \right)14$$

Note that the difference between 33.20 and 33.30 is 10 hundredths and that the difference between 33.20 and 33.23 is 3 hundredths. Note also that the difference between 0.5476 and 0.5490 is 14 ten-thousandths. Make use of the assumption of linearity and note that because 33.23 is $\frac{3}{10}$ of the way from 33.20 to 33.30, $\sin 33.23°$ must be $\frac{3}{10}$ of the way from 0.5476 to 0.5490. This can be expressed with the equation

$$x = \frac{3}{10}(14) \qquad \text{or with the proportion} \qquad \frac{x}{14} = \frac{3}{10}$$

Solve for x.

$$10x = 42$$
$$x = 4.2$$

To get a good estimate of sin 33.23°, you must add 4 ten-thousandths to 0.5476.

$$\sin 33.23° \approx 0.5476 + 0.0004 = 0.5480$$

Verify this result with a calculator. ∎

Example 2   Find cos 15.28°.

Solution   Look up cos 15.20° and cos 15.30° in Table A, compute the differences, and set up a proportion.

$$10\left(8\left(\begin{matrix} \cos 15.20° & \cdots\cdots\cdots\cdots & 0.9650 \\ \cos 15.28° & \cdots\cdots\cdots\cdots & ? \\ \cos 15.30° & \cdots\cdots\cdots\cdots & 0.9646 \end{matrix}\right)x\right)4$$

$$\frac{8}{10} = \frac{x}{4}$$
$$10x = 32$$
$$x = 3.2$$

This time, 3 ten-thousandths must be subtracted from 0.9650.

$$\cos 15.28° \approx 0.9650 - 0.0003 = 0.9647$$

Note that your calculator gives a value of 0.9646. This illustrates that the method of linear interpolation only gives a good estimate. ∎

Example 3   If $\tan\theta = 4.598$, find θ.

Solution   Look in Table A to find 4.598 in the Tan column. Although it does not appear, the numbers 4.586 and 4.625 do.

$$39\left(12\left(\begin{matrix} 4.586 & \cdots\cdots\cdots\cdots & \tan 77.70° \\ 4.598 & \cdots\cdots\cdots\cdots & \tan\theta \\ 4.625 & \cdots\cdots\cdots\cdots & \tan 77.80° \end{matrix}\right)x\right)10$$

Once again, set up a proportion.

$$\frac{12}{39} = \frac{x}{10}$$
$$39x = 120$$
$$x = 3.07$$

To find θ, add 0.03 to 77.70.

$$\theta \approx 77.70° + 0.03° = 77.73°$$ ∎

■ EXERCISE III.1

In Exercises 1–6, use linear interpolation to estimate the required values. Give answers to the nearest thousandth.

1. $\sin 23.35°$ 2. $\sin 83.17°$ 3. $\cos 81.13°$ 4. $\cos 13.21°$

5. $\tan 48.14°$ 6. $\tan 88.23°$

In Exercises 7–12, use linear interpolation to estimate angle θ, $0° \leq \theta < 90°$. Give answers to the nearest hundredth of a degree.

7. $\sin \theta = 0.4444$ 8. $\sin \theta = 0.1234$ 9. $\cos \theta = 0.9876$ 10. $\cos \theta = 0.8642$

11. $\tan \theta = 0.2342$ 12. $\tan \theta = 7.555$

13. Jeff wanted to find $\sin 75°$. So he used linear interpolation.

$$\sin 0° = 0$$
$$\sin 75° = ?$$
$$\sin 150° = 0.5$$

He concluded that $\sin 75° = 0.25$. However, a calculator shows that $\sin 75° \approx 0.97$. What went wrong?

14. Martha wanted to find $\cos 45°$. So she used linear interpolation.

$$\cos 0° = 1$$
$$\cos 45° = ?$$
$$\cos 90° = 0$$

She concluded that $\cos 45° = 0.5$. However, a calculator shows that $\cos 45° \approx 0.7071$. What went wrong?

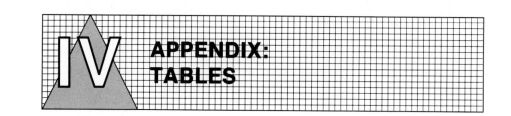

APPENDIX: TABLES

Table A (Continued)

Radians	Degrees	Sin	Cos	Tan	Cot	Degrees	Radians
.0873	5.0°	.0872	.9962	.0875	11.43	85.0°	1.4835
.0890	5.1°	.0889	.9960	.0892	11.20	84.9°	1.4818
.0908	5.2°	.0906	.9959	.0910	10.99	84.8°	1.4800
.0925	5.3°	.0924	.9957	.0928	10.78	84.7°	1.4783
.0942	5.4°	.0941	.9956	.0945	10.58	84.6°	1.4765
.0960	5.5°	.0958	.9954	.0963	10.39	84.5°	1.4748
.0977	5.6°	.0976	.9952	.0981	10.20	84.4°	1.4731
.0995	5.7°	.0993	.9951	.0998	10.02	84.3°	1.4713
.1012	5.8°	.1011	.9949	.1016	9.845	84.2°	1.4696
.1030	5.9°	.1028	.9947	.1033	9.677	84.1°	1.4678
.1047	6.0°	.1045	.9945	.1051	9.514	84.0°	1.4661
.1065	6.1°	.1063	.9943	.1069	9.357	83.9°	1.4643
.1082	6.2°	.1080	.9942	.1086	9.205	83.8°	1.4626
.1100	6.3°	.1097	.9940	.1104	9.058	83.7°	1.4608
.1117	6.4°	.1115	.9938	.1122	8.915	83.6°	1.4591
.1134	6.5°	.1132	.9936	.1139	8.777	83.5°	1.4573
.1152	6.6°	.1149	.9934	.1157	8.643	83.4°	1.4556
.1169	6.7°	.1167	.9932	.1175	8.513	83.3°	1.4539
.1187	6.8°	.1184	.9930	.1192	8.386	83.2°	1.4521
.1204	6.9°	.1201	.9928	.1210	8.264	83.1°	1.4504
.1222	7.0°	.1219	.9925	.1228	8.144	83.0°	1.4486
.1239	7.1°	.1236	.9923	.1246	8.028	82.9°	1.4469
.1257	7.2°	.1253	.9921	.1263	7.916	82.8°	1.4451
.1274	7.3°	.1271	.9919	.1281	7.806	82.7°	1.4434
.1292	7.4°	.1288	.9917	.1299	7.700	82.6°	1.4416
.1309	7.5°	.1305	.9914	.1317	7.596	82.5°	1.4399
.1326	7.6°	.1323	.9912	.1334	7.495	82.4°	1.4382
.1344	7.7°	.1340	.9910	.1352	7.396	82.3°	1.4364
.1361	7.8°	.1357	.9907	.1370	7.300	82.2°	1.4347
.1379	7.9°	.1374	.9905	.1388	7.207	82.1°	1.4329
.1396	8.0°	.1392	.9903	.1405	7.115	82.0°	1.4312
.1414	8.1°	.1409	.9900	.1423	7.026	81.9°	1.4294
.1431	8.2°	.1426	.9898	.1441	6.940	81.8°	1.4277
.1449	8.3°	.1444	.9895	.1459	6.855	81.7°	1.4259
.1466	8.4°	.1461	.9893	.1477	6.772	81.6°	1.4242
.1484	8.5°	.1478	.9890	.1495	6.691	81.5°	1.4224
.1501	8.6°	.1495	.9888	.1512	6.612	81.4°	1.4207
.1518	8.7°	.1513	.9885	.1530	6.535	81.3°	1.4190
.1536	8.8°	.1530	.9882	.1548	6.460	81.2°	1.4172
.1553	8.9°	.1547	.9880	.1566	6.386	81.1°	1.4155
.1571	9.0°	.1564	.9877	.1584	6.314	81.0°	1.4137
.1588	9.1°	.1582	.9874	.1602	6.243	80.9°	1.4120
.1606	9.2°	.1599	.9871	.1620	6.174	80.8°	1.4102
.1623	9.3°	.1616	.9869	.1638	6.107	80.7°	1.4085
.1641	9.4°	.1633	.9866	.1655	6.041	80.6°	1.4067
.1658	9.5°	.1650	.9863	.1673	5.976	80.5°	1.4050
.1676	9.6°	.1668	.9860	.1691	5.912	80.4°	1.4032
.1693	9.7°	.1685	.9857	.1709	5.850	80.3°	1.4015
.1710	9.8°	.1702	.9854	.1727	5.789	80.2°	1.3998
.1728	9.9°	.1719	.9851	.1745	5.730	80.1°	1.3980
		Cos	Sin	Cot	Tan	Degrees	Radians

Table A Values of the Trigonometric Functions

Radians	Degrees	Sin	Cos	Tan	Cot	Degrees	Radians
.0000	0°	.0000	1.0000	.0000	—	90.0°	1.5708
.0017	.1°	.0017	1.0000	.0017	573.0	89.9°	1.5691
.0035	.2°	.0035	1.0000	.0035	286.5	89.8°	1.5673
.0052	.3°	.0052	1.0000	.0052	191.0	89.7°	1.5656
.0070	.4°	.0070	1.0000	.0070	143.2	89.6°	1.5638
.0087	.5°	.0087	1.0000	.0087	114.6	89.5°	1.5621
.0105	.6°	.0105	.9999	.0105	95.49	89.4°	1.5603
.0122	.7°	.0122	.9999	.0122	81.85	89.3°	1.5586
.0140	.8°	.0140	.9999	.0140	71.62	89.2°	1.5568
.0157	.9°	.0157	.9999	.0157	63.66	89.1°	1.5551
.0175	1.0°	.0175	.9998	.0175	57.29	89.0°	1.5533
.0192	1.1°	.0192	.9998	.0192	52.08	88.9°	1.5516
.0209	1.2°	.0209	.9998	.0209	47.74	88.8°	1.5499
.0227	1.3°	.0227	.9997	.0227	44.07	88.7°	1.5481
.0244	1.4°	.0244	.9997	.0244	40.92	88.6°	1.5464
.0262	1.5°	.0262	.9997	.0262	38.19	88.5°	1.5446
.0279	1.6°	.0279	.9996	.0279	35.80	88.4°	1.5429
.0297	1.7°	.0297	.9996	.0297	33.69	88.3°	1.5411
.0314	1.8°	.0314	.9995	.0314	31.82	88.2°	1.5394
.0332	1.9°	.0332	.9995	.0332	30.14	88.1°	1.5376
.0349	2.0°	.0349	.9994	.0349	28.64	88.0°	1.5359
.0367	2.1°	.0366	.9993	.0367	27.27	87.9°	1.5341
.0384	2.2°	.0384	.9993	.0384	26.03	87.8°	1.5324
.0401	2.3°	.0401	.9992	.0402	24.90	87.7°	1.5307
.0419	2.4°	.0419	.9991	.0419	23.86	87.6°	1.5289
.0436	2.5°	.0436	.9990	.0437	22.90	87.5°	1.5272
.0454	2.6°	.0454	.9990	.0454	22.02	87.4°	1.5254
.0471	2.7°	.0471	.9989	.0472	21.20	87.3°	1.5237
.0489	2.8°	.0488	.9988	.0489	20.45	87.2°	1.5219
.0506	2.9°	.0506	.9987	.0507	19.74	87.1°	1.5202
.0524	3.0°	.0523	.9986	.0524	19.08	87.0°	1.5184
.0541	3.1°	.0541	.9985	.0542	18.46	86.9°	1.5167
.0559	3.2°	.0558	.9984	.0559	17.89	86.8°	1.5149
.0576	3.3°	.0576	.9983	.0577	17.34	86.7°	1.5132
.0593	3.4°	.0593	.9982	.0594	16.83	86.6°	1.5115
.0611	3.5°	.0610	.9981	.0612	16.35	86.5°	1.5097
.0628	3.6°	.0628	.9980	.0629	15.89	86.4°	1.5080
.0646	3.7°	.0645	.9979	.0647	15.46	86.3°	1.5062
.0663	3.8°	.0663	.9978	.0664	15.06	86.2°	1.5045
.0681	3.9°	.0680	.9977	.0682	14.67	86.1°	1.5027
.0698	4.0°	.0698	.9976	.0699	14.30	86.0°	1.5010
.0716	4.1°	.0715	.9974	.0717	13.95	85.9°	1.4992
.0733	4.2°	.0732	.9973	.0734	13.62	85.8°	1.4975
.0750	4.3°	.0750	.9972	.0752	13.30	85.7°	1.4957
.0768	4.4°	.0767	.9971	.0769	13.00	85.6°	1.4940
.0785	4.5°	.0785	.9969	.0787	12.71	85.5°	1.4923
.0803	4.6°	.0802	.9968	.0805	12.43	85.4°	1.4905
.0820	4.7°	.0819	.9966	.0822	12.16	85.3°	1.4888
.0838	4.8°	.0837	.9965	.0840	11.91	85.2°	1.4870
.0855	4.9°	.0854	.9963	.0857	11.66	85.1°	1.4853
		Cos	Sin	Cot	Tan	Degrees	Radians

Table A (Continued)

Radians	Degrees	Sin	Cos	Tan	Cot	Degrees	Radians
.1745	10.0°	.1736	.9848	.1763	5.671	80.0°	1.3963
.1763	10.1°	.1754	.9845	.1781	5.614	79.9°	1.3945
.1780	10.2°	.1771	.9842	.1799	5.558	79.8°	1.3928
.1798	10.3°	.1788	.9839	.1817	5.503	79.7°	1.3910
.1815	10.4°	.1805	.9836	.1835	5.449	79.6°	1.3893
.1833	10.5°	.1822	.9833	.1853	5.396	79.5°	1.3875
.1850	10.6°	.1840	.9829	.1871	5.343	79.4°	1.3858
.1868	10.7°	.1857	.9826	.1890	5.292	79.3°	1.3840
.1885	10.8°	.1874	.9823	.1908	5.242	79.2°	1.3823
.1902	10.9°	.1891	.9820	.1926	5.193	79.1°	1.3806
.1920	11.0°	.1908	.9816	.1944	5.145	79.0°	1.3788
.1937	11.1°	.1925	.9813	.1962	5.097	78.9°	1.3771
.1955	11.2°	.1942	.9810	.1980	5.050	78.8°	1.3753
.1972	11.3°	.1959	.9806	.1998	5.005	78.7°	1.3736
.1990	11.4°	.1977	.9803	.2016	4.959	78.6°	1.3718
.2007	11.5°	.1994	.9799	.2035	4.915	78.5°	1.3701
.2025	11.6°	.2011	.9796	.2053	4.872	78.4°	1.3683
.2042	11.7°	.2028	.9792	.2071	4.829	78.3°	1.3666
.2059	11.8°	.2045	.9789	.2089	4.787	78.2°	1.3648
.2077	11.9°	.2062	.9785	.2107	4.745	78.1°	1.3631
.2094	12.0°	.2079	.9781	.2126	4.705	78.0°	1.3614
.2112	12.1°	.2096	.9778	.2144	4.665	77.9°	1.3596
.2129	12.2°	.2113	.9774	.2162	4.625	77.8°	1.3579
.2147	12.3°	.2130	.9770	.2180	4.586	77.7°	1.3561
.2164	12.4°	.2147	.9767	.2199	4.548	77.6°	1.3544
.2182	12.5°	.2164	.9763	.2217	4.511	77.5°	1.3526
.2199	12.6°	.2181	.9759	.2235	4.474	77.4°	1.3509
.2217	12.7°	.2198	.9755	.2254	4.437	77.3°	1.3491
.2234	12.8°	.2215	.9751	.2272	4.402	77.2°	1.3474
.2251	12.9°	.2233	.9748	.2290	4.366	77.1°	1.3456
.2269	13.0°	.2250	.9744	.2309	4.331	77.0°	1.3439
.2286	13.1°	.2267	.9740	.2327	4.297	76.9°	1.3422
.2304	13.2°	.2284	.9736	.2345	4.264	76.8°	1.3404
.2321	13.3°	.2300	.9732	.2364	4.230	76.7°	1.3387
.2339	13.4°	.2317	.9728	.2382	4.198	76.6°	1.3369
.2356	13.5°	.2334	.9724	.2401	4.165	76.5°	1.3352
.2374	13.6°	.2351	.9720	.2419	4.134	76.4°	1.3334
.2391	13.7°	.2368	.9715	.2438	4.102	76.3°	1.3317
.2409	13.8°	.2385	.9711	.2456	4.071	76.2°	1.3299
.2426	13.9°	.2402	.9707	.2475	4.041	76.1°	1.3282
.2443	14.0°	.2419	.9703	.2493	4.011	76.0°	1.3265
.2461	14.1°	.2436	.9699	.2512	3.981	75.9°	1.3247
.2478	14.2°	.2453	.9694	.2530	3.952	75.8°	1.3230
.2496	14.3°	.2470	.9690	.2549	3.923	75.7°	1.3212
.2513	14.4°	.2487	.9686	.2568	3.895	75.6°	1.3195
.2531	14.5°	.2504	.9681	.2586	3.867	75.5°	1.3177
.2548	14.6°	.2521	.9677	.2605	3.839	75.4°	1.3160
.2566	14.7°	.2538	.9673	.2623	3.812	75.3°	1.3142
.2583	14.8°	.2554	.9668	.2642	3.785	75.2°	1.3125
.2601	14.9°	.2571	.9664	.2661	3.758	75.1°	1.3107
		Cos	**Sin**	**Cot**	**Tan**	**Degrees**	**Radians**

Table A (Continued)

Radians	Degrees	Sin	Cos	Tan	Cot	Degrees	Radians
.2618	15.0°	.2588	.9659	.2679	3.732	75.0°	1.3090
.2635	15.1°	.2605	.9655	.2698	3.706	74.9°	1.3073
.2653	15.2°	.2622	.9650	.2717	3.681	74.8°	1.3055
.2670	15.3°	.2639	.9646	.2736	3.655	74.7°	1.3038
.2688	15.4°	.2656	.9641	.2754	3.630	74.6°	1.3020
.2705	15.5°	.2672	.9636	.2773	3.606	74.5°	1.3003
.2723	15.6°	.2689	.9632	.2792	3.582	74.4°	1.2985
.2740	15.7°	.2706	.9627	.2811	3.558	74.3°	1.2968
.2758	15.8°	.2723	.9622	.2830	3.534	74.2°	1.2950
.2775	15.9°	.2740	.9617	.2849	3.511	74.1°	1.2933
.2793	16.0°	.2756	.9613	.2867	3.487	74.0°	1.2915
.2810	16.1°	.2773	.9608	.2886	3.465	73.9°	1.2898
.2827	16.2°	.2790	.9603	.2905	3.442	73.8°	1.2881
.2845	16.3°	.2807	.9598	.2924	3.420	73.7°	1.2863
.2862	16.4°	.2823	.9593	.2943	3.398	73.6°	1.2846
.2880	16.5°	.2840	.9588	.2962	3.376	73.5°	1.2828
.2897	16.6°	.2857	.9583	.2981	3.354	73.4°	1.2811
.2915	16.7°	.2874	.9578	.3000	3.333	73.3°	1.2793
.2932	16.8°	.2890	.9573	.3019	3.312	73.2°	1.2776
.2950	16.9°	.2907	.9568	.3038	3.291	73.1°	1.2758
.2967	17.0°	.2924	.9563	.3057	3.271	73.0°	1.2741
.2985	17.1°	.2940	.9558	.3076	3.251	72.9°	1.2723
.3002	17.2°	.2957	.9553	.3096	3.230	72.8°	1.2706
.3019	17.3°	.2974	.9548	.3115	3.211	72.7°	1.2689
.3037	17.4°	.2990	.9542	.3134	3.191	72.6°	1.2671
.3054	17.5°	.3007	.9537	.3153	3.172	72.5°	1.2654
.3072	17.6°	.3024	.9532	.3172	3.152	72.4°	1.2636
.3089	17.7°	.3040	.9527	.3191	3.133	72.3°	1.2619
.3107	17.8°	.3057	.9521	.3211	3.115	72.2°	1.2601
.3124	17.9°	.3074	.9516	.3230	3.096	72.1°	1.2584
.3142	18.0°	.3090	.9511	.3249	3.078	72.0°	1.2566
.3159	18.1°	.3107	.9505	.3269	3.060	71.9°	1.2549
.3176	18.2°	.3123	.9500	.3288	3.042	71.8°	1.2531
.3194	18.3°	.3140	.9494	.3307	3.024	71.7°	1.2514
.3211	18.4°	.3156	.9489	.3327	3.006	71.6°	1.2497
.3229	18.5°	.3173	.9483	.3346	2.989	71.5°	1.2479
.3246	18.6°	.3190	.9478	.3365	2.971	71.4°	1.2462
.3264	18.7°	.3206	.9472	.3385	2.954	71.3°	1.2444
.3281	18.8°	.3223	.9466	.3404	2.937	71.2°	1.2427
.3299	18.9°	.3239	.9461	.3424	2.921	71.1°	1.2409
.3316	19.0°	.3256	.9455	.3443	2.904	71.0°	1.2392
.3334	19.1°	.3272	.9449	.3463	2.888	70.9°	1.2374
.3351	19.2°	.3289	.9444	.3482	2.872	70.8°	1.2357
.3368	19.3°	.3305	.9438	.3502	2.856	70.7°	1.2339
.3386	19.4°	.3322	.9432	.3522	2.840	70.6°	1.2322
.3403	19.5°	.3338	.9426	.3541	2.824	70.5°	1.2305
.3421	19.6°	.3355	.9421	.3561	2.808	70.4°	1.2287
.3438	19.7°	.3371	.9415	.3581	2.793	70.3°	1.2270
.3456	19.8°	.3387	.9409	.3600	2.778	70.2°	1.2252
.3473	19.9°	.3404	.9403	.3620	2.762	70.1°	1.2235
		Cos	**Sin**	**Cot**	**Tan**	**Degrees**	**Radians**

Table A (Continued)

Radians	Degrees	Sin	Cos	Tan	Cot	Degrees	Radians
.3491	20.0°	.3420	.9397	.3640	2.747	70.0°	1.2217
.3508	20.1°	.3437	.9391	.3659	2.733	69.9°	1.2200
.3526	20.2°	.3453	.9385	.3679	2.718	69.8°	1.2182
.3543	20.3°	.3469	.9379	.3699	2.703	69.7°	1.2165
.3560	20.4°	.3486	.9373	.3719	2.689	69.6°	1.2147
.3578	20.5°	.3502	.9367	.3739	2.675	69.5°	1.2130
.3595	20.6°	.3518	.9361	.3759	2.660	69.4°	1.2113
.3613	20.7°	.3535	.9354	.3779	2.646	69.3°	1.2095
.3630	20.8°	.3551	.9348	.3799	2.633	69.2°	1.2078
.3648	20.9°	.3567	.9342	.3819	2.619	69.1°	1.2060
.3665	21.0°	.3584	.9336	.3839	2.605	69.0°	1.2043
.3683	21.1°	.3600	.9330	.3859	2.592	68.9°	1.2025
.3700	21.2°	.3616	.9323	.3879	2.578	68.8°	1.2008
.3718	21.3°	.3633	.9317	.3899	2.565	68.7°	1.1990
.3735	21.4°	.3649	.9311	.3919	2.552	68.6°	1.1973
.3752	21.5°	.3665	.9304	.3939	2.539	68.5°	1.1956
.3770	21.6°	.3681	.9298	.3959	2.526	68.4°	1.1938
.3787	21.7°	.3697	.9291	.3979	2.513	68.3°	1.1921
.3805	21.8°	.3714	.9285	.4000	2.500	68.2°	1.1903
.3822	21.9°	.3730	.9278	.4020	2.488	68.1°	1.1886
.3840	22.0°	.3746	.9272	.4040	2.475	68.0°	1.1868
.3857	22.1°	.3762	.9265	.4061	2.463	67.9°	1.1851
.3875	22.2°	.3778	.9259	.4081	2.450	67.8°	1.1833
.3892	22.3°	.3795	.9252	.4101	2.438	67.7°	1.1816
.3910	22.4°	.3811	.9245	.4122	2.426	67.6°	1.1798
.3927	22.5°	.3827	.9239	.4142	2.414	67.5°	1.1781
.3944	22.6°	.3843	.9232	.4163	2.402	67.4°	1.1764
.3962	22.7°	.3859	.9225	.4183	2.391	67.3°	1.1746
.3979	22.8°	.3875	.9219	.4204	2.379	67.2°	1.1729
.3997	22.9°	.3891	.9212	.4224	2.367	67.1°	1.1711
.4014	23.0°	.3907	.9205	.4245	2.356	67.0°	1.1694
.4032	23.1°	.3923	.9198	.4265	2.344	66.9°	1.1676
.4049	23.2°	.3939	.9191	.4286	2.333	66.8°	1.1659
.4067	23.3°	.3955	.9184	.4307	2.322	66.7°	1.1641
.4084	23.4°	.3971	.9178	.4327	2.311	66.6°	1.1624
.4102	23.5°	.3987	.9171	.4348	2.300	66.5°	1.1606
.4119	23.6°	.4003	.9164	.4369	2.289	66.4°	1.1589
.4136	23.7°	.4019	.9157	.4390	2.278	66.3°	1.1572
.4154	23.8°	.4035	.9150	.4411	2.267	66.2°	1.1554
.4171	23.9°	.4051	.9143	.4431	2.257	66.1°	1.1537
.4189	24.0°	.4067	.9135	.4452	2.246	66.0°	1.1519
.4206	24.1°	.4083	.9128	.4473	2.236	65.9°	1.1502
.4224	24.2°	.4099	.9121	.4494	2.225	65.8°	1.1484
.4241	24.3°	.4115	.9114	.4515	2.215	65.7°	1.1467
.4259	24.4°	.4131	.9107	.4536	2.204	65.6°	1.1449
.4276	24.5°	.4147	.9100	.4557	2.194	65.5°	1.1432
.4294	24.6°	.4163	.9092	.4578	2.184	65.4°	1.1414
.4311	24.7°	.4179	.9085	.4599	2.174	65.3°	1.1397
.4328	24.8°	.4195	.9078	.4621	2.164	65.2°	1.1380
.4346	24.9°	.4210	.9070	.4642	2.154	65.1°	1.1362
	Cos	Sin	Cot	Tan	Degrees	Radians	

Table A (Continued)

Radians	Degrees	Sin	Cos	Tan	Cot	Degrees	Radians
.4363	25.0°	.4226	.9063	.4663	2.145	65.0°	1.1345
.4381	25.1°	.4242	.9056	.4684	2.135	64.9°	1.1327
.4398	25.2°	.4258	.9048	.4706	2.125	64.8°	1.1310
.4416	25.3°	.4274	.9041	.4727	2.116	64.7°	1.1292
.4433	25.4°	.4289	.9033	.4748	2.106	64.6°	1.1275
.4451	25.5°	.4305	.9026	.4770	2.097	64.5°	1.1257
.4468	25.6°	.4321	.9018	.4791	2.087	64.4°	1.1240
.4485	25.7°	.4337	.9011	.4813	2.078	64.3°	1.1222
.4503	25.8°	.4352	.9003	.4834	2.069	64.2°	1.1205
.4520	25.9°	.4368	.8996	.4856	2.059	64.1°	1.1188
.4538	26.0°	.4384	.8988	.4877	2.050	64.0°	1.1170
.4555	26.1°	.4399	.8980	.4899	2.041	63.9°	1.1153
.4573	26.2°	.4415	.8973	.4921	2.032	63.8°	1.1135
.4590	26.3°	.4431	.8965	.4942	2.023	63.7°	1.1118
.4608	26.4°	.4446	.8957	.4964	2.014	63.6°	1.1100
.4625	26.5°	.4462	.8949	.4986	2.006	63.5°	1.1083
.4643	26.6°	.4478	.8942	.5008	1.997	63.4°	1.1065
.4660	26.7°	.4493	.8934	.5029	1.988	63.3°	1.1048
.4677	26.8°	.4509	.8926	.5051	1.980	63.2°	1.1030
.4695	26.9°	.4524	.8918	.5073	1.971	63.1°	1.1013
.4712	27.0°	.4540	.8910	.5095	1.963	63.0°	1.0996
.4730	27.1°	.4555	.8902	.5117	1.954	62.9°	1.0978
.4747	27.2°	.4571	.8894	.5139	1.946	62.8°	1.0961
.4765	27.3°	.4586	.8886	.5161	1.937	62.7°	1.0943
.4782	27.4°	.4602	.8878	.5184	1.929	62.6°	1.0926
.4800	27.5°	.4617	.8870	.5206	1.921	62.5°	1.0908
.4817	27.6°	.4633	.8862	.5228	1.913	62.4°	1.0891
.4835	27.7°	.4648	.8854	.5250	1.905	62.3°	1.0873
.4852	27.8°	.4664	.8846	.5272	1.897	62.2°	1.0856
.4869	27.9°	.4679	.8838	.5295	1.889	62.1°	1.0838
.4887	28.0°	.4695	.8829	.5317	1.881	62.0°	1.0821
.4904	28.1°	.4710	.8821	.5340	1.873	61.9°	1.0804
.4922	28.2°	.4726	.8813	.5362	1.865	61.8°	1.0786
.4939	28.3°	.4741	.8805	.5384	1.857	61.7°	1.0769
.4957	28.4°	.4756	.8796	.5407	1.849	61.6°	1.0751
.4974	28.5°	.4772	.8788	.5430	1.842	61.5°	1.0734
.4992	28.6°	.4787	.8780	.5452	1.834	61.4°	1.0716
.5009	28.7°	.4802	.8771	.5475	1.827	61.3°	1.0699
.5027	28.8°	.4818	.8763	.5498	1.819	61.2°	1.0681
.5044	28.9°	.4833	.8755	.5520	1.811	61.1°	1.0664
.5061	29.0°	.4848	.8746	.5543	1.804	61.0°	1.0647
.5079	29.1°	.4863	.8738	.5566	1.797	60.9°	1.0629
.5096	29.2°	.4879	.8729	.5589	1.789	60.8°	1.0612
.5131	29.3°	.4894	.8721	.5612	1.782	60.7°	1.0594
.5149	29.4°	.4909	.8712	.5635	1.775	60.6°	1.0577
.5166	29.5°	.4924	.8704	.5658	1.767	60.5°	1.0559
.5184	29.6°	.4939	.8695	.5681	1.760	60.4°	1.0542
.5201	29.7°	.4955	.8688	.5704	1.753	60.3°	1.0524
.5219	29.8°	.4970	.8678	.5727	1.746	60.2°	1.0507
	29.9°	.4985	.8669	.5750	1.739	60.1°	1.0489
	Cos	Sin	Cot	Tan	Degrees	Radians	

Table A (Continued)

Radians	Degrees	Sin	Cos	Tan	Cot	Degrees	Radians
.5236	30.0°	.5000	.8660	.5774	1.732	60.0°	1.0472
.5253	30.1°	.5015	.8652	.5797	1.725	59.9°	1.0455
.5271	30.2°	.5030	.8643	.5820	1.718	59.8°	1.0437
.5288	30.3°	.5045	.8634	.5844	1.711	59.7°	1.0420
.5306	30.4°	.5060	.8625	.5867	1.704	59.6°	1.0402
.5323	30.5°	.5075	.8616	.5890	1.698	59.5°	1.0385
.5341	30.6°	.5090	.8607	.5914	1.691	59.4°	1.0367
.5358	30.7°	.5105	.8599	.5938	1.684	59.3°	1.0350
.5376	30.8°	.5120	.8590	.5961	1.678	59.2°	1.0332
.5393	30.9°	.5135	.8581	.5985	1.671	59.1°	1.0315
.5411	31.0°	.5150	.8572	.6009	1.664	59.0°	1.0297
.5428	31.1°	.5165	.8563	.6032	1.658	58.9°	1.0280
.5445	31.2°	.5180	.8554	.6056	1.651	58.8°	1.0263
.5463	31.3°	.5195	.8545	.6080	1.645	58.7°	1.0245
.5480	31.4°	.5210	.8536	.6104	1.638	58.6°	1.0228
.5498	31.5°	.5225	.8526	.6128	1.632	58.5°	1.0210
.5515	31.6°	.5240	.8517	.6152	1.625	58.4°	1.0193
.5533	31.7°	.5255	.8508	.6176	1.619	58.3°	1.0175
.5550	31.8°	.5270	.8499	.6200	1.613	58.2°	1.0158
.5568	31.9°	.5284	.8490	.6224	1.607	58.1°	1.0140
.5585	32.0°	.5299	.8480	.6249	1.600	58.0°	1.0123
.5603	32.1°	.5314	.8471	.6273	1.594	57.9°	1.0105
.5620	32.2°	.5329	.8462	.6297	1.588	57.8°	1.0088
.5637	32.3°	.5344	.8453	.6322	1.582	57.7°	1.0071
.5655	32.4°	.5358	.8443	.6346	1.576	57.6°	1.0053
.5672	32.5°	.5373	.8434	.6371	1.570	57.5°	1.0036
.5690	32.6°	.5388	.8425	.6395	1.564	57.4°	1.0018
.5707	32.7°	.5402	.8415	.6420	1.558	57.3°	1.0001
.5725	32.8°	.5417	.8406	.6445	1.552	57.2°	.9983
.5742	32.9°	.5432	.8396	.6469	1.546	57.1°	.9966
.5760	33.0°	.5446	.8387	.6494	1.540	57.0°	.9948
.5777	33.1°	.5461	.8377	.6519	1.534	56.9°	.9931
.5794	33.2°	.5476	.8368	.6544	1.528	56.8°	.9913
.5812	33.3°	.5490	.8358	.6569	1.522	56.7°	.9896
.5829	33.4°	.5505	.8348	.6594	1.517	56.6°	.9879
.5847	33.5°	.5519	.8339	.6619	1.511	56.5°	.9861
.5864	33.6°	.5534	.8329	.6644	1.505	56.4°	.9844
.5882	33.7°	.5548	.8320	.6669	1.499	56.3°	.9826
.5899	33.8°	.5563	.8310	.6694	1.494	56.2°	.9809
.5917	33.9°	.5577	.8300	.6720	1.488	56.1°	.9791
.5934	34.0°	.5592	.8290	.6745	1.483	56.0°	.9774
.5952	34.1°	.5606	.8281	.6771	1.477	55.9°	.9756
.5969	34.2°	.5621	.8271	.6796	1.471	55.8°	.9739
.5986	34.3°	.5635	.8261	.6822	1.466	55.7°	.9721
.6004	34.4°	.5650	.8251	.6847	1.460	55.6°	.9704
.6021	34.5°	.5664	.8241	.6873	1.455	55.5°	.9687
.6039	34.6°	.5678	.8231	.6899	1.450	55.4°	.9669
.6056	34.7°	.5693	.8221	.6924	1.444	55.3°	.9652
.6074	34.8°	.5707	.8211	.6950	1.439	55.2°	.9634
.6091	34.9°	.5721	.8202	.6976	1.433	55.1°	.9617
		Cos	Sin	Cot	Tan	Degrees	Radians

Table A (Continued)

Radians	Degrees	Sin	Cos	Tan	Cot	Degrees	Radians
.6109	35.0°	.5736	.8192	.7002	1.428	55.0°	.9599
.6126	35.1°	.5750	.8181	.7028	1.423	54.9°	.9582
.6144	35.2°	.5764	.8171	.7054	1.418	54.8°	.9564
.6161	35.3°	.5779	.8161	.7080	1.412	54.7°	.9547
.6178	35.4°	.5793	.8151	.7107	1.407	54.6°	.9530
.6196	35.5°	.5807	.8141	.7133	1.402	54.5°	.9512
.6213	35.6°	.5821	.8131	.7159	1.397	54.4°	.9495
.6231	35.7°	.5835	.8121	.7186	1.392	54.3°	.9477
.6248	35.8°	.5850	.8111	.7212	1.387	54.2°	.9460
.6266	35.9°	.5864	.8100	.7239	1.381	54.1°	.9442
.6283	36.0°	.5878	.8090	.7265	1.376	54.0°	.9425
.6301	36.1°	.5892	.8080	.7292	1.371	53.9°	.9407
.6318	36.2°	.5906	.8070	.7319	1.366	53.8°	.9390
.6336	36.3°	.5920	.8059	.7346	1.361	53.7°	.9372
.6353	36.4°	.5934	.8049	.7373	1.356	53.6°	.9355
.6370	36.5°	.5948	.8039	.7400	1.351	53.5°	.9338
.6388	36.6°	.5962	.8028	.7427	1.347	53.4°	.9320
.6405	36.7°	.5976	.8018	.7454	1.342	53.3°	.9303
.6423	36.8°	.5990	.8007	.7481	1.337	53.2°	.9285
.6440	36.9°	.6004	.7997	.7508	1.332	53.1°	.9268
.6458	37.0°	.6018	.7986	.7536	1.327	53.0°	.9250
.6475	37.1°	.6032	.7976	.7563	1.322	52.9°	.9233
.6493	37.2°	.6046	.7965	.7590	1.317	52.8°	.9215
.6510	37.3°	.6060	.7955	.7618	1.313	52.7°	.9198
.6528	37.4°	.6074	.7944	.7646	1.308	52.6°	.9180
.6545	37.5°	.6088	.7934	.7673	1.303	52.5°	.9163
.6562	37.6°	.6101	.7923	.7701	1.299	52.4°	.9146
.6580	37.7°	.6115	.7912	.7729	1.294	52.3°	.9128
.6597	37.8°	.6129	.7902	.7757	1.289	52.2°	.9111
.6615	37.9°	.6143	.7891	.7785	1.285	52.1°	.9093
.6632	38.0°	.6157	.7880	.7813	1.280	52.0°	.9076
.6650	38.1°	.6170	.7869	.7841	1.275	51.9°	.9058
.6667	38.2°	.6184	.7859	.7869	1.271	51.8°	.9041
.6685	38.3°	.6198	.7848	.7898	1.266	51.7°	.9023
.6702	38.4°	.6211	.7837	.7926	1.262	51.6°	.9006
.6720	38.5°	.6225	.7826	.7954	1.257	51.5°	.8988
.6737	38.6°	.6239	.7815	.7983	1.253	51.4°	.8971
.6754	38.7°	.6252	.7804	.8012	1.248	51.3°	.8954
.6772	38.8°	.6266	.7793	.8040	1.244	51.2°	.8936
.6789	38.9°	.6280	.7782	.8069	1.239	51.1°	.8919
.6807	39.0°	.6293	.7771	.8098	1.235	51.0°	.8901
.6824	39.1°	.6307	.7760	.8127	1.230	50.9°	.8884
.6842	39.2°	.6320	.7749	.8156	1.226	50.8°	.8866
.6859	39.3°	.6334	.7738	.8185	1.222	50.7°	.8849
.6877	39.4°	.6347	.7727	.8214	1.217	50.6°	.8831
.6894	39.5°	.6361	.7716	.8243	1.213	50.5°	.8814
.6912	39.6°	.6374	.7705	.8273	1.209	50.4°	.8796
.6929	39.7°	.6388	.7694	.8302	1.205	50.3°	.8779
.6946	39.8°	.6401	.7683	.8332	1.200	50.2°	.8762
.6964	39.9°	.6414	.7672	.8361	1.196	50.1°	.8744
		Cos	Sin	Cot	Tan	Degrees	Radians

Table A (Continued)

Radians	Degrees	Sin	Cos	Tan	Cot	Degrees	Radians
.6981	40.0°	.6428	.7660	.8391	1.192	50.0°	.8727
.6999	40.1°	.6441	.7649	.8421	1.188	49.9°	.8709
.7016	40.2°	.6455	.7638	.8451	1.183	49.8°	.8692
.7034	40.3°	.6468	.7627	.8481	1.179	49.7°	.8674
.7051	40.4°	.6481	.7615	.8511	1.175	49.6°	.8657
.7069	40.5°	.6494	.7604	.8541	1.171	49.5°	.8639
.7086	40.6°	.6508	.7593	.8571	1.167	49.4°	.8622
.7103	40.7°	.6521	.7581	.8601	1.163	49.3°	.8604
.7121	40.8°	.6534	.7570	.8632	1.159	49.2°	.8587
.7138	40.9°	.6547	.7559	.8662	1.154	49.1°	.8570
.7156	41.0°	.6561	.7547	.8693	1.150	49.0°	.8552
.7173	41.1°	.6574	.7536	.8724	1.146	48.9°	.8535
.7191	41.2°	.6587	.7524	.8754	1.142	48.8°	.8517
.7208	41.3°	.6600	.7513	.8785	1.138	48.7°	.8500
.7226	41.4°	.6613	.7501	.8816	1.134	48.6°	.8482
.7243	41.5°	.6626	.7490	.8847	1.130	48.5°	.8465
.7261	41.6°	.6639	.7478	.8878	1.126	48.4°	.8447
.7278	41.7°	.6652	.7466	.8910	1.122	48.3°	.8430
.7295	41.8°	.6665	.7455	.8941	1.118	48.2°	.8412
.7313	41.9°	.6678	.7443	.8972	1.115	48.1°	.8395
.7330	42.0°	.6691	.7431	.9004	1.111	48.0°	.8378
.7348	42.1°	.6704	.7420	.9036	1.107	47.9°	.8360
.7365	42.2°	.6717	.7408	.9067	1.103	47.8°	.8343
.7383	42.3°	.6730	.7396	.9099	1.099	47.7°	.8325
.7400	42.4°	.6743	.7385	.9131	1.095	47.6°	.8308
.7418	42.5°	.6756	.7373	.9163	1.091	47.5°	.8290
.7435	42.6°	.6769	.7361	.9195	1.087	47.4°	.8273
.7453	42.7°	.6782	.7349	.9228	1.084	47.3°	.8255
.7470	42.8°	.6794	.7337	.9260	1.080	47.2°	.8238
.7487	42.9°	.6807	.7325	.9293	1.076	47.1°	.8221
.7505	43.0°	.6820	.7314	.9325	1.072	47.0°	.8203
.7522	43.1°	.6833	.7302	.9358	1.069	46.9°	.8186
.7540	43.2°	.6845	.7290	.9391	1.065	46.8°	.8168
.7557	43.3°	.6858	.7278	.9424	1.061	46.7°	.8151
.7575	43.4°	.6871	.7266	.9457	1.057	46.6°	.8133
.7592	43.5°	.6884	.7254	.9490	1.054	46.5°	.8116
.7610	43.6°	.6896	.7242	.9523	1.050	46.4°	.8098
.7627	43.7°	.6909	.7230	.9556	1.046	46.3°	.8081
.7645	43.8°	.6921	.7218	.9590	1.043	46.2°	.8063
.7662	43.9°	.6934	.7206	.9623	1.039	46.1°	.8046
.7679	44.0°	.6947	.7193	.9657	1.036	46.0°	.8029
.7697	44.1°	.6959	.7181	.9691	1.032	45.9°	.8011
.7714	44.2°	.6972	.7169	.9725	1.028	45.8°	.7994
.7732	44.3°	.6984	.7157	.9759	1.025	45.7°	.7976
.7749	44.4°	.6997	.7145	.9793	1.021	45.6°	.7959
.7767	44.5°	.7009	.7133	.9827	1.018	45.5°	.7941
.7784	44.6°	.7022	.7120	.9861	1.014	45.4°	.7924
.7802	44.7°	.7034	.7108	.9896	1.011	45.3°	.7906
.7819	44.8°	.7046	.7096	.9930	1.007	45.2°	.7889
.7837	44.9°	.7059	.7083	.9965	1.003	45.1°	.7871
.7854	45.0°	.7071	.7071	1.0000	1.000	45.0°	.7854
		Cos	Sin	Cot	Tan	Degrees	Radians

Table B Powers and Roots

n	n^2	$\sqrt{n}$	n^3	$\sqrt[3]{n}$	n	n^2	$\sqrt{n}$	n^3	$\sqrt[3]{n}$
1	1	1.000	1	1.000	51	2,601	7.141	132,651	3.708
2	4	1.414	8	1.260	52	2,704	7.211	140,608	3.733
3	9	1.732	27	1.442	53	2,809	7.280	148,877	3.756
4	16	2.000	64	1.587	54	2,916	7.348	157,464	3.780
5	25	2.236	125	1.710	55	3,025	7.416	166,375	3.803
6	36	2.449	216	1.817	56	3,136	7.483	175,616	3.826
7	49	2.646	343	1.913	57	3,249	7.550	185,193	3.849
8	64	2.828	512	2.000	58	3,364	7.616	195,112	3.871
9	81	3.000	729	2.080	59	3,481	7.681	205,379	3.893
10	100	3.162	1,000	2.154	60	3,600	7.746	216,000	3.915
11	121	3.317	1,331	2.224	61	3,721	7.810	226,981	3.936
12	144	3.464	1,728	2.289	62	3,844	7.874	238,328	3.958
13	169	3.606	2,197	2.351	63	3,969	7.937	250,047	3.979
14	196	3.742	2,744	2.410	64	4,096	8.000	262,144	4.000
15	225	3.873	3,375	2.466	65	4,225	8.062	274,625	4.021
16	256	4.000	4,096	2.520	66	4,356	8.124	287,496	4.041
17	289	4.123	4,913	2.571	67	4,489	8.185	300,763	4.062
18	324	4.243	5,832	2.621	68	4,624	8.246	314,432	4.082
19	361	4.359	6,859	2.668	69	4,761	8.307	328,509	4.102
20	400	4.472	8,000	2.714	70	4,900	8.367	343,000	4.121
21	441	4.583	9,261	2.759	71	5,041	8.426	357,911	4.141
22	484	4.690	10,648	2.802	72	5,184	8.485	373,248	4.160
23	529	4.796	12,167	2.844	73	5,329	8.544	389,017	4.179
24	576	4.899	13,824	2.884	74	5,476	8.602	405,224	4.198
25	625	5.000	15,625	2.924	75	5,625	8.660	421,875	4.217
26	676	5.099	17,576	2.962	76	5,776	8.718	438,976	4.236
27	729	5.196	19,683	3.000	77	5,929	8.775	456,533	4.254
28	784	5.292	21,952	3.037	78	6,084	8.832	474,552	4.273
29	841	5.385	24,389	3.072	79	6,241	8.888	493,039	4.291
30	900	5.477	27,000	3.107	80	6,400	8.944	512,000	4.309
31	961	5.568	29,791	3.141	81	6,561	9.000	531,441	4.327
32	1,024	5.657	32,768	3.175	82	6,724	9.055	551,368	4.344
33	1,089	5.745	35,937	3.208	83	6,889	9.110	571,787	4.362
34	1,156	5.831	39,304	3.240	84	7,056	9.165	592,704	4.380
35	1,225	5.916	42,875	3.271	85	7,225	9.220	614,125	4.397
36	1,296	6.000	46,656	3.302	86	7,396	9.274	636,056	4.414
37	1,369	6.083	50,653	3.332	87	7,569	9.327	658,503	4.431
38	1,444	6.164	54,872	3.362	88	7,744	9.381	681,472	4.448
39	1,521	6.245	59,319	3.391	89	7,921	9.434	704,969	4.465
40	1,600	6.325	64,000	3.420	90	8,100	9.487	729,000	4.481
41	1,681	6.403	68,921	3.448	91	8,281	9.539	753,571	4.498
42	1,764	6.481	74,088	3.476	92	8,464	9.592	778,688	4.514
43	1,849	6.557	79,507	3.503	93	8,649	9.644	804,357	4.531
44	1,936	6.633	85,184	3.530	94	8,836	9.695	830,584	4.547
45	2,025	6.708	91,125	3.557	95	9,025	9.747	857,375	4.563
46	2,116	6.782	97,336	3.583	96	9,216	9.798	884,736	4.579
47	2,209	6.856	103,823	3.609	97	9,409	9.849	912,673	4.595
48	2,304	6.928	110,592	3.634	98	9,604	9.899	941,192	4.610
49	2,401	7.000	117,649	3.659	99	9,801	9.950	970,299	4.626
50	2,500	7.071	125,000	3.684	100	10,000	10.000	1,000,000	4.642

Table C Base-10 Logarithms

N	0	1	2	3	4	5	6	7	8	9
1.0	0000	0043	0086	0128	0170	0212	0253	0294	0334	0374
1.1	0414	0453	0492	0531	0569	0607	0645	0682	0719	0755
1.2	0792	0828	0864	0899	0934	0969	1004	1038	1072	1106
1.3	1139	1173	1206	1239	1271	1303	1335	1367	1399	1430
1.4	1461	1492	1523	1553	1584	1614	1644	1673	1703	1732
1.5	1761	1790	1818	1847	1875	1903	1931	1959	1987	2014
1.6	2041	2068	2095	2122	2148	2175	2201	2227	2253	2279
1.7	2304	2330	2355	2380	2405	2430	2455	2480	2504	2529
1.8	2553	2577	2601	2625	2648	2672	2695	2718	2742	2765
1.9	2788	2810	2833	2856	2878	2900	2923	2945	2967	2989
2.0	3010	3032	3054	3075	3096	3118	3139	3160	3181	3201
2.1	3222	3243	3263	3284	3304	3324	3345	3365	3385	3404
2.2	3424	3444	3464	3483	3502	3522	3541	3560	3579	3598
2.3	3617	3636	3655	3674	3692	3711	3729	3747	3766	3784
2.4	3802	3820	3838	3856	3874	3892	3909	3927	3945	3962
2.5	3979	3997	4014	4031	4048	4065	4082	4099	4116	4133
2.6	4150	4166	4183	4200	4216	4232	4249	4265	4281	4298
2.7	4314	4330	4346	4362	4378	4393	4409	4425	4440	4456
2.8	4472	4487	4502	4518	4533	4548	4564	4579	4594	4609
2.9	4624	4639	4654	4669	4683	4698	4713	4728	4742	4757
3.0	4771	4786	4800	4814	4829	4843	4857	4871	4886	4900
3.1	4914	4928	4942	4955	4969	4983	4997	5011	5024	5038
3.2	5051	5065	5079	5092	5105	5119	5132	5145	5159	5172
3.3	5185	5198	5211	5224	5237	5250	5263	5276	5289	5302
3.4	5315	5328	5340	5353	5366	5378	5391	5403	5416	5428
3.5	5441	5453	5465	5478	5490	5502	5514	5527	5539	5551
3.6	5563	5575	5587	5599	5611	5623	5635	5647	5658	5670
3.7	5682	5694	5705	5717	5729	5740	5752	5763	5775	5786
3.8	5798	5809	5821	5832	5843	5855	5866	5877	5888	5899
3.9	5911	5922	5933	5944	5955	5966	5977	5988	5999	6010
4.0	6021	6031	6042	6053	6064	6075	6085	6096	6107	6117
4.1	6128	6138	6149	6160	6170	6180	6191	6201	6212	6222
4.2	6232	6243	6253	6263	6274	6284	6294	6304	6314	6325
4.3	6335	6345	6355	6365	6375	6385	6395	6405	6415	6425
4.4	6435	6444	6454	6464	6474	6484	6493	6503	6513	6522
4.5	6532	6542	6551	6561	6571	6580	6590	6599	6609	6618
4.6	6628	6637	6646	6656	6665	6675	6684	6693	6702	6712
4.7	6721	6730	6739	6749	6758	6767	6776	6785	6794	6803
4.8	6812	6821	6830	6839	6848	6857	6866	6875	6884	6893
4.9	6902	6911	6920	6928	6937	6946	6955	6964	6972	6981
5.0	6990	6998	7007	7016	7024	7033	7042	7050	7059	7067
5.1	7076	7084	7093	7101	7110	7118	7126	7135	7143	7152
5.2	7160	7168	7177	7185	7193	7202	7210	7218	7226	7235
5.3	7243	7251	7259	7267	7275	7284	7292	7300	7308	7316
5.4	7324	7332	7340	7348	7356	7364	7372	7380	7388	7396

Table C (Continued)

N	0	1	2	3	4	5	6	7	8	9
5.5	7404	7412	7419	7427	7435	7443	7451	7459	7466	7474
5.6	7482	7490	7497	7505	7513	7520	7528	7536	7543	7551
5.7	7559	7566	7574	7582	7589	7597	7604	7612	7619	7627
5.8	7634	7642	7649	7657	7664	7672	7679	7686	7694	7701
5.9	7709	7716	7723	7731	7738	7745	7752	7760	7767	7774
6.0	7782	7789	7796	7803	7810	7818	7825	7832	7839	7846
6.1	7853	7860	7868	7875	7882	7889	7896	7903	7910	7917
6.2	7924	7931	7938	7945	7952	7959	7966	7973	7980	7987
6.3	7993	8000	8007	8014	8021	8028	8035	8041	8048	8055
6.4	8062	8069	8075	8082	8089	8096	8102	8109	8116	8122
6.5	8129	8136	8142	8149	8156	8162	8169	8176	8182	8189
6.6	8195	8202	8209	8215	8222	8228	8235	8241	8248	8254
6.7	8261	8267	8274	8280	8287	8293	8299	8306	8312	8319
6.8	8325	8331	8338	8344	8351	8357	8363	8370	8376	8382
6.9	8388	8395	8401	8407	8414	8420	8426	8432	8439	8445
7.0	8451	8457	8463	8470	8476	8482	8488	8494	8500	8506
7.1	8513	8519	8525	8531	8537	8543	8549	8555	8561	8567
7.2	8573	8579	8585	8591	8597	8603	8609	8615	8621	8627
7.3	8633	8639	8645	8651	8657	8663	8669	8675	8681	8686
7.4	8692	8698	8704	8710	8716	8722	8727	8733	8739	8745
7.5	8751	8756	8762	8768	8774	8779	8785	8791	8797	8802
7.6	8808	8814	8820	8825	8831	8837	8842	8848	8854	8859
7.7	8865	8871	8876	8882	8887	8893	8899	8904	8910	8915
7.8	8921	8927	8932	8938	8943	8949	8954	8960	8965	8971
7.9	8976	8982	8987	8993	8998	9004	9009	9015	9020	9025
8.0	9031	9036	9042	9047	9053	9058	9063	9069	9074	9079
8.1	9085	9090	9096	9101	9106	9112	9117	9122	9128	9133
8.2	9138	9143	9149	9154	9159	9165	9170	9175	9180	9186
8.3	9191	9196	9201	9206	9212	9217	9222	9227	9232	9238
8.4	9243	9248	9253	9258	9263	9269	9274	9279	9284	9289
8.5	9294	9299	9304	9309	9315	9320	9325	9330	9335	9340
8.6	9345	9350	9355	9360	9365	9370	9375	9380	9385	9390
8.7	9395	9400	9405	9410	9415	9420	9425	9430	9435	9440
8.8	9445	9450	9455	9460	9465	9469	9474	9479	9484	9489
8.9	9494	9499	9504	9509	9513	9518	9523	9528	9533	9538
9.0	9542	9547	9552	9557	9562	9566	9571	9576	9581	9586
9.1	9590	9595	9600	9605	9609	9614	9619	9624	9628	9633
9.2	9638	9643	9647	9652	9657	9661	9666	9671	9675	9680
9.3	9685	9689	9694	9699	9703	9708	9713	9717	9722	9727
9.4	9731	9736	9741	9745	9750	9754	9759	9763	9768	9773
9.5	9777	9782	9786	9791	9795	9800	9805	9809	9814	9818
9.6	9823	9827	9832	9836	9841	9845	9850	9854	9859	9863
9.7	9868	9872	9877	9881	9886	9890	9894	9899	9903	9908
9.8	9912	9917	9921	9926	9930	9934	9939	9943	9948	9952
9.9	9956	9961	9965	9969	9974	9978	9983	9987	9991	9996

Table D Base-*e* Logarithms

N	0	1	2	3	4	5	6	7	8	9
1.0	.0000	.0100	.0198	.0296	.0392	.0488	.0583	.0677	.0770	.0862
1.1	.0953	.1044	.1133	.1222	.1310	.1398	.1484	.1570	.1655	.1740
1.2	.1823	.1906	.1989	.2070	.2151	.2231	.2311	.2390	.2469	.2546
1.3	.2624	.2700	.2776	.2852	.2927	.3001	.3075	.3148	.3221	.3293
1.4	.3365	.3436	.3507	.3577	.3646	.3716	.3784	.3853	.3920	.3988
1.5	.4055	.4121	.4187	.4253	.4318	.4383	.4447	.4511	.4574	.4637
1.6	.4700	.4762	.4824	.4886	.4947	.5008	.5068	.5128	.5188	.5247
1.7	.5306	.5365	.5423	.5481	.5539	.5596	.5653	.5710	.5766	.5822
1.8	.5878	.5933	.5988	.6043	.6098	.6152	.6206	.6259	.6313	.6366
1.9	.6419	.6471	.6523	.6575	.6627	.6678	.6729	.6780	.6831	.6881
2.0	.6931	.6981	.7031	.7080	.7129	.7178	.7227	.7275	.7324	.7372
2.1	.7419	.7467	.7514	.7561	.7608	.7655	.7701	.7747	.7793	.7839
2.2	.7885	.7930	.7975	.8020	.8065	.8109	.8154	.8198	.8242	.8286
2.3	.8329	.8372	.8416	.8459	.8502	.8544	.8587	.8629	.8671	.8713
2.4	.8755	.8796	.8838	.8879	.8920	.8961	.9002	.9042	.9083	.9123
2.5	.9163	.9203	.9243	.9282	.9322	.9361	.9400	.9439	.9478	.9517
2.6	.9555	.9594	.9632	.9670	.9708	.9746	.9783	.9821	.9858	.9895
2.7	.9933	.9969	1.0006	.0043	.0080	.0116	.0152	.0188	.0225	.0260
2.8	1.0296	.0332	.0367	.0403	.0438	.0473	.0508	.0543	.0578	.0613
2.9	.0647	.0682	.0716	.0750	.0784	.0818	.0852	.0886	.0919	.0953
3.0	1.0986	.1019	.1053	.1086	.1119	.1151	.1184	.1217	.1249	.1282
3.1	.1314	.1346	.1378	.1410	.1442	.1474	.1506	.1537	.1569	.1600
3.2	.1632	.1663	.1694	.1725	.1756	.1787	.1817	.1848	.1878	.1909
3.3	.1939	.1969	.2000	.2030	.2060	.2090	.2119	.2149	.2179	.2208
3.4	.2238	.2267	.2296	.2326	.2355	.2384	.2413	.2442	.2470	.2499
3.5	1.2528	.2556	.2585	.2613	.2641	.2669	.2698	.2726	.2754	.2782
3.6	.2809	.2837	.2865	.2892	.2920	.2947	.2975	.3002	.3029	.3056
3.7	.3083	.3110	.3137	.3164	.3191	.3218	.3244	.3271	.3297	.3324
3.8	.3350	.3376	.3403	.3429	.3455	.3481	.3507	.3533	.3558	.3584
3.9	.3610	.3635	.3661	.3686	.3712	.3737	.3762	.3788	.3813	.3838
4.0	1.3863	.3888	.3913	.3938	.3962	.3987	.4012	.4036	.4061	.4085
4.1	.4110	.4134	.4159	.4183	.4207	.4231	.4255	.4279	.4303	.4327
4.2	.4351	.4375	.4398	.4422	.4446	.4469	.4493	.4516	.4540	.4563
4.3	.4586	.4609	.4633	.4656	.4679	.4702	.4725	.4748	.4770	.4793
4.4	.4816	.4839	.4861	.4884	.4907	.4929	.4951	.4974	.4996	.5019
4.5	1.5041	.5063	.5085	.5107	.5129	.5151	.5173	.5195	.5217	.5239
4.6	.5261	.5282	.5304	.5326	.5347	.5369	.5390	.5412	.5433	.5454
4.7	.5476	.5497	.5518	.5539	.5560	.5581	.5602	.5623	.5644	.5665
4.8	.5686	.5707	.5728	.5748	.5769	.5790	.5810	.5831	.5851	.5872
4.9	.5892	.5913	.5933	.5953	.5974	.5994	.6014	.6034	.6054	.6074
5.0	1.6094	.6114	.6134	.6154	.6174	.6194	.6214	.6233	.6253	.6273
5.1	.6292	.6312	.6332	.6351	.6371	.6390	.6409	.6429	.6448	.6467
5.2	.6487	.6506	.6525	.6544	.6563	.6582	.6601	.6620	.6639	.6658
5.3	.6677	.6696	.6715	.6734	.6752	.6771	.6790	.6808	.6827	.6845
5.4	.6864	.6882	.6901	.6919	.6938	.6956	.6974	.6993	.7011	.7029

Table D (*Continued*)

N	0	1	2	3	4	5	6	7	8	9
5.5	1.7047	.7066	.7084	.7102	.7120	.7138	.7156	.7174	.7192	.7210
5.6	.7228	.7246	.7263	.7281	.7299	.7317	.7334	.7352	.7370	.7387
5.7	.7405	.7422	.7440	.7457	.7475	.7492	.7509	.7527	.7544	.7561
5.8	.7579	.7596	.7613	.7630	.7647	.7664	.7681	.7699	.7716	.7733
5.9	.7750	.7766	.7783	.7800	.7817	.7834	.7851	.7867	.7884	.7901
6.0	1.7918	.7934	.7951	.7967	.7984	.8001	.8017	.8034	.8050	.8066
6.1	.8083	.8099	.8116	.8132	.8148	.8165	.8181	.8197	.8213	.8229
6.2	.8245	.8262	.8278	.8294	.8310	.8326	.8342	.8358	.8374	.8390
6.3	.8405	.8421	.8437	.8453	.8469	.8485	.8500	.8516	.8532	.8547
6.4	.8563	.8579	.8594	.8610	.8625	.8641	.8656	.8672	.8687	.8703
6.5	1.8718	.8733	.8749	.8764	.8779	.8795	.8810	.8825	.8840	.8856
6.6	.8871	.8886	.8901	.8916	.8931	.8946	.8961	.8976	.8991	.9006
6.7	.9021	.9036	.9051	.9066	.9081	.9095	.9110	.9125	.9140	.9155
6.8	.9169	.9184	.9199	.9213	.9228	.9242	.9257	.9272	.9286	.9301
6.9	.9315	.9330	.9344	.9359	.9373	.9387	.9402	.9416	.9430	.9445
7.0	1.9459	.9473	.9488	.9502	.9516	.9530	.9544	.9559	.9573	.9587
7.1	.9601	.9615	.9629	.9643	.9657	.9671	.9685	.9699	.9713	.9727
7.2	.9741	.9755	.9769	.9782	.9796	.9810	.9824	.9838	.9851	.9865
7.3	.9879	.9892	.9906	.9920	.9933	.9947	.9961	.9974	.9988	2.0001
7.4	2.0015	.0028	.0042	.0055	.0069	.0082	.0096	.0109	.0122	.0136
7.5	2.0149	.0162	.0176	.0189	.0202	.0215	.0229	.0242	.0255	.0268
7.6	.0281	.0295	.0308	.0321	.0334	.0347	.0360	.0373	.0386	.0399
7.7	.0412	.0425	.0438	.0451	.0464	.0477	.0490	.0503	.0516	.0528
7.8	.0541	.0554	.0567	.0580	.0592	.0605	.0618	.0631	.0643	.0656
7.9	.0669	.0681	.0694	.0707	.0719	.0732	.0744	.0757	.0769	.0782
8.0	2.0794	.0807	.0819	.0832	.0844	.0857	.0869	.0882	.0894	.0906
8.1	.0919	.0931	.0943	.0956	.0968	.0980	.0992	.1005	.1017	.1029
8.2	.1041	.1054	.1066	.1078	.1090	.1102	.1114	.1126	.1138	.1150
8.3	.1163	.1175	.1187	.1199	.1211	.1223	.1235	.1247	.1258	.1270
8.4	.1282	.1294	.1306	.1318	.1330	.1342	.1353	.1365	.1377	.1389
8.5	2.1401	.1412	.1424	.1436	.1448	.1459	.1471	.1483	.1494	.1506
8.6	.1518	.1529	.1541	.1552	.1564	.1576	.1587	.1599	.1610	.1622
8.7	.1633	.1645	.1656	.1668	.1679	.1691	.1702	.1713	.1725	.1736
8.8	.1748	.1759	.1770	.1782	.1793	.1804	.1815	.1827	.1838	.1849
8.9	.1861	.1872	.1883	.1894	.1905	.1917	.1928	.1939	.1950	.1961
9.0	2.1972	.1983	.1994	.2006	.2017	.2028	.2039	.2050	.2061	.2072
9.1	.2083	.2094	.2105	.2116	.2127	.2138	.2148	.2159	.2170	.2181
9.2	.2192	.2203	.2214	.2225	.2235	.2246	.2257	.2268	.2279	.2289
9.3	.2300	.2311	.2322	.2332	.2343	.2354	.2364	.2375	.2386	.2396
9.4	.2407	.2418	.2428	.2439	.2450	.2460	.2471	.2481	.2492	.2502
9.5	2.2513	.2523	.2534	.2544	.2555	.2565	.2576	.2586	.2597	.2607
9.6	.2618	.2628	.2638	.2649	.2659	.2670	.2680	.2690	.2701	.2711
9.7	.2721	.2732	.2742	.2752	.2762	.2773	.2783	.2793	.2803	.2814
9.8	.2824	.2834	.2844	.2854	.2865	.2875	.2885	.2895	.2905	.2915
9.9	.2925	.2935	.2946	.2956	.2966	.2976	.2986	.2996	.3006	.3016

Use the properties of logarithms and $\ln 10 \approx 2.3026$ to find logarithms of numbers less than 1 or greater than 10.

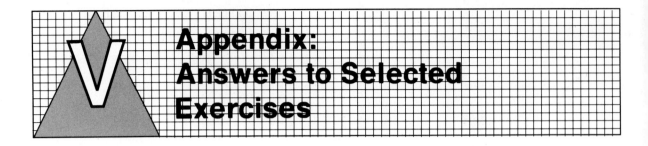

Appendix: Answers to Selected Exercises

Exercise 1.1 (Page 4)

1. 5 **3.** 8 **5.** 12 **7.** 5 **9.** 5 **11.** 13 **13.** 10 **15.** $\sqrt{104} = 2\sqrt{26}$

17. $4\sqrt{5}$ **19.** $d(AB) = 4\sqrt{2}$ and $d(BC) = 4\sqrt{2}$. Hence, the triangle is isosceles.

21. Pick any point P on line CD. Let P have coordinates (x, y). Find the distances PA and PB, set them equal to each other, and simplify. The result is $3x + y = 24$.

23. $(3, 0)$

Exercise 1.2 (Page 10)

1. a function **3.** a function **5.** not a function **7.** a function **9.** not a function

11. Domain is the set of real numbers; range is the set of real numbers.

13. Domain is the set of real numbers except 2; range is the set of real numbers except 0.

15. Domain is the set of real numbers; range is the set of nonnegative real numbers.

17. Domain is the set of real numbers; range is the set of nonnegative real numbers.

19. Domain is the set of real numbers greater than or equal to 2; range is the set of nonnegative real numbers.

21. $f(2) = \frac{3}{2}$ **23.** $f(-3) = \frac{7}{3}$ **25.** $f(0)$ is undefined **27.** $f(h) = \dfrac{2h - 1}{h}$ **29.** $f(1) = 0$

31. $f(0)$ is undefined **33.** $f(2a) = \dfrac{\sqrt{2a - 1}}{2}$ **35.** $f(a + h) = \dfrac{\sqrt{a + h - 1}}{2}$

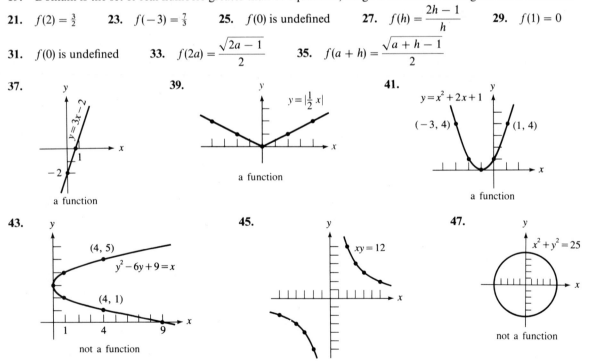

37. $y = 3x - 2$ a function

39. $y = |\frac{1}{2} x|$ a function

41. $y = x^2 + 2x + 1$ $(-3, 4)$ $(1, 4)$ a function

43. $(4, 5)$ $y^2 - 6y + 9 = x$ $(4, 1)$ not a function

45. $xy = 12$ a function

47. $x^2 + y^2 = 25$ not a function

400

49.

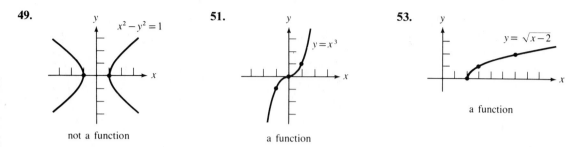

$x^2 - y^2 = 1$

not a function

51.

$y = x^3$

a function

53.

$y = \sqrt{x-2}$

a function

55. Let any point P on the circle have coordinates of (x, y). Find the distance r between point P and the origin: $r = \sqrt{(x - 0)^2 + (y - 0)^2} = \sqrt{x^2 + y^2}$. Square both sides to get $x^2 + y^2 = r^2$.

Exercise 1.3   (Page 18)

1. Domain is $\{2, 3, 4\}$; range is $\{1, 2, 3\}$; a function   **3.** Domain is $\{2\}$; range is $\{3, 4, 5\}$; not a function

5. Domain is $\{5, 4, 3, 2\}$; range is $\{1\}$; a function   **7.** $\{(1, 1), (8, 2), (27, 3), (64, 4)\}$; a function

9. $y = -\dfrac{x + 2}{2}$; a function   **11.** $y = 3 - x$; a function   **13.** $y = \pm\sqrt{x - 2}$; not a function

15. $y = \dfrac{4x + 2}{x}$; a function   **17.** $y = \pm\sqrt{x + 1}$; not a function   **19.** $x = |y - 2|$; not a function

21. $y = \sqrt{x + 4}$; a function   **23.** one-to-one   **25.** not one-to-one

27.

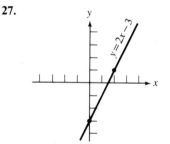

$y = 2x - 3$

a one-to-one function

29.

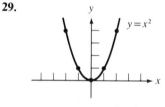

$y = x^2$

not a one-to-one function

31.

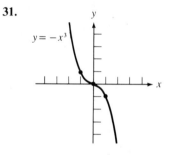

$y = -x^3$

a one-to-one function

33.

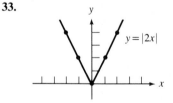

$y = |2x|$

not a one-to-one function

35.

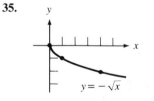

$y = -\sqrt{x}$

a one-to-one function

37.

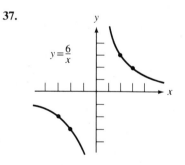

$y = \dfrac{6}{x}$

a one-to-one function

39.

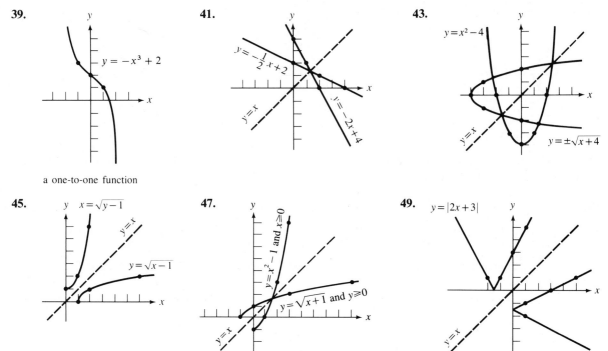

$y = -x^3 + 2$

a one-to-one function

41.

$y = -\frac{1}{2}x + 2$

$y = x$

$y = -2x + 4$

43.

$y = x^2 - 4$

$y = x$

$y = \pm\sqrt{x+4}$

45.

$y \quad x = \sqrt{y-1}$

$y = x$

$y = \sqrt{x-1}$

47.

$y = x^2 - 1$ and $x \geqslant 0$

$y = x$

$y = \sqrt{x+1}$ and $y \geqslant 0$

49.

$y = |2x + 3|$

$y = x$

$x = |2y + 3|$

51. Range is the set of all real numbers except 1. **53.** Range is the set of all real numbers except 3.

REVIEW EXERCISES (Page 19)

1. 7 **2.** 10 **3.** 17 **4.** 35 **5.** 5 **6.** 13 **7.** 5 **8.** 5 **9.** $5\sqrt{2}$
10. $5\sqrt{2}$ **11.** $\sqrt{170}$ **12.** $2\sqrt{17}$ **13.** $d(BC) = \sqrt{5}$ and $d(CA) = \sqrt{5}$. Hence, triangle ABC is isosceles.
14. Because $d(AB) + d(BC) = d(CA)$, points A, B, and C lie on a line.
15. Domain $= \{2, 4, 5, 6, 7\}$; range $= \{1, 2, 6, 7, 8\}$ **16.** Domain $= \{-2, 1, 3\}$; range $= \{-2, 2, 3, 4, 5\}$
17. Domain is the set of real numbers; range is the set of real numbers.
18. Domain is the set of real numbers; range is the set of all real numbers greater than or equal to -10.
19. Domain is the set of all real numbers greater than or equal to 10; range is the set of nonnegative real numbers.
20. Domain is the set of all real numbers; range is the set of nonnegative real numbers. **21.** not a function
22. a function **23.** not a function **24.** a function **25.** not a function **26.** not a function
27. -4 **28.** -1 **29.** $h^2 + 2h - 4$ **30.** $2ah + 2h + h^2$

31.

$y = \frac{1}{2}x - 3$

a function

32.

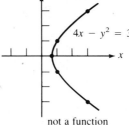

$4x - y^2 = 3$

not a function

33.

$x^2 - y = 4$

a function

34.

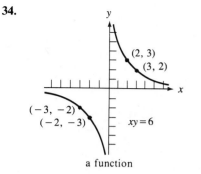

a function

35. $\{(1,2),(1,3),(1,4)\}$; not a function
36. $\{(10,1),(20,2),(30,3),(40,4)\}$; a function
37. $y = \dfrac{x-12}{7}$; a function   **38.** $y = 4x - 17$; a function
39. $y = \pm\sqrt{3x}$; not a function   **40.** $yx = 36$; a function
41. $y = \dfrac{x^2+1}{2}$, $x \geq 0$; a function   **42.** $y = \sqrt{x-4}$; a function

43.

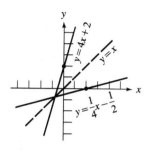

44.

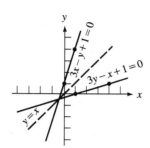

45.

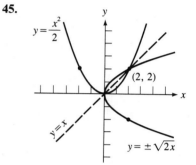

46.

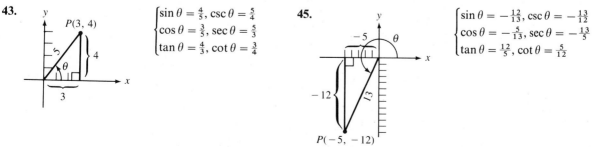

47. one-to-one   **48.** not one-to-one   **49.** not one-to-one
50. one-to-one   **51.** one-to-one   **52.** one-to-one

Exercise 2.1   (Page 28)

1. $84.607°$ **3.** $124.880°$ **5.** $23°8'24''$ **7.** $96°30'$ **9.** yes; positive   **11.** no; positive
13. no; negative   **15.** yes; negative   **17.** no; positive   **19.** yes; positive   **21.** QII
23. QIV   **25.** QII   **27.** QIII   **29.** QIII   **31.** QIV   **33.** QIII   **35.** yes
37. no   **39.** yes   **41.** yes

43.

$\begin{cases} \sin\theta = \frac{4}{5},\ \csc\theta = \frac{5}{4} \\ \cos\theta = \frac{3}{5},\ \sec\theta = \frac{5}{3} \\ \tan\theta = \frac{4}{3},\ \cot\theta = \frac{3}{4} \end{cases}$

45.

$\begin{cases} \sin\theta = -\frac{12}{13},\ \csc\theta = -\frac{13}{12} \\ \cos\theta = -\frac{5}{13},\ \sec\theta = -\frac{13}{5} \\ \tan\theta = \frac{12}{5},\ \cot\theta = \frac{5}{12} \end{cases}$

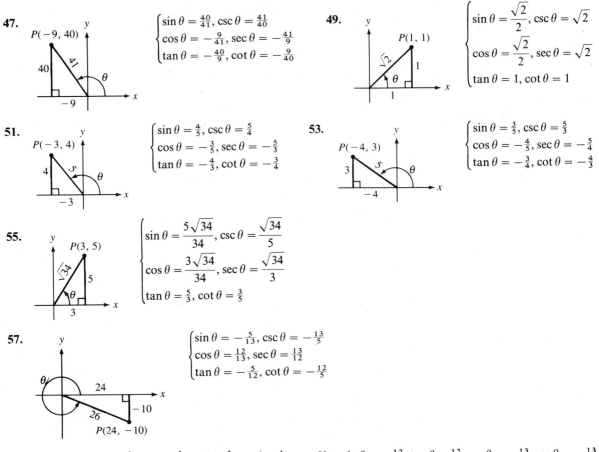

47. $P(-9, 40)$

$\begin{cases} \sin\theta = \frac{40}{41}, \csc\theta = \frac{41}{40} \\ \cos\theta = -\frac{9}{41}, \sec\theta = -\frac{41}{9} \\ \tan\theta = -\frac{40}{9}, \cot\theta = -\frac{9}{40} \end{cases}$

49. $P(1, 1)$

$\begin{cases} \sin\theta = \frac{\sqrt{2}}{2}, \csc\theta = \sqrt{2} \\ \cos\theta = \frac{\sqrt{2}}{2}, \sec\theta = \sqrt{2} \\ \tan\theta = 1, \cot\theta = 1 \end{cases}$

51. $P(-3, 4)$

$\begin{cases} \sin\theta = \frac{4}{5}, \csc\theta = \frac{5}{4} \\ \cos\theta = -\frac{3}{5}, \sec\theta = -\frac{5}{3} \\ \tan\theta = -\frac{4}{3}, \cot\theta = -\frac{3}{4} \end{cases}$

53. $P(-4, 3)$

$\begin{cases} \sin\theta = \frac{3}{5}, \csc\theta = \frac{5}{3} \\ \cos\theta = -\frac{4}{5}, \sec\theta = -\frac{5}{4} \\ \tan\theta = -\frac{3}{4}, \cot\theta = -\frac{4}{3} \end{cases}$

55. $P(3, 5)$

$\begin{cases} \sin\theta = \dfrac{5\sqrt{34}}{34}, \csc\theta = \dfrac{\sqrt{34}}{5} \\ \cos\theta = \dfrac{3\sqrt{34}}{34}, \sec\theta = \dfrac{\sqrt{34}}{3} \\ \tan\theta = \frac{5}{3}, \cot\theta = \frac{3}{5} \end{cases}$

57. $P(24, -10)$

$\begin{cases} \sin\theta = -\frac{5}{13}, \csc\theta = -\frac{13}{5} \\ \cos\theta = \frac{12}{13}, \sec\theta = \frac{13}{12} \\ \tan\theta = -\frac{5}{12}, \cot\theta = -\frac{12}{5} \end{cases}$

59. $\cos\theta = \frac{4}{5}, \tan\theta = \frac{3}{4}, \csc\theta = \frac{5}{3}, \sec\theta = \frac{5}{4}, \cot\theta = \frac{4}{3}$ **61.** $\sin\theta = -\frac{12}{13}, \tan\theta = \frac{12}{5}, \csc\theta = -\frac{13}{12}, \sec\theta = -\frac{13}{5}$

63. $\cos\theta = \frac{40}{41}, \tan\theta = -\frac{9}{40}, \csc\theta = -\frac{41}{9}, \sec\theta = \frac{41}{40}, \cot\theta = -\frac{40}{9}$

65. $\sin\theta = \frac{4}{5}, \cos\theta = -\frac{3}{5}, \tan\theta = -\frac{4}{3}, \cot\theta = -\frac{3}{4}$ **67.** $\sin\theta = -\frac{40}{41}, \csc\theta = -\frac{41}{40}, \sec\theta = \frac{41}{9}, \cot\theta = -\frac{9}{40}$

69. Because the tangent and cotangent functions cannot differ in sign for a specific value of θ.

71. If $\sin\theta = 2$, then $\sin\theta = \frac{2}{1} = \frac{x}{r}$ and a leg of a right triangle would be greater than the hypotenuse.

73. If $\sec\theta = \frac{1}{2}$, then $\cos\theta = 2$, which is impossible.

Exercise 2.2 (Page 36)

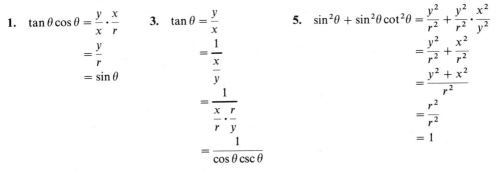

1. $\tan\theta\cos\theta = \dfrac{y}{x} \cdot \dfrac{x}{r}$

$= \dfrac{y}{r}$

$= \sin\theta$

3. $\tan\theta = \dfrac{y}{x}$

$= \dfrac{1}{\dfrac{x}{y}}$

$= \dfrac{1}{\dfrac{x}{r} \cdot \dfrac{r}{y}}$

$= \dfrac{1}{\cos\theta\csc\theta}$

5. $\sin^2\theta + \sin^2\theta\cot^2\theta = \dfrac{y^2}{r^2} + \dfrac{y^2}{r^2} \cdot \dfrac{x^2}{y^2}$

$= \dfrac{y^2}{r^2} + \dfrac{x^2}{r^2}$

$= \dfrac{y^2 + x^2}{r^2}$

$= \dfrac{r^2}{r^2}$

$= 1$

7. $\cot^2\theta + \sin^2\theta = \dfrac{x^2}{y^2} + \dfrac{y^2}{r^2}$

$\qquad = \dfrac{r^2 - y^2}{y^2} + \dfrac{r^2 - x^2}{r^2}$

$\qquad = \dfrac{r^2}{y^2} - 1 + 1 - \dfrac{x^2}{r^2}$

$\qquad = \dfrac{r^2}{y^2} - \dfrac{x^2}{r^2}$

$\qquad = \csc^2\theta - \cos^2\theta$

9. $\sin^2\theta + \dfrac{\cot^2\theta}{\csc^2\theta} = \dfrac{y^2}{r^2} + \dfrac{\tfrac{x^2}{y^2}}{\tfrac{r^2}{y^2}}$

$\qquad = \dfrac{y^2}{r^2} + \dfrac{x^2}{r^2}$

$\qquad = \dfrac{y^2 + x^2}{r^2}$

$\qquad = \dfrac{r^2}{r^2}$

$\qquad = 1$

11. $\tan\theta\cos\theta = \dfrac{\sin\theta}{\cos\theta}\cos\theta$

$\qquad = \sin\theta$

13. $\tan\theta = \dfrac{\sin\theta}{\cos\theta}$

$\qquad = \dfrac{1}{\csc\theta} \cdot \dfrac{1}{\cos\theta}$

$\qquad = \dfrac{1}{\cos\theta\csc\theta}$

15. $\sin^2\theta + \sin^2\theta\cot^2\theta = \sin^2\theta(1 + \cot^2\theta)$

$\qquad = \sin^2\theta\csc^2\theta$

$\qquad = 1$

17. $\cot^2\theta + \sin^2\theta = \csc^2\theta - 1 + 1 - \cos^2\theta$

$\qquad = \csc^2\theta - \cos^2\theta$

19. $\sin^2\theta + \dfrac{\cot^2\theta}{\csc^2\theta} = \sin^2\theta + \dfrac{\cos^2\theta}{\sin^2\theta} \cdot \dfrac{\sin^2\theta}{1}$

$\qquad = \sin^2\theta + \cos^2\theta$

$\qquad = 1$

21. $\dfrac{\sec\theta}{\csc\theta} = \sec\theta \cdot \dfrac{1}{\csc\theta}$

$\qquad = \dfrac{1}{\cos\theta} \cdot \dfrac{\sin\theta}{1}$

$\qquad = \dfrac{\sin\theta}{\cos\theta}$

$\qquad = \tan\theta$

23. $\cos\theta\tan\theta = \dfrac{\cos\theta}{1} \cdot \dfrac{\sin\theta}{\cos\theta}$

$\qquad = \sin\theta$

25. $\dfrac{\sin^2\theta}{\sec^2\theta} + \sin^4\theta = \sin^2\theta\left(\dfrac{1}{\sec^2\theta} + \sin^2\theta\right)$

$\qquad = \sin^2\theta(\cos^2\theta + \sin^2\theta)$

$\qquad = \sin^2\theta(1)$

$\qquad = \sin^2\theta$

27. $(\sec\theta + \tan\theta)(\sec\theta - \tan\theta) = \sec^2\theta - \tan^2\theta$

$\qquad = 1 + \tan^2\theta - \tan^2\theta$

$\qquad = 1$

29. $(1 + \cos\theta)(1 - \cos\theta) = 1 - \cos^2\theta$

$\qquad = \sin^2\theta + \cos^2\theta - \cos^2\theta$

$\qquad = \sin^2\theta$

31. $(\cos\theta\sec\theta + \cos\theta)(\cos\theta\sec\theta - \cos\theta) = \cos^2\theta\sec^2\theta - \cos^2\theta$

$\qquad = 1 - \cos^2\theta$

$\qquad = \sin^2\theta + \cos^2\theta - \cos^2\theta$

$\qquad = \sin^2\theta$

33. $\cos\theta = \tfrac{3}{5}$, $\tan\theta = \tfrac{4}{3}$, $\csc\theta = \tfrac{5}{4}$, $\sec\theta = \tfrac{5}{3}$, $\cot\theta = \tfrac{3}{4}$

35. $\sin\theta = \tfrac{12}{13}$, $\tan\theta = -\tfrac{12}{5}$, $\csc\theta = \tfrac{13}{12}$, $\sec\theta = -\tfrac{13}{5}$, $\cot\theta = -\tfrac{5}{12}$

37. $\sin\theta = \tfrac{4}{5}$, $\cos\theta = -\tfrac{3}{5}$, $\csc\theta = \tfrac{5}{4}$, $\sec\theta = -\tfrac{5}{3}$, $\cot\theta = -\tfrac{3}{4}$

39. $\sin\theta = \tfrac{40}{41}$, $\cos\theta = \tfrac{9}{41}$, $\tan\theta = \tfrac{40}{9}$, $\csc\theta = \tfrac{41}{40}$, $\sec\theta = \tfrac{41}{9}$

41. $\sin\theta = -\tfrac{12}{13}$, $\cos\theta = -\tfrac{5}{13}$, $\tan\theta = \tfrac{12}{5}$, $\csc\theta = -\tfrac{13}{12}$, $\cot\theta = \tfrac{5}{12}$ **43.** odd   **45.** even   **47.** odd

49. neither   **51.** neither   **53.** $y = \csc\theta = 1/\sin\theta$. Because $\sin\theta$ is an odd function, so is $1/\sin\theta = y$.

55. yes   **57.** no

59.

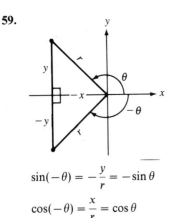

$$\sin(-\theta) = -\frac{y}{r} = -\sin\theta$$

$$\cos(-\theta) = \frac{x}{r} = \cos\theta$$

$$\tan(-\theta) = -\frac{y}{x} = -\tan\theta$$

61.

$$\sin(-\theta) = -\frac{y}{r} = -\sin\theta$$

$$\cos(-\theta) = \frac{x}{r} = \cos\theta$$

$$\tan(-\theta) = -\frac{y}{x} = -\tan\theta$$

63. negative

65. positive

Exercise 2.3 (Page 45)

1.

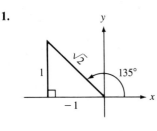

$$\begin{cases} \sin 135° = \dfrac{1}{\sqrt{2}} = \dfrac{\sqrt{2}}{2} \\[2mm] \cos 135° = \dfrac{-1}{\sqrt{2}} = -\dfrac{\sqrt{2}}{2} \\[2mm] \tan 135° = \dfrac{1}{-1} = -1 \end{cases}$$

3.

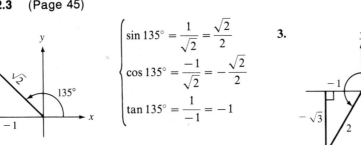

$$\begin{cases} \sin 240° = -\dfrac{\sqrt{3}}{2} \\[2mm] \cos 240° = -\dfrac{1}{2} \\[2mm] \tan 240° = \sqrt{3} \end{cases}$$

5.

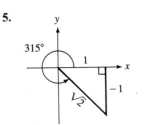

$$\begin{cases} \sin 315° = \dfrac{-1}{\sqrt{2}} = -\dfrac{\sqrt{2}}{2} \\[2mm] \cos 315° = \dfrac{1}{\sqrt{2}} = \dfrac{\sqrt{2}}{2} \\[2mm] \tan 315° = \dfrac{-1}{1} = -1 \end{cases}$$

7.

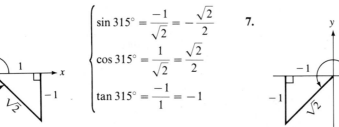

$$\begin{cases} \csc 225° = -\sqrt{2} \\[2mm] \sec 225° = -\sqrt{2} \\[2mm] \cot 225° = 1 \end{cases}$$

9.

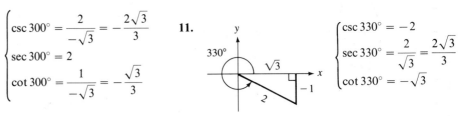

$$\begin{cases} \csc 300° = \dfrac{2}{-\sqrt{3}} = -\dfrac{2\sqrt{3}}{3} \\[2mm] \sec 300° = 2 \\[2mm] \cot 300° = \dfrac{1}{-\sqrt{3}} = -\dfrac{\sqrt{3}}{3} \end{cases}$$

11.

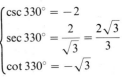

$$\begin{cases} \csc 330° = -2 \\[2mm] \sec 330° = \dfrac{2}{\sqrt{3}} = \dfrac{2\sqrt{3}}{3} \\[2mm] \cot 330° = -\sqrt{3} \end{cases}$$

13. $\sin 390° = \dfrac{1}{2}$, $\cos 390° = \dfrac{\sqrt{3}}{2}$, $\tan 390° = \dfrac{\sqrt{3}}{3}$ **15.** $\sin 510° = \dfrac{1}{2}$, $\cos 510° = -\dfrac{\sqrt{3}}{2}$, $\tan 510° = -\dfrac{\sqrt{3}}{3}$

17. $\sin(-45°) = -\dfrac{\sqrt{2}}{2}$, $\cos(-45°) = \dfrac{\sqrt{2}}{2}$, $\tan(-45°) = -1$ **19.** 1   **21.** 0   **23.** 2   **25.** $-\dfrac{23}{4}$

27. $\dfrac{\sqrt{6}+4}{4}$ **29.** $\dfrac{8\sqrt{3}-\sqrt{6}}{3}$ **31.** $-\sqrt{3}$ **33.** 1   **35.** $\theta = 30°$ **37.** $\theta = 300°$

39. $\theta = 210°$ **41.** $\theta = 315°$ **43.** $\theta = 135°$ **45.** $\theta = 90°, 270°$ **47.** $\theta = 330°$

49. $30°, 150°$ **51.** $30°, 210°$ **53.** $120°, 240°$ **55.** impossible   **57.** $210°$

59. $30°, 60°, 210°, 240°$

Exercise 2.4   (Page 51)

1. $\sin 50° \approx 0.77$, $\cos 50° \approx 0.64$, $\tan 50° \approx 1.2$ **3.** $\sin 235° \approx -0.82$, $\cos 235° \approx -0.57$, $\tan 235° \approx 1.4$

5. $\sin 12° \approx 0.21$, $\cos 12° \approx 0.98$, $\tan 12° \approx 0.21$ **7.** $\sin 115° \approx 0.91$, $\cos 115° \approx -0.42$, $\tan 115° \approx -2.1$

9. $\sin(-260°) \approx 0.98$, $\cos(-260°) \approx -0.17$, $\tan(-260°) \approx -5.7$

11. $\sin 17° \approx 0.2924$, $\cos 17° \approx 0.9563$, $\tan 17° \approx 0.3057$

13. $\sin 62° \approx 0.8829$, $\cos 62° \approx 0.4695$, $\tan 62° \approx 1.881$

15. $\sin 119° \approx 0.8746$, $\cos 119° \approx -0.4848$, $\tan 119° \approx -1.804$

17. $\sin(-233°) \approx 0.7986$, $\cos(-233°) \approx -0.6018$, $\tan(-233°) \approx -1.327$

19. $\sin 1723° \approx -0.9744$, $\cos 1723° \approx 0.2250$, $\tan 1723° \approx -4.331$ **21.** $\sin 20° \approx 0.3420$, $\cos 70° \approx 0.3420$

23. $\cos 5° \approx 0.9962$, $\sin 85° \approx 0.9962$ **25.** $\sec 84° \approx 9.567$, $\csc 6° \approx 9.567$ **27.** false   **29.** true

31. true   **33.** true   **35.** false   **37.** 0.3923   **39.** -0.6909 **41.** 0.9490

43. -1.3542 **45.** 1.4774   **47.** -0.1871 **49.** $14.0°$ **51.** $110.0°$ **53.** $207.0°$

55. $280.0°$ **57.** $49.0°$ **59.** $220.0°$

Exercise 2.5   (Page 58)

1. angle $B = 53°$, $a = 12$, $b = 16$ **3.** angle $B = 21.3°$, $a = 206$, $c = 221$ **5.** 15 ft   **7.** 721 ft

9. $4.0°$ **11.** 28.95 ft   **13.** 319 mi   **15.** S 62.9° W   **17.** 5.8 mi   **19.** 685 mph

21. 60.7 ft   **23.** 9180 ft; 1.74 mi   **25.** 22 ft   **27.** 29.9 ft   **29.** $48.3°$

Exercise 2.6   (Page 64)

1. 1450 ft   **3.** 556 ft   **5.** 631 ft   **7.** 143 ft   **9.** 123 ft   **11.** $\dfrac{r}{\tan\frac{\theta}{2}}$ **13.** $b = 2k\cos\alpha$

15. $H = k(\tan\alpha + \tan\beta)$ **17.** $D(\tan\phi - \tan\theta)$ **19.** $a = d\tan\beta$

Exercise 2.7   (Page 69)

1. 3 mph; 7 mph   **3.** $63°$ **5.** 352 mph   **7.** N 11.6° W   **9.** 323 lb   **11.** 384 lb

13. 1160 lb   **15.** $13.7°$ **17.** 224 lb; 48.1°   **19.** $13°$ **21.** $(F_1)_y = 80$ lb; $(F_1)_x = 95$ lb

23. N 79.9° E   **25.** 58 lb

Exercise 2.8   (Page 74)

1. $45°$ **3.** 1.31   **5.** 130 ohms   **7.** $46.4°$ **9.** 120 ohms   **11.** $\dfrac{\sqrt{3}k}{2}$ **13.** 6% less

REVIEW EXERCISES   (Page 77)

1. yes   **2.** no   **3.** no   **4.** yes   **5.** yes   **6.** no

7. $\cos\theta = -\dfrac{\sqrt{51}}{10}$, $\tan\theta = \dfrac{7\sqrt{51}}{51}$, $\csc\theta = -\dfrac{10}{7}$, $\sec\theta = -\dfrac{10\sqrt{51}}{51}$, $\cot\theta = \dfrac{\sqrt{51}}{7}$

8. $\sin\theta = -\dfrac{7\sqrt{130}}{130}$, $\cos\theta = -\dfrac{9\sqrt{130}}{130}$, $\csc\theta = -\dfrac{\sqrt{130}}{7}$, $\sec\theta = -\dfrac{\sqrt{130}}{9}$, $\cot\theta = \dfrac{9}{7}$

9. $\sin\theta = \dfrac{\sqrt{51}}{10}$, $\tan\theta = -\dfrac{\sqrt{51}}{7}$, $\csc\theta = \dfrac{10\sqrt{51}}{51}$, $\sec\theta = -\dfrac{10}{7}$, $\cot\theta = -\dfrac{7\sqrt{51}}{51}$

10. $\sin\theta = \dfrac{-8\sqrt{145}}{145}$, $\cos\theta = \dfrac{9\sqrt{145}}{145}$, $\tan\theta = \dfrac{-8}{9}$, $\csc\theta = -\dfrac{\sqrt{145}}{8}$, $\sec\theta = \dfrac{\sqrt{145}}{9}$

11. $\dfrac{1}{\sec\theta} = \cos\theta = \cos\theta\dfrac{\sin\theta}{\sin\theta} = \sin\theta\cot\theta$ 12. $\cos\theta\csc\theta = \cos\theta\dfrac{1}{\sin\theta} = \dfrac{\cos\theta}{\sin\theta} = \cot\theta$ 13. $\dfrac{\sqrt{6}}{4}$

14. $\dfrac{1}{2}$ 15. $\dfrac{9}{16}$ 16. 0   17. $\sin 930° = \dfrac{-1}{2}$, $\cos 930° = \dfrac{-\sqrt{3}}{2}$, $\tan 930° = \dfrac{\sqrt{3}}{3}$

18. $\sin 1380° = -\dfrac{\sqrt{3}}{2}$, $\cos 1380° = \dfrac{1}{2}$, $\tan 1380° = -\sqrt{3}$

19. $\sin(-300°) = \dfrac{\sqrt{3}}{2}$, $\cos(-300°) = \dfrac{1}{2}$, $\tan(-300°) = \sqrt{3}$

20. $\sin(-585°) = \dfrac{\sqrt{2}}{2}$, $\cos(-585°) = \dfrac{-\sqrt{2}}{2}$, $\tan(-585°) = -1$

21. $\sin 15° \approx 0.26$, $\cos 15° \approx 0.97$, $\tan 15° \approx 0.27$ 22. $\sin 160° \approx 0.34$, $\cos 160° \approx -0.94$, $\tan 160° \approx -0.36$
23. $\sin 265° \approx -0.99$, $\cos 265° \approx -0.09$, $\tan 265° \approx 11.4$
24. $\sin 340° \approx -0.34$, $\cos 340° \approx 0.94$, $\tan 340° \approx -0.36$
25. $\sin 15° \approx 0.2588$, $\cos 15° \approx 0.9659$, $\tan 15° \approx 0.2679$
26. $\sin 160° \approx 0.3420$, $\cos 160° \approx -0.9397$, $\tan 160° \approx -0.3640$
27. $\sin 265° \approx -0.9962$, $\cos 265° \approx -0.0872$, $\tan 265° \approx 11.430$
28. $\sin 340° \approx -0.3420$, $\cos 340° \approx 0.9397$, $\tan 340° \approx -0.3640$
29. $\sin(-160°) \approx -0.3420$, $\cos(-160°) \approx -0.9397$, $\tan(-160°) \approx 0.3640$
30. $\sin(-340°) \approx 0.3420$, $\cos(-340°) \approx 0.9397$, $\tan(-340°) \approx 0.3640$ 31. 119°   32. 211°
33. 317°   34. 57.7°   35. 100°   36. 287°   37. 17.3°   38. 59.7 ft   39. 53 mi
40. 150 mi   41. 10 lb   42. 11.5°

Exercise 3.1   (Page 86)

1. $\frac{1}{12}\pi$ 3. $\frac{2}{3}\pi$ 5. $\frac{7}{6}\pi$ 7. $\frac{5}{3}\pi$ 9. $\frac{13}{3}\pi$ 11. $-\frac{26}{9}\pi$ 13. 135°   15. 450°
17. 240°   19. $1080°/\pi \approx 343.77°$ 21. $-1800°/\pi \approx -572.96°$ 23. $2250°/\pi \approx 716.20°$
25. $\sqrt{3}/2$ 27. 1   29. $-\sqrt{2}/2$ 31. $-\frac{1}{2}$ 33. -2 35. $\sqrt{3}/3$ 37. 39 cm
39. 76°   41. 2970 mi   43. 38.6° N   45. 104.72 sq. units
47. 221,000 mi   49. 74 m   51. 1.15 mi

Exercise 3.2   (Page 90)

1. $2\pi\dfrac{\text{rads}}{\text{hr}}$ 3. $\dfrac{\pi}{1800}\dfrac{\text{rads}}{\text{sec}}$ 5. $\dfrac{4\pi}{59}\dfrac{\text{rads}}{\text{day}}$ 7. $\dfrac{176}{5\pi} \approx 11\dfrac{\text{rev}}{\text{sec}}$ 9. $\dfrac{1716}{\pi} \approx 550$ rpm

11. $\dfrac{75}{16}\pi \approx 14.7$ ft/sec   13. $\dfrac{300}{\pi} \approx 95$ rpm   15. $\dfrac{50}{3}$ rpm   17. $\dfrac{225}{88}\pi \approx 8$ mph

19. approximately 933 mph   21. 44.9° N

23. Use the results of Exercise 22. $W = \dfrac{R_1}{R}\cdot W_1$. Hence, $W_2 = \dfrac{R}{R_2}\cdot W = \dfrac{R}{R_2}\cdot\dfrac{R_1}{R}\cdot W_1 = \dfrac{R_1}{R_2}\cdot W_1$

Exercise 3.3   (Page 95)

1. $\dfrac{1}{2}$ **3.** $-\dfrac{\sqrt{3}}{2}$ **5.** 1   **7.** 2   **9.** $-\dfrac{\sqrt{3}}{2}$ **11.** -1 **13.** 0.9093   **15.** -0.1455

17. 0.8163   **19.** 3.5253   **21.** 0.6421   **23.** -0.1411 **25.** $(0, -1)$ **27.** $(-1, 0)$

29. $(-1, 0)$ **31.** $(0, 1)$ **33.** $\left(\dfrac{\sqrt{2}}{2}, \dfrac{\sqrt{2}}{2}\right)$ **35.** $\left(-\dfrac{\sqrt{2}}{2}, -\dfrac{\sqrt{2}}{2}\right)$ **37.** $\left(\dfrac{1}{2}, \dfrac{\sqrt{3}}{2}\right)$

39. $\left(\dfrac{-1}{2}, \dfrac{-\sqrt{3}}{2}\right)$ **41.** $\left(\dfrac{-1}{2}, \dfrac{-\sqrt{3}}{2}\right)$ **43.** $\left(\dfrac{1}{2}, \dfrac{\sqrt{3}}{2}\right)$ **45.** $\left(\dfrac{\sqrt{3}}{2}, \dfrac{-1}{2}\right)$ **47.** $\left(\dfrac{\sqrt{3}}{2}, \dfrac{-1}{2}\right)$

49. The formula for area of a sector of a circle is $A = \frac{1}{2}r^2\theta$. In the unit circle, $r = 1$. Hence, $A = \frac{1}{2}(1)^2\theta = \frac{1}{2}\theta$.
51. $A = \frac{1}{2}bh$ with $b = 1$ and $h = \tan\theta$. Hence, $A = \frac{1}{2}\tan\theta$.

Exercise 3.4   (Page 102)

1. $2; 2\pi$ **3.** $1; \dfrac{2\pi}{9}$ **5.** $1; 6\pi$ **7.** $1; 10\pi$ **9.** $3; 4\pi$ **11.** $\dfrac{1}{2}; 2$ **13.** $3; 1$ **15.** $\dfrac{1}{3}; \dfrac{2\pi^2}{3}$

17. **19.** **21.**

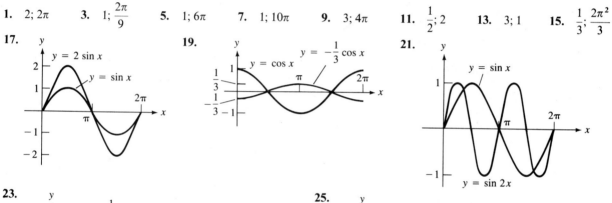

23. **25.**

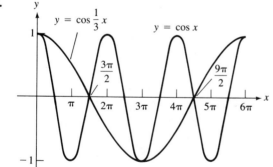

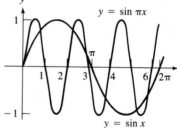

27. **29.** **31.**

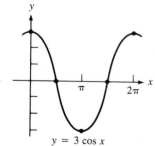

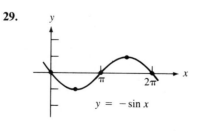

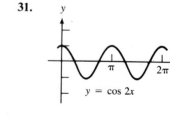

33.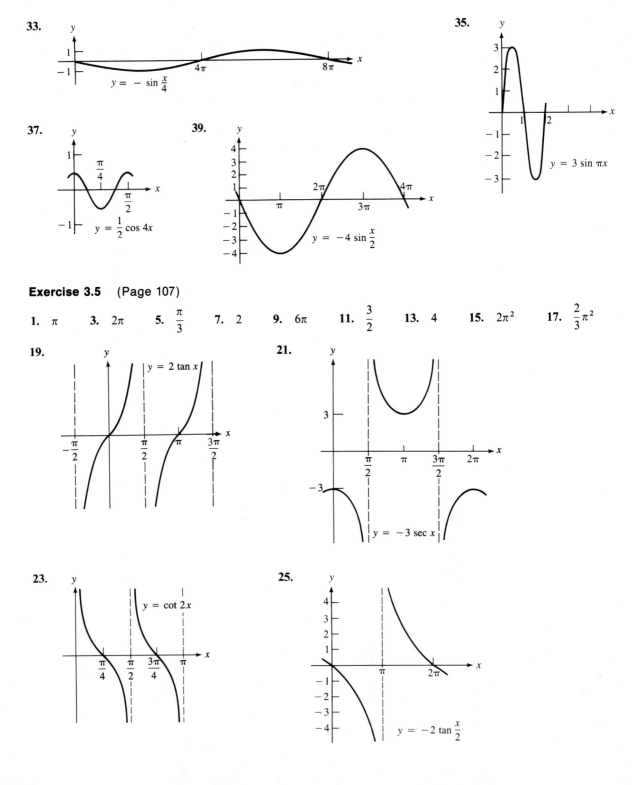

35.

37.

39.

Exercise 3.5 (Page 107)

1. π **3.** 2π **5.** $\dfrac{\pi}{3}$ **7.** 2   **9.** 6π **11.** $\dfrac{3}{2}$ **13.** 4   **15.** $2\pi^2$ **17.** $\dfrac{2}{3}\pi^2$

19.

21.

23.

25.

27.

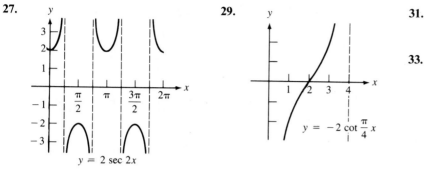

$y = 2 \sec 2x$

29.

$y = -2 \cot \dfrac{\pi}{4} x$

31. The value of y can be as large as you want.

33. $\ldots, -\pi, 0, \pi, 2\pi, 3\pi, \ldots$

Exercise 3.6 (Page 113)

1. 2 units up; 2π **3.** 1 unit down; π **5.** 7 units up; $\dfrac{2\pi}{5}$ **7.** 3 units up; 2π **9.** 5 units down; π

11. 6 units up; 1   **13.** 2π; $\dfrac{\pi}{3}$ to the right   **15.** 2π; $\dfrac{\pi}{6}$ to the left   **17.** 1; no phase shift

19. π; π to the right   **21.** 2π; $\dfrac{\pi}{4}$ to the left   **23.** π; $\dfrac{\pi}{2}$ to the left   **25.** 2; $\dfrac{1}{2}$ to the left

27. 6π; 18π to the right   **29.** $\dfrac{\pi}{7}$; $\dfrac{3}{2}\pi$ to the right

31.

$y = -4 + \sin x$

33.

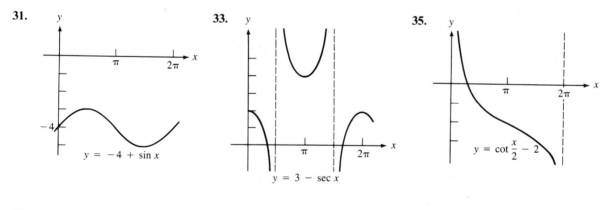

$y = 3 - \sec x$

35.

$y = \cot \dfrac{x}{2} - 2$

37.

$y = \sin\left(x + \dfrac{\pi}{2}\right)$

39.

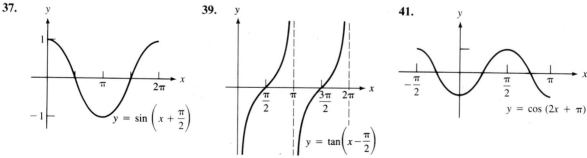

$y = \tan\left(x - \dfrac{\pi}{2}\right)$

41.

$y = \cos(2x + \pi)$

43.

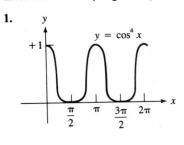

$y = \sec\left(3x + \dfrac{\pi}{2}\right)$

45.

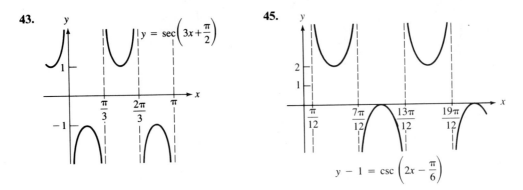

$y - 1 = \csc\left(2x - \dfrac{\pi}{6}\right)$

Exercise 3.7 (Page 116)

1.

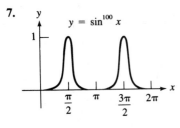

$y = \cos^4 x$

3.

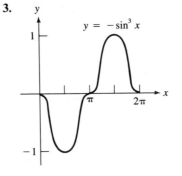

$y = -\sin^3 x$

5.

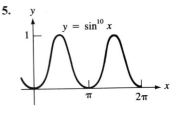

$y = \sin^{10} x$

7.

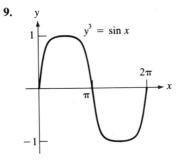

$y = \sin^{100} x$

9.

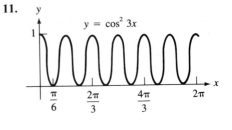

$y^3 = \sin x$

11.

$y = \cos^2 3x$

13.

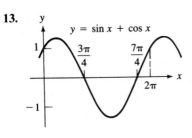

$y = \sin x + \cos x$

15.

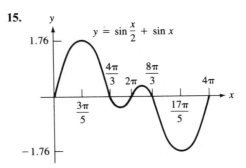

$y = \sin \dfrac{x}{2} + \sin x$

17.

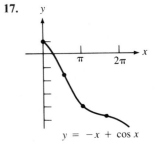

$y = -x + \cos x$

19.

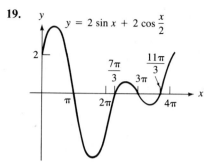

$y = 2 \sin x + 2 \cos \dfrac{x}{2}$

21.

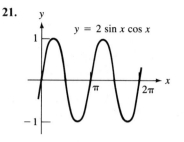

$y = 2 \sin x \cos x$

23.

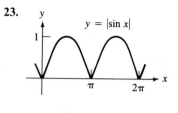

$y = |\sin x|$

25.

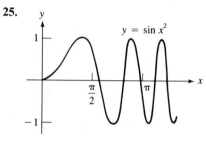

$y = \sin x^2$

27.

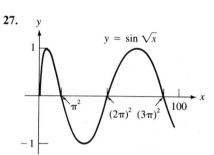

$y = \sin \sqrt{x}$

Exercise 3.8   (Page 119)

1. $V = 310 \sin 100\pi t$ **3.** 4π sec   **5.**   48 dynes per cm

7.   period of 2 seconds per cycle; frequency of $\frac{1}{2}$ cycle per second   **9.**   286 hertz

11.

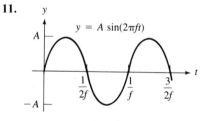

$y = A \sin(2\pi f t)$

REVIEW EXERCISES   (Page 121)

1. $\frac{7}{12}\pi$ **2.** $\frac{65}{36}\pi$ **3.** $\frac{53}{30}\pi$ **4.** $\frac{-7}{12}\pi$ **5.** $570°$ **6.** $-150°$

7. $1260°$ **8.** $458.366°$ **9.** $\frac{1}{2}$ **10.** $\dfrac{\sqrt{3}}{2}$ **11.** $-\sqrt{3}$ **12.** 2

13. 2750 mi   **14.** $41.5°$ **15.** 19 sq cm   **16.** $\dfrac{\pi}{43,200}\dfrac{\text{rads}}{\text{sec}}$

17. $\dfrac{1815}{\pi}$ rpm   **18.** about 12 mph   **19.** $\left(\dfrac{-\sqrt{3}}{2}, \dfrac{-1}{2}\right)$ **20.** $\left(-\dfrac{\sqrt{2}}{2}, -\dfrac{\sqrt{2}}{2}\right)$

21. 0.0000 **22.** 0.7539 **23.** -2.1850 **24.** undefined   **25.** 1.8508 **26.** -0.5774

27. -0.3983 **28.** -0.2003 **29.** $4; \dfrac{2\pi}{3}$ **30.** $\dfrac{1}{8}; \dfrac{\pi}{2}$ **31.** $\dfrac{1}{3}; 6\pi$ **32.** $0.875; 8\pi$

33. 2 units up   **34.** $\dfrac{\pi}{6}$ to the left   **35.** 4 units up; $\dfrac{21}{2}$ to the left   **36.** 1 unit down; $\dfrac{5}{2}\pi$ to the right

37.

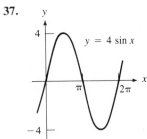

$y = 4 \sin x$

38.

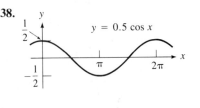

$y = 0.5 \cos x$

39.

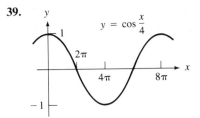

$y = \cos \dfrac{x}{4}$

40.

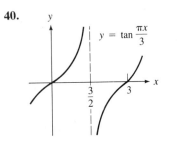

$y = \tan \dfrac{\pi x}{3}$

41.

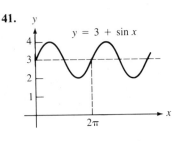

$y = 3 + \sin x$

42.

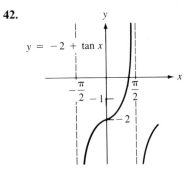

$y = -2 + \tan x$

43.

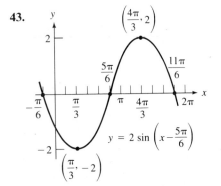

$\left(\dfrac{4\pi}{3}, 2\right)$

$y = 2 \sin\left(x - \dfrac{5\pi}{6}\right)$

$\left(\dfrac{\pi}{3}, -2\right)$

44.

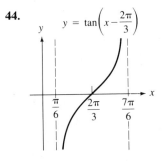

$y = \tan\left(x - \dfrac{2\pi}{3}\right)$

45.

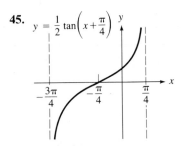

$y = \dfrac{1}{2} \tan\left(x + \dfrac{\pi}{4}\right)$

46.

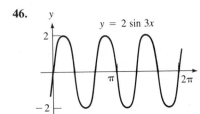

$y = 2 \sin 3x$

47.

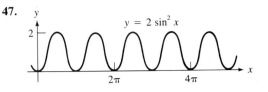

$y = 2 \sin^2 x$

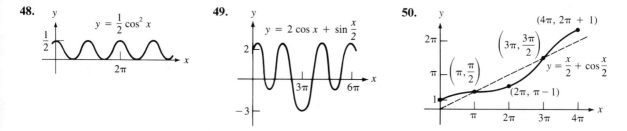

48. $y = \frac{1}{2}\cos^2 x$

49. $y = 2\cos x + \sin\dfrac{x}{2}$

50. $y = \dfrac{x}{2} + \cos\dfrac{x}{2}$; $(4\pi, 2\pi + 1)$; $\left(3\pi, \dfrac{3\pi}{2}\right)$; $\left(\pi, \dfrac{\pi}{2}\right)$; $(2\pi, \pi - 1)$

Exercise 4.1   (Page 128)

1. yes   3. no   5. no   7. yes   9. no   11. a meaningless statement   13. no
15. yes   17. no   19. no   21. yes   23. yes   25. no   27. no   29. no

Exercise 4.2   (Page 136)

1. $\sin 195° = \sin(45° + 150°) = \dfrac{-\sqrt{6} + \sqrt{2}}{4}$ 3. $\tan 195° = \tan(225° - 30°) = \dfrac{3 - \sqrt{3}}{3 + \sqrt{3}}$

5. $\cos\dfrac{11\pi}{12} = \cos\left(\dfrac{\pi}{6} + \dfrac{3\pi}{4}\right) = \dfrac{-\sqrt{6} - \sqrt{2}}{4}$ 7. $\cos\dfrac{19\pi}{12} = \cos\left(\dfrac{11\pi}{6} - \dfrac{\pi}{4}\right) = \dfrac{\sqrt{6} - \sqrt{2}}{4}$

9. $\sin 75° = \sin(30° + 45°) = \dfrac{\sqrt{2} + \sqrt{6}}{4}$ 11. $\tan 105° = \tan(60° + 45°) = \dfrac{\sqrt{3} + 1}{1 - \sqrt{3}}$

13. $\sin 105° = \sin(60° + 45°) = \dfrac{\sqrt{6} + \sqrt{2}}{4}$ 15. $\sin\dfrac{\pi}{12} = \sin\left(\dfrac{\pi}{3} - \dfrac{\pi}{4}\right) = \dfrac{\sqrt{6} - \sqrt{2}}{4}$

17. $\tan\dfrac{5\pi}{12} = \tan\left(\dfrac{2\pi}{3} - \dfrac{\pi}{4}\right) = \dfrac{-\sqrt{3} - 1}{1 - \sqrt{3}}$ 19. $\cos\dfrac{21\pi}{12} = \cos\left(2\pi - \dfrac{\pi}{4}\right) = \dfrac{\sqrt{2}}{2}$

21. $\sec 75° = \dfrac{1}{\cos 75°} = \dfrac{1}{\cos(30° + 45°)} = \dfrac{1}{\dfrac{\sqrt{6} - \sqrt{2}}{4}} = \sqrt{6} + \sqrt{2}$

23. $\cot(-165°) = -\cot 165° = -\dfrac{1}{\tan(120° + 45°)} = -\dfrac{1}{\dfrac{1 - \sqrt{3}}{1 + \sqrt{3}}} = 2 + \sqrt{3}$

25. $\csc\dfrac{7\pi}{12} = \dfrac{1}{\sin\left(\dfrac{\pi}{4} + \dfrac{\pi}{3}\right)} = \dfrac{1}{\dfrac{\sqrt{2} + \sqrt{6}}{4}} = \sqrt{6} - \sqrt{2}$

27. $\sec\left(-\dfrac{\pi}{12}\right) = \dfrac{1}{\cos\left(\dfrac{\pi}{4} - \dfrac{\pi}{3}\right)} = \dfrac{1}{\dfrac{\sqrt{2} + \sqrt{6}}{4}} = \sqrt{6} - \sqrt{2}$

29. $\sin(60° + \theta) = \sin 60° \cos\theta + \cos 60° \sin\theta = \dfrac{\sqrt{3}}{2}\cos\theta + \dfrac{1}{2}\sin\theta$

31. $\tan(\pi + x) = \dfrac{\tan\pi + \tan x}{1 - \tan\pi\tan x} = \dfrac{\tan x}{1} = \tan x$ 33. $\cos(\pi - x) = \cos\pi\cos x + \sin\pi\sin x = -\cos x$

35. $\sin(10° + 30°) = \sin 40°$ 37. $\tan(75° + 40°) = \tan 115°$ 39. $\cos(120° - 40°) = \cos 80°$
41. $\sin(x + 2x) = \sin 3x$ 43. $\sin(\alpha + \beta) = -\frac{56}{65}$; $\cos(\alpha - \beta) = \frac{63}{65}$
45. $\tan(\alpha + \beta) = \frac{140}{171}$; $\tan(\alpha - \beta) = -\frac{220}{21}$ 47. $\sin\alpha = \frac{416}{425}$; $\cos\alpha = \frac{87}{425}$

Exercise 4.3 (Page 141)

1. $\sin 2\alpha$ **3.** $\sin 6\theta$ **5.** $\cos 2\beta$ **7.** $\cos \beta$ **9.** $2 \sin 2\theta$ **11.** $(\sin 4\theta)^2 = \sin^2 4\theta$

13. $\frac{1}{2}\cos 2\alpha$ **15.** $\cos 18\theta$ **17.** $\sin^2 10\theta$ **19.** $\tan 8C$ **21.** $\tan A$ **23.** $\cos 8x$

25. $-\cos 10x$ **27.** $\frac{\sqrt{3}}{2}$ **29.** undefined   **31.** -0.5 **33.** 0 **35.** $\frac{\sqrt{3}}{2}$ **37.** 0.5

39. $\sin 2\theta = \frac{120}{169}, \cos 2\theta = -\frac{119}{169}, \tan 2\theta = -\frac{120}{119}$ **41.** $\sin 2\theta = \frac{120}{169}, \cos 2\theta = -\frac{119}{169}, \tan 2\theta = -\frac{120}{119}$

43. $\sin 2\theta = \frac{24}{25}, \cos 2\theta = \frac{7}{25}, \tan 2\theta = \frac{24}{7}$ **45.** $\sin 2\theta = -\frac{336}{625}, \cos 2\theta = -\frac{527}{625}, \tan 2\theta = \frac{336}{527}$

47. $\sin 2\theta = \frac{720}{1681}, \cos 2\theta = -\frac{1519}{1681}, \tan 2\theta = -\frac{720}{1519}$ **49.** $\sin 2\theta = -\frac{720}{1681}, \cos 2\theta = \frac{1519}{1681}, \tan 2\theta = -\frac{720}{1519}$

Exercise 4.4 (Page 149)

1. $\frac{\sqrt{2+\sqrt{3}}}{2}$ **3.** $-2-\sqrt{3}$ **5.** $\frac{\sqrt{2-\sqrt{2}}}{2}$ **7.** $\frac{\sqrt{2+\sqrt{3}}}{2}$ **9.** $\sqrt{3}-2$ **11.** 1

13. $\sin\frac{\theta}{2} = \frac{\sqrt{10}}{10}, \cos\frac{\theta}{2} = \frac{3\sqrt{10}}{10}, \tan\frac{\theta}{2} = \frac{1}{3}$ **15.** $\sin\frac{\theta}{2} = \frac{2\sqrt{5}}{5}, \cos\frac{\theta}{2} = -\frac{\sqrt{5}}{5}, \tan\frac{\theta}{2} = -2$

17. $\sin\frac{\theta}{2} = \frac{3\sqrt{34}}{34}, \cos\frac{\theta}{2} = -\frac{5\sqrt{34}}{34}, \tan\frac{\theta}{2} = -\frac{3}{5}$ **19.** $\sin\frac{\theta}{2} = \frac{\sqrt{82}}{82}, \cos\frac{\theta}{2} = \frac{9\sqrt{82}}{82}, \tan\frac{\theta}{2} = \frac{1}{9}$

21. $\sin\frac{\theta}{2} = \frac{4\sqrt{17}}{17}, \cos\frac{\theta}{2} = \frac{\sqrt{17}}{17}, \tan\frac{\theta}{2} = 4$ **23.** $\sin\frac{\theta}{2} = \frac{\sqrt{6}}{6}, \cos\frac{\theta}{2} = \frac{-\sqrt{30}}{6}, \tan\frac{\theta}{2} = \frac{-\sqrt{5}}{5}$ **25.** $\cos 15°$

27. $\sin 40°$ **29.** $-\sin 200°$ **31.** $\cos 275°$ **33.** $\tan 100°$ **35.** $\tan 40°$ **37.** $\tan \pi = 0$

39. $\tan\frac{x}{4}$ **41.** $\tan 5A$

Exercise 4.5 (Page 155)

1. $\frac{1}{4}$ **3.** $\frac{1}{4}$ **5.** $\frac{\sqrt{2}}{4}$ **7.** $\frac{1}{4}$ **9.** $\frac{\sqrt{3}-2}{4}$ **11.** $\frac{\sqrt{6}}{2}$ **13.** $-\frac{\sqrt{2}}{2}$ **15.** $-\frac{\sqrt{2}}{2}$

17. $\frac{\sqrt{6}}{2}$ **19.** 0 **21.** $\frac{1}{2}\cos 10° - \frac{1}{2}\cos 70°$ **23.** $\frac{1}{2}\cos\frac{6\pi}{7} + \frac{1}{2}\cos\frac{4\pi}{7}$ **25.** $\frac{1}{2}\cos 8\theta + \frac{1}{2}\cos 2\theta$

27. $\frac{1}{2}\sin\frac{3A}{2} + \frac{1}{2}\sin\frac{A}{2}$ **29.** $\frac{1}{2}\cos\frac{x}{6} - \frac{1}{2}\cos\frac{5x}{6}$ **31.** $2\sin 40° \cos 10°$ **33.** $-2\sin\frac{2\pi}{3}\sin\frac{\pi}{9}$

35. $2\cos 4\theta \cos \theta$ **37.** $2\cos\frac{3A}{4}\sin\frac{A}{4}$ **39.** $2\cos\frac{5x}{12}\cos\frac{x}{12}$

Exercise 4.6 (Page 159)

1. $10\sin(x + 53.1°)$ **3.** $10\sin(x - 53.1°)$ **5.** $\sqrt{5}\sin(x + 26.6°)$ **7.** $\sqrt{2}\sin(x + 45°)$

9. $\sqrt{26}\sin(x + 101.3°)$ or $-\sqrt{26}\sin(x - 78.7°)$ **11.** $2\sqrt{3}\sin(x - 60°)$

13. $\left(\frac{A}{\sqrt{A^2 + B^2}}\right)^2 + \left(\frac{B}{\sqrt{A^2 + B^2}}\right)^2 = \frac{A^2}{A^2 + B^2} + \frac{B^2}{A^2 + B^2} = \frac{A^2 + B^2}{A^2 + B^2} = 1$

15.

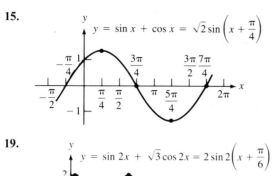

$$y = \sin x + \cos x = \sqrt{2}\sin\left(x + \frac{\pi}{4}\right)$$

17.

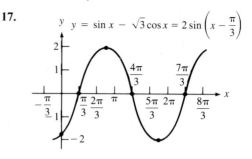

$$y = \sin x - \sqrt{3}\cos x = 2\sin\left(x - \frac{\pi}{3}\right)$$

19.

$$y = \sin 2x + \sqrt{3}\cos 2x = 2\sin 2\left(x + \frac{\pi}{6}\right)$$

Exercise 4.7   (Page 162)

1. 8.1°

3. If l_1 is perpendicular to l_2, then the angle between the lines is 90°. Because tan 90° is undefined, the denominator of $\dfrac{m_2 - m_1}{1 + m_2 m_1}$ must be zero. Hence, $1 + m_2 m_1 = 0$, or $m_1 = -\dfrac{1}{m_2}$.

5. 13   **7.** $\frac{5}{2}$ cycles per second apart.

REVIEW EXERCISES   (Page 164)

1. $\dfrac{\sqrt{2} - \sqrt{6}}{4}$ **2.** $\dfrac{\sqrt{2} + \sqrt{6}}{4}$ **3.** $\sqrt{3} - 2$ **4.** $\dfrac{-\sqrt{6} - \sqrt{2}}{4}$ **5.** $\dfrac{\sqrt{2} - \sqrt{6}}{4}$

6. $\dfrac{-\sqrt{3} - 1}{1 - \sqrt{3}} = 2 + \sqrt{3}$ **7.** $-\dfrac{\sqrt{3}}{2}$ **8.** $-\dfrac{1}{2}$ **9.** $\sqrt{3}$ **10.** $\dfrac{\sqrt{2} - \sqrt{2}}{2}$ **11.** $\dfrac{\sqrt{2} + \sqrt{2}}{2}$

12. $\sqrt{\dfrac{2 - \sqrt{2}}{2 + \sqrt{2}}} = \sqrt{2} - 1$ **13.** $\sin 71°$ **14.** $\tan(-31°) = -\tan 31°$ **15.** $\cos 20°$ **16.** $\cos \frac{4}{11}\pi$

17. $\sin\dfrac{\pi}{7}$ **18.** $\sin 34°$ **19.** $\cos 34°$ **20.** $\tan 434° = \tan 74°$ **21.** $2\sin\dfrac{3\pi}{4} = \sqrt{2}$

22. $\cos\dfrac{2\pi}{3} = -\dfrac{1}{2}$ **23.** $\cos\dfrac{2\pi}{13}$ **24.** $2\sin^2\theta$ **25.** $\sin\theta\cos\alpha + \cos\theta\sin\alpha$

26. $\cos\theta\cos\alpha - \sin\theta\sin\alpha$ **27.** $\dfrac{\tan\theta + \tan\alpha}{1 - \tan\theta\tan\alpha}$ **28.** $\sin\theta\cos\alpha - \cos\theta\sin\alpha$

29. $\cos\theta\cos\alpha + \sin\theta\sin\alpha$ **30.** $\dfrac{\tan\theta - \tan\alpha}{1 + \tan\theta\tan\alpha}$ **31.** $2\sin\theta\cos\theta$

32. $\cos^2\theta - \sin^2\theta$, or $2\cos^2\theta - 1$, or $1 - 2\sin^2\theta$ **33.** $\dfrac{2\tan\theta}{1 - \tan^2\theta}$ **34.** $\pm\sqrt{\dfrac{1 - \cos\theta}{2}}$

35. $\pm\sqrt{\dfrac{1 + \cos\theta}{2}}$ **36.** $\dfrac{\sin\theta}{1 + \cos\theta}$ or $\dfrac{1 - \cos\theta}{\sin\theta}$

37. $\cos(60° + \theta) = \cos 60° \cos \theta - \sin 60° \sin \theta = \frac{1}{2}(\cos \theta - \sqrt{3} \sin \theta)$

38. $\sin\left(\frac{3\pi}{2} - \theta\right) = \sin\frac{3\pi}{2} \cos \theta - \cos\frac{3\pi}{2} \sin \theta = -\cos \theta$　　　**39.** $\tan(180° - \theta) = \frac{\tan 180° - \tan \theta}{1 + \tan 180° \tan \theta} = -\tan \theta$

40. $\sin(120° + \theta) = \sin 120° \cos \theta + \cos 120° \sin \theta = \frac{1}{2}(\sqrt{3} \cos \theta - \sin \theta)$

41. $\cos(300° - \theta) = \cos 300° \cos \theta + \sin 300° \sin \theta = \frac{1}{2}(\cos \theta - \sqrt{3} \sin \theta)$

42. $\tan\left(\frac{\pi}{4} + \theta\right) = \dfrac{\tan\frac{\pi}{4} + \tan \theta}{1 - \tan\frac{\pi}{4}\tan \theta} = \dfrac{1 + \tan \theta}{1 - \tan \theta}$　　　**43.** $\dfrac{\sin 2x}{2 \cos x} = \dfrac{2 \sin x \cos x}{2 \cos x} = \sin x$

44. $\pm\sqrt{\sin^2\theta + \cos 2\theta} = \pm\sqrt{\sin^2\theta + \cos^2\theta - \sin^2\theta} = \pm\sqrt{\cos^2\theta} = \pm\cos \theta$

45. $\sin 2\theta = -\frac{120}{169}, \cos 2\theta = \frac{119}{169}, \tan 2\theta = -\frac{120}{119}$　　　**46.** $\sin 2\theta = \frac{120}{169}, \cos 2\theta = \frac{119}{169}, \tan 2\theta = \frac{120}{119}$

47. $\sin 2\theta = -\frac{840}{841}, \cos 2\theta = \frac{41}{841}, \tan 2\theta = -\frac{840}{41}$　　　**48.** $\sin 2\theta = \frac{840}{841}, \cos 2\theta = \frac{41}{841}, \tan 2\theta = \frac{840}{41}$

49. $\sin 2\theta = \frac{24}{25}, \cos 2\theta = \frac{-7}{25}, \tan 2\theta = \frac{-24}{7}$　　　**50.** $\sin 2\theta = -\frac{24}{25}, \cos 2\theta = -\frac{7}{25}, \tan 2\theta = \frac{24}{7}$

51. $\sin\frac{\theta}{2} = \frac{\sqrt{26}}{26}, \cos\frac{\theta}{2} = \frac{5\sqrt{26}}{26}, \tan\frac{\theta}{2} = \frac{1}{5}$　　　**52.** $\sin\frac{\theta}{2} = \frac{\sqrt{26}}{26}, \cos\frac{\theta}{2} = -\frac{5\sqrt{26}}{26}, \tan\frac{\theta}{2} = -\frac{1}{5}$

53. $\sin\frac{\theta}{2} = \frac{\sqrt{26}}{26}, \cos\frac{\theta}{2} = -\frac{5\sqrt{26}}{26}, \tan\frac{\theta}{2} = -\frac{1}{5}$　　　**54.** $\sin\frac{\theta}{2} = \frac{2\sqrt{5}}{5}, \cos\frac{\theta}{2} = \frac{\sqrt{5}}{5}, \tan\frac{\theta}{2} = 2$

55. $\sin\frac{\theta}{2} = \frac{2\sqrt{5}}{5}, \cos\frac{\theta}{2} = \frac{\sqrt{5}}{5}, \tan\frac{\theta}{2} = 2$　　　**56.** $\sin\frac{\theta}{2} = \frac{2\sqrt{5}}{5}, \cos\frac{\theta}{2} = -\frac{\sqrt{5}}{5}, \tan\frac{\theta}{2} = -2$　　　**57.** $-\dfrac{2 + \sqrt{3}}{4}$

58. $\frac{1}{2}\left(-\frac{\sqrt{3}}{2} + 1\right)$　　　**59.** $\dfrac{1 - \sqrt{3}}{4}$　　　**60.** $\frac{1}{2}\left(-\frac{1}{2} - \frac{\sqrt{3}}{2}\right)$　　　**61.** $2 \sin 6° \cos 1°$

62. $2 \cos 226° \sin 86°$　　　**63.** $-2 \sin\frac{2\pi}{5} \sin\frac{\pi}{5}$　　　**64.** $2 \cos\frac{5\pi}{14} \cos\frac{\pi}{14}$　　　**65.** $-\dfrac{\sqrt{2}}{2}$　　　**66.** $\dfrac{\sqrt{6}}{2}$

67. $-\dfrac{\sqrt{2}}{2}$　　　**68.** $\dfrac{\sqrt{6}}{2}$

Exercise 5.1　(Page 174)

1. $\frac{\pi}{6} + 2n\pi, \frac{5\pi}{6} + 2n\pi$　　　**3.** $\frac{\pi}{3} + n\pi$　　　**5.** $\frac{\pi}{2} + n\pi, 0 + 2n\pi$　　　**7.** $n\pi, \frac{\pi}{4} + n\pi$　　　**9.** $\frac{\pi}{6} + n\pi, \frac{5\pi}{6} + n\pi$

11. $60° + n360°, 120° + n360°$　　　**13.** $135° + n180°$　　　**15.** $n90°$　　　**17.** $n180°, 45° + n180°$

19. $45° + n180°$　　　**21.** $120°$　　　**23.** $0°, 90°, 180°, 270°$　　　**25.** $90°, 270°$　　　**27.** $0°, 180°$

29. $45°, 135°, 225°, 315°$　　　**31.** $0°, 120°, 240°$　　　**33.** $210°, 330°$　　　**35.** $60°, 240°, 120°, 300°$

37. no solution　　　**39.** $0°, 180°, 30°, 210°$　　　**41.** $60°, 300°, 180°$　　　**43.** $90°, 210°, 270°, 330°$

45. $45°, 225°$　　　**47.** $0°, 240°$　　　**49.** $0°, 120°, 240°$　　　**51.** $0°, 180°$　　　**53.** $30°, 150°, 210°, 330°$

55. $45°$　　　**57.** $0°, 30°, 150°, 180°$　　　**59.** $90°, 120°, 240°, 270°$

61. $0°, 30°, 90°, 150°, 180°, 210°, 270°, 330°$　　　**63.** $0°, 240°$　　　**65.** $60°, 120°, 240°, 300°$　　　**67.** $0°, 180°$

69. $45°, 135°, 315°$　　　**71.** $0°, 120°$　　　**73.** $30°$　　　**75.** $0°, 240°$　　　**77.** $\frac{\pi}{4}, \frac{5\pi}{4}$　　　**79.** $0, \frac{\pi}{2}, \pi, \frac{3\pi}{2}$

81. $0, \frac{\pi}{4}, \frac{\pi}{2}, \frac{3\pi}{4}, \pi, \frac{5\pi}{4}, \frac{3\pi}{2}, \frac{7\pi}{4}$　　　**83.** $0, \frac{\pi}{3}, \frac{\pi}{2}, \frac{2\pi}{3}, \pi, \frac{4\pi}{3}, \frac{3\pi}{2}, \frac{5\pi}{3}$

85. $\frac{\pi}{12}, \frac{\pi}{6}, \frac{\pi}{4}, \frac{5\pi}{12}, \frac{\pi}{2}, \frac{7\pi}{12}, \frac{3\pi}{4}, \frac{5\pi}{6}, \frac{11\pi}{12}, \frac{13\pi}{12}, \frac{7\pi}{6}, \frac{5\pi}{4}, \frac{17\pi}{12}, \frac{3\pi}{2}, \frac{19\pi}{12}, \frac{7\pi}{4}, \frac{11\pi}{6}, \frac{23\pi}{12}$　　　**87.** $\frac{\pi}{4}, \frac{3\pi}{4}, \frac{5\pi}{4}, \frac{7\pi}{4}$

89. $\frac{\pi}{6}, \frac{\pi}{2}, \frac{5\pi}{6}, \frac{7\pi}{6}, \frac{3\pi}{2}, \frac{11\pi}{6}$　　　**91.** $\frac{\pi}{2}, \frac{3\pi}{2}$　　　**93.** $\frac{7\pi}{6}, \frac{11\pi}{6}$

95. $\dfrac{\pi}{20}, \dfrac{3\pi}{20}, \dfrac{\pi}{4}, \dfrac{7\pi}{20}, \dfrac{9\pi}{20}, \dfrac{11\pi}{20}, \dfrac{13\pi}{20}, \dfrac{3\pi}{4}, \dfrac{17\pi}{20}, \dfrac{19\pi}{20}, \dfrac{21\pi}{20}, \dfrac{23\pi}{20}, \dfrac{5\pi}{4}, \dfrac{27\pi}{20}, \dfrac{29\pi}{20}, \dfrac{31\pi}{20}, \dfrac{33\pi}{20}, \dfrac{7\pi}{4}, \dfrac{37\pi}{20}, \dfrac{39\pi}{20}$

97. $0, \dfrac{\pi}{4}, \pi, \dfrac{5\pi}{4}$ **99.** $\dfrac{\pi}{6}, \dfrac{5\pi}{6}, \dfrac{7\pi}{6}, \dfrac{11\pi}{6}$ **101.** $63.4°, 116.6°, 243.4°, 296.6°$ **103.** $193.7°, 346.3°$

105. $20.7°, 159.3°$ **107.** $9.2°, 99.2°, 189.2°, 279.2°$

Exercise 5.2   (Page 178)

1. $30°, 150°$ **3.** $135°, 315°$ **5.** $135°, 225°$ **7.** $30°, 210°$ **9.** $0°, 180°$ **11.** no values

13. $270°$ **15.** $60°, 300°$ **17.** $\dfrac{\pi}{6}, \dfrac{11\pi}{6}$ **19.** $\dfrac{\pi}{6}, \dfrac{5\pi}{6}$ **21.** $\dfrac{\pi}{3}, \dfrac{4\pi}{3}$ **23.** $\dfrac{2\pi}{3}, \dfrac{5\pi}{3}$

25. no values   **27.** $\dfrac{\pi}{2}$ **29.** There are no numbers with a sine of 2.   **31.** $\dfrac{\pi}{3}, \dfrac{5\pi}{3}$ **33.** $1.000, 2.142$

35. $1.000, 4.141$ **37.** $2.000, 4.283$ **39.** $2.000, 5.142$ **41.** $0.142, 3.000$ **43.** $3.000, 6.142$

45. no values   **47.** $55°, 125°$ **49.** $125°, 305°$ **51.** $55°, 305°$ **53.** $130°, 310°$

55. $50°, 310°$ **57.** $15°, 165°$ **59.** no values

Exercise 5.3   (Page 187)

1. $\dfrac{\pi}{6}$ **3.** $\dfrac{\pi}{2}$ **5.** $\dfrac{\pi}{4}$ **7.** no value   **9.** $\dfrac{3\pi}{4}$ **11.** $-\dfrac{\pi}{3}$

13. $\sin\dfrac{\pi}{6} = \dfrac{1}{2}, \cos\dfrac{\pi}{6} = \dfrac{\sqrt{3}}{2}, \tan\dfrac{\pi}{6} = \dfrac{\sqrt{3}}{3}$ **15.** $\sin 0 = 0, \cos 0 = 1, \tan 0 = 0$

17. $\sin\dfrac{5\pi}{6} = \dfrac{1}{2}, \cos\dfrac{5\pi}{6} = -\dfrac{\sqrt{3}}{2}, \tan\dfrac{5\pi}{6} = -\dfrac{\sqrt{3}}{3}$ **19.** $\sin\dfrac{\pi}{2} = 1, \cos\dfrac{\pi}{2} = 0, \tan\dfrac{\pi}{2}$ is undefined

21. $\sin\pi = 0, \cos\pi = -1, \tan\pi = 0$ **23.** $\sin\dfrac{\pi}{4} = \dfrac{\sqrt{2}}{2}, \cos\dfrac{\pi}{4} = \dfrac{\sqrt{2}}{2}, \tan\dfrac{\pi}{4} = 1$ **25.** $\frac{1}{2}$ **27.** 1

29. 1 **31.** $\frac{1}{2}$ **33.** $-\sqrt{3}$ **35.** $\dfrac{\sqrt{2}}{2}$ **37.** $\frac{3}{5}$ **39.** $\frac{12}{13}$ **41.** $-\frac{4}{3}$ **43.** $\frac{12}{5}$

45. $\frac{12}{13}$ **47.** $\frac{40}{41}$ **49.** $\dfrac{2\sqrt{2}}{3}$ **51.** $-\dfrac{3\sqrt{7}}{7}$ **53.** $\sin\left(\dfrac{\pi}{6} + \dfrac{\pi}{3}\right) = 1$ **55.** 1 **57.** $\frac{120}{169}$

59. $\frac{720}{1681}$ **61.** $\frac{3}{5}$ **63.** $\frac{5}{13}$ **65.** 1 **67.** $\frac{24}{25}$ **69.** $-\frac{119}{169}$ **71.** undefined   **73.** 7

75. $\dfrac{x}{\sqrt{1 + x^2}}$ **77.** $\dfrac{x}{\sqrt{1 - x^2}}$ **79.** $\sqrt{1 - x^2}$ **81.** $2x\sqrt{1 - x^2}$ **83.** $\dfrac{2x}{1 - x^2}$

85. $1 - 2x^2$ **87.** $\sqrt{\dfrac{1 + x}{2}}$ **89.** $\sqrt{\dfrac{1 - x}{2}}$ **91.** $-\dfrac{\pi}{6}$ **93.** $\dfrac{\pi}{4}$ **103.** $-1 \le x \le 1$

105. $-\dfrac{\pi}{2} \le x \le \dfrac{\pi}{2}$

Exercise 5.4   (Page 193)

1. $\dfrac{\pi}{6}$ **3.** π **5.** $-\dfrac{\pi}{2}$ **7.** 3 **9.** $\dfrac{4\sqrt{15}}{15}$ **11.** $\frac{40}{9}$ **13.** $\frac{5}{3}$ **15.** $\frac{24}{25}$ **17.** $-\frac{33}{65}$

19. 0 **21.** $\dfrac{1}{x}$ **23.** $\dfrac{\sqrt{1 + x^2}}{x}$ **25.** $\pm\sqrt{x^2 - 1}$

Exercise 5.5   (Page 197)

1. 48.8°   **3.** 38.7°   **5.** 54.7°

REVIEW EXERCISES   (Page 198)

1. 30°, 330°   **2.** 150°, 330°   **3.** 45°, 225°   **4.** 90°, 270°   **5.** no solutions in the given interval
6. 300°   **7.** 0°, 180°, 270°   **8.** 0°, 180°   **9.** 0°, 90°   **10.** 180°, 210°, 330°
11. 0°, 45°, 225°   **12.** 0°, 60°, 180°, 300°   **13.** 0°, 30°, 150°, 180°   **14.** 90°, 210°, 330°
15. 60°, 180°, 300°   **16.** 0°, 120°, 240°   **17.** 120°, 240°   **18.** 45°, 225°, 135°, 315°
19. 45°, 135°, 225°, 315°   **20.** an identity   **21.** no solutions   **22.** 60°, 180°, 300°   **23.** 45°

24. 330°   **25.** $\dfrac{\pi}{3}, \dfrac{2\pi}{3}$ **26.** $\dfrac{\pi}{6}, \dfrac{7\pi}{6}$ **27.** $\dfrac{2\pi}{3}, \dfrac{4\pi}{3}$ **28.** $\dfrac{4\pi}{3}, \dfrac{5\pi}{3}$ **29.** $\dfrac{2\pi}{3}, \dfrac{5\pi}{3}$

30. no values   **31.** no values   **32.** $\dfrac{\pi}{6}, \dfrac{7\pi}{6}$ **33.** $\dfrac{\pi}{2}, \dfrac{3\pi}{2}$ **34.** $\dfrac{\pi}{6}, \dfrac{5\pi}{6}$ **35.** no values

36. $\dfrac{2\pi}{3}, \dfrac{4\pi}{3}$ **37.** 0.629, 2.513   **38.** no values   **39.** 1.557, 4.699   **40.** 0.013, 3.155

41. no values   **42.** no values   **43.** 0.013, 3.155   **44.** 0.003, 3.145   **45.** no values

46. 1.369, 4.914   **47.** 1.047, 5.236   **48.** no values   **49.** $-\dfrac{\pi}{6}$ **50.** $\dfrac{2\pi}{3}$ **51.** $\dfrac{\pi}{6}$

52. $-\dfrac{\pi}{6}$ **53.** $\dfrac{\pi}{2}$ **54.** $-\dfrac{\pi}{2}$ **55.** $-\dfrac{\pi}{4}$ **56.** $\dfrac{\pi}{3}$ **57.** 0   **58.** no values   **59.** 0

60. $\dfrac{\pi}{2}$ **61.** $\dfrac{4}{5}$ **62.** $\dfrac{1}{3}$ **63.** $\dfrac{\sqrt{3}}{2}$ **64.** $\dfrac{\sqrt{3}}{2}$ **65.** 1   **66.** $-\sqrt{3}$ **67.** $\dfrac{3}{4}$

68. $\frac{3}{4}$ **69.** $\frac{13}{12}$ **70.** $\frac{13}{12}$ **71.** $-\frac{5}{12}$ **72.** $-\frac{5}{12}$ **73.** $\dfrac{2\sqrt{10}}{7}$ **74.** $\dfrac{3\sqrt{5}}{7}$ **75.** 1

76. $\frac{1}{2}$ **77.** $\frac{7}{25}$ **78.** $\frac{120}{169}$ **79.** 7   **80.** $\frac{4}{3}$ **81.** 1   **82.** $\frac{24}{25}$ **83.** 1

84. $-\frac{169}{119}$ **85.** $\dfrac{\sqrt{3}}{2}$ **86.** $\frac{1}{2}$ **87.** $\sqrt{3}$ **88.** $-\dfrac{\sqrt{3}}{3}$ **89.** $\frac{24}{25}$ **90.** $\frac{119}{169}$

91. $-\frac{7}{25}$ **92.** $\frac{120}{169}$ **93.** $\sqrt{1 - u^2}$ **94.** $\sqrt{1 - u^2}$ **95.** $\dfrac{u}{\sqrt{1 - u^2}}$ **96.** $\dfrac{u}{\sqrt{1 + u^2}}$

97. $\dfrac{\sqrt{1 + u^2}}{1 + u^2}$ **98.** $\dfrac{\sqrt{1 - u^2}}{u}$ **99.** $2u\sqrt{1 - u^2}$ **100.** $2u^2 - 1$ **101.** $\dfrac{2u}{1 - u^2}$

102. $2u\sqrt{1 - u^2}$ **103.** $\sqrt{\dfrac{1 + \sqrt{1 - u^2}}{2}}$ **104.** $\pm\sqrt{\dfrac{1 + u}{2}}$ **105.** $\dfrac{\pi}{3}$ **106.** $\dfrac{\pi}{4}$ **107.** $\dfrac{2\pi}{3}$

108. $\dfrac{5\pi}{6}$ **109.** $\frac{4}{5}$ **110.** $\frac{12}{13}$ **111.** $\frac{12}{13}$ **112.** $\frac{40}{41}$ **113.** $\dfrac{2\sqrt{10}}{3}$ **114.** $\dfrac{2\sqrt{21}}{21}$

Exercise 6.1   (Page 205)

1. 61.0 cm   **3.** 1410 km   **5.** 65.9 cm   **7.** 54°   **9.** 90°   **11.** 37.85°   **13.** 210 lb
15. 5°, if 210 lbs is used for the resultant force.   **17.** 1090 lb   **19.** 36.7 nautical miles   **21.** 131 m
23. 69.3°, 64.4°, 46.3°   **25.** $x = \sqrt{2R^2 - 2R^2 \cos\theta}$ **27.** 5.9 and 14 ft   **33.** 96°   **35.** 71.6°
37. 107°   **39.** 120 m

Exercise 6.2   (Page 212)

1. 67 km   **3.** 256 m   **5.** 2.55 m   **7.** 305 m   **9.** 49.2 cm   **11.** 1.0 mi   **13.** 218

15. 3.97   **17.** 180 yd   **19.** 420 ft   **21.** 7.1°   **23.** 2.5 nautical miles
25. approximately 2:50 P.M.   **27.** 38.9 ft   **29.** 75 ft

Exercise 6.3 (Page 219)

1. 31.5°   **3.** 61.6°   **5.** no triangle   **7.** 136°   **9.** 156.19° or 4.09°   **11.** no triangle
13. 12 m or 2.1 m   **15.** no triangle   **17.** 2900 ft or 780 ft   **19.** $h = 957$ ft
21. 7.2 in. or 12.7 in.

Exercise 6.4 (Page 222)

1. 74.1   **3.** 86   **5.** 44.39   **7.** 10.3   **9.** 19   **11.** 850   **13.** 70.5 cm

Exercise 6.5 (Page 227)

1. 190 sq ft   **3.** 301 sq cm   **5.** 6.5 sq units   **7.** 6 sq units   **9.** 42 sq units
11. 960 sq units   **13.** 0.001 sq mm   **15.** 195,000 sq km   **17.** 31,400 sq ft   **19.** 72 sq m
21. $\sqrt{2295}$ sq cm   **23.** 126,000 sq ft
25. $\dfrac{s(s-a)}{bc} = \dfrac{(a+b+c)(b+c-a)}{4bc} = \dfrac{1}{2}\left(\dfrac{(a+b+c)(b+c-a)}{2bc}\right) = \dfrac{1}{2}(1 + \cos A) = \cos^2\dfrac{A}{2}$
27. Area of $\triangle ABC = \frac{1}{2}(AC)\frac{1}{2}(DB)\sin\alpha = \frac{1}{4}(AC)(DB)\sin\alpha$
$\underline{\text{Area of } \triangle ACD = \frac{1}{2}(AC)\frac{1}{2}(DB)\sin\alpha = \frac{1}{4}(AC)(DB)\sin\alpha}$
Area of $\square\ ABCD = \frac{1}{2}(AC)(DB)\sin\alpha$
29. $A = \dfrac{1}{4}b^2 \cot\dfrac{\alpha}{2} = \dfrac{b^2\cos^2(\alpha/2)}{2\sin\alpha}$ **31.** $\frac{1}{4}b^2\sqrt{15}$

Exercise 6.6 (Page 237)

1. $\langle 7, -5\rangle$ **3.** $\langle 6, -9\rangle$ **5.** $\langle 9, -8\rangle$ **7.** $\sqrt{13}$ **9.** $\sqrt{5}$ **11.** $\sqrt{13} + \sqrt{2}$
13. $\dfrac{3}{5}i + \dfrac{4}{5}j$ **15.** $\dfrac{\sqrt{3}}{2}i - \dfrac{1}{2}j$ **17.** $8i + 8j$ **19.** $5\sqrt{3}i + 5j$ **21.** $18.6i + 14.1j$ **23.** 9
25. 7   **27.** 0   **29.** 45°   **31.** 150°   **33.** 36.9°   **35.** perpendicular
37. not perpendicular   **39.** perpendicular   **41.** $\dfrac{63}{13}$ **43.** 0   **45.** $\dfrac{315}{169}i + \dfrac{756}{169}j$ **47.** 0
49. $16.7i + 2.06j$; 16.9 lbs

REVIEW EXERCISES (Page 240)

5. 7.6   **6.** 32   **7.** 0.6   **8.** 11.1   **9.** 25°   **10.** 37°, 143°   **11.** 18.6°, 161.4°
12. 51.31°, 128.69°   **13.** 65.7   **14.** 70.8   **15.** 14°, 82°   **16.** 42.4°   **17.** 640 sq units
18. 1200 sq units   **19.** 39 sq units   **20.** 110 sq units   **21.** 1.4 sq units   **22.** 12,000 sq units
23. 2400 sq units   **24.** 67.3 sq units   **25.** 558 mi   **26.** 180 ft   **27.** 611 ft   **31.** 112.2°
32. about 27.2°, 48.8°, and 104.0°   **33.** 280 sq m   **34.** 24 sq units   **35.** $\langle 0, 29\rangle$
36. $3\sqrt{58} - \sqrt{29}$ **37.** $\langle 25, 10\rangle$ **38.** $\sqrt{377}$ **39.** 90°   **40.** 0°   **41.** 75°   **42.** 120°
43. $\dfrac{1}{3}i + \dfrac{2\sqrt{2}}{3}j$ **44.** $-j$ **45.** $2\sqrt{2}j$ **46.** $-2\sqrt{2}$

Exercise 7.1 (Page 249)

1. i **3.** -1 **5.** $-i$ **7.** i **9.** 1   **11.** i **13.** $x = -\frac{1}{2}; y = -\frac{1}{2}$

15. $x = 0;\ y = 0$ **17.** $x = \frac{2}{3};\ y = -\frac{2}{9}$ **19.** $5 - 6i$ **21.** $-2 - 10i$ **23.** $4 + 10i$

25. $6 - 17i$ **27.** $52 + 56i$ **29.** $-5 + 12i$ **31.** $-6 + 17i$ **33.** $2 - 11i$ **35.** $-\frac{3}{4} + 0i$

37. $\frac{2}{5} - \frac{1}{5}i$ **39.** $\frac{1}{25} + \frac{7}{25}i$ **41.** $\frac{1}{2} + \frac{1}{2}i$ **43.** $\dfrac{-2 - 2\sqrt{5}}{17} + \dfrac{16 - \sqrt{5}}{34}i$

45. $\dfrac{6 + \sqrt{3}}{10} + \dfrac{3\sqrt{3} - 2}{10}i$ **47.** $(1 - i)(1 - i) = 1 - i - i - 1 = -2i$

49. $(1 + 2i)(1 + 2i) = 1 + 2i + 2i - 4 = -3 + 4i$ **53.** $-1 + i, -1 - i$ **55.** $-2 + i, -2 - i$

57. $\frac{1}{3} + \frac{1}{3}i, \frac{1}{3} - \frac{1}{3}i$ **59.** $34.87 + 32.69i$ **61.** approximately $-6.92 + 9.26i$

Exercise 7.2   (Page 252)

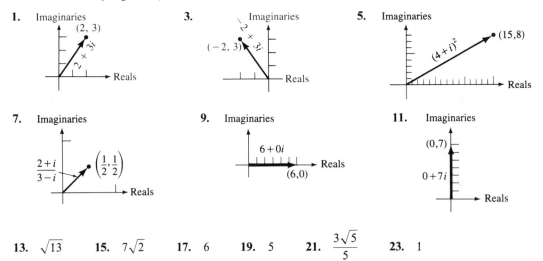

1. Imaginaries $(2, 3)$, $2 + 3i$

3. Imaginaries $(-2, 3)$, $-2 + 3i$

5. Imaginaries $(15,8)$, $(4 + i)^2$

7. Imaginaries $\left(\frac{1}{2}, \frac{1}{2}\right)$, $\dfrac{2 + i}{3 - i}$

9. Imaginaries $6 + 0i$, $(6,0)$

11. Imaginaries $(0,7)$, $0 + 7i$

13. $\sqrt{13}$ **15.** $7\sqrt{2}$ **17.** 6 **19.** 5 **21.** $\dfrac{3\sqrt{5}}{5}$ **23.** 1

Exercise 7.3   (Page 256)

1. $6(\cos 0° + i \sin 0°)$ **3.** $3(\cos 270° + i \sin 270°)$ **5.** $\sqrt{2}(\cos 225° + i \sin 225°)$

7. $6(\cos 60° + i \sin 60°)$ **9.** $2(\cos 240° + i \sin 240°)$ **11.** $2(\cos 210° + i \sin 210°)$ **13.** $\sqrt{3} + i$

15. $0 + 7i$ **17.** $1 - \sqrt{3}\,i$ **19.** $-\frac{1}{2} + 0i$ **21.** $\dfrac{3\sqrt{2}}{2} + \dfrac{3\sqrt{2}}{2}i$ **23.** $\dfrac{11\sqrt{3}}{2} - \dfrac{11}{2}i$

25. $8(\cos 90° + i \sin 90°)$ **27.** $\cos 300° + i \sin 300°$ **29.** $6(\cos 2\pi + i \sin 2\pi)$

31. $6\left(\cos \dfrac{\pi}{2} + i \sin \dfrac{\pi}{2}\right)$ **33.** $30 \operatorname{cis} 116°$ **35.** $36 \operatorname{cis} \dfrac{13\pi}{12}$ **37.** $6(\cos 30° + i \sin 30°)$

39. $\dfrac{3}{2}\left(\cos \dfrac{\pi}{2} + i \sin \dfrac{\pi}{2}\right)$ **41.** $\dfrac{12}{5} \operatorname{cis} 130°$ **43.** $\dfrac{1}{2} \operatorname{cis} \dfrac{\pi}{2}$ **45.** $\operatorname{cis} 40°$ **47.** $4 \operatorname{cis} \dfrac{5\pi}{6}$

Exercise 7.4   (Page 262)

1. $27(\cos 90° + i \sin 90°)$ **3.** $\cos 180° + i \sin 180°$ **5.** $3125 \operatorname{cis} 10°$ **7.** $81(\cos \pi + i \sin \pi)$

9. $256(\cos 12 + i \sin 12)$ **11.** $\dfrac{1}{27} \operatorname{cis} \dfrac{3\pi}{2}$ **13.** $1 + \sqrt{3}\,i$ **15.** $\dfrac{1}{2} + \dfrac{\sqrt{3}}{2}i$ **17.** $1 + \sqrt{3}\,i$

19. $1 + \sqrt{3}\,i, -2, 1 - \sqrt{3}\,i;$

21. $\dfrac{\sqrt{2}}{2} + \dfrac{\sqrt{2}}{2}i, -\dfrac{\sqrt{2}}{2} - \dfrac{\sqrt{2}}{2}i;$

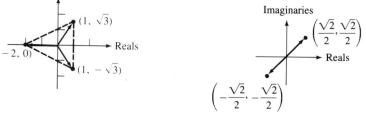

23. $\cos 54° + i\sin 54°, \cos 126° + i\sin 126°, \cos 198° + i\sin 198°, \cos 270° + i\sin 270°, \cos 342° + i\sin 342°;$

25. $\sqrt{3} + i, -1 + i\sqrt{3}, -\sqrt{3} - i, 1 - i\sqrt{3};$

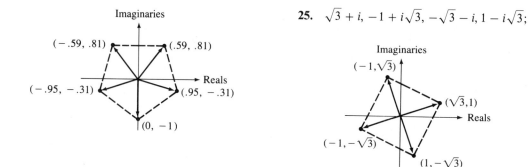

Exercise 7.5   (Page 269)

1. $(\sqrt{3}, 1)$ **3.** $\left(\dfrac{7}{2}, -\dfrac{7\sqrt{3}}{2}\right)$ **5.** $\left(\dfrac{-3}{2}, \dfrac{-3\sqrt{3}}{2}\right)$ **7.** $(0, 2)$ **9.** $(-\sqrt{3}, -1)$

11. $\left(\dfrac{-5\sqrt{2}}{2}, \dfrac{-5\sqrt{2}}{2}\right)$ **13.** $(\sqrt{3}, -1)$ **15.** $(0, 10)$ **17.** $(0, 0)$ **19.** $(-3\sqrt{3}, 3)$

21. $(\sqrt{2}, 45°)$ **23.** $(4, 330°)$ **25.** $(2, 210°)$ **27.** $(2, 150°)$ **29.** $(0, 0°)$ **31.** $(5, 180°)$

33. $(3\sqrt{2}, 315°)$ **35.** $(14, 60°)$ **37.** $r\cos\theta = 3$ **39.** $r(3\cos\theta + 2\sin\theta) = 3$ **41.** $r = 9\cos\theta$

43. $r^2 = 4\cos^2\theta\sin^2\theta$ **45.** $r = 2\cos\theta$ **47.** $r^2\cos^2\theta - 2r\sin\theta = 1$ **49.** $x^2 + y^2 = 9$

51. $x = 5$ **53.** $\sqrt{x^2 + y^2} + y = 1$ **55.** $(x^2 + y^2)^2 = 2xy$ **57.** $y = 0$

59. $2\sqrt{x^2 + y^2} - x = 2$

61.   **63.**   **65.**

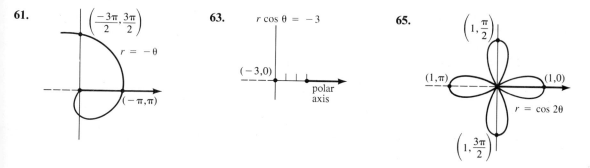

67.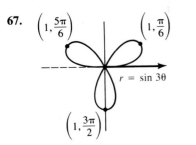

69.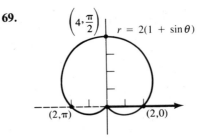

71.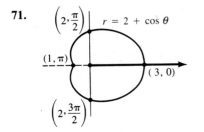

73.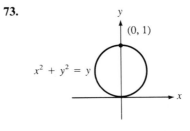

Exercise 7.6 (Page 275)

1.

3.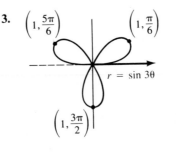

5.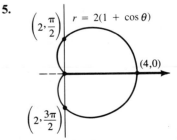

7.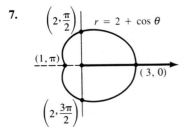

9.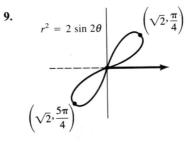

11.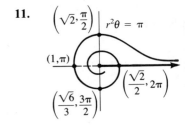

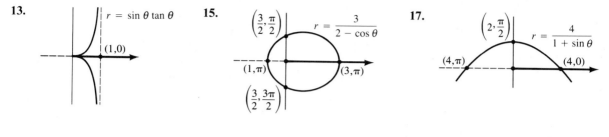

13. $r = \sin\theta\tan\theta$ (1,0)

15. $\left(\frac{3}{2}, \frac{\pi}{2}\right)$ $r = \dfrac{3}{2 - \cos\theta}$ (1, π)   (3, π) $\left(\frac{3}{2}, \frac{3\pi}{2}\right)$

17. $\left(2, \frac{\pi}{2}\right)$ $r = \dfrac{4}{1 + \sin\theta}$ (4, π)   (4, 0)

Exercise 7.7   (Page 278)

1. 42.9 lbs; 24.0°   **3.** 8 units

REVIEW EXERCISES   (Page 279)

1. $-i$ **2.** $-i$ **3.** $-i$ **4.** $\dfrac{1}{i^{1812}} = 1$ **5.** $\left(\frac{4}{3}, \frac{2}{3}\right)$ **6.** $(-11, -18)$ **7.** $-4 + i$

8. $-5 + i$ **9.** $-5 - 5i$ **10.** $21 + i$ **11.** $0 - \frac{1}{5}i$ **12.** $0 + \frac{13}{6}i$ **13.** $\frac{8}{17} - \frac{2}{17}i$

14. $-\frac{3}{2} - \frac{1}{2}i$ **15.** $0 + i$ **16.** $\frac{2}{5} - \frac{1}{5}i$ **17.** $\dfrac{2 - 3\sqrt{2}}{3} + \dfrac{3 + 2\sqrt{2}}{3}i$ **18.** $\dfrac{3 + \sqrt{3}}{4} + \dfrac{3\sqrt{3} - 1}{4}i$

19. Imaginaries   Reals $4 - 5i$ $(4, -5)$

20. Imaginaries $(-7, 2)$ $-7 + 2i$ Reals

21. Imaginaries   Reals   6   (6, 0)

22. Imaginaries   (0, 3) $3i$ Reals

23. $\sqrt{73}$ **24.** $10\sqrt{2}$ **25.** $\dfrac{3\sqrt{10}}{10}$ **26.** 1

27. $2\sqrt{2}(\cos 135° + i\sin 135°)$ **28.** $5\sqrt{2}(\cos 315° + i\sin 315°)$

29. $6(\cos 60° + i\sin 60°)$ **30.** $4(\cos 0° + i\sin 0°)$ **31.** $\dfrac{3}{2} + \dfrac{3\sqrt{3}}{2}i$

32. $\sqrt{3} - i$ **33.** $-\dfrac{3}{2} - \dfrac{3\sqrt{3}}{2}i$ **34.** $-\dfrac{7\sqrt{3}}{2} + \dfrac{7}{2}i$ **35.** cis 110°   **36.** 6 cis 570°

37. $6\left(\cos\dfrac{\pi}{4} + i\sin\dfrac{\pi}{4}\right)$ **38.** $21(\cos\pi + i\sin\pi)$ **39.** 2 cis 50°   **40.** $\dfrac{2}{3}(\cos 10° + i\sin 10°)$

41. $\cos 20° + i\sin 20°$ **42.** $\sqrt[4]{14}\left(\cos\dfrac{45°}{4} + i\sin\dfrac{45°}{4}\right)$ **43.** $5, \dfrac{-5 + 5\sqrt{3}i}{2}, \dfrac{-5 - 5\sqrt{3}i}{2}$

44. $3, 3i, -3, -3i$ **45.** $\left(\dfrac{5}{2}, \dfrac{5\sqrt{3}}{2}\right)$ **46.** $(-\sqrt{3}, -1)$ **47.** $\left(\dfrac{\sqrt{3}}{2}, \dfrac{1}{2}\right)$ **48.** $(-5\sqrt{2}, 5\sqrt{2})$

49. $(2, 135°)$ **50.** $(2, 150°)$ **51.** $(1, 0°)$ **52.** $(2, 300°)$ **53.** $r^2\cos\theta\sin\theta = 1$

54. $r(\cos\theta + 2\sin\theta) = 2$ **55.** $r\cos^2\theta = 3\sin\theta$ **56.** $r^2 = 4\cos\theta\sin\theta$

57. $(x^2 + y^2)^2 = 9x^2 - 9y^2$ **58.** $x^2 + y^2 = 5y$ **59.** $4\sqrt{x^2 + y^2} + y = 1$

60. $\sqrt{x^2 + y^2} - x = 2$

61.

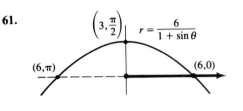

$r = \dfrac{6}{1 + \sin\theta}$

$\left(3, \dfrac{\pi}{2}\right)$ $(6,\pi)$ $(6,0)$

62.

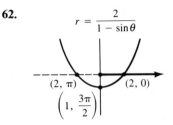

$r = \dfrac{2}{1 - \sin\theta}$

$(2, \pi)$ $(2, 0)$ $\left(1, \dfrac{3\pi}{2}\right)$

63.

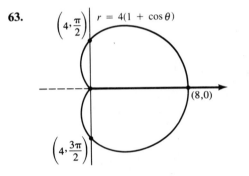

$\left(4, \dfrac{\pi}{2}\right)$ $r = 4(1 + \cos\theta)$ $(8,0)$ $\left(4, \dfrac{3\pi}{2}\right)$

64.

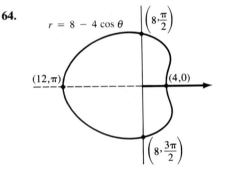

$r = 8 - 4\cos\theta$ $\left(8, \dfrac{\pi}{2}\right)$ $(12,\pi)$ $(4,0)$ $\left(8, \dfrac{3\pi}{2}\right)$

Exercise 8.1   (Page 292)

1.

$4x + 3y = 12$

3. 5   **5.** -2 **7.** -1 **9.** $\frac{1}{2}$ **11.** $\frac{2}{3}$ **13.** 0
15. perpendicular   **17.** parallel   **19.** perpendicular
21. parallel   **23.** perpendicular   **25.** They do not lie on a straight line.
27. They lie on a straight line.   **29.** None are perpendicular.
31. PQ and PR are perpendicular.   **33.** $y = 2x$ **35.** $y = 2x + \frac{7}{2}$
37. $3x - y = 2$ **39.** $ax - y = -\dfrac{1}{a}$ **41.** $m = \frac{42}{5}; b = \frac{21}{5}$
43. no defined slope; no y-intercept   **45.** $y = \frac{3}{2}x + \frac{5}{2}$

47. $y = \frac{13}{8}x - \frac{1}{4}$ **49.** $2x + 3y = 1$ **51.** $8x + 13y = 55$ **53.** $x = 1$ **55.** $y = 5$
57. $m = -1; b = 2$ **59.** $m = 3; b = 11$ **61.** $m = \frac{1}{3}; b = \frac{7}{3}$

Exercise 8.2   (Page 297)

1. $x^2 + y^2 = 1$ **3.** $(x - 6)^2 + (y - 8)^2 = 16$ **5.** $(x + 5)^2 + (y - 3)^2 = 25$
7. $(x - 3)^2 + (y + 4)^2 = 2$ **9.** $(x - 3)^2 + (y - 3)^2 = 25$ **11.** $(x + 5)^2 + (y - 1)^2 = 65$
13. $(x + 3)^2 + (y - 4)^2 = 25$ **15.** $(x + 2)^2 + (y + 6)^2 = 40$ **17.** $x^2 + (y + 3)^2 = 157$
19. $(x - 5)^2 + (y - 8)^2 = 338$ **21.** $(x + 4)^2 + (y + 2)^2 = 98$ **23.** $(x - 1)^2 + (y + 2)^2 = 36$
25. $x^2 + (y + 12)^2 = 10$ **27.** no

29.

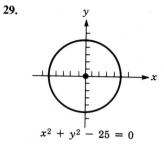

$$x^2 + y^2 - 25 = 0$$

31.

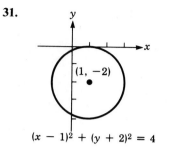

$$(x - 1)^2 + (y + 2)^2 = 4$$

33.

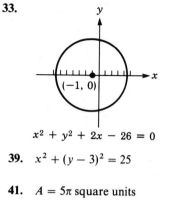

$$x^2 + y^2 + 2x - 26 = 0$$

35.

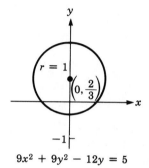

$$9x^2 + 9y^2 - 12y = 5$$

37.

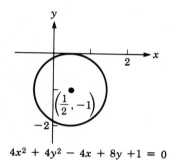

$$4x^2 + 4y^2 - 4x + 8y + 1 = 0$$

39. $x^2 + (y - 3)^2 = 25$

41. $A = 5\pi$ square units

43. $A = 2a\sqrt{r^2 - a^2}$ square units

Exercise 8.3   (Page 303)

1. $x^2 = 12y$ **3.** $y^2 = 12x$ **5.** $(x - 3)^2 = -12(y - 5)$ **7.** $(x - 3)^2 = -28(y - 5)$

9. $(x - 2)^2 = -2(y - 2)$ or $(y - 2)^2 = -2(x - 2)$ **11.** $(x + 4)^2 = -\frac{16}{3}(y - 6)$ or $(y - 6)^2 = \frac{9}{4}(x + 4)$

13. $(y - 8)^2 = -4(x - 6)$ **15.** $(x - 3)^2 = \frac{1}{2}(y - 1)$

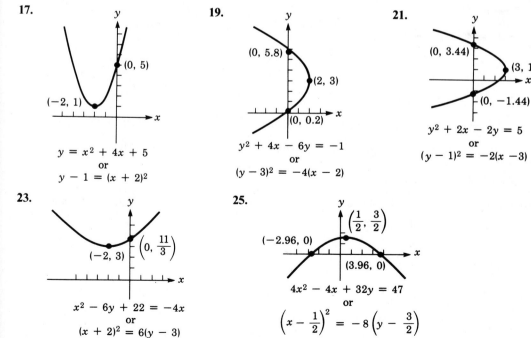

17.

$$y = x^2 + 4x + 5$$
or
$$y - 1 = (x + 2)^2$$

19.

$$y^2 + 4x - 6y = -1$$
or
$$(y - 3)^2 = -4(x - 2)$$

21.

$$y^2 + 2x - 2y = 5$$
or
$$(y - 1)^2 = -2(x - 3)$$

23.

$$x^2 - 6y + 22 = -4x$$
or
$$(x + 2)^2 = 6(y - 3)$$

25.

$$4x^2 - 4x + 32y = 47$$
or
$$\left(x - \frac{1}{2}\right)^2 = -8\left(y - \frac{3}{2}\right)$$

27. $x^2 = -\frac{45}{2}y$ **29.** 10 and 10   **31.** 8 cabins   **33.** $\frac{80}{7}$ meters   **35.** $\frac{1}{36}$ foot
39. $y = x^2 + 4x + 3$

Exercise 8.4   (Page 311)

1. $\dfrac{x^2}{25} + \dfrac{y^2}{16} = 1$ **3.** $\dfrac{9x^2}{16} + \dfrac{9y^2}{25} = 1$ **5.** $\dfrac{x^2}{7} + \dfrac{y^2}{16} = 1$ **7.** $\dfrac{(x-3)^2}{4} + \dfrac{(y-4)^2}{9} = 1$

9. $\dfrac{(x-3)^2}{9} + \dfrac{(y-4)^2}{4} = 1$ **11.** $\dfrac{(x-3)^2}{41} + \dfrac{(y-4)^2}{16} = 1$ **13.** $\dfrac{x^2}{36} + \dfrac{(y-4)^2}{20} = 1$ **15.** $\dfrac{x^2}{100} + \dfrac{y^2}{64} = 1$

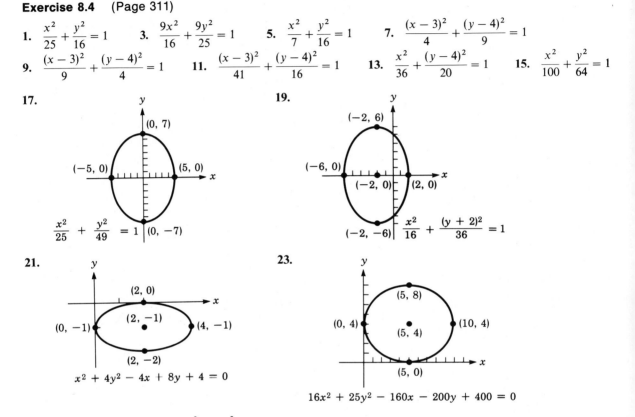

17.

$\dfrac{x^2}{25} + \dfrac{y^2}{49} = 1$

19.

$\dfrac{x^2}{16} + \dfrac{(y+2)^2}{36} = 1$

21.

$x^2 + 4y^2 - 4x + 8y + 4 = 0$

23.

$16x^2 + 25y^2 - 160x - 200y + 400 = 0$

25. 199,000 miles   **27.** $\dfrac{x^2}{2500} + \dfrac{y^2}{900} = 1$; 36 meters

29. In the ellipse, $b^2 = a^2 - c^2$, or $a = \sqrt{b^2 + c^2}$. Using the distance formula, you have $d(FB) = \sqrt{b^2 + c^2} = a$.

31. Substitute c for x in the equation $\dfrac{x^2}{a^2} + \dfrac{y^2}{b^2} = 1$. Use the fact that $c^2 = a^2 - b^2$, and solve for y to determine $d(FA') = \dfrac{b^2}{a}$. Because $d(AA') = 2y$, $d(AA') = \dfrac{2b^2}{a}$.

33. The thumbtacks are the foci at $(\pm 1, 0)$. Hence, $c = 1$. The string is 6 meters, so $2a = 6$, or $a^2 = 9$. Because $b^2 = a^2 - c^2$, you have $b^2 = 8$. The equation is $\dfrac{x^2}{9} + \dfrac{y^2}{8} = 1$.

Exercise 8.5   (Page 317)

1. $\dfrac{x^2}{25} - \dfrac{y^2}{24} = 1$ **3.** $\dfrac{(x-2)^2}{4} - \dfrac{(y-4)^2}{9} = 1$ **5.** $\dfrac{(y-3)^2}{9} - \dfrac{(x-5)^2}{9} = 1$ **7.** $\dfrac{y^2}{9} - \dfrac{x^2}{16} = 1$

9. $\dfrac{(x-1)^2}{4} - \dfrac{(y+3)^2}{16} = 1$ or $\dfrac{(y+3)^2}{4} - \dfrac{(x-1)^2}{16} = 1$ **11.** $\dfrac{x^2}{10} - \dfrac{3y^2}{20} = 1$ **13.** 24 square units

15. 12 square units   **17.** $\dfrac{(x+2)^2}{4} - \dfrac{4(y+4)^2}{81} = 1$ **19.** $\dfrac{x^2}{36} - \dfrac{16y^2}{25} = 1$

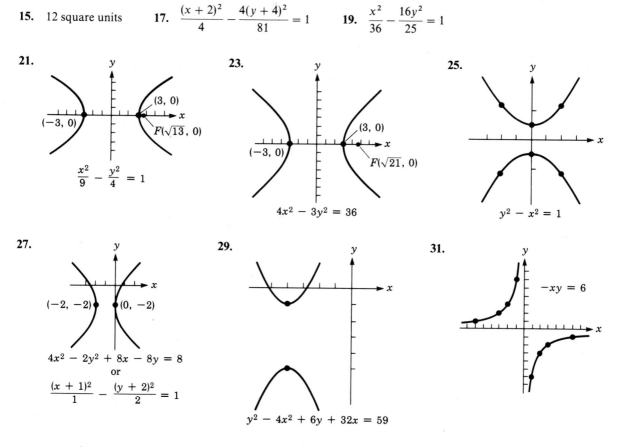

21.

$\dfrac{x^2}{9} - \dfrac{y^2}{4} = 1$

$(-3, 0)$ $(3, 0)$ $F(\sqrt{13}, 0)$

23.

$4x^2 - 3y^2 = 36$

$(-3, 0)$ $(3, 0)$ $F(\sqrt{21}, 0)$

25.

$y^2 - x^2 = 1$

27.

$4x^2 - 2y^2 + 8x - 8y = 8$
or
$\dfrac{(x+1)^2}{1} - \dfrac{(y+2)^2}{2} = 1$

$(-2, -2)$ $(0, -2)$

29.

$y^2 - 4x^2 + 6y + 32x = 59$

31.

$-xy = 6$

33. $\dfrac{(x-3)^2}{9} - \dfrac{(y-1)^2}{16} = 1$ **35.** $4x^2 - 5y^2 - 60y = 0$

37. From the geometry of Figure 8-27, $d(PF) + d(F'F) > d(F'P)$. This is equivalent to $d(F'P) - d(PF) < d(F'F)$, or $2a < 2c$, which implies that $c > a$.

Exercise 8.6   (Page 326)

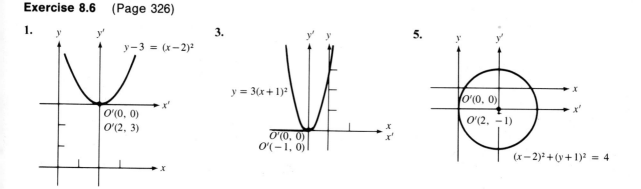

1.

$y - 3 = (x - 2)^2$

$O'(0, 0)$
$O'(2, 3)$

3.

$y = 3(x + 1)^2$

$O'(0, 0)$
$O'(-1, 0)$

5.

$O'(0, 0)$
$O'(2, -1)$

$(x - 2)^2 + (y + 1)^2 = 4$

7.

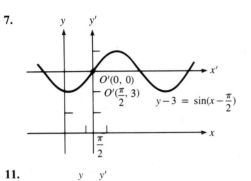

$O'(0, 0)$
$O'(\frac{\pi}{2}, 3)$
$y - 3 = \sin(x - \frac{\pi}{2})$

9.

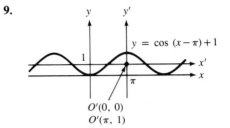

$y = \cos(x - \pi) + 1$

$O'(0, 0)$
$O'(\pi, 1)$

11.

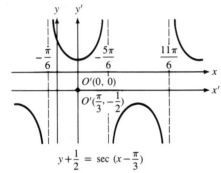

$-\frac{\pi}{6}$ $\frac{5\pi}{6}$ $\frac{11\pi}{6}$

$O'(0, 0)$
$O'(\frac{\pi}{3}, -\frac{1}{2})$

$y + \frac{1}{2} = \sec(x - \frac{\pi}{3})$

13.

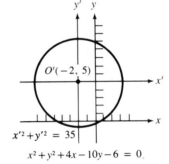

$O'(-2, 5)$

$x'^2 + y'^2 = 35$

$x^2 + y^2 + 4x - 10y - 6 = 0$

15.

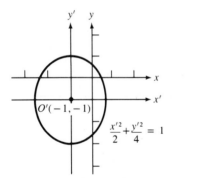

$O'(-1, -1)$

$\frac{x'^2}{2} + \frac{y'^2}{4} = 1$

$2x^2 + y^2 + 4x + 2y - 1 = 0$

17.

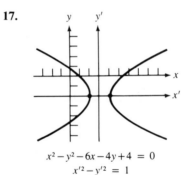

$x^2 - y^2 - 6x - 4y + 4 = 0$
$x'^2 - y'^2 = 1$

19. $(3, \sqrt{3})$

21. $(-2\sqrt{2}, 4\sqrt{2})$

23. $\left(\dfrac{7 - 4\sqrt{3}}{2}, \dfrac{7\sqrt{3} + 4}{2}\right)$

25. $60°$ **27.** $15°$

29.

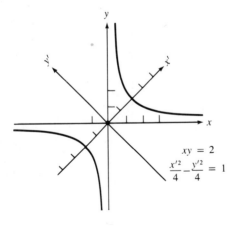

$xy = 2$
$\dfrac{x'^2}{4} - \dfrac{y'^2}{4} = 1$

31.

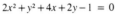

$x^2 + 3xy + y^2 = 2$
$\dfrac{5x'^2}{4} - \dfrac{y'^2}{4} = 1$

33.

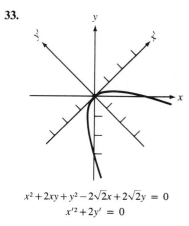

$$x^2 + 2xy + y^2 - 2\sqrt{2}x + 2\sqrt{2}y = 0$$
$$x'^2 + 2y' = 0$$

35.

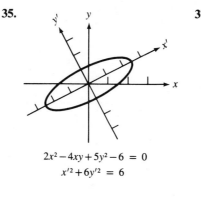

$$2x^2 - 4xy + 5y^2 - 6 = 0$$
$$x'^2 + 6y'^2 = 6$$

37.

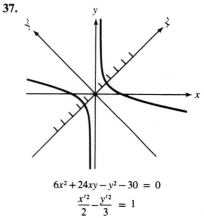

$$6x^2 + 24xy - y^2 - 30 = 0$$
$$\frac{x'^2}{2} - \frac{y'^2}{3} = 1$$

Exercise 8.7   (Page 333)

1. a parabola

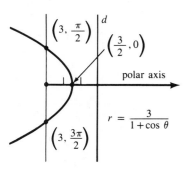

$\left(3, \frac{\pi}{2}\right)$ $\left(\frac{3}{2}, 0\right)$

polar axis

$\left(3, \frac{3\pi}{2}\right)$ $r = \dfrac{3}{1 + \cos \theta}$

3. an ellipse

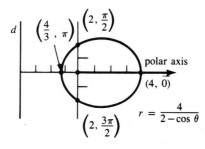

$\left(\frac{4}{3}, \pi\right)$ $\left(2, \frac{\pi}{2}\right)$

polar axis

(4, 0)

$\left(2, \frac{3\pi}{2}\right)$ $r = \dfrac{4}{2 - \cos \theta}$

5. a hyperbola

$\left(0.8, \frac{3\pi}{2}\right)$

polar axis

$(2, \pi)$ $(2, 0)$

d $\left(-4, \frac{\pi}{2}\right)$ $r = \dfrac{4}{2 - 3 \sin \theta}$

7. $r = \dfrac{10}{3 + 2\cos \theta}$

9. $r = \dfrac{14}{12 + 8\cos \theta} = \dfrac{7}{6 + 4\cos \theta}$

11. major axis $= \dfrac{ep}{1 + e} + \dfrac{ep}{1 - e}$
$$= \dfrac{2ep}{1 - e^2}$$
minor axis $= 2ep$

13. $\left(\dfrac{p}{2}, \pi\right)$

15. $2a = \dfrac{2ep}{1 - e^2}; c = \dfrac{e^2 p}{1 - e^2}; \dfrac{c}{a} = e$

Exercise 8.8   (Page 339)

1.

$$\begin{cases} x = 2t \\ y = t - 5 \end{cases}$$

3.

$$\begin{cases} x = 2(t - 3) \\ y = 3(t - 2) \end{cases}$$

5.

$$\begin{cases} x = t^2 \\ y = t - 2 \end{cases}$$

$(4, 0)$

$(0, -2)$

$(4, -4)$

7.

$$\begin{cases} x = \sin t \\ y = \cos t \end{cases}$$

$(1, 0)$

9. $2x - 3y = 4$; line

11. $y = 5$; line

13. $\dfrac{x^2}{9} + \dfrac{y^2}{25} = 1$; ellipse

15. $xy = 1$;   hyperbola

17.

$$\begin{cases} x = t^2 \\ y = 2t^2 + 1 \end{cases}$$

19.

$$\begin{cases} x = 2t^2 - 1 \\ y = 1 - t^2 \end{cases}$$

$(-1, 1)$

21.

$$\begin{cases} x = \sqrt{t} \\ y = t \end{cases}$$

$(1, 1)$

23.

$$\begin{cases} x = |t| \\ y = |t| \end{cases}$$

25.

$(2, 3)$

$(-2, -3)$

$$\begin{cases} x = 2 \sin t \\ y = 3 \sin t \end{cases}$$

27.

$$\begin{cases} x = 3 \sin^2 t \\ y = 4 \cos^2 t \end{cases}$$

29.

$$\begin{cases} x = \sin t \\ y = \sin 2t \end{cases}$$

31. $\begin{cases} x = a \cos t \\ y = a \sin t \end{cases}$

33. $\begin{cases} x = a \cos t \\ y = b \sin t \end{cases}$

35. $y = x \tan \theta - \dfrac{16x^2}{v_0^2 \cos^2 \theta}$; parabola

37. $\begin{cases} x = x_1 + t(x_2 - x_1) \\ y = y_1 + t(y_2 - y_1) \end{cases}$

REVIEW EXERCISES   (Page 344)

1.

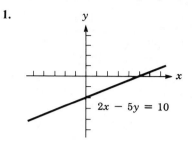

$2x - 5y = 10$

2.

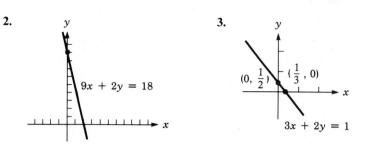

$9x + 2y = 18$

3.

$(0, \frac{1}{2})$ $(\frac{1}{3}, 0)$

$3x + 2y = 1$

4.

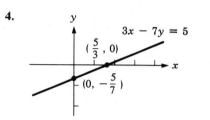

$3x - 7y = 5$

$(\frac{5}{3}, 0)$

$(0, -\frac{5}{7})$

5. -6 **6.** 2 **7.** -1 **8.** 1 **9.** $7x + 5y = 0$

10. $4x + y = -7$ **11.** $2x + y = 9$ **12.** $2x - 3y = -9$

13. $y = 17$ **14.** $x = -5$ **15.** $7x + y = 54$ **16.** $x - 7y = 22$

17. $3x - 4y = 6$ **18.** $3x - y = 21$ **19.** $x^2 + y^2 = 50$

20. $x^2 + y^2 = 100$ **21.** $(x - 5)^2 + (y - 10)^2 = 85$

22. $(x - 2)^2 + (y - 2)^2 = 89$

23.

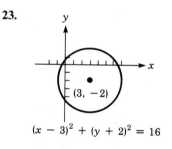

$(3, -2)$

$(x - 3)^2 + (y + 2)^2 = 16$

24.

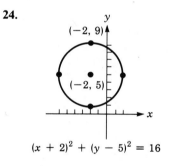

$(-2, 9)$

$(-2, 5)$

$(x + 2)^2 + (y - 5)^2 = 16$

25. $y^2 = -2x$ **26.** $x^2 = 16y$

27. $\left(x + \dfrac{b}{2a}\right)^2 = \dfrac{1}{a}\left(y + \dfrac{b^2 - 4ac}{4a}\right)$;

Thus, the x-coordinate of the vertex is $-\dfrac{b}{2a}$.

28. $(x + 2)^2 = -\dfrac{4}{11}(y - 3)$

29.

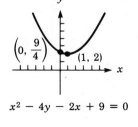

$\left(0, \dfrac{9}{4}\right)$ $(1, 2)$

$x^2 - 4y - 2x + 9 = 0$

30.

$(1, 3)$

$\left(\dfrac{13}{4}, 0\right)$

$y^2 - 6y = 4x - 13$

or

$(y - 3)^2 = 4(x - 1)$

31.

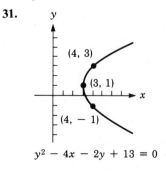

$(4, 3)$

$(3, 1)$

$(4, -1)$

$y^2 - 4x - 2y + 13 = 0$

32. $\dfrac{x^2}{36} + \dfrac{y^2}{16} = 1$

33. $\dfrac{(x+2)^2}{16} + \dfrac{(y-3)^2}{9} = 1$

34.

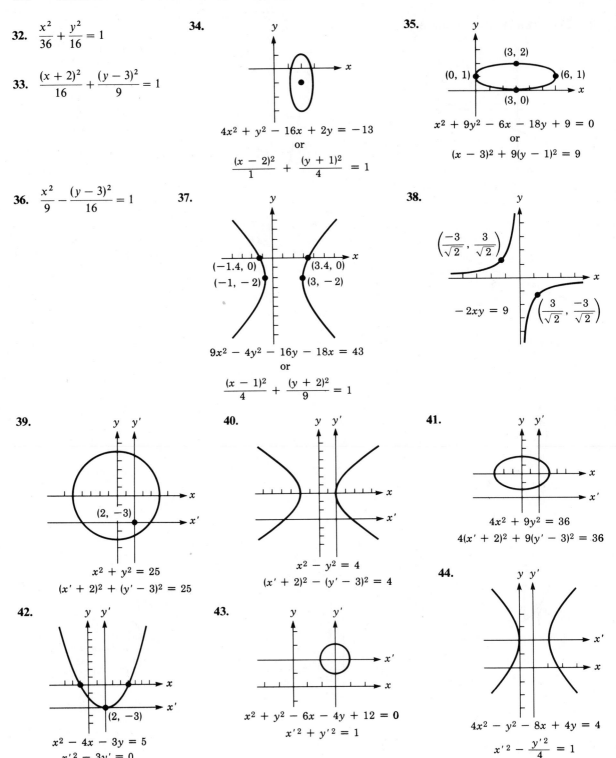

$4x^2 + y^2 - 16x + 2y = -13$

or

$\dfrac{(x-2)^2}{1} + \dfrac{(y+1)^2}{4} = 1$

35.

$x^2 + 9y^2 - 6x - 18y + 9 = 0$

or

$(x-3)^2 + 9(y-1)^2 = 9$

36. $\dfrac{x^2}{9} - \dfrac{(y-3)^2}{16} = 1$

37.

$9x^2 - 4y^2 - 16y - 18x = 43$

or

$\dfrac{(x-1)^2}{4} + \dfrac{(y+2)^2}{9} = 1$

38.

$-2xy = 9$

39.

$x^2 + y^2 = 25$

$(x'+2)^2 + (y'-3)^2 = 25$

40.

$x^2 - y^2 = 4$

$(x'+2)^2 - (y'-3)^2 = 4$

41.

$4x^2 + 9y^2 = 36$

$4(x'+2)^2 + 9(y'-3)^2 = 36$

42.

$x^2 - 4x - 3y = 5$

$x'^2 - 3y' = 0$

43.

$x^2 + y^2 - 6x - 4y + 12 = 0$

$x'^2 + y'^2 = 1$

44.

$4x^2 - y^2 - 8x + 4y = 4$

$x'^2 - \dfrac{y'^2}{4} = 1$

45.

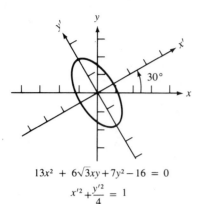

$$13x^2 + 6\sqrt{3}xy + 7y^2 - 16 = 0$$

$$x'^2 + \frac{y'^2}{4} = 1$$

46.

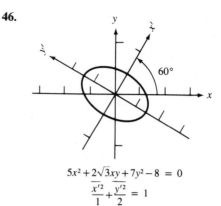

$$5x^2 + 2\sqrt{3}xy + 7y^2 - 8 = 0$$

$$\frac{x'^2}{1} + \frac{y'^2}{2} = 1$$

47.

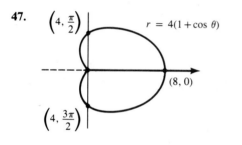

48.

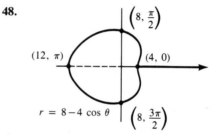

49.

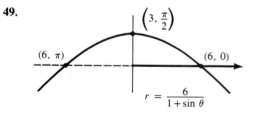

50.

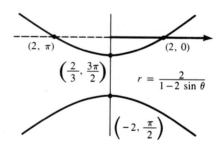

51.   **52.**   **53.**

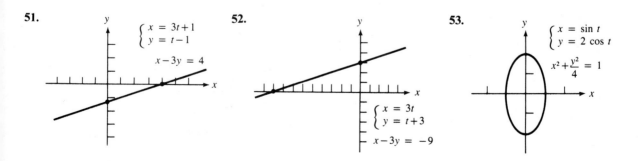

54.

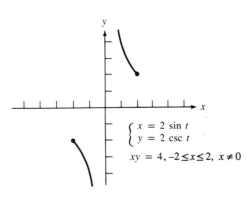

$$\begin{cases} x = 2 \sin t \\ y = 2 \csc t \end{cases}$$
$$xy = 4, -2 \le x \le 2, \; x \ne 0$$

55.

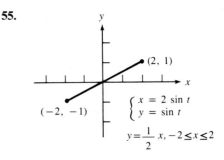

$$\begin{cases} x = 2 \sin t \\ y = \sin t \end{cases}$$
$$y = \frac{1}{2} x, -2 \le x \le 2$$

56.

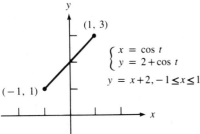

$$\begin{cases} x = \cos t \\ y = 2 + \cos t \end{cases}$$
$$y = x + 2, -1 \le x \le 1$$

Exercise I.1   (Page 351)

1. angle ABC or GBE **3.** angle A, GAC, CAG, or CAB **5.** angle BDE or EDB **7.** acute
9. obtuse   **11.** right   **13.** none   **15.** straight   **17.** $130°$ **19.** $96°$ **21.** $63°$
23. $60°$ **25.** $123°$ **27.** yes   **29.** no   **31.** yes   **33.** 50 m   **35.** $\sqrt{5}$ cm
37. $\sqrt{a^2 + b^2}$ ft   **39.** $BC = 2$ cm; $AB = 2\sqrt{2}$ cm   **41.** $AC = BC = 5\sqrt{2}$ in.
43. $BC = 3\sqrt{3}$ km; $AB = 3\sqrt{6}$ km   **45.** $BC = 3\sqrt{3}$ cm; $AB = 6$ cm
47. $AC = \dfrac{5\sqrt{3}}{3}$ mi; $AB = \dfrac{10\sqrt{3}}{3}$ mi   **49.** $AC = 5$ cm; $BC = 5\sqrt{3}$ cm   **51.** $BC = 6$ cm; $AB = 4\sqrt{3}$ cm
53. $AC = 2$ ft; $AB = 4$ ft   **55.** $AC = \sqrt{3}$ cm; $BC = 3$ cm   **57.** $4|x|$ **59.** $25x^4$ **61.** x
63. $|x|$

Exercise I.2   (Page 355)

1. $\frac{60}{7}$ cm   **3.** $\frac{168}{25}$ cm   **5.** $37\frac{1}{2}$ ft   **7.** 60 ft   **9.** 43 ft $+ 5\frac{1}{2}$ ft $= 48\frac{1}{2}$ ft   **11.** 528 ft
13. 40 ft   **15.** $40°$ **17.** Since angle A + angle 2 = $90°$ and angle 1 + angle 2 = $90°$, angle A = angle 1.
Use same argument to show angle B = angle 2.
19. Triangles are similar because two angles of triangle ACD equal two angles of triangle ACB.   **21.** 4.5

Exercise I.3   (Page 358)

1. 12π in.   **3.** 9 units   **5.** $\dfrac{81}{2}$ units   **7.** $\sqrt{\dfrac{63}{\pi}} = \dfrac{\sqrt{63\pi}}{\pi}$ units   **9.** 2 units   **11.** 12 units

13. $10\sqrt{2}\,\pi$ cm   **15.** $\dfrac{2\sqrt{6\pi}}{\pi}$ sq in.   **17.** about 94 plants   **19.** about \$177 million

21. about 747 revolutions   **23.** about \$2094   **25.** 24 m   **27.** 54.76 cm   **29.** 3979 mi

Exercise II.1   (Page 366)

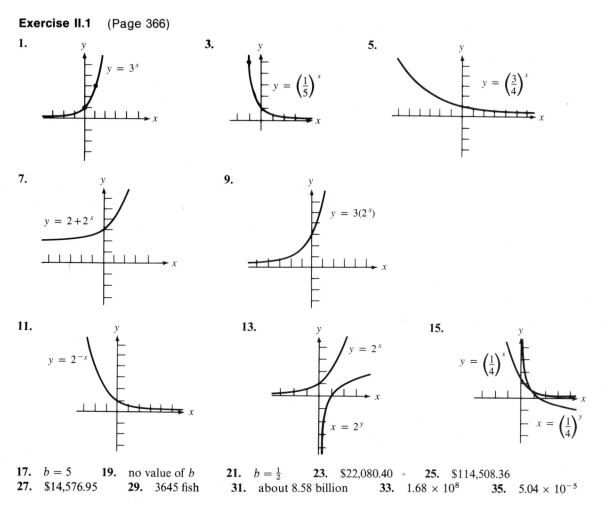

1. $y = 3^x$

3. $y = \left(\frac{1}{5}\right)^x$

5. $y = \left(\frac{3}{4}\right)^x$

7. $y = 2 + 2^x$

9. $y = 3(2^x)$

11. $y = 2^{-x}$

13. $y = 2^x$ $x = 2^y$

15. $y = \left(\frac{1}{4}\right)^x$ $x = \left(\frac{1}{4}\right)^y$

17. $b = 5$ 19.   no value of b 21. $b = \frac{1}{2}$ 23.   \$22,080.40   25.   \$114,508.36
27.   \$14,576.95   29.   3645 fish   31.   about 8.58 billion   33. 1.68×10^8 35. 5.04×10^{-5}

Exercise II.2   (Page 374)

1. $3^4 = 81$ 3. $\left(\frac{1}{2}\right)^3 = \frac{1}{8}$ 5. $4^{-3} = \frac{1}{64}$ 7. $x^z = y$ 9. $\log_8 64 = 2$ 11. $\log_4 \frac{1}{16} = -2$
13. $\log_{1/2} 32 = -5$ 15. $\log_x z = y$ 17.   0   19.   2   21.   1   23.   3   25.   3
27.   3   29.   16   31.   81   33.   2   35.   2   37.   2   39.   3   41.   81   43.   25
45. -2 47. $\frac{1}{3}$ 49.   4

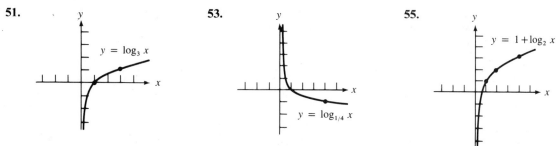

51. $y = \log_3 x$

53. $y = \log_{1/4} x$

55. $y = 1 + \log_2 x$

57.   **59.**

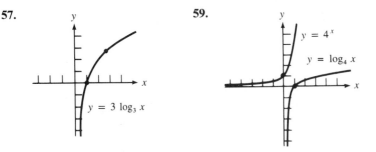

61. no value of b **63.** $b = 3$ **65.** no value of b

67. $\log_b x + \log_b y + \log_b z$ **69.** $2 \log_b x - 2 \log_b y$ **71.** $\dfrac{3}{2}\log_b x + \dfrac{1}{2}\log_b z$

73. $\dfrac{3}{4}\log_b x + \dfrac{1}{2}\log_b y - \log_b z$ **75.** $\log_b \dfrac{x+1}{x}$ **77.** $\log_b x^2 \sqrt[3]{y}$ **79.** $\log_b \dfrac{\sqrt{z}}{x^3 y^2}$

81. $\log_b \dfrac{x}{y}$ **83.** false   **85.** true   **87.** false   **89.** true   **91.** true   **93.** false

95. false   **97.** false   **99.** 1.4472   **101.** 0.3521   **103.** 1.1972   **105.** 2.4014
107. 0.3011   **109.** 0.7782   **111.** 1.6990   **113.** -0.3010

Exercise II.3   (Page 381)

1. 0.5119   **3.** 0.3854   **5.** 1949.8446   **7.** 0.0001   **9.** 1.3610   **11.** -3.6109
13. 0.0002   **15.** 16.9455   **17.** 2.3282   **19.** pH = 4.77   **21.** 6.31×10^{-14} gram-ions per L
23. from 2.5119×10^{-8} to 1.585×10^{-7} **25.** 0.71 V   **27.** 49.54 db   **29.** 4.4
31. 2500 microns   **33.** 12.2 years   **35.** 23.1 years
37. The original loudness L_0 is increased by $k \ln 2$ because $L = L_0 + k \ln 2$, where $L_0 = k \ln I$.
39. The original intensity must be cubed.

Exercise II.4   (Page 388)

1. $\dfrac{\log 5}{\log 4}$ **3.** $\dfrac{1}{\log 2} + 1$ **5.** -2 **7.** $\pm\sqrt{2}$ **9.** 0   **11.** $\dfrac{\log 3}{\log 4 - \log 3}$ **13.** 9
15. 8   **17.** 2   **19.** 3   **21.** 7   **23.** 4   **25.** $10, -10$ **27.** 50   **29.** 20
31. 10   **33.** 10   **35.** 2   **37.** 10^{10} **39.** 1.771   **41.** -1 **43.** $\frac{1}{6}$ **45.** 0.3028
51. 5.146 years   **53.** about 42.7 days   **55.** about 4200 years old   **57.** It will be 2.828 times larger.
59. about 9.46 years   **61.** about 9.694 years

Exercise III.1   (Page 392)

1. 0.3963   **3.** 0.1542   **5.** 1.116   **7.** 26.39°   **9.** 9.03°   **11.** 13.18°
13. The graph of $y = \sin x$ cannot be assumed to be linear over so large an interval.

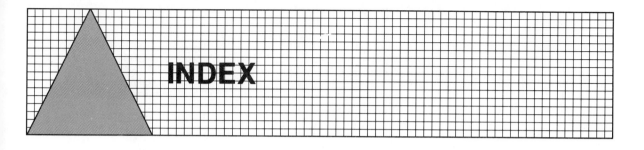

INDEX

Absolute value of a complex number, 251
Acute angle, 347
Acute triangle, 348
Addition:
  of complex numbers, 246
  of ordinates, 115
  of vectors, 67, 230
Alternating current, 116
Ambiguous case, 215
Amplitude, 101, 107, 116
Angle, 21, 347
  acute, 347
  between two lines, 159
  central, 357
  complementary, 348
  coterminal, 22
  critical, 195
  of depression, 55
  of elevation, 55
  of incidence, 72
  obtuse, 348
  phase, 73
  quadrantal, 22
  reference, 43
  reflex, 348
  of refraction, 72
  right, 347
  straight, 348
  supplementary, 348
  trigonometric, 22
  vertex of, 21, 347
Angle-of-rotation formula, 324
Angular velocity, 88
Arccosecant function, 191
Arccosine function, 183
Arccotangent function, 190
Archimedean spiral, 266
Arc length, 82
Arc of a circle, 357
Arcsecant function, 191
Arcsine function, 181
Arctangent function, 183
Area:
  of a circle, 357
  of a sector of a circle, 83
  of a triangle, 222–227

Argument:
  of a complex number, 253
  of a function, 96
Asymptote, 104
  of a hyperbola, 315
Axis (axes), 1
  of an ellipse, 305
  of a hyperbola, 312, 316
  of a parabola, 298
  polar, 263
  rotation of the coordinate, 322, 323
  translation of the coordinate, 319, 323

Base:
  of the exponential function, 361
  of the logarithmic function, 368
Bearing, 55
Beat frequencies, 161

Calculators, 49–50
Carbon dating, 386
Cardioid, 274
Center:
  of a circle, 294
  of an ellipse, 305
Central angle, 357
Change-of-base formula, 385
Chord of a circle, 357
Circle, 281, 294, 357
  arc of, 357
  area of, 357
  area of a sector of, 83
  center of, 294
  central angle of an arc of, 357
  chord of, 357
  circumference of, 357
  diameter of, 357
  equation of, 295
  radius of, 294, 357
  sector of, 358
  unit, 91
Circumference of a circle, 357
Cis θ, 253
Cofunctions, 48
Common logarithms, 377
Complementary angles, 348

Complex number(s), 245
  absolute value of, 251
  addition of, 246
  argument of, 253
  conjugate, 247
  division of, 248, 255
  equality of, 246
  graphing, 250
  imaginary part of, 245
  modulus of, 253
  multiplication of, 246, 254
  real part of, 245
  trigonometric form of, 253
Complex plane, 250
Components of a vector, 68, 235, 236
Compound interest, 364, 365
Conditional equation, 123
Congruent triangles, 353
Conics, polar equations of, 329
Conic sections, 281
Conjugate axis of a hyperbola, 316
Conjugate complex numbers, 247
Continuous compound interest, 365
Coordinate(s), 1, 2
Coordinate axes, 1
  polar, 263
  rotation of, 322
  translation of, 319
Coordinate vectors, 232
Cosecant function, 24, 92
  inverse of, 191
Cosine function, 24, 92
  inverse of, 176
Cosines, law of, 202
Cotangent function, 24, 92
  inverse of, 190
Coterminal angles, 22
Course, 67
Critical angle, 195
Current, alternating, 116
Curves, polar, 275
Cycloid, 337

Decibel voltage gain, 378
Decreasing functions, 362
Degree, 21
De Moivre, Abraham, 258

De Moivre's theorem, 258
Dependent variable, 5
Depression, angle of, 55
Descartes, René, 1, 243, 281
Diameter, of a circle, 357
Difference of two angle identities,
    133–135
Digits, significant, 53
Directrix, of a parabola, 298
Distance formula, 3
Division, of complex numbers, 248, 255
Domain, 5
    of exponential function, 361
    of a function, 5
    of the logarithmic function, 368
    of a relation, 6
Dot product of vectors, 233
Double-angle identities, 137

Eccentricity, 309, 328
Elevation, angle of, 55
Eliminating the parameter, 335
Ellipse, 281, 305
    center of, 305
    eccentricity of, 309
    equation of, 307
    foci of, 305
    major axis of, 305
    minor axis of, 305
    vertices of, 305
Equality:
    of complex numbers, 246
    of vectors, 66, 229
Equation(s), 123
    of a circle, 295
    conditional, 123
    of the conics, 295, 299, 307, 314, 329
    of an ellipse, 307
    exponential, 382
    of a hyperbola, 314
    impossible, 123
    of a line, 287, 289, 292
    logarithmic, 382
    of a parabola, 299, 300
    parametric, 334
    for rotation of axes, 323
    for translation of axes, 323
Equiangular triangle, 348
Equilateral triangle, 348
Eratosthenes, 358
Even function, 35
Exponential equation, 382
Exponential function, 361
    base of, 361
    domain of, 361
    range of, 361

First-quadrant angle, 22
Focal width, 311
Focus (foci):
    of an ellipse, 305

Focus (foci) (*continued*)
    of a hyperbola, 312
    of a parabola, 298
Formulas:
    angle-of-rotation, 324
    change-of-base, 385
    distance, 3
    Heron's (Hero's), 227
    polar to rectangular conversion, 264
    rectangular to polar conversion, 265
    translation-of-axis, 320
Fourth-quadrant angle, 22
Frequencies of oscillation, 117, 161
Function(s), 5
    arccosecant, 191
    arccosine, 183
    arccotangent, 190
    arcsecant, 191
    arcsine, 181
    arctangent, 183
    argument of, 96
    decreasing, 362
    domain of, 5
    even, 35
    exponential, 361
    inverse, 12
    inverse cosecant, 191
    inverse cosine, 176
    inverse cotangent, 190
    inverse secant, 191
    inverse sine, 181
    inverse tangent, 183
    inverse trigonometric, 179
    logarithmic, 368
    odd, 35
    one-to-one, 15
    periodic, 96
    range of, 5
    trigonometric, 24, 92
Function notation, 7
Fundamental rectangle, 315
Fundamental trigonometric
    relationships, 32, 124

General form:
    of the equation of a line, 292
    of a second-degree equation, 281
Graphing complex numbers, 250
Graphing functions and relations, 8
Ground speed, 67

Half-angle identities, 143
Harmonic motion, 116, 118, 160
Heading, 67
Heron's formula, 227
Horizontal line, 291
    test, 15
Horizontal translation, 108, 110
Hyperbola, 281, 312
    asymptotes of, 315
    conjugate axis of, 316

Hyperbola (*continued*)
    equation of, 314
    foci of, 312
    fundamental rectangle of, 315
    transverse axis of, 312
    vertices of, 312
Hyperbolic spiral, 274
Hypotenuse of a right triangle, 348

Identity(ies), 123
    difference-of-two-angle, 133–135
    double-angle, 137
    half-angle, 143
    product-to-sum, 154
    sum-to-product, 154
    sum-of-two-angle, 132–135
Imaginary number, 243
Imaginary part of a complex number,
    245
Impedance, 73
Impossible equation, 123
Incidence, angle of, 72
Independent variable, 5
Index, refractive, 72
Initial side of a trigonometric angle, 22
Interest, compound, 364, 365
Internal reflection, 195
Interpolation, linear, 390
Inverse:
    cosecant function, 191
    cosine function, 176
    cotangent function, 190
    function, 12
    relation, 12
    secant function, 191
    sine function, 181
    tangent function, 183
    trigonometric function, 179
Involute, 341
Isosceles triangle, 348

Land, Edwin H., 196
Law:
    of cosines, 202
    of sines, 209
    of tangents, 221
Length of an arc, 82
Line(s):
    angle between two, 159
    general form of the equation of, 292
    horizontal, 291
    parallel, 284, 354
    perpendicular, 285
    point-slope form of the equation
        of, 287
    slope of, 71, 283
    slope-intercept form of the equation
        of, 289
    x-intercept of, 282
    y-intercept of, 282
Linear interpolation, 390

Linear velocity, 87
Logarithmic equation, 382
Logarithmic function, 368
    base of, 368
    domain of, 368
    range of, 368
Logarithms:
    change-of-base formula, 385
    common, 377
    natural, 379
    properties of, 370

Magnitude of a vector, 66, 229, 230
Major axis, 305
Minor axis, 305
Minute, 21
Modulus, 253
Multiplication of complex numbers, 246, 254

Natural logarithm, 379
Negative reciprocals, 285
Norm of a vector, 66, 229, 230
Number(s):
    absolute value of a complex, 251
    addition of complex, 246
    argument of a complex, 253
    complex, 245
    conjugate complex, 247
    division of complex, 248, 255
    equality of complex, 246
    graphing complex, 250
    imaginary, 243
    imaginary part of a complex, 245
    modulus of a complex, 253
    multiplication of complex, 246, 254
    real part of a complex, 245
    trigonometric form of a complex, 253

Oblique triangle, 201, 348
Obtuse angle, 348
Obtuse triangle, 348
Odd function, 35
One-to-one function, 15
Ordered pair, 2
Ordinates, addition of, 115
Origin of a rectangular coordinate system, 1
Oscillation:
    frequency of, 117
    period of, 101, 107, 117

Parabola, 281, 298
    axis of, 298
    directrix of, 298
    equation of, 299, 300
    focus of, 298
    vertex of, 298
Parallel lines, 284, 354
Parallelogram law, 67, 230
Parametric equations, 334

Parametric equations (*continued*)
    eliminating the parameter, 335
Pascal, Blaise, 281
Period, 101, 107
Periodic function, 96
Period of oscillation, 117
Perpendicular lines, 285
Perpendicular vectors, 234
Phase angle, 73
Phase shift, 110, 112, 113, 231
pH scale, 378
Plane curves, 333
Point, coordinates of, 1, 263
Point-slope form of the equation of a line, 287
Polar axis, 263
Polar coordinate system, 263
Polar curves, 275
Polar equations of the conics, 329
Polarized light, 196
Polar-to-rectangular conversion formulas, 264
Pole, 263
Population growth, 365, 387
Principal square root, 349
Product-to-sum identities, 154
Projection, of vectors, 235, 236
Properties, of logarithms, 370
Pythagorean theorem, 349

Quadrant, 22
Quadrantal angle, 22

Radian measure, 79
Radioactive decay, 385
Radius of a circle, 294, 357
Range, 5
    of an exponential function, 361
    of a function, 5
    of a logarithmic function, 368
    of a relation, 6
Reactance, 73
Real part of a complex number, 245
Reciprocal relationships, 30
Rectangle, fundamental, 315
Rectangular coordinate system, 1
Rectangular-to-polar conversion formulas, 265
Reference angle, 43
Reference triangle, 42
Reflection, total internal, 195
Reflex angle, 348
Refraction, angle of, 72
Refractive index, 72
Relation(s), 6, 179
    domain of, 6
    inverse, 12
    range of, 6
Resistance, 73
Resolving a vector into components, 68, 232

Resultant vector, 67, 230
Richter scale, 379
Right angle, 347
Right triangle, 348
    hypotenuse of, 348
    trigonometry, 52, 61
Roses, 273
Rotated coordinate system, 322
Rotation of axes, 323

Scalar:
    multiplication, 231
    projection, 235, 236
    quantities, 66
Seasons, 74
Secant function, 24, 92
    inverse of, 191
Second, 21
Second-degree equation, general form, 281
Second-quadrant angle, 22
Sector:
    area of a, 83
    of a circle, 358
Semiperimeter, 227
Side of a trigonometric angle, 22
Significant digits, 53
Similar triangles, 353
Simple harmonic motion, 116, 118, 160
Sine function, 24, 92
    inverse of, 181
Sine wave, 97
Slope-intercept form of the equation of a line, 289
Slope of a line, 71, 283
Snell, Willebrord, 72
Snell's law, 72, 195
Solstice, 74
Solving right triangles, 54
Solving trigonometric equations, 166
Sound waves, 118
Spiral:
    Archimedean, 266
    hyperbolic, 274
Spring constant, 116
Square root, 349
Standard position of a trigonometric angle, 22
Straight angle, 348
Sum-of-two-angle identities, 132–135
Sum of vectors, 230
Sum-to-product identities, 154
Supplementary angles, 348

Tables:
    base-*e* logarithms, 399
    base-10 logarithms, 398
    powers and roots, 397
    trigonometric, 47, 393–397
Tangent function, 24, 92
    inverse of, 183

Tangents, law of, 221
Terminal side of a trigonometric angle, 22
Third-quadrant angle, 22
Total internal reflection, 195
Translated coordinate system, 319
Translation:
  horizontal, 108, 110
  vertical, 109
Translation of axes, 319, 323
Translation-of-axis formulas, 320
Transversal, 354
Transverse axis, 312
Triangle(s):
  acute, 348
  area of, 222–227
  congruent, 353
  equiangular, 348
  equilateral, 348
  hypotenuse of a right, 348
  isosceles, 348
  oblique, 201, 348
  obtuse, 348
  reference, 42
  right, 348
  similar, 353
  solving right, 54
Trigonometric angle, 22
  initial side of, 22
  standard position of, 22

Trigonometric angle (*continued*)
  terminal side of, 22
Trigonometric equations, 166
Trigonometric form of a complex number, 253
Trigonometric functions, 24, 92
  inverses of, 179
Trigonometric relationships, 32, 124
Trigonometric tables, 47

Unit circle, 91
Unit coordinate vectors, 232
Using calculators, 49–50
Using a trigonometric table, 47

Value, absolute, 251
Variable:
  dependent, 5
  independent, 5
Vector(s), 66, 229
  addition of, 67, 231
  components of, 232, 235, 236
  dot product of, 233
  equal, 66
  equality of, 229
  magnitude of, 66, 229, 230
  norm of, 66, 229, 230
  perpendicular, 234
  projection, 235, 236
  quantities, 66

Vector(s) (*continued*)
  resultant of, 67, 230
  scalar multiplication, 231
  scalar projection of, 235, 236
  sum of, 67, 230
  unit coordinate, 232
Velocity:
  angular, 88
  linear, 87
Vertex (vertices):
  of an angle, 21, 347
  of an ellipse, 305
  of a hyperbola, 312
  of a parabola, 298
Vertical line test, 10
Vertical translation, 109

Wavelength, 118
Weber-Fechner law, 380

x-axis, 1
x-component of a vector, 232
x-coordinate, 2
x-intercept of a line, 282

y-axis, 1
y-component of a vector, 232
y-coordinate, 2
y-intercept of a line, 282